Agroforestry in Sustainable Agricultural Systems

ADVANCES IN AGROECOLOGY

CLIVE EDWARDS, SERIES EDITIOR

Soil Ecology in Sustainable Agricultural Systems
Lijbert Brussaard and Ronald Ferrera-Cerrato

Biodiversity in Agroecosystems
Wanda Williams Collins and Calvin O. Qualset

Agroforestry in Sustainable Agricultural Systems
Louise E. Buck, James P. Lassoie, and Erick C.M. Fernandes

Agroforestry in Sustainable Agricultural Systems

Edited by

Louise E. Buck • James P. Lassoie
Erick C.M. Fernandes

CRC Press
Boca Raton London New York Washington, D.C.

LEWIS PUBLISHERS
Boca Raton London New York Washington, D.C.

Acquiring Editor: John Sulzycki
Project Editor: Albert W. Starkweather
Cover design: Dawn Boyd

Library of Congress Cataloging-in-Publication Data

Agroforestry in sustainable agricultural systems / edited by Louise E. Buck, James P. Lassoie, and Erick C. M. Fernandes
p. cm. -- (Advances in agroecology)
Includes bibliographical references and index.
ISBN 1-56670-294-1 (alk. paper)
1. Agroforestry. I. Buck, Louise. II. Lassoie, J. P. III. Fernandes, E. C. M.(Erick C. M.) IV. Series
S494.5.A5 A375123 1998
634.9′9—ddc21 98-31470
CIP

Direct all inquiries to CRC Press LLC, 2000 Corporate Blvd., N.W., Boca Raton, Florida 33431.

Lewis Publishers is an imprint of CRC Press LLC

International Standard Book Number 1-56670-294-1
Library of Congress Card Number 98-31470
Printed in the United States of America 1 2 3 4 5 6 7 8 9 0
Printed on acid-free paper

The Editors

Louise E. Buck is a Senior Extension Associate in the Department of Natural Resources at Cornell University, Ithaca, NY, and a Senior Associate Scientist for the Center for International Forestry Research (CIFOR) with social science expertise in the fields of agroforestry and natural resources management. She is the leader of the Cornell Agroforestry Working Group (CAWG) of the Cornell International Institute for Food, Agriculture and Development (CIIFAD). She instructs and coordinates a graduate level course in agroforestry and manages several agroforestry research and outreach projects. Buck also coordinates CIIFAD's Protected Area Management Working Group and the organization's collaborative program in Madagascar. Prior to joining Cornell, she worked for a decade in eastern and southern Africa with research and development organizations including the International Center for Research in Agroforestry (ICRAF), CARE International, and the Beijer Institute of the Swedish Royal Academy of Sciences on social dimensions of agroforestry science and practice. During the past decade Buck has been active in temperate zone agroforestry, serving as a founding member of the Association for Temperate Agroforestry (AFTA) and collaborating with the USDA's National Agroforestry Center on policy and research development initiatives. She has worked with small-holder farmers and agroforestry systems in Costa Rica, the Dominican Republic, Ethiopia, Indonesia, Kenya, Madagascar, Rwanda, Somalia, Sudan, Tanzania, and Uganda, and with landowners in northeastern North America to evaluate the potential for agroforestry practice.

She received a B.A. in Sociology and Psychology and a B.S. in Outdoor Recreation from the University of Colorado in 1973, a M.S. in Natural Resources from Colorado State University in 1977, and is a Ph.D. candidate in Natural Resources at Cornell University.

Her current research activities include being Principal Investigator of an action research project in Madagascar with CIFOR, supported by USAID, to pilot and test community-based resource use agreements in fragile forest margins where agroforestry can be a viable land use alternative; Principal Investigator for a National Assessment of the Potential of Specialty Forest Products in Agroforestry Systems, funded by the USDA National Agroforestry Center; Principal Investigator for an action research project in the northeastern U.S. on Advancing Woods Cultivated Ginseng Knowledge and Practice: A Forest Farming Approach to Sustainable Agriculture, funded by USDA's Hatch Research program; and Project Leader for the International Study Group on the Devolution of Forest Governance and Management (USAID).

James P. Lassoie, Ph.D., is a Professor of Forest Science and chairman of the Department of Natural Resources, College of Agriculture and Life Sciences, at Cornell University. He completed his undergraduate and graduate education at the College of Forest Resources, University of Washington in Seattle, receiving a B.S. in 1968 and a Ph.D. in 1975. Before joining Cornell in 1976, he was a research assistant for Weyerhaeuser Company in Washington state and completed a two-year, postdoctoral fellowship at the School of Forestry, Fisheries, and Wildlife, University of Missouri in Columbia. As an assistant professor at Cornell he was responsible for developing a Cooperative Extension and research program in forest ecology and management. Lassoie spent 1984 as a Visiting Associate Professor at the University of Washington in Seattle. Upon returning to Cornell, he served as the Extension Leader for the Department of Natural Resources until January 1988 when he became the chairman of that department, a position he held until assuming the directorship of the Cornell Center for the Environment in July 1993. He returned to the faculty in the Department of Natural Resources and reassumed responsibilities as department chairman in August 1996.

Lassoie's past research activities have been focused on the ecology, management, and physiology of temperate forest tree species. He was particularly interested in the physiological impacts of air pollution and acidic precipitation on forest trees and has worked jointly with scientists at Boyce Thompson Institute where he held an adjunct scientist appointment for many years. He also was interested in the physiological basis for sap production and sugar yield from sugar maple trees. More recently, Lassoie has extended his research interests to environmental and conservation problems, sustainable development, and the role of trees in agricultural systems of developing countries and rural North America. Lassoie has published 35 peer-reviewed research papers, nine book chapters, three monographs, and has more than 85 other scholarly research publications, including symposium proceedings, project reports, book reviews, and abstracts.

Lassoie helped initiate the Cornell Tree Crops Research Project in 1978. In 1985, he started the Cornell Agroforestry Study Group that continues to coordinate agroforestry activities within the College of Agriculture and Life Sciences as the Cornell Agroforestry Working Group under the leadership of Louise E. Buck. In 1991, Lassoie developed an interdisciplinary graduate minor in conservation and sustainable development that included agroforestry and social forestry components. He was instrumental in establishing the Cornell International Institute for Food, Agriculture, and Development in 1990 and served on its program committee for many years. In addition, he helped develop the Cornell Center for the Environment in 1991 and served as its second director. Dr. Lassoie has developed graduate-level courses in agroforestry and an interdisciplinary course on conservation and sustainable development, which included a research field practicum in Latin America.

Erick C. M. Fernandes, Ph.D., is an assistant professor of tropical cropping systems and agroforestry in the Department of Soil Crop and Atmospheric Science at Cornell University, having joined the faculty in 1995. Fernandes is the leader of the Management of Organic Inputs in Soils of the Tropics program of the Cornell International Institute for Food, Agriculture and Development. In 1997, he was appointed Global Coordinator of the Alternatives to Slash and Burn (ASB) Program. The ASB program was initiated by the International Centre for Research in Agroforestry (ICRAF) with funding from the Global Environment Facility (GEF). Prior to joining Cornell, he was based in the Brazilian Amazon as Leader of a Rockefeller Foundation-sponsored Tropical Soils and Agroforestry Program that undertook research aimed at improving cropping systems to reduce deforestation and rehabilitate abandoned pasture land in the Brazilian Amazon (1991–1995). He has worked at the International Centre for Research in Agroforestry (ICRAF) and was involved in the global inventory and study of tropical agroforestry systems. Fernandes was a consulting editor of the journals *Agroforestry Systems* and *Acta Amazonica* and has worked with small holder farmers and agroforestry systems in Brazil, Cameroon, India, Indonesia, Kenya, Madagascar, and Peru.

He received a B.S. in Forestry from the University of Aberdeen, Scotland, in 1982 and a Ph.D. in Soil Science from North Carolina State University in 1991.

His current research activities include being Principal Investigator and Member of the Science Team of NASA's contribution to Large Scale Biosphere Atmosphere (LBA-Ecology) research program in the Brazilian Amazon, whose key research focus is Carbon and Nutrient Stocks and Soil Water Dynamics in Abandoned Pastures and Agroforestry Systems in the Central Amazon. He is also Principal Investigator of a G7 grant to Evaluate Agroforestry System Prototypes to Rehabilitate Abandoned Pasture Land in the Brazilian Amazon and leader of the Management of Organic Inputs in Soils of the Tropics (MOIST) program that has field and on-farm research activities in Honduras and Costa Rica.

Contributors

J. E. Michael Arnold
Consulting Forester
Oxford, UK

Michael R. Barnhart
Department of Entomology
University of Nebraska
Lincoln, NE

David M. Bates
Bailey Hortarium
Cornell University
Ithaca, NY

Badege Bishaw
Department of Forest Science
Oregon State University
Corvallis, OR

Eric B. Brennan
Department of Pomology
University of California at Davis
Davis, CA

John W. Bruce
Land Tenure Center
University of Wisconsin
Madison, WI

Louise E. Buck
Department of Natural Resources
Cornell University
Ithaca, NY

Roland J. Buresh
International Centre for Research in Agroforestry (ICRAF)
Nairobi, Kenya

Hubert de Foresta
International Centre for Research in Agroforestry (ICRAF)
Bogor, Indonesia

Peter A. Dewees
Environmental Technical Group
Africa Region – World Bank
Washington, DC

Anne M. Dix
Universidad del Vallé
Guatemala City, Guatemala

Mary Ellen Dix
Forest Health Protection
USDA Forest Service
Washington, DC

Clive A. Edwards
Entomology Department
Ohio State University
Columbus, OH

Erick C. M. Fernandes
Department of Soil, Crop and Atmospheric Science
Cornell University
Ithaca, NY

Louise Fortmann
Department of Environmental Science, Policy and Management
University of California at Berkeley
Berkeley, CA

Harold E. Garrett
School of Natural Resources
University of Missouri
Columbia, MO

Larry S. Harper
Editor
Missouri Ruralist
Columbia, MO

Donald C. L. Kass
Watershed Management and Agroforestry Systems
Tropical Agriculture Research and Higher Education Center (CATIE)
Turrialba, Costa Rica

Ned B. Klopfenstein
USDA Forest Service
National Agroforestry Center
University of Nebraska
Lincoln, NE

James P. Lassoie
Department of Natural Resources
Cornell University
Ithaca, NY

Christopher R. Latt
School of Forest Resources and Conservation
University of Florida
Gainesville, FL

Roger R. B. Leakey
Institute of Terrestrial Ecology
Penicuik, Midlothian, UK

Beverly D. McIntyre
Kawanda Banana Research Center
Kampala, Uganda

Genevieve Michon
ORSTOM
Bogor, Indonesia

Kenneth W. Mudge
Department of Floriculture and Ornamental Horticulture
Cornell University
Ithaca, NY

Daniel N. Mugendi
Kenya Forestry Research Institute
Nairobi, Kenya

P. K. Ramachandran Nair
School of Forest Resources and Conservation
University of Florida
Gainesville, FL

Alice N. Pell
Department of Animal Science
Cornell University
Ithaca, NY

David Pimentel
Department of Entomology
Cornell University
Ithaca, NY

Susan J. Riha
Department of Soil, Crop and Atmospheric Science
Cornell University
Ithaca, NY

Dianne Rochelau
Graduate School of Geography
Clark University
Worcester, MA

Kenneth Schlather
Department of Soil, Crop and Atmospheric Science
Cornell University
Ithaca, NY

Steven H. Sharrow
Department of Rangeland Resources
Oregon State University
Corvallis, OR

Fergus L. Sinclair
School of Agriculture/Forest Sciences
University of Wales
Gwynedd, UK

Charles Staver
Tropical Crops Program
Tropical Agriculture Research and Higher Education Center (CATIE)
Turrialba, Costa Rica

H. David Thurston
Department of Plant Pathology
Cornell University
Ithaca, NY

Thomas P. Tomich
International Center for Research in Agroforestry SE Asia (ICRAF)
Bogor, Indonesia

Daniel H. Walker
CSIRO Davies Lab
Aitkenvale, Australia

Aaron Wightman
Department of Forestry
Virginia Polytechnical Institute
Blacksburg, VA

Sarah W. Workman
USDA National Agroforestry Center
University of Nebraska
Lincoln, NE

Foreword

Both the terms agroforestry and sustainable agriculture are value-laden and cause two general reactions. One, both are intrinsically good and worthy goals, without challenge. The opposite reaction is that both terms must be defined in testable ways and scrutinized in a rigorous scientific manner, just like any land use option. Agroforestry was recently redefined by Roger R. B. Leakey as "a dynamic, ecologically based, natural resources management system that, through the integration of trees in farms and in the agricultural landscape, diversifies and sustains production for increased social, economic and environmental benefits for land users at all levels." Sustainable agriculture, a much overused term, implies the productive use of natural resources for agricultural production in a way that the resource base is maintained for future generations.

The evolution of agroforestry from a highly descriptive and value-laden practice into an applied, multidisciplinary science is now in full swing, and this book is a significant contribution in this direction. The chapters provide excellent coverage. Some describe broad issues, while others focus on highly specific topics. Geographical coverage is wide, including both tropical and temperate regions. Multidisciplinarity is evident, with chapters focusing on biophysical and social sciences. Interdisciplinarity, the harder aspect, is evident in some of the chapters.

The leadership in developing this book comes from Cornell University. In 1997 ICRAF and Cornell entered into a long-term strategic alliance. This book, in a way, is a product of such an alliance, which includes partners from other leading institutions in agroforestry research. I commend the authors and editors on a job well done. Enjoy reading it.

Pedro A. Sanchez
Director General
International Centre for Research in Agroforestry
Nairobi, Kenya

Preface

Sustainable agricultural systems depend on agroecological processes that promote soil fertility and pest resistance through biologically acquired inputs, and social processes that generate knowledge and incentives for producing a variety of foods and fibers within locally affordable means. Agroforestry practice, the cultivation of trees or other woody plants with crops or pasture for multiple benefits, can contribute substantially to advancing a sustainable agriculture through its influence on ecological and social processes. To realize this potential a better understanding is needed of how agroforestry practice affects soil, water, plant, animal and atmospheric relations, and the roles of management in bringing about desirable outcomes. This volume is designed to help meet the challenge and to demonstrate how perspectives of key scientists and various works in progress are shaping the field of agroforestry.

The contributions highlight advances in characterizing and evaluating agroforestry practice, including incentives and constraints for managing these complex land use systems. The collection aims to generate insight from domains of knowledge that are becoming consolidated and to foster speculation about new directions for practice and science. Its depth of analysis demonstrates a substantial maturation of the field since the publication a decade ago of ICRAF's well-known and largely descriptive "decade of development," a volume similarly designed to reflect advances across a broad spectrum of agroforestry inquiry.

The first eight chapters of this volume focus on biophysical properties and processes in agroforestry systems, and roles of management in bringing about desired characteristics and benefits. Nair and his co-authors describe tree-mediated processes that determine the extent and rate of nutrient cycling in agroforestry systems, and review interaction effects in four major tropical agroforestry systems. They conclude that agroforestry systems can meet the nitrogen requirements to sustain crop production, though are less of an advantage in P nutrition. Concerned with nutrient cycling processes in agrosilvopastoral systems in the tropics, Pell identifies the roles of these systems in minimizing unintentional nutrient loss and ensuring that natural resources and agricultural productivity are maintained while farmers earn a viable living. She also examines effects of tree legumes in animal nutrition, and elaborates their important role in the provision of protein. Processes of water movement and water use by plants are the focus of Riha and McIntyre's examination of functions that hedgerow agroforestry systems perform, in comparison to more conventional agricultural systems. The authors demonstrate how hedgerow components combine with climate, soil and landscape factors to affect water movement and use processes.

Staver draws on research advances and development experience with weed management in Central American coffee systems to suggest treatments and hypothesize effects of improved weed control and soil conservation measures in agroforestry-based coffee systems. He then describes the use of pilot methods for strengthening farmer capacity in ecologically based decision-making for the management of ground cover heterogeneity in perennial crops under trees. Garrett and Harper focus on management as well in their description of a successful 20-year-old alleycropping development and research program in the state of Missouri. Unreliable nut production from wild populations stimulated land owners to plant and manage black walnut as a reforestation measure on abandoned agricultural lands, resulting in the "Missouri system of agroforestry." The authors identify the biological and financial methods that are used to evaluate trade-offs and guide management decisions for these intensive systems. Demonstrating that intensively managed systems of trees, livestock, and pasture is the most common form of agroforestry practiced in developed countries, Sharrow examines the effects of interactions among these components on ecosystems processes such as succession, facilitation, competition and herbivory.

Dix and her co-authors break new ground in providing a thoroughly researched review of current and potential integrated pest management practices in agroforestry systems. Drawing on literature from wide-ranging sources, they demonstrate that these approaches and their effectiveness depend on the intensity of energy or labor inputs within the system and discuss techniques that are applicable to both types. To complete this section, Mudge and Brennan provide an in-depth review of clonal propagation as a traditional and an increasingly important contemporary approach to plant propagation. They stress the importance of clonal approaches for improving the availability of high quality planting stock of appropriate multipurpose and fruit trees, and opening up numerous new options for the domestication of tropical trees.

Chapters nine through 14 illuminate social perspectives, processes and issues in advancing agroforestry practice and science. The material marks a shift in emphasis from the comparatively static basic needs approach that dominated earlier social science activity in the field of agroforestry toward an orientation to understand and manage complex and dynamic decision-making environments that encompass multiple actors with competing as well as overlapping objectives and constraints. Rocheleau sets the stage for this section in elaborating a hybrid, coalition model of agroforestry science. Arguing that diverse and complex social and ecological systems demand more robust and flexible agroforestry sciences, she maps out a fourfold mandate for social science that expands its role in improving the science and practice of agroforestry.

Evaluating the social content of agroforestry technologies is an element of Rocheleau's mandate for social science in agroforestry. Bruce and Fortmann exemplify it in drawing our attention to property rights in trees as potentially powerful and flexible policy instruments. The authors illuminate ways in which tree tenure provides policy levers to encourage social relationships and technical practices that lead to sustainable agroforestry. A second element of Rocheleau's social science mandate, facilitating participatory processes and practices, is portrayed by Sinclair and Walker who pose a method for rigorously analyzing the local knowledge held by practitioners that underpins the management of their practice. The investigators illustrate how explicit treatment of rural people's ecological knowledge is possible and useful in complementing participatory approaches to agroforestry development. Arnold and Dewees examine tree management in terms of farmer livelihood strategies and the dynamics of rural change. Drawing on farmers' perceptions they explore patterns in relationships between tree management and land use characteristics as well as contemporary development strategies, and how various policies may affect them. The piece reflects a third element in Rocheleau's mandate, articulating the social context of agroforestry decision-making and practice.

The final two chapters in the cluster on social dimensions assess the potential for agroforestry development in economic as well as environmental terms, reflecting a forth element in Rocheleau's social science mandate, bridging multiple sciences across scales, disciplines or publics. Pimentel and Wightman adopt macro and local perspectives in reviewing the potential of agroforestry in soil and water conservation as well as crop protection. They highlight a broad range of economic and environmental benefits of agroforestry that can help address major world issues of poverty and natural resource degradation. Leaky and Tomich recommend the further domestication and commercialization of trees for timber and non-timber forest products through agroforestry. Providing a thorough review of expected gains in product yield and quality through domestication they identify the biological and economic possibilities, priorities and pitfalls.

The last three chapters of the volume are exploratory in approach, offering insights into future directions for agroforestry science and practice. Bates seeks to integrate agroforestry into an ethnobotanical framework by characterizing ethnobotanical thought and activity and drawing parallels from the field of agroforestry. From this perspective he evaluates the types and extent of plant diversity in existing agroforestry systems, and explores prospects for maintaining or increasing the plant species and genetic diversity of these systems in the future. Kass and his co-authors focus on the important roles of mulch in sustainable agricultural systems. Using a land use intensification framework they trace the evolution of mulch-based cropping systems with trees and anticipate which of these agroforestry practices are likely to become more prominent. Michon and de Foresta paint a compelling vision of agroforestry modeled on forest architecture and function, urging the creation of space for this concept alongside agoforestry's more common depiction as an agricultural field oriented enterprise. The authors highlight the qualities of agro-forests and propose initial approaches for their analysis.

We are grateful to the authors of this unique collection of material for their creativity and persistent hard work in bringing the chapters to fruition. We also appreciate the extraordinary efforts of our scientific peers who reviewed the chapters to ensure their state-of-the-art quality. We hope the payoff is in satisfaction that the reader receives in contemplating relationships between agroforestry and sustainable agriculture, and in stimulating his or her preparedness to participate in the challenge of further improving our understanding and practice.

The Editors

Contents

1 Nutrient Cycling in Tropical Agroforestry Systems: Myths and Science

P. K. Ramachandran Nair, Roland J. Buresh, Daniel N. Mugendi, and Christopher R. Latt

1. INTRODUCTION

One of the conceptual foundations upon which tropical agroforestry research was initiated two decades ago is that trees help maintain soil fertility and support the growth of associated crops (Mongi and Huxley, 1979; Nair, 1984). Various hypotheses emanating from this concept have been tested vigorously in a large number of situations (Young, 1989; 1997), and such efforts have formed an important part of scientific research in tropical agroforestry during these years (Sanchez, 1995; Nair, 1997). Today, a substantial body of knowledge is available on the role of nutrient cycling in the maintenance of soil fertility in agroforestry systems (Rao et al., 1997), and this continues to be a major tenet of the study of tree-crop-soil interactions. Although many of the claims about nutrient cycling and its importance in tropical agroforestry systems have been verified or disproved by scientific research, it seems an aura of mysticism still surrounds most discussions. The objective of this paper is to review the current state of knowledge of nutrient cycling in tropical agroforestry systems in the hope of separating the grain from the chaff — science from myth.

In the following sections, we discuss nutrient cycling in several types of agroforestry systems — alley cropping, trees in cropland (parkland) systems, improved fallows, and shaded perennial-crop systems. The first two are simultaneous systems in which trees and crops occur together in the

1-56670-294-1/99/$0.00+$.50

same field at the same time; improved fallow is a sequential planting system in which the crops and the trees do not occupy the same piece of land at the same period of time; and, in shaded perennial-crop systems, both components are perennial species and are usually grown simultaneously.

Of the agroforestry systems discussed in this paper, alley cropping is the most extensively studied. Briefly, it involves the growing of crops between hedgerows of regularly coppiced woody species. Both N_2-fixing and non-N_2-fixing tree species are used (Rao et al., 1997). In "trees in cropland" systems, such as the "parkland system" of West Africa (CTFT, 1988) and in seasonally dry areas in East Africa (Laike, 1992) and southern Africa (Saka et al., 1994), trees are scattered or dispersed in agricultural fields. The best known systems are those that include *Faidherbia albida*, néré (*Parkia biglobosa*), karité *(Vitellaria paradoxa)*, melia *(Melia volkensii)* or neem *(Azadirachta indica)* in the semiarid tropics (SAT) of Africa (Kater et al., 1992; Kessler, 1992; Tilander et al., 1995), and neem or *Prosopis cineraria* ("khejri") in the SAT of India (Tejwani, 1994). Trees in these systems are rarely planted, but are derived from natural regeneration, and their stand densities vary considerably from two to three trees ha^{-1} for néré (Kater et al., 1992) to 10 to 45 trees ha^{-1} for *Prosopis* (Tejwani, 1994).

In improved fallow systems, cropping and fallow phases are alternated. During the fallow phase, woody (or herbaceous) species with beneficial qualities are either planted or retained from natural regeneration as a means of improving soil fertility to support food crop production during the subsequent cropping phase. Short-rotation fallow cycles involving fast-growing leguminous trees such as *Sesbania* spp., pigeonpea (*Cajanus cajan*), leucaena (*Leucaena leucocephala*), and *Tephrosia* spp. are being studied in Africa (Kwesiga and Coe, 1994; ICRAF, 1995, 1996).

In shaded perennial-crop systems, shade trees, often leguminous ones, are retained over perennial crops such as coffee (*Coffea* sp.) and cacao (*Theobroma cacao*). The shade trees, up to 300 trees ha^{-1}, are pollarded two or three times a year and the pruned biomass is left around the trees, spread on the ground, or removed for use as fodder or firewood (Beer et al., 1997).

2. NUTRIENT CYCLING — THE GENERAL CONCEPT

In a soil-plant system, plant nutrients are in a state of continuous, dynamic transfer. Plants take up nutrients from the soil and use them for metabolic processes. In turn, plants return nutrients to the soil either naturally as litterfall in unmanaged systems, deliberately as prunings in some agroforestry systems, or through root senescence in both managed and unmanaged systems. These plant parts are decomposed by soil microorganisms, releasing the nutrients bound in them into the soil. The nutrients then become available for plant uptake once again. The term nutrient cycling, as used in most agroforestry discussions, refers to the continuous transfer of nutrients that are already present within a soil-plant system, such as a farmer's field (Nair, 1993; Nair et al., 1995; Sanchez and Palm, 1996; Buresh and Tian, 1997). However, in a broader sense, nutrient cycling involves the continuous transfer of nutrients within and between different components of an ecosystem and includes processes such as weathering of minerals, activities of soil biota, and other transformations occurring in the biosphere, lithosphere, and hydrosphere (Jordan, 1985).

A generalized model of nutrient cycling in an ecosystem is presented in Figure 1 (DeAngelis, 1992). The model consists of a soil-plant-animal system partitioned into several compartments (pools), with inputs into the system (gains), outputs from the system (losses), and internal turnover or transfer within the system (cycling). Inputs into the system come through fertilizer, rain, dust, organic materials from outside the system, N_2 fixation, and weathering of rocks. Outputs result from erosion, leaching, plant harvest, denitrification, volatilization of N, and burning (Nair et al., 1995).

Nutrients entering the soil compartment contribute to the soil-nutrient pool. Water can remove (leach) nutrients from the soil nutrient pool, with the level of nutrient loss determined by the flow rate of the percolating water and the soil properties. Nutrients, such as nitrates, that dissolve readily

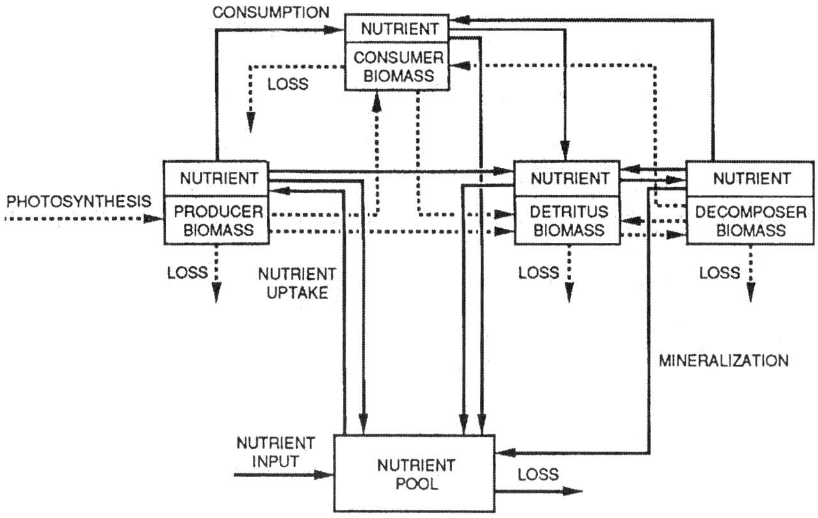

FIGURE 1. General representation of energy and nutrient flows in an ecosystem. Dashed lines represent energy flows and solid lines represent nutrient cycles and inputs and outputs. Outputs of nutrients include losses by water transport of dissolved nutrients and drift or migration of living organisms. Source: DeAngelis (1992).

in water and are weakly held by the soil matrix have a greater likelihood of being leached than do nutrients, such as phosphates, that have very low solubility and mobility in soils. On the other hand, the loss of cations such as potassium depends on the exchange capacity of the soils.

The autotroph (producer) component of an ecosystem (mostly plants) produces biomass through photosynthesis, a process which also involves transpiration of water and uptake of nutrients from the soil. Nutrients that are taken up are either stored within the plant or used in the production of new biomass. Some of these nutrients are subsequently returned to the soil through litter fall, root turnover, crown drip and stem flow. Decomposers in the soil mineralize nutrients back to inorganic forms that can be used again by autotrophs, but also use available nutrients, decreasing nutrient availability for autotrophs. Within-system movement of nutrients by water, wind and organisms, as well as inputs to and losses from the ecosystem, are essential processes (DeAngelis, 1992).

Natural forest ecosystems of the tropics represent self-sustaining and efficient nutrient cycling systems. These are "closed" nutrient cycling systems with relatively little loss or gain of the actively cycling nutrients, and high rates of nutrient turnover within the system. In contrast, most agricultural systems represent "open" or "leaky" systems with comparatively high nutrient losses. Nutrient cycling in agroforestry systems falls between these "extremes." (Nair et al., 1995).

Figure 2, originally proposed by Nair (1984), presents a generalized model of nutrient cycling in an agroforestry system, in comparison to cycling in monocrop agricultural and natural forest systems. The figure emphasizes that the major difference between agroforestry and other agricultural production systems is the greater possibility of managing the agroforestry system or its components to facilitate increased rates of nutrient turnover or transfer within different compartments of the system (Nair, 1993; Nair et al., 1995). In order to "exploit" these nutrient-cycling advantages of agroforestry systems, we need to understand the processes involved. Several recent reviews have addressed the topic (e.g., Sanchez, 1995; Rhoades, 1997; Buresh and Tian, 1997; Mafongoya et al., 1997a; Khanna, 1997; Young, 1997). Based on the current level of understanding, there appear to be three main tree-mediated processes that determine nutrient cycling in tropical agroforestry systems: (1) increased input of N through biological N_2 fixation (BNF) by trees; (2) enhanced availability of nutrients resulting from production and decomposition of substantial quantities of tree biomass; and (3) greater uptake and utilization of nutrients from deeper layers of soil by trees. Additionally, agroforestry systems offer the possibility for reducing the loss of soil — and therefore

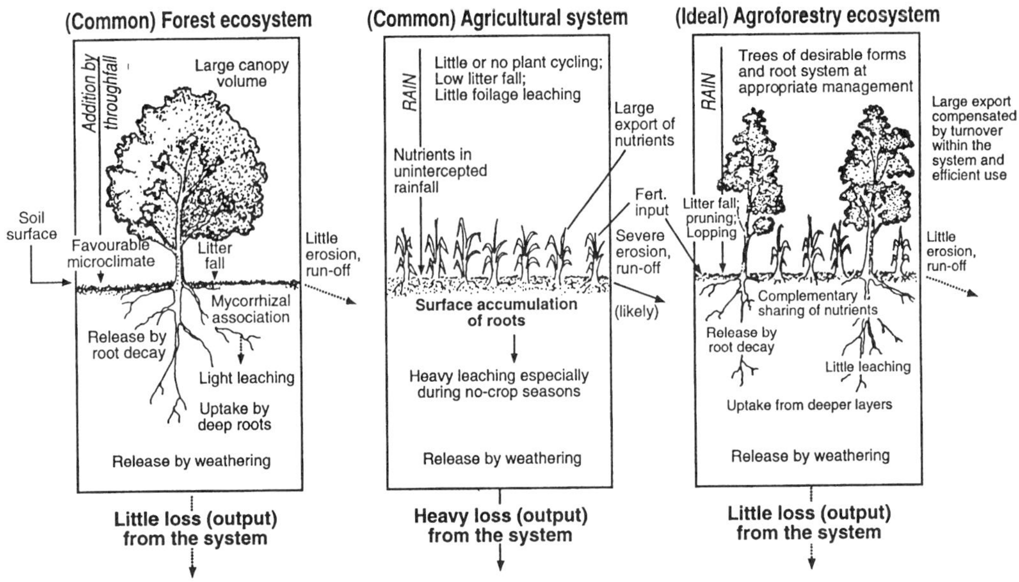

FIGURE 2. Schematic representation of nutrient relations and advantages of "ideal" agroforestry systems in comparison with common agricultural and forestry systems. Source: Nair (1984).

nutrients — through erosion. This aspect, although relevant in the broad discussion on nutrient cycling, is not considered here. In some discussions, input, output, and turnover phases of nutrient cycling are discussed separately. We feel that it is more useful and realistic to include all three phases in a discussion organized according to the three main processes identified above.

3. TREE-MEDIATED PROCESSES THAT AFFECT NUTRIENT CYCLING IN AGROFORESTRY SYSTEMS

3.1 Biological Nitrogen Fixation by Trees

Both science and myth are involved in discussions on the role of BNF in nutrient cycling in tropical agroforestry systems. The fact is that some trees that are, or potentially can be, used in agroforestry systems have the ability to add N to the soil through BNF. The main myth concerns the amount of N_2 fixed by trees and shrubs and the extent to which it is actually used or potentially available to the associated crop during various periods of time.

Among the 650 woody species belonging to nine families that are capable of fixing atmospheric N_2, 515 belong to the family Leguminosae (320 in Mimosoideae, 170 in Papilionoideae and 25 in Caesalpinoideae). Several genera of nonleguminous N_2-fixing trees (NFTs) are also important in tropical agroforestry systems; examples include *Alnus* and *Casuarina*. Among some 120 genera of NFTs (MacDicken, 1994), only a few are used directly as human food — fruits, flowers, leaves — (examples include the genera *Erythrina, Inga, Leucaena, Parkia, Pterocarpus,* and *Sesbania*); many are used for timber, fuelwood, or fodder; and most, if not all, for soil improvement. This last aspect — soil improvement — is achieved through several processes: (1) the direct contribution by trees to the soil N pool through transfer of biologically fixed N, (2) increased nutrient turnover and availability due to increased production and decomposition of biomass, and (3) improved erosion control via appropriate tree planting arrangements and mulching with tree prunings; however, the last aspect is not discussed in this paper.

Among the various agroforestry systems in the tropics (Nair, 1989), the most widely studied in terms of N_2 fixation are two simultaneous systems: alley cropping and shaded perennial-crop systems. Lately, sequential systems such as improved fallows are also being studied more rigorously (Rao et al., 1997). The present knowledge on BNF in the two simultaneous systems has been reviewed extensively by Sanginga et al. (1995) (alley cropping) and Beer et al. (1997) (shaded perennial-crop systems).

Some early reports on alley cropping claimed that enormously large quantities of N were fixed by some fast-growing tree species used as hedgerows, especially *Leucaena leucocephala* and *Gliricidia sepium*. For example, Sanginga et al. (1995), in their review, cited N_2 fixation levels of 100 to 300 kg — and sometimes up to 500 kg — N ha^{-1} yr^{-1}. But such estimates are subject to a number of variables such as soil, climate, and plant management conditions. Furthermore, it has lately been found that high variability exists among provenances or isolines of NFTs in the percentage of total plant N derived from atmospheric N_2 (% Ndfa) (Sanginga et al., 1995; 1996). Yet another problem is the difficulty in assessing the extent to which the N_2 fixed by NFTs becomes available (N recovery rates) to crops that are associated with the NFTs during current and subsequent seasons. The extent of N recovery is dependent on the rate of organic matter decomposition and N mineralization (discussed in Section 3.2). To complicate the issue further, little is known about the effects that management practices can have on N_2 fixation, for example, the effect of pruning on N_2-fixing, hedgerow trees. Thus, even though many of the tree species used in alley cropping systems are active N_2 fixers, we do not yet clearly understand the extent and time sequence of the benefits that these species will deliver under actual field conditions.

The extent of N_2 fixation by trees in cropland systems is not clearly known. Several of the widely used trees in these systems (e.g., néré, karité, and neem) are not NFTs, and even the NFTs (e.g., *F. albida* and most *Acacia* species) are known to have low N_2 fixation rates (Dommergues, 1995). Most observations and discussions on soil fertility improvement by trees in these systems (for example, see Section 4.2) agree that the "tree effect" results from the cumulative benefits of several factors, such as BNF, nutrient cycling, and nutrient accumulation under trees via excreta from animals and birds that take shelter under or in the trees.

The present state of knowledge on BNF in sequential agroforestry systems, such as improved fallows, is somewhat similar to that in simultaneous systems. Here again, exact information on BNF by the NFTs and the rate and extent to which the fixed N becomes available to succeeding crops are not clearly known.

Nitrogen fixation has been evaluated in a number of studies involving shaded perennial-crop systems as well. Studies during the 1980s in unfertilized coffee and cacao plantations shaded with *Inga jinicuil, Gliricidia sepium*, and *Erythrina poeppigiana* (Escalante et al., 1984; Roskowski and Van Kessel, 1985) and in fertilized plantations under *E. poeppigiana* (Lindblad and Russo, 1986) estimated N_2 fixation levels of 35 to 60 kg N ha^{-1} yr^{-1}, using acetylene reduction assay. These could well be underestimates because the acetylene reduction method measures only short-term nitrogenase activity (Peoples and Herridge, 1990). Comparing nutrient balances of leguminous and non-leguminous shade-tree/coffee associations, Fassbender (1987) also estimated 60 kg N ha^{-1} yr^{-1} as N_2 fixation by *E. poeppigiana*. Nygren and Ramirez (1995) found that *E. poeppigiana* nodules disappeared almost completely for 10 weeks after pruning, which suggests that there may be 20 weeks in the year during which these biannually pruned trees would not fix N_2 and hence compete with the associated crop for soil N. Herrera et al. (1987) reported that the nodules of unpruned *E. poeppigiana* shade trees in cacao plantations in Venezuela disappeared during the dry season. These reports show that contributions of N by N_2-fixing shade trees in coffee and cacao plantations are relatively low.

Because of its relevance to the present discussion, it is worthwhile to mention another simultaneous system involving perennial species although it may not be considered an agroforestry system: the mixed planting of leguminous and nonleguminous, fast-growing tree species, such as *Acacia* s and *Eucalyptus* s, in short-rotation industrial plantations. This practice is gaining popularity in some parts of the tropics, especially in southeast Asia (Pandey, 1995). Comparing nutrient cycling under such mixed stands with nutrient cycling under pure stands, Khanna (1997) concludes that the addition of NFTs to pure stands of non-NFTs may alter nutrient cycling in the system through (1) direct effects from the addition of N by NFTs, (2) indirect effects due to interactions caused by the addition of N by NFTs, and (3) increased competition for nutrients. In six-year-old mixed plantings of *Eucalyptus* sp. and *Albizia* sp. in the proportion of 34:66 in Hawaii, Binkley (1992) and Binkley et al. (1992) observed higher biomass production, above-ground net primary production, and annual growth increment in the mixed-species stand than in the respective single-species stands. Similarly, in Thailand, Foelster and Khanna (1997) observed higher mean tree basal area and basal area increments in four-year-old mixtures of *Eucalyptus globulus* and *Acacia mearnsii* than in the pure stands of these species. Both studies concluded that N contribution from the NFTs in the mixtures was substantial enough to not only compensate for, but even to outweigh the detrimental effects of possible competition between the species for light, water, and other nutrients, especially P. Although detailed studies that describe the processes leading to such enhanced growth of species in mixtures are lacking, facilitation through N additions to the systems is strongly suggested.

NFTs are a valuable resource in agroforestry systems. However, some of the widely held assumptions about their benefits could be wrong or information about them may be inadequate. Because of methodological difficulties in quantifying N_2 fixation under field conditions, especially in older tree-stands (Danso et al., 1992; Sanginga et al., 1996), quantitative information on the extent of benefit that is actually realized by using NFTs in agroforestry systems is far from satisfactory. Furthermore, it is not clearly understood what proportion of the N_2 that is fixed by a NFT is actually utilized by, or potentially made available to, an associated crop during the current crop cycle, and what proportion goes into the soil's N store for eventual use by subsequent crops. If the N is transferred continuously from the NFT to the soil, the inclusion of the NFT should enhance the soil N status in the long run. Obviously, rigorous, long-term monitoring of these aspects is essential.

3.2 Tree Biomass and its Decomposition

One of the major recognized avenues of soil fertility improvement in tropical agroforestry systems is the recycling of nutrients through decomposition of tree biomass — mainly leaf litter or prunings, but also roots — that is added to the soil. Obviously, the extent of benefits derived will depend on the quantity and nutrient content of the biomass added, and the rate at which it is decomposed. Voluminous information is available on the nutrient content and quantity of biomass produced by different trees and shrubs used in agroforestry systems under a variety of conditions, especially in systems such as alley cropping and improved fallows where soil fertility improvement is a major objective. As is to be expected, considerable variation exists in such data. Most reports on nutrient content of tree biomass deal with N; other elements such as P and K are less commonly reported. The C-to-N ratios of the leaf biomass of 17 N_2 fixers included in Table 1 range from 10 to 25, whereas for the 10 non-N_2-fixing species, the range is from 14 to 32. If the C content of the leaf biomass is assumed to range from 45 to 50%, the N content of leaf biomass will range from 2 to 5% for these N_2-fixing species and 1.4 to 3.5% for the non-N_2-fixers. As for P and K, information of a general nature available in the literature shows a range of 0.15 to 0.29% for P and 0.9 to 1.52% for K in leaf biomass of common agroforestry tree species (Nair, 1993: 293; Palm, 1995). Added to these variations in nutrient contents of the materials, there is enormous variation in the reported quantities of biomass production by different species under various situations (Table 2). Therefore, the extent of soil fertility improvement caused by nutrient cycling via tree biomass decomposition in agroforestry systems is very site-specific.

While trees in agroforestry systems may supply N to associated crops, their ability to supply P is very limited. Many tropical soils have very low native P levels (Sanchez and Palm, 1996; Buresh et al., 1997). Indeed, the low native soil P, high P fixation by soils with high iron and aluminum contents, and the nutrient-depleting effects of long-term cropping without additions of adequate external inputs have contributed to P deficiencies in many tropical soils (Jama et al.,

1997). Nevertheless, application of tree biomass to the soil has been shown to increase crop available P especially in the highly weathered tropical soils. This is achieved either directly by the process of decomposition and release of P from the biomass or indirectly by the production of organic acids (by-products of decomposition) that chelate iron and aluminum, reducing P fixation (Coleman et al., 1983; Nziguheba et al., 1998). However, as reported by Palm (1995), the quantity of P contained in the biomass of most multipurpose tree species used in agroforestry systems is insufficient to supply the associated crop's P demand, though the biomass may contain sufficient N to meet the immediate crop N requirements. Jama et al. (1997) concluded that it could be economically attractive to integrate an inorganic P source with the organic material, whereby the organic material would provide the required N for the crop and the inorganic P source would meet the additional requirement of P.

Roots constitute a "black hole" in our understanding of nutrient cycling in agroforestry systems: compared to above-ground tree biomass (i.e., leaves and twigs), very little is known about their dynamics. Almost all past reviewers of the subject (e.g., Young, 1989; Anderson and Sinclair, 1993; Nair, 1993; Schroth, 1995) are unanimous about the importance of roots in nutrient cycling and soil fertility maintenance in agroforestry systems, as well as about the unsatisfactory level of knowledge on the subject. Widely varying estimates of root biomass addition have been reported from different agroforestry systems. For example, Alpizar et al. (1986) and Fassbender et al. (1991) reported that in a five-year-old stand of *Theobroma cacao* and *Cordia alliodora* in Costa Rica with an above-ground biomass store of 45.9 Mg ha^{-1}, the fine- and small-root biomass constituted 4.2 Mg ha^{-1} (9% of above-ground biomass). The corresponding figures for a 10-year-old stand were 95.4 and 9.8 Mg ha^{-1} (10%). These percentage figures are on the lower end of the range (three to 33) reported by Vogt et al. (1997) for a wide variety of tropical forest- and forest-plantation ecosystems. Schroth and Zech (1995) reported that an alley cropping system involving *Gliricidia sepium* with maize (*Zea mays*) and groundnut (*Arachis hypogaea*) in the West African rainforest zone produced 1.1 Mg ha^{-1} yr^{-1} of root biomass in the 0-50 cm soil layer, equivalent to 8.8% of aboveground biomass production (13.6 Mg ha^{-1} yr^{-1}). Govindarajan et al. (1996) reported from semiarid highlands of Kenya that in an alley cropping system of *Leucaena leucocephala* and maize, fine-root-biomass production by *L. leucocephala* was only 510 kg ha^{-1} during a cropping season of about 120 days.

It is believed that root tissues are continuously sloughed off and replaced, and that these sloughed-off tissues, along with senescent and dead roots constitute a significant avenue of addition of organic matter (and nutrient) addition to the soil ecosystem. Furthermore, aboveground management of plants (such as pruning in hedgerow intercropping systems) might influence root dynamics. Studies on these aspects have been very limited and inconclusive. A major problem is methodological: sound procedures for determining the quantities and dynamics of roots, especially fine roots, are not yet available.

Decomposition of organic materials and the rate at which their nutrients are released are determined by the "quality" of the material, the environment, and the decomposer organisms that are present (Swift et al., 1979). Since many recent studies have focused on the quality of plant biomass available in agroforestry systems, a discussion on the current status of this topic will be useful.

3.2.1 Plant Litter Quality

For this discussion, the quality of an organic material refers to its (organic) constituents and nutrient content (Mafongoya et al., 1997c; Cadish and Giller, 1997). Organic (C) constituents are important because the energy available to decomposer organisms depends on the proportion of soluble C, cellulose and hemicellulose, and lignin. Soluble C includes metabolic and storage C, and is primarily responsible for promoting microbial growth and activity (Smith, 1994). Green foliage usually contains 20% to 30% soluble C. Cellulose and hemicellulose, which constitute 30%

TABLE 1
Leaf and litter characteristics and resulting decomposition patterns of multipurpose agroforestry trees. Direct values are not given due to difficulties in comparing results obtained with different analytical methods. Instead, the likely relative leaf qualities and the probable decomposition characteristics are given, based on extrapolation from data for fully opened leaves (including pinnae). A higher number of crosses (+) indicates a higher relative value. Source: Mafongoya et al. (1997)

Species	N_2-fixer	C-to-N ratio	Lignin content	Polyphenol content	Protein-binding capacity	Decomposition rate	%N released in first month	Ref.[a]
Acacia angustissima	yes	18	+++	+++	++++	slow	ND[c]	7
A. auriculiformis	yes	20	++	+++	ND	ND	immobilizes	1
Apuleia leiocarpa	no	18	+	+	+	moderate	immobilizes	13
[b]*Dactyladenia barteri*	no	25-30	++++	+++	ND	slow	ND	11
Cajanus cajan	yes	10-15	+/++	+/++	+	moderate	ND	7,9
Calliandra calothyrsus	yes	10-18	++	+++	++++	slow	15-30	3,7,8
Centrolobium ochroxylum	no	18	+	+++	+++	moderate	immobilizes	13
Crotalaria anagyroides	yes	12	+	ND	ND	fast	ND	5
[b]*Croton macrostachyus*	no	14	++	++	ND	ND	20	6
Erythrina s	yes	11	++	+	ND	fast	25	9
Flemingia macrophylla	yes	19-25	++++	+++	++	slow	ND	7,10
Gliricidia sepium	yes	10-20	+	+	+	fast	35-60	1,2,3,7,9,11
[b]*Grevillia robusta*	no	32	+++	++	ND	slow	immobilizes	6

Inga edulis	yes	14-19	+++	+++	ND	slow	0-10	1,8,9
Leucaena leucocephala	yes	10-16	++	+++	+	fast	30	2,3,7,9,11
Myroxylon balsamum	yes	16	+	ND	+	moderate	ND	13
Peltophorum dasyrrachis	no	20	+++	ND	+++	slow	12-15	2,3
Piptadenia buchtienii	yes	16	+	+++	++++	slow	ND	13
Pithecellobium saman	yes	14	+	+	ND	fast	ND	9
Schizolobium amazonicum	yes	15	+	++	+	fast	ND	13
Senna reticulata	no	18	+	+	ND	ND	0-20	8,9
Senna spectabilis	no	14	+	+	+	moderate	ND	4,10
Senna siamea	no	16-18	+	++	+	moderate	immobilizes	1,8,12
Sesbania sesban	yes	15-16	+	+/++	+	fast	55	1,7
Tephrosia candida	yes	13-20	+	+	ND	fast	ND	5
Tephrosia vogelii	yes	15	+	+	ND	ND	ND	10
[b]*Tithonia diversifolia*	no	14	+	+	ND	ND	30	6

[a]1, Constantinides and Fownes (1994b); 2, Handayanto et al. (1996); 3, Handayanto et al. (1994; 1995);, 4, Itimu, O.A., Cadisch, G. and Giller, K.E. (unpublished results); 5, Joachim and Kandiah (1936); 6, Kwabiah, and Palm, C. (unpublished results); 7, Mafongoya (1995); 8, Oglesby and Fownes (1992); 9, Palm and Sanchez (1991); 10, Palm, C. and Wangari, N. (unpublished results); 11, Tian et al. (1992a,b); 12, Tian et al. (1995); 13, Vargas, E. and Giller, K.E. (unpublished results).

[b] non-legumes

[c]ND = no data

TABLE 2
Biomass production by tree species in some alley cropping and biomass transfer systems in the humid, subhumid, and semiarid zones of the tropics.

Site description and species	Trees (nos/ha)	Tree age (months)	Prunings (no/yr)	Dry matter (Mg/ha/yr)	Ref.
Hyderabad, India; rainfall 750 mm/yr; alfisol, pH 7.0, P = 8 ppm (Olsen)					
Leucaena leucocephala	2 000	48	ND[1]	1.4 l	1
	833	48	ND	7.4 l+w[2]	2
Ibadan, SW Nigeria; rainfall 1280 mm/yr; alfisol, pH 6.2, P = 25 ppm (Olsen)					
Leucaena leucocephala	10 000	36	6	6.5 l	3
Yurimaguas, Peru; rainfall 2200 mm/yr; ultisol, pH 4.2-4.6, P = 8 ppm (Olsen)					
Inga edulis	8 888	11	3	9.6 l+w	4
Gliricidia sepium 14/84	5 000	11	3	8.1 l+w	
Gliricidia sepium 34/85	5 000	11	3	1.8 l+w	
Onne, SE Nigeria; rainfall 2400 mm/yr; ultisol, pH 4.0, P = 50 ppm (Bray-1)					
Acioa barteri	2 500	48	ND	13.8 l+w	5
Alchornea cordifolia	2 500	48	ND	14.9 l+w	
Cassia siamea (Serna Siamea)	2 500	48	ND	12.2 l+w	
Gmelina arborea	2 500	48	ND	12.3 l+w	
Sumatra; rainfall 2575 mm/yr; oxisol, pH 4.1, P = 4.8-6.8 mg/kg (Melich I)					
Paraserianthes falcataria	19 900	09	4	4.9 l+w	6
		21	4	9.7 l+w	
Calliandra calothyrsus	19 900	09	4	6.8 l+w	
		21	4	10.7 l+w	
Gliricidia sepium	10 000	09	4	0.6 l+W	
		21	4	1.4 l+w	
Costa Rica; rainfall 2640 mm/yr; inceptisol, pH 4.3-4.8, P = 15 ppm (Olsen)					
Gliricidia sepium	6 666	24	2	9.6 l+w	7
		60	2	15.2 l+w	
Erythrina poeppigiana	555	24	2	7.4 l+w	
		60	2	11.1 l+w	
Western Samoa; rainfall 3000 mm/yr; mod. fertile inceptisol, no soil data					
Calliandra calothyrsus	5 000	48	3	12.1 l+w	8
	3 333	48	3	7.6 l+w	
Gliricidia sepium	5 000	48	3	10.7 l+w	
	3 333	48	3	6.5 l+w	
Embu, Central Kenya; rainfall 1500 mm/yr; ultisol, pH 5.5, P = 7 mg/kg (Olsen), N = 0.25 g/100g					
Calliandra calothyrsus	4 444	48	2	4.0 l	9
Leucaena leucocephala	4 444	48	2	3.9 l	
Katherine, Northern Australia; rainfall 948 mm/yr; alfisol, pH 6.6, P = 5 mg/kg (bicarb), N = 1.1%					
Leucaena leucocephala	ND	60	ND	1.7 l+w	10
Chitedze, Malawi; rainfall 750-785 mm/yr; alfisol, pH 5.7, P = 2.4 mg/kg (Olsen), N = 0.18%					
Leucaena leucocephala	ND	60	2	3.1 l	11

TABLE 2 (continued)
Biomass production by tree species in some alley cropping and biomass transfer systems in the humid, subhumid, and semiarid zones of the tropics.

Site description and species	Trees (nos/ha)	Tree age (months)	Prunings (no/yr)	Dry matter (Mg/ha/yr)	Ref.
Machakos (NDFRC), Kenya; rainfall; 700 mm/yr; alfisol, pH 6.0-6.5, P = 10-16 ppm (Olsen), N = 0.08%					
Senna siamea	11 111	60	2	3.1 l	12
Machakos (ICRAF), Kenya; rainfall; 700 mm/yr; alfisol, pH 6.0-6.5, P = 10-16 ppm (Olsen), N = 0.08%					
Leucaena leucocephala	2 998	48	4	2.2 l	13
Senna siamea	2 998	48	4	2.7 l	

[1] ND = not determined
[2] l = leaves and green shoots; w = woody material
Source: Updated from Fernandes et al. (1994):
References;
1. Singh et al. (1989)
2. Hocking and Rao (1990)
3. Kang et al. (1981)
4. Fernandes (1990)
5. Ruhigwa et al. (1992)
6. Evenson (1989)
7. Kass et al. (1989)
8. Rosecrance et al. (1992)
9. Mugendi (1997)
10. Xu et al. (1993)
11. Jones et al. (1996) and Wendt et al. (1996)
12. Mugendi et al. (1994)
13. Jama et al. (1995)

to 70% of plant C (12% to 30% of total plant material) are structural polysaccharides of "intermediate" quality; they are attacked by the decomposer microbes after soluble carbohydrates have been depleted (Swift et al., 1979). Lignin, which intertwines the cell wall and physically protects cellulose and other cell wall constituents from degradation (Chesson, 1997), is the "lowest"-quality C constituent, providing little or no energy to the decomposers until the last stages of decomposition. Thus, the lignin content of the organic material is considered to be the most important factor determining the rate of decomposition (Jama and Nair, 1996; Mafongoya et al., 1997a; Mafongoya and Nair, 1997; Mugendi and Nair, 1997). Lignin content of agroforestry tree species varies from 5% to 20% of dry weight in green foliage and 10% to 40% in senescent foliage or leaf litters (Constantinoides and Fownes, 1994a). It has been suggested that 15% is a critical level, above which decomposition is impaired (Mafongoya et al., 1997c).

Many recent studies with agroforestry tree species have shown that polyphenols, which comprise a relatively small percentage of the organic material, have a disproportionately large negative influence on decomposition and N release (Palm and Sanchez, 1990; Constantinides and Fownes, 1994a; Handayanto et al., 1994; Mafongoya et al., 1997a;b). Although polyphenols can serve as a C substrate for decomposers, in general, they inhibit the growth or function of decomposers (Swift et al., 1979). Condensed tannins, also known as proanthocyanidins, are the polyphenols most noted for their effects on decomposition and N release. Some condensed tannins are soluble in polar extractants, while others are insoluble and bound to protein or the cell wall (Jackson et al., 1996). The effect of the bound condensed tannins (insoluble) is similar to that of lignin: they make the cell wall and proteins physically or chemically inaccessible to decomposer organisms and thus slow decomposition. In general, total soluble polyphenol content of green foliage of agroforestry species can be as high as 10%, but is usually less than 5% (Constantinides and Fownes, 1994a). There is no apparent correlation between total polyphenols and condensed tannins. Some species such as *Senna (*syn. *Cassia) siamea* have a fairly high total soluble polyphenol content (Constantinides and Fownes, 1994b) but a very low level of condensed tannins (Jackson et al., 1996), whereas *Calliandra calothyrsus* has high levels of both fractions (Handayanto et al., 1995). In addition to the C quality, nutrient — especially N — content of plant materials is a major determinant of litter quality. Generally, materials with N content higher than 20 mg g^{-1} are considered to be of high quality, although this can be modified by lignin and polyphenol contents (Mafongoya et al., 1997a;c).

Several recent studies have related the rate of biomass decomposition with a number of these plant (litter)-quality indices. These include ratios of C to N, polyphenol to N, lignin to N, and polyphenols + lignin to N (Palm and Sanchez, 1991; Oglesby and Fownes, 1992; Handayanto et al., 1994; 1995; Constantinides and Fownes, 1994b; Tian et al., 1995; Jama and Nair, 1996; Mafongoya and Nair, 1997; Mafongoya et al., 1997a; 1997b; Mugendi and Nair, 1997). All of these are valid indicators, but each has its own advantages and disadvantages (Mafongoya et al., 1997b).

3.2.2 Litter (Biomass) Decomposition

In addition to quality variations in biomass from different species, other factors such as the methods of analyses and the plant parts that were examined complicate comparisons of the abundant decomposition data that are becoming available. Nevertheless, it is possible to make reasonably accurate predictions about decomposition rates of plant materials that are commonly used in agroforestry systems. Mafongoya et al. (1997c) summarized these for a number of agroforestry species, as reported in Table 1. They concluded that leaves that are high in N, low in lignin, and low in polyphenols (e.g., those of *Gliricidia sepium* and *Sesbania* spp.) will decompose quickly and release a large proportion of their N. Well-lignified leaves (e.g., those of *Dactyladenia barteri* and *Flemingia macrophylla*) will decompose slowly and may cause immobilization of soil N for a fairly long period (several weeks) after they are added to the soil. The decomposition pattern of biomass of species with high N and polyphenol contents may be governed by the protein-binding capacity of the polyphenols: decomposition will be rapid when protein-binding capacity is low (as in *Leucaena leucocephala*), whereas decomposition will be slow when protein-binding capacity is high (as in *Calliandra calothyrsus*). Furthermore, even species with narrow C-to-N ratio and low lignin and polyphenol contents may decompose slowly if large amounts of N are bound to condensed tannins, as is the case with *Senna siamea.* Thus, the large variations in decomposition patterns for biomass from several agroforestry species, reported from a wide variety of situations, can largely be interpreted based on the chemical quality parameters of the materials, and this information can be — and is now being — used for making appropriate management decisions.

3.2.3 Management of Decomposition and Nutrient-use Efficiency

Biomass decomposition can be "manipulated" to improve the efficiency of uptake and utilization of nutrients by growing plants, especially in simultaneous agroforestry systems. Mafongoya et al. (1997c) suggest two strategies for this: (1) regulate the rates of release of nutrients to improve the synchrony of nutrient supply with plant (crop) demand, and (2) provide a more favorable environment for plant growth. While the former is of "immediate" (short-term) nature, the latter involves longer-term improvements, often mediated through improvements in soil organic matter (SOM) status.

Green foliage composes the bulk of plant biomass that is available for decomposition in agroforestry systems, as opposed to senescent material (litter) that dominates the biomass input in natural and agricultural systems. Because mobile nutrients are translocated from senescent leaves to other plant parts before litterfall, litter differs from green foliage in quality and, therefore, decomposition rates (Constantinides and Fownes, 1994a). Conditions prevailing during tree growth can also result in plant biomass of differing quality. Nitrogen limitation, in particular, can enhance concentration of polyphenols in leaves. For example, Handayanto et al. (1995) produced litters of *C. calothyrsus* and *G. sepium* of widely varying quality by altering N supply to the plants. Increased N concentration and reduced polyphenol concentration in leaves consequent to enhanced N supply to the plants resulted in faster decomposition and uptake of released N by maize (Handayanto et al., 1997a;b). These results indicate that higher N_2 fixation by trees could result in the production of "better-quality" leaf biomass compared with biomass from trees growing under N-starved conditions.

A number of field management operations can alter biomass quality or the rate of its decay: (1) the duration and temperature of drying the material before applying it to the soil (fresh prunings decompose faster than sun-dried prunings), (2) the physical size of the material (smaller-sized or ground materials decompose faster than larger and coarser materials), (3) the mixing of biomass of differing compositions, and (4) the method of applying the materials (incorporating materials into the soil results in faster decomposition than surface placement) (Mafongoya et al., 1997b;c).

When tree biomass is used as a source of nutrients for crops, it is important to ensure synchrony between the release of nutrients (via decomposition) and their uptake by the crop (Swift, 1987; Nair, 1993). Improved synchrony will enhance nutrient-use efficiency by minimizing the loss of nutrients (Myers et al., 1994). Synchrony can be achieved (1) by manipulating the crop's demand for nutrients through adjustments in the time of planting and crop selection, and (2) by manipulating nutrient release through adjustments in biomass management, as described earlier in this section. A schematic presentation of manipulation of synchrony is provided in Figure 3.

An important point to consider in this context, as in the case of biological N_2 fixation, is the so-called nutrient recovery, which indicates the extent to which the nutrients that are released from biomass decomposition are taken up by the current (and subsequent) seasons' crops. Many leguminous tree species used in agroforestry systems, especially alley cropping and biomass transfer systems, are capable of producing substantial quantities of biomass (see Table 2), through which nutrients, with the notable exception of P, are recycled in quantities sufficient to support crop growth (Young, 1989; Szott et al., 1991; Palm, 1995). For example, Table 3 shows Palm's (1995) comparison of the nutrient requirement of an "average" crop of maize and the nutrient addition provided in 4 Mg ha^{-1} of leaves of some commonly used MPTs. However, the reported rates of nutrient recovery by the current season's crop from decomposing tree biomass are highly variable and, in general, low (from about 10% to 40%, with some estimates as high as 60%) (Haggar et al., 1993; Handayanto et al., 1994; Palm, 1995; Mafongoya and Nair, 1997). In subhumid Kenya, Mugendi et al. (1997b) used ^{15}N to estimate N recovery from tree biomass applied to the soil in an alley cropping experiment. Only 9% to 13% of the initial ^{15}N was recovered by the first season's maize crop, while 55% to 69% was recovered in the soil organic N pool after the cropping season. The remaining 20% to 30% of the ^{15}N could not be accounted for (Table 4). Haggar et al. (1993) also reported the amount of N left in the soil after the first crop to be as high as 80% of the initial N applied in the tree biomass. Although low recovery by a crop of N released from decomposing organic material does not necessarily imply a corresponding build-up of SOM, these studies suggest that a considerable portion of the N added as tree biomass to crop production fields can be retained in SOM.

In shaded perennial-crop systems, transfer of N from the N_2-fixing leguminous shade trees to non-N_2-fixing associated crops has generally been assumed to occur largely through the decomposition of aboveground pruning residues and litterfall (e.g., Fassbender, 1993). Studies carried out in Latin American coffee and cacao plantations, with 120 to 560 leguminous shade trees per hectare pollarded 0 to 3 times annually, showed that these inputs could vary from 3 to 14 Mg ha^{-1} yr^{-1} of dry matter containing 60 to 340 kg N ha^{-1} yr^{-1} (Beer, 1988). Escalante et al. (1984) calculated that 57 to 66 kg N ha^{-1} yr^{-1} was released through nodule senescence and decomposition, with no difference in nodule N content (22 to 23 kg N ha^{-1}) between fertilized and unfertilized plots. Nygren

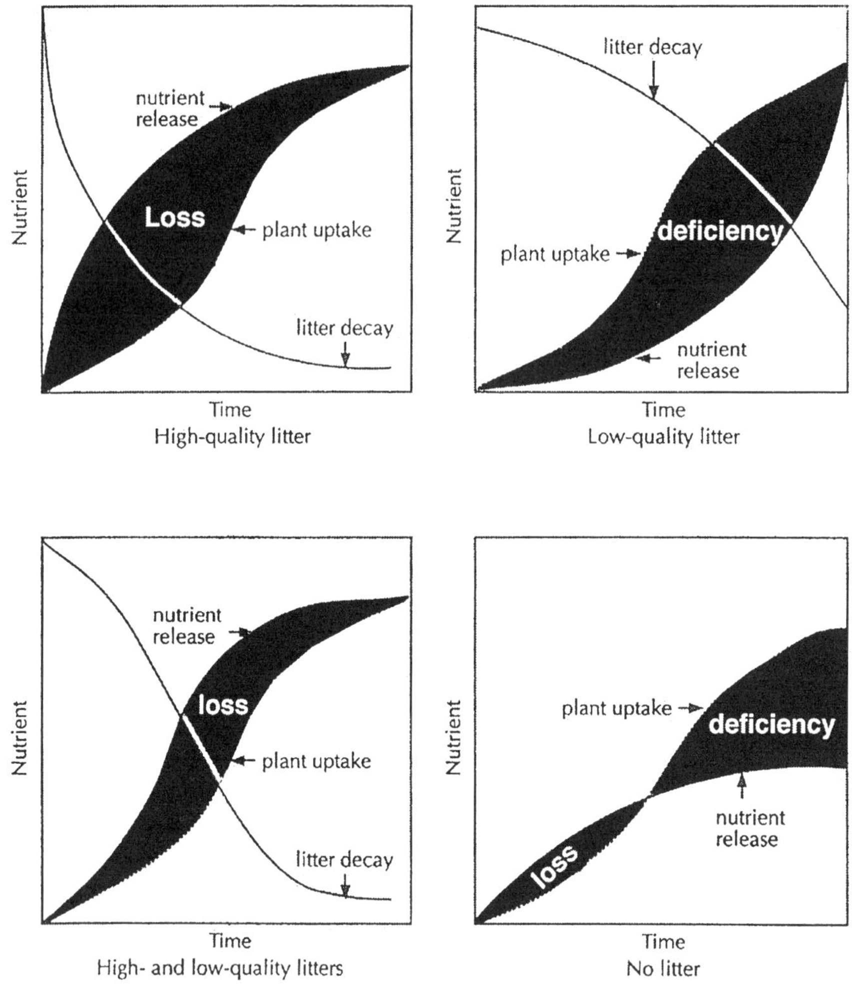

FIGURE 3. Hypothetical patterns of nutrient availability in four treatments of an experiment to test the synchrony principle. Source: Myers et al. (1994).

and Ramírez (1995) found a turnover of 6.8 to 35.4 g N tree^{-1} in a 23-week pruning cycle (9.6 to 50.0 kg N ha^{-1} yr^{-1}) through *E. poeppigiana* nodule senescence and decomposition. These studies suggest that a significant proportion of N_2 fixed by shade trees may be transferred below ground to non-N_2-fixing plants. As regards the extent of N cycling, Babbar and Zak (1994, 1995) found higher rates of N mineralization in Costa Rican coffee plantations shaded by *E. poeppigiana* (148 kg N ha^{-1} yr^{-1}) than in plantations without shade trees (111 kg N ha^{-1} yr^{-1}; both sites were heavily fertilized with mineral N at rates up to 300 kg N ha^{-1} yr^{-1}). They concluded that N cycling was more efficient in shaded plantations because, despite the greater availability of mineralized N, less N was lost through leaching.

3.3 Tree Uptake of Nutrients from Deeper Soil Layers

It has long been recognized that in some tree species, roots extend far deeper into the soil than the rooting depth of common agricultural crops (Stone and Kalisz, 1991). Recent research on agroforestry trees has focused on this deep-rooting attribute of trees, with a view to understanding the spatial distribution and temporal patterns of root growth (Jonsson et al., 1988; Ruhigwa et al.,

TABLE 3
Comparison between nutrient requirements of a maize crop[1] and nutrient addition by MPTs leaves[2].

	Nutrients kg ha^{-1}				
Species	**N**	**P**	**K**	**Ca**	**Mg**
Crop requirement	80	18	66	15	10
Leucaena leucocephala	154	8	84	52	13
Erythrina poeppigiana	132	7	46	61	—
Inga edulis	127	9	50	30	7
Senna siamea	105	6	44	110	7
Dactyladenia barteri	60	4	31	40	8
Grevillea robusta	52	2	24	60	7

[1] A maize crop yielding 2 Mg/ha grain and 3 Mg/ha stover.
[2] MPT leaves at the rate of 4 Mg/ha.

Source: Adapted from Palm (1995).

1992; van Noordwijk et al., 1996) and relating such information to nutrient uptake by tree roots from deeper soil layers (Mekonnen et al., 1997; Buresh and Tian, 1997). Reviewing the current level of knowledge in this area of research, Buresh and Tian (1997) concluded that the potential of trees to take up subsoil nutrients is generally greatest when the trees have deep root systems and a high demand for nutrients, and when they are grown in locations with water and/or nutrient stress in the surface soil but considerable reserves of plant-available nutrients or weatherable minerals in the subsoil.

In western Kenya, researchers have noted the accumulation of fairly large quantities of nitrate (70 to 315 kg N ha^{-1}) in acid soils (Kandiudalfic Eutrudox) at 0.5- to 2.0-m depth under unfertilized maize and have attributed it to the formation of nitrate by mineralization of SOM and the sorption and retention of nitrates by clay minerals (Mekonnen et al., 1997). Fast-growing trees such as *C. calothyrsus*, *Sesbania sesban*, and *Eucalyptus grandis* grown in rotation with maize rapidly root into this nitrate-accumulation zone and take up the sorbed nitrate that is inaccessible to maize (Hartemink et al., 1996; Jama et al., 1998). Mekonnen et al. (1997) observed that nitrate-N to 4-m soil depth was only 51 kg N ha^{-1} in a 15-month *S. sesban* fallow, as compared to 199 kg N ha^{-1} under fertilized maize. The maximum rooting depth was 1.2 m for maize, whereas roots of 15-month-old *S. sesban* extended to below 4 m. In another study on an acid soil in western Kenya, Jama et al. (1998) further showed that fast-growing trees with high root length densities can rapidly utilize subsoil nitrate on soil with no chemical and physical barriers to rooting. Fast-growing *S. seban* and *C. calothyrsus* rooted to >4 m depth in 11 months, but slower growing *G. robusta* had few roots below 3-m depth (Figure 4). *S. sesban* and *C. calothyrsus* had root-length densities of >0.1 cm cm^{-3} to below 1.5-m depth, and they reduced soil nitrate throughout the 2-m-deep soil profile (Figure 5). The reduction in soil nitrate in the top 2 m (150 to 200 kg N ha^{-1}) corresponded to large accumulation of N in aboveground biomass for *S. sesban* (336 kg N ha^{-1}) and *C. calothyrsus* (312 kg N ha^{-1}). Slower growing *G. robusta* only accumulated 107 kg N ha^{-1} in aboveground biomass, and soil nitrate increased rather than decreased during the 11 months after establishment (Figure 5).

The potential for nutrient uptake from deeper soils is much greater for water soluble nutrients such as nitrate than for immobile nutrients such as P. There is typically little potential for trees to take up and recycle P from below the rooting depth of annual crops because plant extractable P is normally low in subsoil and the phosphate ion is relatively immobile in soil (Buresh and Tian,

TABLE 4
Labeled N (^{15}N) recovered in alley-cropped maize and trees supplied with biomass of *ex situ* grown trees that had been treated with ^{15}N in the subhumid highlands of Kenya.

	Labeled N in plants (kg/ha)					Recovery of labeled N (%) at maize harvest							
	Maize					Maize							
Trt⁻	Grain	Cob	Stover	Roots	Tree leaves	Grain	Cob	Stover	Roots	Total	Tree leaves	Soil	Unaccounted?
Call	24.0 b	1.4 a	17.3 b	3.3 b	8.2 a	5.3 a	0.3 a	2.9 b	0.8 b	9.3 b	2.7 b	69.3 a	19.6 b
Leuc	32.7 a	1.8 a	21.8 a	4.7 a	6.4 a	6.5 a	0.3 a	5.0 a	1.4 a	13.0 a	2.1 a	55.0 b	29.7 a

Means followed by the same letter within a column are not significantly different at $p < 0.05$.
Trt⁻ = treatments with enriched ^{15}N calliandra (call) and leucaena (leuc) prunings.

Source: Mugendi (1997).

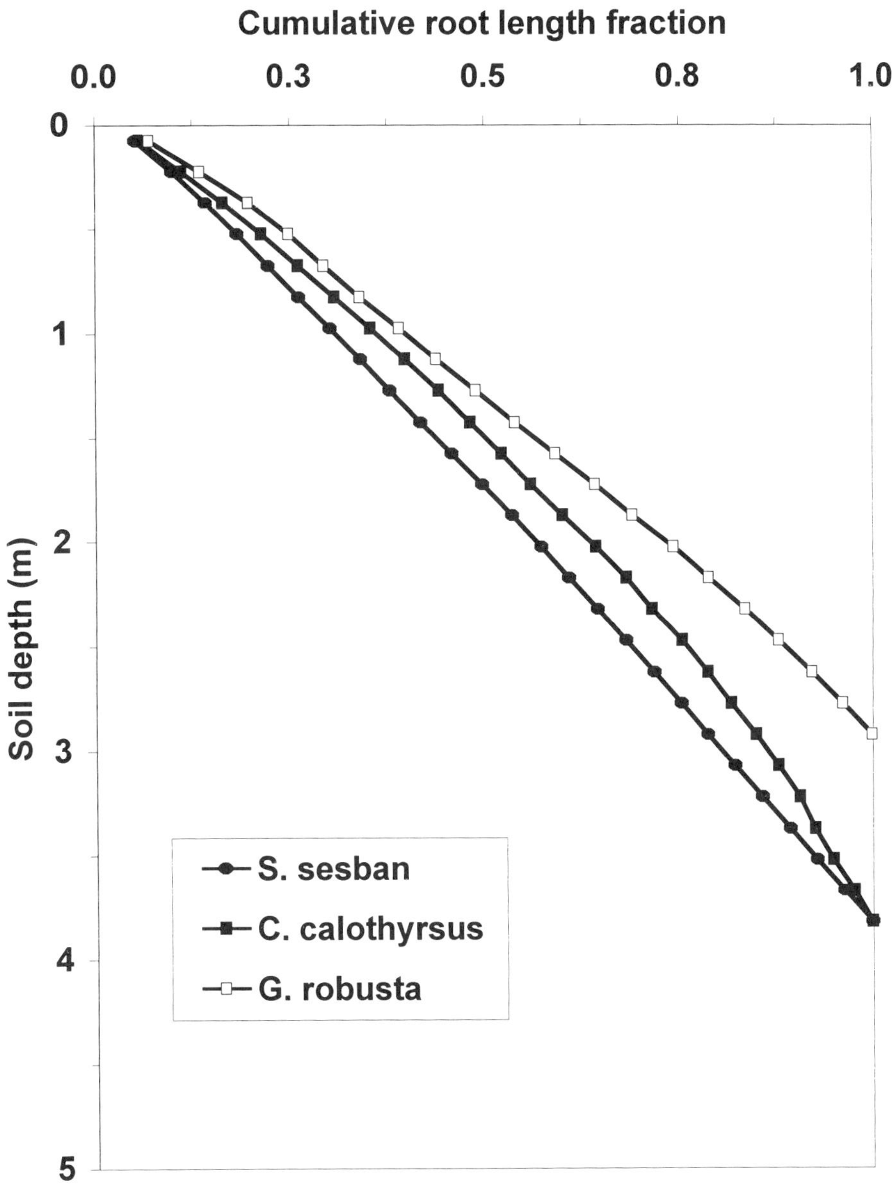

FIGURE 4. Cumulative root-length densities of three agroforestry tree species in western Kenya. Source: Jama et al. (1998)

1997). The role of trees in nutrient uptake from deeper soil layers for nutrients other than N and P is, in general, little studied.

Soil physical and chemical barriers to rooting will reduce the potential of trees to retrieve and take up subsoil nutrients. Mobile nutrients in acid soils of the humid tropics can be leached by high rains into subsoil where high aluminum saturation restricts the rooting of crops. In such systems, the roots of trees with a horizontal spread in the subsoil may act as a "safety net," which intercepts nutrients as they leach down the soil profile (van Noordwijk et al., 1996).

In semiarid areas, the lateral extension of tree roots is considered to be more important in nutrient uptake than the penetration of roots to deeper soil horizons (Breman and Kessler, 1995).

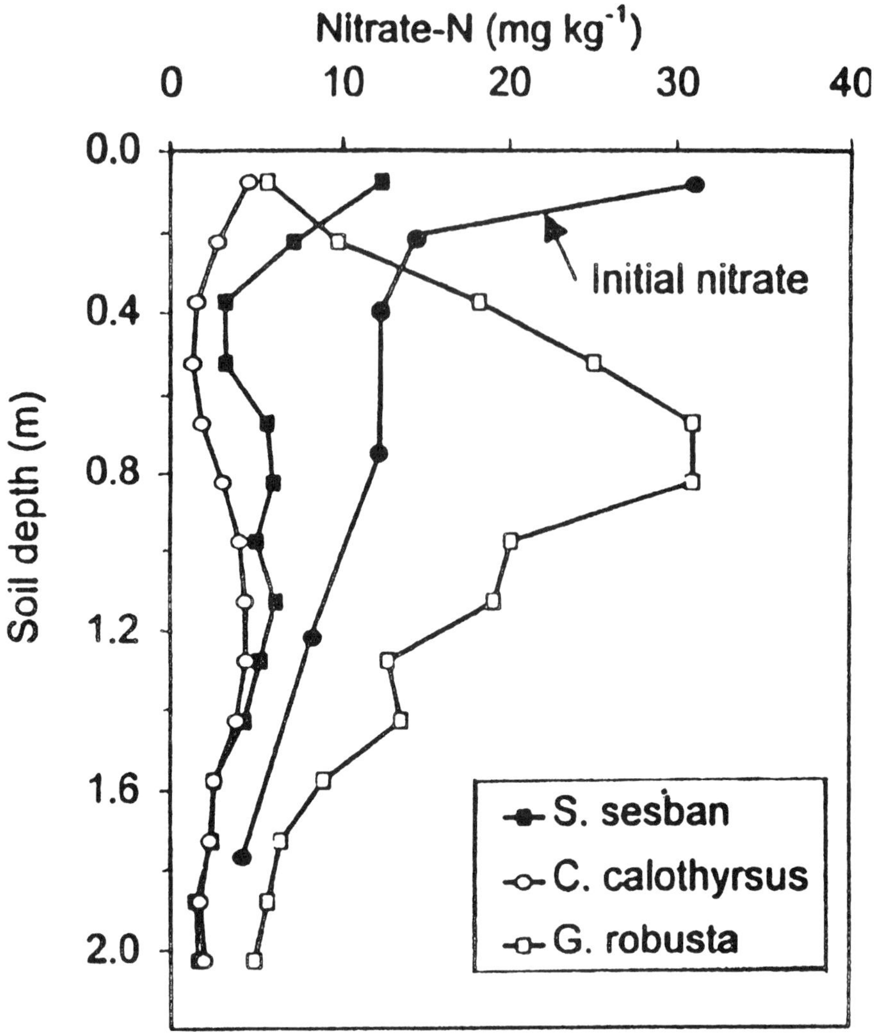

FIGURE 5. Effect of trees on soil nitrate eleven months after planting as seedlings in western Kenya. Source: Jama et al. (1998).

As noted by Sanchez and Palm (1996), nutrients taken up by tree roots from below or beyond the root zone of interplanted annual crops can be an important input in agroforestry systems when these nutrients are transferred to the crop-rooting zone and made available to crops through biomass addition and decomposition.

4. NUTRIENT CYCLING IN TROPICAL AGROFORESTRY SYSTEMS: THE REALITY

Most discussions on nutrient cycling in agroforestry systems fail to connect the seemingly convincing scientific evidence for nutrient cycling — deep root uptake, decomposition processes, improved soil nutrient status — with practically convincing field examples that demonstrate nutrient cycling in action and the benefits derived from this process. Unfortunately, it is difficult to identify field examples that specifically portray the effects of nutrient cycling in farm systems. The effects of nutrient cycling are so inseparably mixed with other tree-crop interaction effects that it becomes complicated — if not impossible and even unrealistic — to isolate them as independent factors and processes. Because of this difficulty, it is often necessary to present the measurable effects of nutrient cycling as an inseparable part of the "package" of cumulative effects of all tree-crop interaction processes. Measures of plant productivity, often expressed as economic yields, provide the most widely used, easily understood, and perhaps practically most important means of evaluating these processes.

In a comprehensive review of biophysical interactions in tropical agroforestry systems, Rao et al. (1997) presented data on crop performances in a number of simultaneous and sequential systems. In the absence of better measures of productivity, we believe that it is reasonable to accept these crop performance data as partial expressions of nutrient cycling. At the same time, we acknowledge that the relative contribution of nutrient cycling will vary among different systems and situations. In this section, we present several examples of agroforestry systems in which nutrient cycling is arguably an important factor.

4.1 Hedgerow Intercropping

In order to assess the long-term performance of hedgerow intercropping (alley cropping), Rao et al. (1997) reviewed the results of 29 trials conducted for four or more years, mostly with small plots, over a wide range of soils and climates across the tropics. Experiments on sloping lands, where the primary benefit is likely to be soil conservation, were excluded. In 28 of the trials, no N fertilizer was applied to crops, but in most trials P fertilizer was applied; the remaining trial evaluated hedgerows for fodder production, with crop fertilization at the recommended rate (Rao et al., 1991b). When the trials involved multiple hedgerow species or different alley widths, results of the best treatment system were considered. The average yields of annual crops from intercropping systems relative to those from sole-crop systems are presented in Figure 6. Yields of sequential crops in bimodal rainfall sites are presented separately if different crops were used in different seasons (hence more than 29 observations in Figure 6). Tree species used in these trials were *L. leucocephala* (n = 12), *Senna siamea* (n = 3), *L. diversifolia* (n = 1), *Gliricidia sepium* alone or mixed with *C. calothyrsus* (n = 2) and *Albizia lebbeck* (n = 1) in the semiarid and subhumid climates; and *C. calothyrsus* (n = 3), *Erythrina* sp. (n = 4), *Inga* sp., *Peltophorum* sp. and *Paraserianthes falcataria* (one site each) in the humid tropics.

The results show both positive effects (n = 15 for cereals, n = 8 for noncereal crops such as beans (*Phaseolus* sp.), cowpea (*Vigna unguiculata*) and cassava (*Manihot esculenta*)) and negative effects (n = 13 for cereal crops and n = 1 for sweet potato (*Ipomoea batatas*) and taro (*Colocasia esculenta*)) of hedgerow intercropping on crop yields, indicating that the system performance is location-specific and sensitive to management, so generalizations may be difficult if not misleading. Ignoring <15% yield increases as unattractive to farmers, only two of 10 studies in semiarid sites (<1000 mm rainfall) gave substantial yield increases. In subhumid environments (rainfall between 1000 and 1600 mm), significant positive yield responses were observed in seven of eleven studies where the soils were either inherently fertile (Kang, 1993; Hauser and Kang, 1993; Mureithi et al., 1994; Shannon and Vogel, 1994) or the acid infertile sites received external nutrients and lime (Akyeampong et al., 1995;1996; Matthews et al., 1992a;b). In the humid tropics (rainfall >2000 mm), maize and taro did not benefit from hedgerow intercropping in four out of eight trials, but interestingly bean and cowpea yields invariably increased.

Whereas low yield of hedgerow prunings (2 to 3 Mg ha^{-1} yr^{-1}) and competition of hedgerows for water were the major reasons for the negative results in water-limited areas (Ong et al., 1991; Rao et al., 1991a; Mathuva et al., 1998), low yield of prunings and competition of hedgerows for nutrients were responsible for negative results in poor soils (Matthews et al., 1992a, b; Fernandes et al., 1993). Inadequate water limited the response of crops even though hedgerow intercropping improved soil fertility in certain sites of the semiarid tropics. In favorable environments, hedgerows produced 8 to 12 Mg ha^{-1} yr^{-1} of prunings (dry weight) which increased SOM and ensured adequate supply of nutrients, especially N, to the alleycrops.

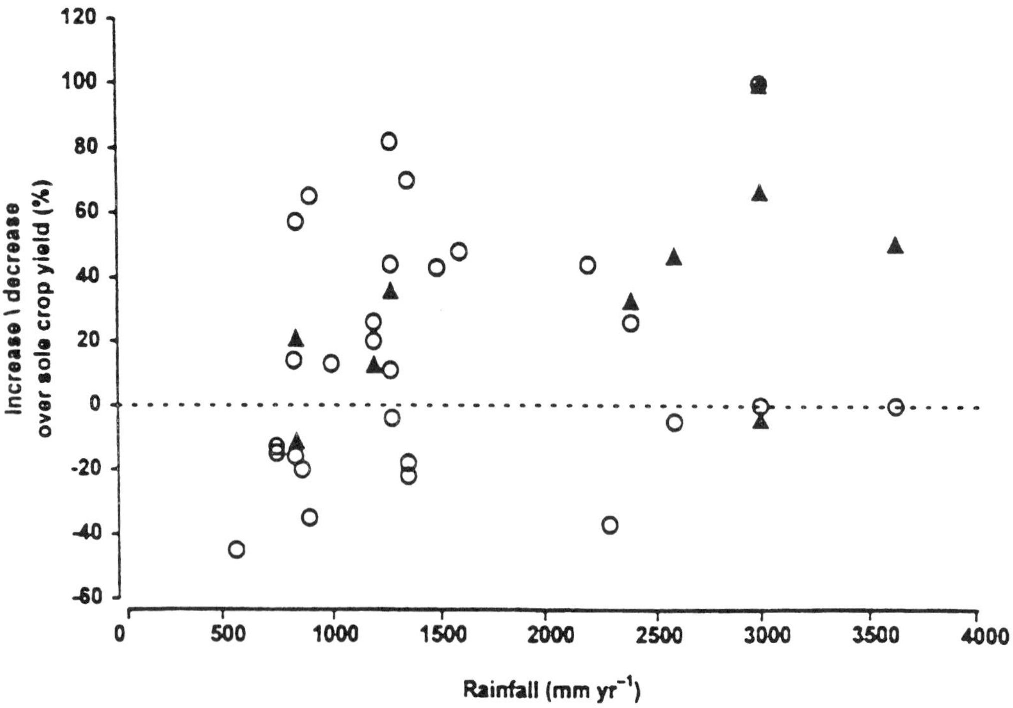

FIGURE 6. Crop yields in hedgerow intercropping expressed as a percent of yields in sole annual cropping in 29 experiments conducted throughout the tropics. Open circles represent the average relative yields of cereal alley-crops (maize (*Zea mays*), sorghum (*Sorghum bicolor*), millet or rice (*Oryza sativa*)) grown singly or in mixtures with other crops. Closed triangles represent the average relative yields of non-cereal alley-crops grown throughout the experimental period such as taro (*Colocasia esculenta*) or in the second season where two crops were grown per year such as beans (*Phaseolus vulgaris*) or cowpea (*Vigna unguiculata*). (Sources: Akondé et al., 1996; Akyeampong et al., 1996; Balasubramanian and Sekayange, 1991; Bundersen, 1992; Chiyenda and Materechera, 1989; Duguma et al., 1994; Evenson et al., 1995; Fernandes et al., 1993; Hauser and Kang, 1993; ICRAF, 1995; Kang, 1993; Kang et al., unpublished; Kass et al., 1995; Lal, 1989a; Mathews et al., 1992a, b; Mathuva et al., 1997; Mittal and Singh, 1989; Mureithi et al.,1994; Mugendi et al. 1997; Rao et al., 1991a; Rosecrance et al., 1992; Schroth et al., 1995b; Shannon et al., 1994; Singh et al., 1989; Tilander et al., 1995; van Noordwijk et al., 1995) Source: Rao et al. (1997)

In Peru, on Typic Paleudults characterized by high acidity and Al toxicity, annual crops did not produce economic yields beyond two or three years without applications of lime and fertilizer. On this site, hedgerow intercropping maintained yields of Al-tolerant rice and cowpea crops at 1 and 0.5 Mg ha^{-1}, respectively, for many seasons (ICRAF, 1995; Palm et al., 1995). Despite declining maize yields in sole-crop systems, hedgerow intercropping produced higher yields over a long period in acid soils of Indonesia (van Noordwijk et al., 1995; ICRAF, 1995) and higher yields in subsequent years after a one-season fallow in acid soils of Burundi (Akyeampong and Hitimana, 1996). In both sites, rainfall was adequate, P was not limiting or was added through fertilizer, and 5 to 12 Mg ha^{-1} of prunings were harvested per year. Reduced Al toxicity to maize, in addition to increased supply of N, could be responsible for the favorable response to hedgerow intercropping in these situations. Many studies have indicated reduction of Al toxicity with the addition of large amounts of organic residues because their decomposition products bind Al and reduce Al saturation (Wong et al., 1995).

4.2 Trees in Cropland (Parkland) Systems

Rhoades (1997) has reviewed parkland systems and summarized the influence of parkland trees on crops and soils (Table 5). Crop-yield increases have been widespread under open and well-managed canopies of fully grown trees. Under *Faidherbia albida*, it has been reported that maize yields were increased by more than 100% in Malawi (Saka et al., 1994) and 76% in Ethiopia (Poschen, 1986); sorghum (*Sorghum bicolor*) yields were increased by 36% in Ethiopia (Poschen, 1986) and 125% in Burkina Faso (Depommier et al., 1992). These yield increases under *F. albida* (often referred to as the 'albida effect') and other tree species are attributed to the combined effects of improved soil fertility, soil water retention and microclimate amelioration. A fertilizer experiment at a low N and P site (Bray 1 P = 3.7 mg kg^{-1} and total N = 0.02%) in the Sahel (N'dounga near Niamey, Niger) indicated that about 60% of the 'albida effect' was due to increased N availability and 40% was due to increased P availability (ICRAF, 1997); however, these results have to be considered as specific to the study site.

Crop yield declines, relative to yields in open, treeless fields, have been noted mostly under the canopies of large, evergreen, unmanaged trees. In Burkina Faso, sorghum yields under karité and néré trees were reduced on average by 50% and 70% respectively (Kessler, 1992). Sorghum yields were reduced by up to 60% under canopies of both these species in southern Mali (Kater et al., 1992). In India, wheat yields were reduced by up to 60% (Puri and Bangarwa, 1992) and mustard (*Brassica* sp.) yields by up to 65% (Yadav et al., 1993) under *A. nilotica* trees. Obviously, detailed analyses of the effect of site-specific factors are essential before generalizations can be drawn from these reports. But, it seems that with favorable tree-shade conditions (achieved either naturally as in the reverse-phenology of *F. albida* trees or deliberately as with the managed canopies of néré and karité trees), tree-mediated soil-improvement — including nutrient cycling — is an important factor that can support crop productivity in the parkland systems.

4.3 Improved Fallows

Sequential systems using planted tree fallows have not received the same level of attention as hedgerow intercropping, but the few studies undertaken so far illustrate the positive effect of tree fallows on crops. Short-duration fallows with herbaceous legumes have been examined widely and found to increase yields of subsequent crops compared with grass fallows or continuous cropping systems (Drechsel et al., 1996; van Noordwijk et al., 1995). Tree fallows, however, have distinct advantages over herbaceous fallows, particularly in seasonally dry climates, because they may take up nutrients from deep soil layers, and accumulate a large quantity of biomass through which nutrients can be recycled (Szott et al., 1994). Furthermore, N_2-fixing trees (as well as herbaceous legumes) may add N to the system through BNF.

On Vertisols in semiarid India, one-year monocrops of pigeonpea, harvested for grain, increased the subsequent maize yield by 57% compared with the yield after a bare fallow. The beneficial effect of pigeonpea, equivalent to the effect of 38 kg ha^{-1} of fertilizer N, was attributed to the enhanced mineralization of its post-harvest residues (Kumar Rao et al., 1983). In Malawi, maize yield following a three-year-old pigeonpea fallow was 55% higher than yield after a natural fallow (Prinz, 1986). However, on Ultisols in Nigeria, while two-year fallow of *Tephrosia candida* increased the yield of subsequent maize by 1.5 Mg ha^{-1} (or 157%) compared with bush fallow, a two-year pigeonpea fallow showed no benefit (Gichuru, 1991). *Tephrosia vogelii* was found to be promising in Rwanda, where a one-year fallow of this species increased yield of the first sequential crop (maize) by 72% over control and the second crop (bean) by 96% (Balasubramanian and Sekayange, 1992). In a review of on-farm trials in Rwanda, Drechsel et al. (1996) found that yields of crops following improved fallows increased up to 74% in the first season and up to 46% in the second season compared with yields of the "no-fallow" control plots; however, yield increases on farms following fallows did not compensate for the loss of crops during the fallow period.

TABLE 5
Influence of parkland trees on soil and crops.

Site description/source	Tree species	Condition	Total C (%)	Total N (%)	Ca (cmol$_c$ kg^{-1})	Mg (cmol$_c$ kg^{-1})	K	pH	Crop response[a]
North-central Senegal	*Faidherbia albida*	Canopy	3.70	0.40	1.61	0.71	0.10	5.7	Groundnuts: +37%
(Dancette and Poulain, 1969)		Open	2.70	0.30	1.13	0.62	0.07	5.5	Sorghum: +200%
Soil depth 0–20 cm									
Central Plateau, Burkina Faso	*Faidherbia albida*	Canopy	0.90	0.13	5.80	2.08	0.65	6.7	Sorghum grain: +115%
(Depommier et al., 1992)		Open	0.78	0.90	5.05	2.00	0.38	6.6	
Soil depth 0–20 cm									
Lakeshore Plain, Malawi	*Faidherbia albida*	Canopy	2.50	0.22	5.71	1.50	0.98	6.3	Maize grain: +100–400%
(Rhoades, 1995; Saka et al., 1994)		Open	2.20	0.19	6.84	1.78	0.87	6.3	
Soil depth 0–15 cm									
South Central Mexico	*Prunus capuli*	Canopy							Maize grain
(Farrell, 1990)	*Juniperus deppeana*	Prunus	1.34	0.09	6.50	1.29	0.58	6.6	Prunus: –50%
Soil depth 0–15 cm									
		Juniperus	0.63	0.04	6.60	1.24	0.84	7.4	Juniperus: –50%
		Open	0.45	0.03	3.45	0.62	0.33	6.2	
South Eastern Mali	*Vitellaria paradoxa* (VP)	Canopy	0.66	0.06	1.68	0.67	0.27	6.0	Cotton: VP: –2%; PB: –66%;
(Kater et al., 1992)									Sorghum: VP: –44%; PB: –65%
Soil depth: 0–20	*Parkia biglobosa* (PB)	Open	0.51	0.05	1.45	0.42	0.16	6.0	Millet: VP & PB: –60%

[a] Yield difference under tree canopy and open sites.

Source: Rhoades (1997).

In Zambia, one-year *S. sesban* fallows increased the yield of subsequent maize by 50% to 80%, and two-year fallows by 150% to 270% over maize yields after a grass fallow or continuous cropping (Kwesiga and Coe, 1994; Torquebiau and Kwesiga, 1996). The residual effect of both one- and two-year fallows was observed even four years after clearing the fallow, with three times higher yield than in monocropped maize (Kwesiga et al., 1998). Such a large residual effect more than compensates for the loss of production during the fallow period. In western Kenya, 18 months of *S. sesban* fallow produced 9.7 Mg ha^{-1} of maize in the subsequent three seasons, compared with 6.9 Mg ha^{-1} after a grass fallow and 4.9 Mg ha^{-1} after continuous maize cropping (B.A. Jama, pers. comm., 1997). The fallow benefit at this P-deficient site was substantially higher after applying P fertilizer to maize (ICRAF, 1996), and application of P fertilizer to maize increased the economic attractiveness of the *S. sesban* fallow (ICRAF, 1997).

Recent results on an acid soil in western Kenya (pH = 5.1, clay = 29%) indicated that rotation of maize with *S. sesban* fallows can lessen losses in maize yield resulting from potassium deficiency (ICRAF, 1998). Yield of maize without a preceding *S. sesban* fallow averaged 1.2 Mg ha^{-1} without K fertilizer and 4.8 Mg ha^{-1} with fertilizer. Yield of maize following *S. sesban* (in the season before maize) was 4.6 Mg ha^{-1} without added K fertilizer. Sesbania presumably increased K availability to maize through cycling via its leaf biomass that was incorporated into the soil after the fallow.

Pruning the trees during the fallow phase might increase the production of foliar biomass for greater *in situ* nutrient cycling at the expense of wood, which is normally taken out of the field. In western Kenya, Onim et al. (1990) pruned fallows of *S. sesban*, *L. leucocephala* and pigeonpea at two-month intervals and incorporated the prunings into the soil. Following a one-year fallow, they recorded yield increases of 67% in the first crop of maize, as compared to yields after a natural fallow. There was no residual effect in the second crop of maize, but a 26% yield increase was noted in the third crop of beans after *L. leucocephala* and *S. sesban* fallows; this could be an effect of the tree/crop rotation. Root biomass of fallow trees might contribute substantially to the residual effect of fallows over the years. Sanginga et al. (1988) estimated the N contribution of roots, nodules and, probably, a small percent of leaf litter of *L. leucocephala* fallow to be equivalent to 32 kg N ha^{-1} on maize, similar to that of aboveground prunings in hedgerow intercropping.

Based on these and other promising results, "improved fallows" is now being promoted as a promising technology for soil improvement in N-deficient soils of Africa (ICRAF, 1997). Improved fallows may have potential to alleviate K deficiency (ICRAF, 1998), but they do not eliminate the need for P inputs on severely P-deficient soils (ICRAF, 1997). The benefits of this technology derive largely from the nutrient-cycling capability (and other soil-improvement capabilities) of shrubs and trees that are used during short-rotation fallows.

4.4 Shaded Perennial-Crop Systems

Direct relationships between crop yields and nutrient cycling are far more difficult to establish under shaded perennial-crop systems compared with sequential and simultaneous systems involving annual crops. Yield responses of shaded perennial species such as coffee and cacao are generally reported in terms of the effect of shade-management practices (involving pruning, pollarding, and spacing of overstory species). Such practices alter the light and radiation environment of understory species more than their nutrient relations. Furthermore, yields of shaded perennial species are known to be affected by the interaction between light availability and soil fertility (Beer et al., 1997).

Reviewing the biophysical interactions in shaded perennial systems, Beer et al. (1997) concluded that most aspects of nutrient cycling in these systems were directly affected by the choice of the shade tree species. A comparison of two shade-tree species that are widely used in coffee and cacao plantations found that accumulation of Ca and Mg was greater in the stems and branches of unpruned, nonleguminous *Cordia alliodora* than in pruned, leguminous *E. poeppigiana* (Fassbender, 1993). However, transfer of N, P, K, Ca, and Mg to the soil was greater from *E. poeppigiana*.

TABLE 6
Biomass production, turnover and recycling index for four shaded perennial-crop systems in Costa Rica.[a]

Component	Shaded production system			
	Erythrina + coffee	Cordia + coffee	Erythrina + coffee	Cordia +coffee
Fine root biomass, Mg ha^{-1}	2.6	4.5	3.8	7.0
Standing biomass, Mg ha^{-1}	35.4	37.1	44.5	63.7
Turnover, Mg ha^{-1}	20.0	5.7	22.9	11.4
Recycling, %	56.4	15.3	51.4	17.9
N in standing biomass, kg ha^{-1}	522	286	357	400
N turnover, kg ha^{-1}	461	114	447	169
N Recycling, %	88.3	40.0	125.0	42.3
P in standing biomass, kg ha^{-1}	46.0	34.0	38.0	50.0
P turnover, kg ha^{-1}	35.0	7.0	40.0	24.0
P recycling, %	76.1	21.5	105	48.0
K in standing biomass, kg ha^{-1}	338	229	428	346
K turnover, kg ha^{-1}	260	54.0	177	73.0
K recycling, %	77.0	23.5	41.3	21.1

[a] Calculated from Fassbender et al. (1991) and Fassbender (1993).

Source: Nair et al. (1995).

Working in the same plots, Muñoz (1993) found that the combined fine roots in the *E. poeppigiana*-cacao association decomposed more quickly than the leaves of either species. Although total biomass and nutrient levels were greater in leaf litter, he pointed out that the more rapid turnover of smaller amounts of nutrients in fine roots may be of considerable ecological significance. A higher total nutrient content was found in the *C. alliodora* associations (Alpizar et al., 1986), which had a greater biomass of fine roots than the *E. poeppigiana* associations (Fassbender, 1993; Muñoz, 1993). Greater leaching losses of Ca and Mg occurred in these *E. poeppigiana* associations (Imbach et al., 1989a;b) possibly because the return of nutrients to the soil surface was concentrated in green (pruned) biomass which decomposed rapidly (Heuveldop et al., 1985). Despite the greater nutrient losses, cycling indices (relation of nutrient turnover to nutrients in the biomass) were higher in the pruned *E. poeppigiana* association (Beer et al., 1990; Nair et al., 1995) (see Table 6).

Beer et al. (1997) surmise that shade management, especially pruning, has a critical influence on nutrient cycling and, hence, in addition to its use in managing the microclimate of the underlying crop, provides a tool to manipulate the timing and quantity of nutrient transfer from tree to soil. Furthermore, the authors state that although some information exists on the below-ground processes of these systems (see above), this stratum is still a "black box" whose internal biological and chemical mechanisms are poorly quantified and little understood.

5. CONCLUSION

The results from the four major categories of agroforestry systems that have been reviewed here occur across the four major agroecological/geographical zones of the tropics as categorized by Nair (1992): hedgerow intercropping (at least, where results have been favorable) for humid and subhumid tropical lowlands with seasonal rainfall, parkland systems for the semiarid regions, improved fallow systems for the subhumid and humid lowlands and subhumid highlands, and shaded perennial-crop systems for tropical subhumid highlands. In all of these systems, there are encouraging examples which demonstrate that agroforestry systems can provide N for crop production. However, as already stated, not all of the benefits of the systems to crop production can be attributed to nutrient cycling, nor can positive results be expected in all situations. In contrast to the provision of N, it seems that agroforestry systems are not capable of providing sufficient amounts of P to maintain crop yields. Little can be said about other nutrients in agroforestry systems because experimental data are scarce.

A major problem in nutrient cycling studies in tropical agroforestry systems is the lack of appropriate research methodologies. There are strong indications that nutrient cycling is operational, that it contributes to improved soil fertility — especially with respect to N and SOM — and that it enhances the productivity of most tropical agroforestry systems. The major tree-mediated processes of the mechanism have been identified: N_2 fixation, production and decomposition of tree biomass, and nutrient uptake from deep soil horizons. However, rigorous methodologies do not exist for accurately determining the dynamics of these processes under field conditions. Therefore, it remains difficult to predict the success of management strategies involving the application of the processes. Until this deficiency is corrected — obviously through more research — the rate of success in exploiting the seemingly attractive nutrient-cycling potential in tropical agroforestry systems will remain uncertain.

REFERENCES

Akondé, T. P., Leihner, D. E., and Steinmüller, N., 1996, "Alley cropping on an Ultisol in subhumid Benin. Part 1: Long-term effect on maize, cassava and tree productivity," *Agrofor. Syst.*, 34, 1.

Akyeampong, A., Duguma, B., Heineman, A. M., Kamara, C. S., Kiepe, P., Kwesiga, F., Ong, C. K., Otieno, H. J., and Rao, M. R., 1995a, "A synthesis of ICRAF's research on alley cropping," in *Alley Farming Research and Development*, Kang, B. T., Osiname, A. O., and Larbi, A. Eds., International Institute of Tropical Agriculture (IITA), 40–51.

Akyeampong, A. and Hitimana, L., 1996, "Agronomic and economic appraisal of alley cropping with *Leucaena diversifolia* on an acid soil in the highlands of Burundi," *Agrofor. Syst.*, 33, 1.

Akyeampong, A., Hitimana, L., Franzel, S. and Munyemana, P. C., 1995b, "The agronomic and economic performance of banana, bean and tree intercropping in the highlands of Burundi: An interim assessment," *Agrofor. Syst.*, 31, 199.

Alpizar, L., Fassbender, H. W., Heuveldop, J., Folster, H., and Enriquez, G., 1986, "Modelling agroforestry systems of cacao (*Theobroma cacao*) with laurel (*Cordia alliodora*) and poro (*Erythrina poeppigiana*) in Costa Rica, I, Inventory of organic matter and nutrients," *Agrofor. Syst.*, 4, 3.

Anderson, L. S. and Sinclair, F. L., 1993, "Ecological interactions in agroforestry systems." *Agrofor. Abs.* 6, 5.

Babbar, L. I. and Zak, D. R., 1994, "Nitrogen cycling in coffee agroecosystems: Net nitrogen mineralization and nitrification in the presence and absence of shade trees," *Agri. Ecosystems Environ.*, 48, 107.

Babbar, L. I. and Zak, D. R., 1995, "Nitrogen loss from coffee agroecosystems in Costa Rica. Leaching and denitrification in the presence and absence of shade trees," *J. Envir. Qual.*, 24, 227.

Balasubramanian, V. and Sekayange, L., 1991, "Effect of tree legumes in hedgerows on soil fertility changes and crop performance in the semi-arid highlands of Rwanda." *Biol., Ag. and Hort.*, 8, 17.

Balasubramanian, V. and Sekayange, L., 1992, "Five years of research on improved fallow in the semi-arid highlands of Rwanda," in *Biological Nitrogen Fixation and Sustainability of Tropical Agriculture*, Mulongoy, K., Gueye, M. and Spencer, D. S. C., Eds., John Wiley & Sons and Sayce Publishing, Chichester, U.K., 405–422.

Beer, J. W., 1988, "Litter production and nutrient cycling in coffee (*Coffea arabica*) or cacao (*Theobroma cacao*) plantations with shade trees," *Agro. Syst.*, 7, 103.

Beer, J. W., Bonneman, A., Chavez, W., Fassbender, H. W., Imbach, A. C., and Martel, I., 1990, "Modeling agroforestry systems of cacao with *Cordia alliodora* and *Erythrina poeppigiana* in Costa Rica. V. Productivity indices, organic matter models and sustainability over 10 years," *Agrofor. Syst.*, 12, 229.

Beer, J., Muschler, R., Kass, D., and Somarriba, E., 1997, "Shade management in coffee and cacao plantations," *Agrofor. Syst.*, 38, 139-164.

Binkley, D., 1992, "Mixtures of N_2-fixing and non-N_2-fixing species," in *The Ecology of Mixed-Species Stands of Trees*, Cannell, M. G. R., Malcolm, D. C., and Robertson, P. A., Eds., Blackwell Sci. Publi., London.

Binkley, D., Dunkin, K. A., DeBell, D., and Ryan, M. G., 1992, "Production and nutrient cycling in mixed plantations of *Eucalyptus* and *Albizia* in Hawaii," *Forest Sci.*, 38, 393.

Breman, H. and Kessler, J. J., 1995. *Woody Plants in Agro-ecosystems of Semi-arid Regions*, Springer-Verlag, Berlin Heidelberg, 340.

Bunderson, W. T., 1992, *Final report of the agroforestry technical assistance to the national agroforestry team, MARE project*, Chitedze Agricultural Research Station, Malawi.

Buresh, R. J., Smithson, P. C., and Hellums, D. T., 1997, "Building soil phosphorus capital in Africa," in *Replenishing Soil Fertility in Africa*, Buresh, R.J., Sanchez, P. A. and Calhoun, F. Eds., Soil Sci. Soc. Am. Spec. Publ. 51, Soil Sci. Soc. Am. and Am. Soc. Agron., Madison, WI.

Buresh, R. J. and Tian, G., 1997, "Soil improvement by trees in sub-Saharan Africa," *Agrofor. Syst.*, 38, 51-76.

Cadisch, G. and Giller, K. E., 1997, Eds., *Driven by Nature: Plant Litter Quality and Decomposition*, CAB International, Wallingford, U.K., 409.

Chesson, A., 1997, "Plant degradation by ruminants: parallels with litter decomposition in soils," in *Driven by Nature: Plant Litter Quality and Decomposition*, Cadisch, G. and Giller, K. E., Eds., CAB International, Wallingford, U.K., Ch.3.

Chiyenda, S. and Materechera, S.A., 1989, "Some results from alley cropping *Leucaena leucocephala*, *Cassia siamea* and *Cajanus cajan* with maize at Bunda College of Agriculture," in *Trees for Development in sub-Saharan Africa*, IFS, Stockholm, Sweden, 135–142.

Coleman, D. C., Reid, C. P. P. and Cole, C. V., 1983, "Biological strategies of nutrient cycling in soil systems," *Adv. Eco. Res.*, 13, 1.

Constantinides, M. and Fownes, J. H., 1994a, "Nitrogen mineralization from leaves and litter of tropical plants: relationship to nitrogen, lignin and soluble polyphenol concentrations," *Soil. Biol. Biochem.* 26, 49.

Constantinides, M. and Fownes, J. H., 1994b, "Tissue-to-solvent ratio and other factors affecting determination of soluble polyphenols in tropical leaves," *Comm. Soil Sci. Pl. Anal.*, 25, 3221.

CTFT, 1988, "Faidherbia albida *(Del.) A. Chev. (synonyme* Acacia albida *Del.)*," Centre Technique Forestier Tropical, Nogent-sur-Marne, France.

Dancette, C. and Poulain, J. F., 1969, "Influence of *Acacia albida* on pedoclimatic factors and crop yields," *African Soils/Sols Africains* 14, 143.

Danso, S. K. A., Bowen, G. D., and Sanginga, N., 1992, "Biological nitrogen fixation in trees in agro-ecosystems," *Pl. Soil*, 141, 177.

DeAngelis, D. L., 1992, *Dynamics of Nutrient Cycling and Food Webs*, Chapman & Hall, London, 270.

Depommier, D., Janodet, E., and Oliver, R., 1992, "*Faidherbia albida* parks and their influence on soils and crops at Watinoma, Burkina Faso," in *Faidherbia albida in the West African Semi-arid Tropics*, Vandenbeldt, R. J., Ed., ICRISAT, Patancheru-502 324, India.

Dommergues, Y. R., 1995, "Nitrogen fixation by trees in relation to soil nitrogen economy," *Fert. Res.*, 42, 215.

Drechsel, P., Steiner, K. G., and Hagedorn, F., 1996, "A review on the potential of improved fallows and green manure in Rwanda," *Agrofor. Syst.*, 33, 109.

Duguma, B., Mollet, M., and Tiki Manga, T., 1994, *Annual Progress Report, Institute of Agronomic Research (IRA) and International Centre for Research in Agroforestry (ICRAF)*, AFRENA Report, ICRAF, Nairobi, Kenya.

Escalante, G., Herrera, R., and Aranguren, J., 1984, "Nitrogen fixation in shade trees (*Erythrina poeppigiana*) in cacao plantations in North Venezuela," *Pesq. Agropec. Bras.* 19 (s/n), 223.

Evenson, C. L. I., 1989, *Alley Cropping and Green Manuring for Upland Crop Production in West Sumatra*, Ph.D. dissertation, University of Hawaii.

Evenson, C. I., Dierof, T. S., and Yost, R. S., 1995, "Decreasing rice and cowpea yields in alley cropping on highly weathered Oxisol in West Sumatra, Indonesia," *Agrofor. Syst.*, 31, 1.

Farrell, J., 1990, "The influence of trees in selected agroecosystems in Mexico," in *Agroecology: Researching the Ecological Basis for Sustainable Agriculture*, Gliessman, S. R. Ed., Springer-Verlag, New York, 167–183.

Fassbender, H. W., 1987, "Nutrient cycling in agroforestry systems of coffee (*Coffea arabica*) with shade trees in the Central Experiment of CATIE," in *Advances in Agroforestry Research*, Beer, J., Fassbender, H. and Heuveldop, J., Eds., CATIE, Turrialba.

Fassbender, H. W., 1993, *Modelos Edafológicos de Sistemas Agroforestales*, 2nd Ed., CATIE, Turrialba, Costa Rica.

Fassbender, H. W., Beer, J., Heuveldop, J., Imbach, A., Enriquez, G., and Bonnemann, A., 1991, "Ten-year balances of organic matter and nutrients in agroforestry systems at CATIE," Costa Rica. *For. Ecol. Mgt.*, 45, 173-183.

Fernandes, E. C. M., 1990, *Alley Cropping on Acid Soils*, Ph.D. dissertation, North Carolina State University, Raleigh (Diss. Abstr. 52-03 B).

Fernandes, E. C. M., Davey, C. B., and Nelson, L. A., 1993, "Alley cropping on an acid soil in the upper Amazon: mulch, fertilizer, and hedgerow root pruning effects," in *Technologies for Sustainable Agriculture in the Tropics*, ASA Special Publication 56, American Society of Agronomy, Madison, WI.

Fernandes, E. C. M., Garrity, D. P., Szott, L. T., and Palm, C. A., 1994, "Use and potential of domesticated trees for soil improvement," in *Tropical Trees: The Potential for Domestication and Rebuilding of Forest Resources*, Leakey, R. R. P. and Newton, A. C., Eds., ITE/ECTF symposium, HMSO, London.

Foelster, H. and Khanna, P. K., 1997, "Dynamics of nutrient supply in plantation soils," in *Better Management of Soil, Nutrients, and Water in Tropical Plantation Forests*, Nambiar, E. K. S. and Brown, A., Eds., ACIAR, Canberra, Australia, 339-378.

Gichuru, M. P., 1991, "Residual effects of natural bush, *Cajanus cajan* and *Tephrosia candida* on the productivity of an acid soil in southeastern Nigeria," *Pl. Soil*, 134, 31.

Govindarajan, M., Rao, M. R., Mathua, M. N., and Nair, P. K. R., 1996, "Soil-water and root dynamics under hedgerow intercropping in semiarid Kenya." *Agron. J.* 88, 513-520.

Haggar, J., Tanner, E. V. E., Beer, J., and Kass, D. B. L., 1993, "Nitrogen dynamics of tropical agroforestry and annual cropping systems," *Soil Biol. Biochem.*, 25, 1363.

Handayanto, E., Cadisch, G. and Giller, K. E., 1994, "Nitrogen release from prunings of legume hedgerow trees in relation to quality of the prunings and incubation method," *Pl. Soil*, 160, 237.

Handayanto, E., Cadisch, G., and Giller, K. E., 1995, "Manipulation of quality and mineralization of tropical legume tree prunings by varying nitrogen supply," *Pl. Soil*, 176, 149.

Handayanto, E., Cadisch, G., and Giller, K. E., 1997a, "Regulating N mineralization from plant residues by manipulation of quality," in *Driven by Nature: Plant Litter Quality and Decomposition*, Cadisch, G. and Giller, K. E. Eds., CAB International, Wallingford, U.K., Ch.14.

Handayanto, E., Giller, K. E., and Cadisch, G., 1997b, "Nitrogen mineralization from mixtures of legume tree prunings of different quality and recovery of nitrogen by maize," *Soil Biol. Biochem.*, 29, 1417–1426.

Hartemink, A. E., Buresh, R. J., Jama, B., and Janssen, B. H., 1996, "Soil nitrate and water dynamics in sesbania fallow, weed fallows, and maize," *Soil Sci. Soc. Am. J.*, 60, 568.

Hauser, S. and Kang, B. T., 1993, "Nutrient dynamics, maize yield and soil organic matter dynamics in alley cropping with *Leucaena leucocephala*," in *Soil Organic Matter Dynamics and Sustainability of Tropical Agriculture*, Mulongoy, K., and Merckx, R., Eds., John Wiley & Sons and Sayce Publishing, Chichester, Ch. 3.6.

Herrera, R., Accardi, A., Aranguren, J., Navidad, E., Escalante, G, Toro, M., and Cuenca, G., 1987, "Coffee and cacao plantations under shade trees in Venezuela," in *Advances in Agroforestry Research*, Beer, J., Fassbender, H. W., Heuveldop, J., Eds., CATIE, Turrialba, Costa Rica.

Heuveldop, J., Alpizar, L., Fassbender, H. W., Enriquez, G., and Folster, H., 1985, *Agroforestales de café (Coffea arabica) con laurel (Cordia alliodora) y café con poró (Erythrina poeppigiana) en Turrialba, Costa Rica., II, Producción agrícola, maderable y de residuos vegetales*, Sistemas, Turrialba, 35, 347.

Hocking, D. and Rao, D. G., 1990, "Canopy management possibilities for arboreal *Leucaena* in mixed sorghum and livestock small farm production systems in semi-arid India," *Agrofor. Syst.*, 10, 135.

ICRAF, 1995. *Annual Report for 1994*, International Centre for Research in Agroforestry, Nairobi, Kenya.

ICRAF, 1996, *Annual Report for 1995*, International Centre for Research in Agroforestry, Nairobi, Kenya.

ICRAF, 1997, *Annual Report for 1996*, International Centre for Research in Agroforestry, Nairobi, Kenya.

ICRAF, *Annual Report for 1997*, International Centre for Research in Agroforestry, Nairobi, Kenya, 1998,

Imbach, A. C., Fassbender, H. W., Beer, J., Borel, R., and Bonneman, A., 1989a. *Agroforestales de café (Coffea arabica) con laurel (Cordia alliodora) y café con poro (Erythrina poeppigiana) en Turrialba, Costa Rica, IV, Balances hídricos y lixiviación de elementos nutritivos,* Sistemas, Turrialba, 39, 400.

Imbach, A. C., Fassbender, H. W., Borel, R., Beer, J., and Bonnemann, A., 1989b, "Modeling agroforestry systems of cacao (*Theobroma cacao*) with laurel (*Cordia alliodora*) and *Erythrina poeppigiana* in Costa Rica, IV, Water balances, nutrient inputs, and leaching," *Agrofor. Syst.*, 8, 267.

Jackson, F. S., Barry, T. N., and Lascano, C., 1996, "The extractable and bound condensed tannin content of leaves from tropical trees, shrubs and forage legumes," *J. Sci. Food Agri.*, 71, 103.

Jama, B. A. and Nair P. K. R., 1996, "Decomposition- and nitrogen-mineralization patterns of *Leucaena leucocephala* and *Cassia siamea* mulch under tropical semiarid conditions in Kenya," *Pl. Soil* 179, 275.

Jama, B., Nair, P. K. N. and Rao, M. R., 1995, "Productivity of hedgerow shrubs and maize under alleycropping and block planting systems in semiarid Kenya," *Agrofor. Syst.*, 31, 257.

Jama, B., Swinkels, R. A., and Buresh, R., 1997, "Agronomic and economic evaluation of organic and inorganic sources of phosphorus in western Kenya," *Agron. J.* 89, 597.

Jama, B., Buresh, R.J., Ndufa, J. K., and Shepherd, K. D., 1998, "Vertical distribution of roots and soil nitrate: Tree species and phosphorus effects," *Soil Sci. Soc. Am. J.*

Joachim, A. W. R. and Kandiah, S., 1936, "The change in composition and decomposability of tropical Ceylon green manures with age." *The Tropical Agriculturalist*, 82, 3-20.

Jones, R. B., Wendt, J. W., Bunderson, W. T., and Itimu, O. A., 1996, "Leucaena and maize alley cropping in Malawi, part 1, Effects of N, P, and leaf application on maize yields and soil properties," *Agrofor. Syst.*, 33, 281.

Jonsson, K., Fidjeland, L., Maghembe, J. A., and Högberg, P., 1988, "The vertical distribution of fine roots of five tree species and maize in Morogoro, Tanzania," *Agrofor. Syst.*, 6, 63.

Jordan, C. F., 1985, *Nutrient Cycling in Tropical Forest Ecosystems*, John Wiley, New York, 190.

Kang, B. T., 1993, "Alley cropping: Past achievements and future directions," *Agrofor. Syst.*, 23, 141.

Kang, B. T., Wilson, G. F., and Sipkens, L., 1981, "Alley cropping maize (*Zea mays* L.) and *Leucaena leucocephala* in Southern Nigeria," *Pl. Soil*, 63, 165.

Kass, D. L., Araya, J. S., Sanchez, J. O., Pinto, L. S., and Ferreira, P., 1995, "Ten years experience with alley farming in Central America," in *Alley Farming Research and Development*, Kang, B. T., Osiname, A. O. and Larbi, A. Eds. IITA, Ibadan, Nigeria, 393–402.

Kass, D. C. L., Barrantes, A., Bermudez, W., Campos, W., Jiminez, M., and Sanchez, J., 1989, "Resultados de investigaciốn de cultivo en callejones (alley cropping) en 'La Montana'," *Turrialba,* 19, 5-14.

Kater, L. J. M., Kante. S., and Budelman, A., 1992, "Karité (*Vitellaria paradoxa*) and néré (*Parkia biglobosa*) associated with crops in South Mali," *Agrofor. Syst.*, 18, 89.

Kessler, J. J., 1992, "The influence of karité (*Vitellaria paradoxa*) and néré (*Parkia biglobossa)* trees on sorghum production in Burkina Faso." *Agrofor. Syst.*, 17, 97.

Khanna, P. K., 1997, "Nutrient cycling under mixed tree systems in southeast Asia," *Agrofor. Syst.,* 38, 99-120.

Kumar Rao, J. V. D. K., Dart, P. J., and Sastry, P. V. S. S., 1983, "Residual effect of pigeonpea (Cajanus cajan) on yield and nitrogen response of maize," *Exp. Agri.*, 19, 131.

Kwesiga, F. and Coe, R., 1994, "The effect of short rotation *Sesbania sesban* planted fallows on maize yield," *For. Ecol. Mgt.*, 64, 199.

Kwesiga, F. R., Franzel, S., Place, F., Phiri, D., and Simwanza, C. P., 1998, "*Sesbania sesban* improved fallows in eastern Zambia: Their inception, development and farmer enthusiasm," *Agrofor Sys*, in press.

Laike, A., 1992, "*Faidherbia albida* in the traditional farming systems of central Ethiopia," in *Faidherbia albida in the West African Semi-arid Tropics*, Vandenbeldt, R. J., Ed., ICRISAT, Patancheru, India.

Lal, R., 1989, "Agroforestry systems and soil surface management of a tropical Alfisol. I. Soil moisture and crop yield," *Agrofor. Syst.*, 8, 7.

Lindblad, P. and Russo, R., 1986, "C_2H_2 reduction by *Erythrina poeppigiana* in a Costa Rican coffee plantation," *Agrofor. Syst.*, 4, 33.

MacDicken, K. G., 1994, *Selection and Management of Nitrogen-Fixing Trees*, Winrock International, Morrilton, AR, and FAO, Bangkok, 272.

Mafongoya, P. L., 1995. *Multipurpose Tree Prunings as a Source of Nitrogen to Maize* (Zea mays) *under Semiarid Conditions in Zimbabwe*, Ph.D. Dissertation, University of Florida, Gainesville, FL.

Mafongoya, P. L. and Nair, P. K. R., 1997, "Multipurpose tree prunings as a source of nitrogen to maize under semiarid conditions in Zimbabwe, Part 1, Nitrogen-recovery rates as influenced by pruning quality and method of application," *Agrofor. Syst.*, 35, 31.

Mafongoya, P. L., Nair, P. K. R., and Dzowela, B. H., 1997a, "Multipurpose tree prunings as a source of nitrogen to maize under semiarid conditions in Zimbabwe, Part 2, Nitrogen-recovery rates and crop growth as influenced by mixtures and prunings," *Agrofor. Syst.*, 35, 47.

Mafongoya, P. L., Nair, P. K. R., and Dzowela, B. H., 1997b, "Multipurpose tree prunings as a source of nitrogen to maize under semiarid conditions in Zimbabwe, Part 3, Interactions of pruning quality and time and method of application on nitrogen recovery by maize in two soil types," *Agrofor. Syst.*, 35, 57.

Mafongoya, P. L., Giller, K. E., and Palm, C. A., 1997c, "Decomposition and nutrient release patterns of prunings and litter of agroforestry trees," *Agrofor. Syst. 38,* 77–97.

Mathuva, M. N., Rao, M. R., Smithson, P. C., and Coe, R., 1998, "Improving maize (*Zea mays*) yields in semiarid highlands of Kenya: agroforestry or inorganic fertilizers?" *Field Crops Res*, 55, 57.

Matthews, R. B., Holden, S. T., Volk, J., and Lungu, S., 1992a, "The potential of alley cropping in improvement of cultivation systems in the high rainfall areas of Zambia, I," *Chitemene and Fundikila, Agrofor. Syst.*, 17, 219.

Matthews, R. B., Lungu, S., Volk, J., Holden, S. T., and Solberg, K., 1992b, "The potential of alley cropping in improvement of cultivation systems in the high rainfall areas of Zambia, II, Maize production," *Agrofor. Syst.*, 17.

Mekonnen, K., Buresh, R. J., and Jama, B., 1997, "Root and inorganic nitrogen distributions in sesbania fallow, natural fallow and maize," *Pl. Soil* 188, 319.

Mittal, S. P. and Singh, P., 1989, "Intercropping field crops between rows of *Leucaena leucocephala* under rainfed conditions in northern India," *Agrofor. Syst.*, 8, 165.

Mongi, H. O. and Huxley, P. A., Eds., 1979. *Soils Research in Agroforestry: Proceedings of an Expert Consultation*, ICRAF, Nairobi, Kenya, 584.

Mugendi, D. N., Mochoge, B. O., Coulson, C. L. and Sang, F. K., 1997a, "Effect of intercropping *Cassia siamea* prunings on maize yield in semiarid Kenya," *Af. Crop Sci. J.*, 5, 271.

Mugendi, D. N., Mochoge, B. O., Coulson, C. L., Stigter, C. J. and Sang, F. K., 1994, "Decomposition of *Cassia siamea* loppings in Semi-arid Machakos," *Arid Soil Res. Rehab.*, 8, 363.

Mugendi, D. N. and Nair, P. K. R., 1997, "Predicting the decomposition patterns of tree biomass in the tropical highland microregions of Kenya," *Agrofor. Syst.*, 35, 187.

Mugendi, D. N., Nair, P. K. R., and Graetz, D. A., 1997b, "Nitrogen recovery by alley-cropped maize and trees from ^{15}N-labeled tree biomass," *Paper for the ASA 1997 Annual Conference, Anaheim, California*; *Agronomy Abstracts*, 45.

Mugendi, D. N., 1997, *Tree-biomass Decomposition, Nitrogen Dynamics, and Maize Growth under Agroforestry Conditions in the Subhumid Highlands of Kenya*, Ph.D. Dissertation, University of Florida, Gainesville, FL, 131.

Muñoz, F., 1993, *Dinámica de Raíces Finas en los Sistemas Agroforestales de Cacao con Sombras de Poró o Laurel en Turrialba, Costa Rica*, M.S. Thesis, CATIE, Turrialba, Costa Rica.

Mureithi, J. G., Taylor, R. S., and Thorpe, W., 1994, "The effects of alley cropping with *Leucaena leucocephala* and of different management practices on the productivity of maize and soil chemical properties in lowland coastal Kenya," *Agrofor. Syst.*, 27, 31.

Myers, R. J. K., Palm, C.A., Cuevas, E., Gunatilleke, I. U. N., and Brossard, M., 1994, "The synchronisation of nutrient mineralization and plant nutrient demand," in *The Biological Management of Soil Fertility*, Woomer, P. L. and Swift, M. J., Eds., John Wiley & Sons, Chichester, Ch. 4.

Nair, P. K. R., 1984. *Soil Productivity Aspects of Agroforestry*, ICRAF, Nairobi, Kenya, 85.

Nair, P. K. R., 1989, Ed., *Agro. Syst. in the Tropics*, Kluwer, Dordrecht, Netherlands.

Nair, P. K. R., 1992, "Agroforestry systems design: an ecozone approach," in *Managing the World's Forests: Looking for Balance Between Conservation and Development,* Sharma, N. P., Ed., Kendall/Hunt Publ., Falls Church, VA, for The World Bank, Washington, DC, Ch. 15.

Nair, P. K. R., 1993, *An Introduction to Agroforestry*, Kluwer, Dordrecht, 499.

Nair, P. K. R., Kang, B. T., and Kass, D. B. L., 1995, "Nutrient cycling and soil-erosion control in agroforestry system," in *Agriculture and Environment: Bridging Food Production in Developing Countries*, ASA special publication no. 60, American Society of Agronomy, Madison, WI, Ch. 7.

Nair, P. K. R. 1997, Directions in tropical agroforestry. *Agrofor. Syst.*, 38, 223-246.

Nygren, P. and Ramírez, C., 1995, "Production and turnover of N_2 fixing nodules in relation to foliage development in periodically pruned *Erythrina poeppigiana* (Leguminosae) tree," *For. Ecol. Mgt.,* 73, 59.

Nziguhebu, G, Palm, C. A., Buresh, R. J., and Smithson, P. C., 1998, "Soil phosphorus fractions and adsorption as affected by organic and inorganic sources." *Pl. Soil*, In press.

Oglesby, K. A. and Fownes, J. H., 1992, "Effects of chemical-composition on nitrogen mineralization from green manures of 7 tropical leguminous trees," *Pl. Soil*, 127.

Ong, C. K., Corlett, J. E., Singh, R. P., and Black, C. R., 1991, "Above and below ground interactions in agroforestry systems," *For. Ecol. Mgt.*, 45, 45.

Onim, J. F. M., Mathuva, M., Otieno, K., and Fitzhugh, H. A., 1990, "Soil fertility changes and response of maize and beans to green manures of leucaena, sesbania and pigeonpea," *Agrofor. Syst.*, 12, 197.

Palm, C. A. and Sanchez P. A., 1990, "Decomposition and nutrient release patterns of the leaves of three tropical legumes," *Biotropica*, 22, 330.

Palm, C. A. and Sanchez, P. A., 1991, "Nitrogen release from the leaves of some tropical legumes as affected by their lignin and polyphenolic contents," *Soil Biol. Biochem.*, 23, 83.

Palm, C. A., 1995, "Contribution of agroforestry trees to nutrient requirements of intercropped plants," *Agrofor. Syst.*, 30, 105.

Palm, C. A., Salazar, A. A., Szott, L. T., and Fernandes, E. C. M., 1995, "Long-term performance of alley cropping on acid soils of the Amazon basin," in *Alley Farming Research and Development*, Kang, B. T., Osiname, A. O., and Larbi, A., Eds., IITA, Ibadan, Nigeria.

Pandey, D., 1995, *Forest Resources Assessment 1990, Tropical Forest Plantation Resources*, FAO Forestry paper 128. FAO/SUAS/SIDA, Rome, 81.

Peoples, M. B. and Herridge, D. F., 1990, "Nitrogen fixation by legumes in tropical and subtropical agriculture," *Adv. Agron.*, 44, 155.

Poschen, P., 1986, "An evaluation of the *Acacia albida*-based agroforestry practices in the Hararghe highlands of eastern Ethiopia," *Agrofor. Syst.*, 4, 129.

Prinz, D., 1986, "Increasing the productivity of smallholder farming systems by introduction of planted fallows," *Plant Res. Dev.*, 24, 31.

Puri, S. and Bangarwa, K. S., 1992, "Effects of trees on the yield of irrigated wheat crop in semi-arid regions," *Agrofor. Syst.* 20, 229.

Rao, M. R., Ong, C. K., Pathak, P., and Sharma, M. M., 1991a, "Productivity of annual cropping and agroforestry systems on a shallow Alfisol in semi-arid India," *Agrofor. Syst.*, 15, 51.

Rao, M. R., Sharma, M. M., and Ong, C. K., 1991b, "A tree/crop interface design and its use for evaluating the potential of hedgerow intercropping," *Agrofor. Syst.*, 13, 143.

Rao, M. R., Nair, P. K. R., and Ong, C. K., 1997, "Biophysical interactions in tropical agroforestry systems," *Agrofor. Syst.*, 38, 3-49.

Rhoades, C. C., 1995, "Seasonal pattern of nitrogen mineralization and soil moisture beneath *Faidherbia albida* (syn. *Acacia albida*) in central Malawi," *Agrofor. Syst.*, 29, 133.

Rhoades, C. C., 1997, "Single-tree influences on soil properties in agroforestry: Lessons from natural forest and savanna ecosystems," *Agrofor. Syst.*, 35, 71.

Rosecrance, R. C., Brewbaker, J. L., and Fownes, J. H., 1992, "Alley cropping of maize with nine leguminous trees," *Agrofor. Syst.*, 17, 159.

Roskowski, J. P. and Van Kessel, C., 1985, "Annual seasonal and daily variation in nitrogen fixing activity by *Inga jinicuil*, a tropical leguminous tree," *Oikos*, 44, 306.

Ruhigwa, B. A., Gichuru, M. P., Mambani, B., and Tariah, N. M., 1992, "Root distribution of *Acioa barteri*, *Alchornea cordifolia*, *Cassia siamea* and *Gmelina arborea* in an acid Ultisol," *Agrofor. Syst.*, 19, 67.

Saka, A. R., Bunderson, W. T., Itimu, O. A., Phombeya, H. S. K., and Mbekeani, Y., 1994, "The effects of *Acacia albida* on soils and maize grain yields under smallholder farm conditions in Malawi," *For. Ecol. Mgt.*, 64, 217.

Sanginga, N. K., Mulongoy, K., and Ayanaba, A., 1988, "Nitrogen contribution of leucaena/rhizobium symbiosis to soil and subsequent maize crop." *Pl. Soil*, 112, 137.

Sanginga, N., Vanlauwe, B., and Danso, S. K. A., 1995, "Management of biological N_2 fixation in alley cropping systems: Estimation and contribution to N balance," *Pl. Soil*, 174, 119.

Sanginga, N., Danso, S. K. A., and Zapata, F., 1996, "Field requirements of nitrogen fixation in leguminous trees used in agroforestry systems: Influence of ^{15}N-labeling approaches and reference trees," *Biol Fert Soils*, 23, 26.

Sanchez, P. A., 1995, "Science in agroforestry," *Agrofor. Syst.*, 30, 5.

Sanchez, P. A. and Palm, C. A., 1996, "Nutrient cycling and agroforestry in Africa," *Unasylva* 185 (47), 24.

Schroth, G., 1995, "Tree-root characteristics as criteria for species selection and systems design in agroforestry." *Agrofor. Syst.*, 30, 125.

Schroth, G., Oliver, R., Balle, P., Gnahaua, G. M., Kanchenkanti, N., Leduc, B., Peltier, R., and Zech, W., 1995, "Alley cropping with *Gliricidia sepium* on a high base status soil following forest clearing: Effect of soil conditions, plant nutrition and crop yields." *Agrofor. Syst.*, 32, 261.

Schroth, G. and Zech, W., 1995, "Above- and below-ground biomass dynamics in a sole cropping and alley cropping system with *Gliricidia sepium* in the semideciduous rainforest zone of West Africa." *Agrofor. Syst.*, 31, 181.

Shannon, D. A. and Vogel, W. O., 1994, "The effects of alley cropping and fertilizer application on continuously cropped maize," *Trop Agric*, 71, 163.

Singh, R. P., Ong, C. K., and Saharan, N., 1989, "Above and below ground interactions in alley cropping in semi-arid India," *Agrofor. Syst.*, 9, 59.

Smith, J. L., 1994, "Cycling of nitrogen through microbial activity," in *Soil Biology: Effects on Soil Quality, Advances in Soil Sci.,* Hatfield, J. L., and Stewart, B. A., Eds., CRC Press, Boca Raton, FL.

Stone, E. L. and Kalisz, P. J., 1991, "On the maximum extent of tree roots," *For. Ecol. Mgt.*, 46, 59.

Swift, M. J., Heal, O. W., and Anderson, J. M., 1979. *Decomposition in Terrestrial Ecosystems, Studies in Ecology Vol. 5,* University of California Press, Berkeley, CA.

Swift, M. J., 1987, "Tropical soil biology and fertility: interregional research planning workshop, special issue 13," in *Biology International,* IUBS, Paris.

Szott, L. T., Fernandes, E. C. M., and Sanchez, P. A., 1991, "Agroforestry in acid soils of the humid tropics," *Adv. Agron.*, 45, 275.

Szott, L. T., Palm, C. A., and Davey, C. B., 1994, "Biomass and litter accumulation under managed and natural tropical fallows," *For. Ecol. Mgt.*, 67, 177.

Tejwani, K. G., 1994, *Agroforestry in India,* Oxford & IBH Publishing Co. Pvt. Ltd., New Delhi, India, 233.

Tian, G., Brussaard, L., and Kang, B. T., 1995, "An index for assessing the quality of plant residues and evaluating their effects on soil and crop in the (sub-)humid tropics," *A Soil Ecol.*, 2, 25.

Tian, G., Kang, B. T., and Brussard, L., 1992a, "Effects of chemical composition on N, Ca, and Mg release during incubation of leaves from selected agroforestry and fallow plant species," *Biogeochemistry,* 16, 103-119.

Tian, G., Kang, B. T., and Brussard, L., 1992b, "Biological effects of plant residues with contrasting chemical composition under humid tropical conditions: Decomposition and nutrient release," *Soil Biol. Biochem.*, 24, 1051-1060.

Tilander, Y., Ouedraogo, G. and Yougma, F., 1995, "Impact of tree coppicing on tree–crop competition in parkland and alley farming systems in semiarid Burkina Faso," *Agrofor. Syst.*, 30, 363.

Torquebiau, E. F. and Kwesiga, F., 1996, "Root development in *Sesbania sesban* fallow-maize system in eastern Zambia," *Agrofor. Syst.*, 34, 193.

van Noordwijk, M., Lawson, G., Soumare, A., and Groot, J. J. R., 1996, "Root distribution of trees and crops: competition and/complementarity," in *Tree crop Interactions, A Physiological Approach,* Ong, C. K. and Huxley, P., Eds., CAB International, Wallingford, U.K., Ch. 9.

van Noordwijk, M., Sitompul, S. M., Hairiah, K., Listyarini, E., and Syekhfani, M. S., 1995, "Nitrogen supply from rotational or spatially zoned inclusion of Leguminosae for sustainable maize production on an acid soil in Indonesia," in *Plant-Soil Interactions at Low pH,* R. A., Grundon, N. J., Rayment, G. E., and Probert, M. E. Eds. Kluwer, Dordrecht, Netherlands, 779–784.

Vogt, K., Asbjornsen, H., Ercelawn, A., Montagnini, M., and Valdes, M., 1997, "Roots and mycorrhizas in plantation ecosystems," in *Management of Soil, Nutrients and Water in Tropical Plantation Forests,* Nambiar, E. K. S. and Brown, A. G. Eds., Australian Centre for International Agricultural Research, Canberra, Australia, Ch. 8.

Wendt, J. W., Jones, R. B., Bunderson, W. T., and Itimu, O. A., 1996, "Leucaena and maize alley cropping in Malawi, part 2, Residual P and leaf management effects on maize nutrition and soil properties," *Agrofor. Syst.*, 33, 295.

Wong, M. T. F., Akyeampong, A., Nortcliff, S., Rao, M. R., and Swift, R. S., 1995, "Initial responses of maize and beans to decreased concentrations of monomeric inorganic aluminium with application of manure or tree prunings to an Oxisol in Burundi," *Pl. Soil*, 171, 275.

Xu, Z. H., Saffigna, P. G., Myers, R. J. K., and Chapman, A. L., 1993, "Nitrogen cycling in leucaena (*Leucaena leucocephala*) alley cropping in semiarid tropics. I. Mineralization of nitrogen from leucaena residues." *Pl. Soil*, 148, 63.

Yadav, J. P., Sharma, K. K., and Khanna, P., 1993, "Effect of *Acacia nilotica* on mustard crop," *Agrofor. Syst.*, 21.

Young, A., 1989, *Agroforestry for Soil Conservation,* CAB International, Wallingford, U.K., 276.

Young, A., 1997, *Agroforestry for Soil Management,* CAB International, Wallingford, U.K., and ICRAF, Nairobi, Kenya.

2 Animals and Agroforestry in the Tropics

Alice N. Pell

1. INTRODUCTION

The role of animals in tropical farming systems is controversial: they have been considered both to be essential components of a sustainable farming system and to be the cause of deforestation of the rain forests, encroachment of the Sahara and other types of environmental degradation. While there is no question that animals that are poorly managed or that are maintained in fragile ecosystems can cause environmental damage, such degradation is not a necessary by-product of livestock production. In many cases, animals provide the manure that furnishes many nutrients, especially nitrogen and phosphorus, needed for crop production. Throughout this paper, the term *manure* will refer to a mixture of urine, feces, and bedding or uneaten feed according to the definition used by Strauch (1987).

In semi-arid areas particularly, the economic role of livestock is important: in one study in Ethiopia, animal-related products provided about 40% of the gross farm income, with other estimates as high as 70% (Reed and Bert, 1993). Animals are mobile bank accounts that can contribute needed cash for expenses like school fees or to buy essential food during lean times. In addition, animals provide traction power, fiber, and soil amendments (manure). Animal products are an important dietary source of protein and often-deficient micronutrients like iron. The pertinent question is how animals can best be incorporated into sustainable, integrated farming systems.

The need for "closed systems" in which recycling of nutrients is maximized has been emphasized (Stangel, 1993). The intended objective in closing the system is to prevent agricultural mining of the soils so that they are no longer capable of reasonable crop yields. Because replenishment of soil nutrients through excessive reliance on purchased soil amendments is ill-advised economically

1-56670-294-1/99/$0.00+$.50

and biologically (Power, 1994), the idea of creating a closed system in which few nutrients escape has merit. However, complete closure of the system contradicts the primary goal of both crop and livestock production systems: to harvest nutrients for sale or consumption.

Returning nutrients to the site of production to maintain agricultural productivity and soil quality will become increasingly difficult as societies become more urbanized, an accelerating trend throughout the developed and developing worlds. As of 1991, 71% of Mexicans, 32% of Ghanaians and 31% of Indonesians lived in cities (NRC, 1993). More than 54% of all Africans will live in urban areas by 2025 (Cooper et al., 1996). Although forecasting population growth rate is difficult, in many areas the population is doubling every 25 years (NRC, 1993). The need for additional food drives the cultivation of more marginal land and shortens or eliminates the fallow period needed for soil rejuvenation. Extensive systems that rely on transhumanance (human and animal migration) to optimize forage utilization which is often practiced in fragile, arid areas of Africa are at particular risk. The dangers of soil degradation and erosion are increased especially in the oxisols, alfisols, and ultisols, three highly weathered soil types that cover about 50% of the land in the tropics (Lal, 1987). A major challenge in the development of sustainable agricultural systems is to ensure that they can meet the food needs of both the rural and urban members of a society without degradation of the agricultural areas. Recognizing that complete closure of the system is neither possible nor desirable, we will explore the integration of crop, forestry, and livestock (agro-silvo-pastoral (ASP)) systems to minimize unintentional nutrient loss and to ensure that natural resources and agricultural productivity are maintained while farmers earn a viable living.

Over the last 20 years, there has been recognition of the role that both herbaceous and tree legumes play in nitrogen fixation and in meeting the protein requirements of animals. Tree legumes also are important in providing fuelwood and building materials, and for use as living fences (Gutteridge and Shelton, 1994). The first section of this paper will consider how nutrients can be used to maximum advantage in an ASP system with a focus on the livestock component. Nitrogen and phosphorus will be emphasized because they are the first limiting nutrients in most cropping and animal systems (Williams et al., 1995) although other minerals, especially sulfur, copper, cobalt, sodium, and iodine, are likely to be deficient and limit animal performance (McDowell et al., 1984). Because ruminants digest fiber better than monogastrics and are thus the primary forage consumers, and because they are better able to tolerate secondary compounds present in many legumes, we will focus on ruminants. Although poultry and swine are often found in the humid tropics, livestock generally and ruminants in particular are less important in moist environments because of disease and parasite problems. In sub-Saharan Africa, the semi-humid and semi-arid areas contain approximately 80% of the livestock (Agboola and Kintomo, 1993) so we will emphasize semi-humid and semi-arid areas. Second, we will address some problems encountered in ASP systems including increased labor requirements, toxicity of secondary compounds in many tropical legumes and other unresolved environmental issues.

2. NUTRIENT CYCLING IN AN ASP SYSTEM

2.1 Nitrogen and Phosphorus Transformations

2.1.1 Nitrogen

Ideally, the nutrient requirements of a crop should be met but not greatly exceeded during all stages of the crop production cycle (Power, 1994). Achievement of this goal is difficult because many complex processes are involved and many factors are beyond human control. Understanding the chemical transformations of nutrients alone without consideration of the impacts of management practices, weather conditions, and differences in soils and environmental conditions is difficult.

Nitrogen used by crops comes from several sources including soil organic matter, biological nitrogen fixation, fertilizers, crop residues, and manures, so that N is present in the soil as ammonium, nitrate, and organic forms of varying availability (Power, 1994). These dynamic N pools are in constant flux. By attempting to integrate animal, tree, and crop production while maintaining soil fertility, we make the system much more complicated because interactions among all the different processes must be considered. For example, animal diet affects the rate of soil mineralization of manure nitrogen and thus nutrient availability to crops.

When we look at just one small part of the nutrient cycling system, manure composition, it would seem that our task should become easier. Surely after decades of animal science research we know the quantity of manure produced and its nutrient composition. Unfortunately, even this problem is difficult. Manure composition is dependent on animal species and genotype, diet, feed availability, physiological status, and level of intake of the animal, and manure handling practices (Brower and Powell, 1993; Somda et al., 1993). To give an indication of the variation with which we are dealing, the amount of fecal dry matter excreted daily by stall-fed cattle in the tropics varied from 764 to 7222 g/day and the comparable figures for sheep ranged from 39 to 898 g/day (Somda et al., 1993). Even when these data were adjusted to account for differences in metabolic body size ($BW^{0.75}$), there was still a four-fold difference among the cattle and a ten-fold difference among the sheep in the amount of dry matter excreted. Manure composition varied as much as the quantity produced.

As in soils, N in manure is present in several forms: (1) urinary nitrogen that is subject to volatilization and leaching, (2) fecal nitrogen of metabolic origin that is readily available for soil mineralization, and (3) fecal nitrogen that consists of the undigested feed. Urinary N includes ammonia, urea, allantoin, hippuric, and uric acids (Van Soest, 1994), most of which are likely to be lost through volatilization and leaching (Russelle, 1992). Despite these losses, urine is beneficial because it raises soil pH thus increasing the availability of soil P (Somda et al., 1993). The amount of urinary N excreted is strongly influenced by diet: if animals consume more N than they require, urinary N excretion increases. Increased urinary excretion also occurs if ruminally available or ruminally undegradable protein levels are excessive even if the amount of crude protein is close to recommended levels.

The N in the undigested feed residue is either associated with lignin or consists of tannin-bound protein. This fraction is insoluble in neutral detergent solution while the metabolic fraction is soluble (Van Soest, 1994) indicating that the metabolic fecal nitrogen is more readily mineralized in the soil than the nitrogen associated with the indigestible feed (Somda et al., 1993). Inclusion of browse in animal diets generally decreases urinary N and increases the amount of neutral detergent insoluble N in the feces (Powell et al., 1994). These effects are more pronounced when feeds high in polyphenols or with a high polyphenol:N ratio are fed (Somda et al., 1993).

2.1.2 Phosphorus

The chemical transformations of phosphorus are less complex than those of nitrogen, but bioavailability must be considered for both plants and animals. Almost all P is excreted in the feces so we need not consider the form and fate of urinary P. Manure P includes both inorganic and organic fractions with inorganic P comprising from 47 to 63% of the total P in manure from animals raised on commercial farms in temperate regions (Barnett, 1994). The animals from which these data were collected all received adequate or excess dietary P and the range would have been broader had P-deficient animals been included in the database. The organic fraction includes residual P from nucleic acids, acid-soluble P from inositol hexaphosphates and small amounts of phospholipids (Barnett, 1994). The availability of the various forms of organic P to plants is variable (Sharpley, 1996) and the linkages between P transformations and organic matter decomposition are not well understood (Gressel and McColl, 1997).

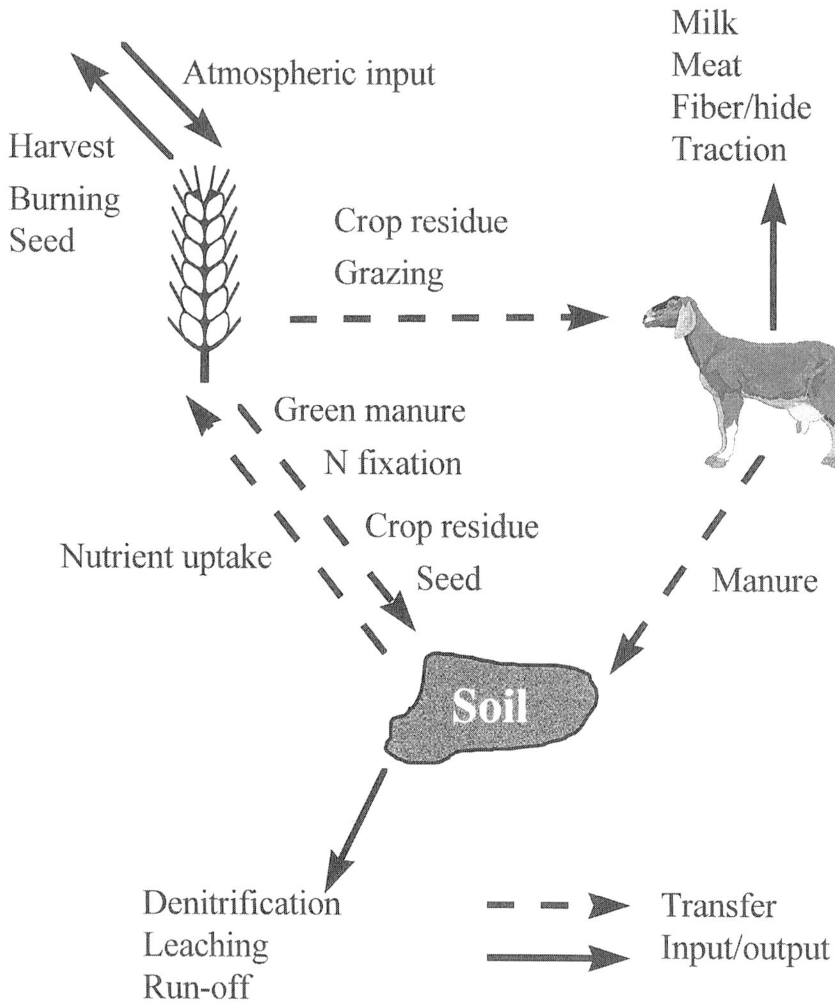

FIGURE 1. Inputs, outputs and nutrient flows in an extensive grazing system. Solid lines denote imports or exports and dashed lines indicate transfers. Adapted from (Agboola and Kintomo, 1993).

2.2 Nutrient Transfers In Cropland-Rangeland Systems

2.2.1 Use of Animals for Nutrient Transfer

The role of animals in nutrient cycling in ASP systems is important. Low efficiency of use of nutrients is not viewed favorably by animal nutritionists, but environmentally this low efficiency may be an advantage. Estimates of the amount of nitrogen harvested by grazing animals that is actually retained vary between 10 to 20% with the remainder being returned to the land via excreta and plant litter (Thomas and Lascano, 1993). When these low retention figures are combined with biological nitrogen fixation, it often is possible to maintain soil organic matter and nitrogen at levels capable of sustaining crop yields. Without a P supply comparable to biological N fixation, phosphorus often is deficient for plants and animals, so P supplementation may be essential in many ASP systems. Figure 1 (adapted from Agboola and Kintomo, 1993) shows some of the inputs, outputs, and transfers in an ASP grazing system.

If animals range freely during the day and are confined at night, they harvest nutrients from rangelands and excrete some of these nutrients in the overnight confinement areas for later spreading onto cropland. During the dry season, animals often are enclosed in crop-growing areas to permit grazing of crop residues and to manure the cropland. In this situation, manure is directly deposited on the cropland minimizing labor input. This nutrient transfer system is sufficiently well developed in parts of West Africa that exchanges of manure for crop residues or water are common (de Leeuw et al., 1993). However, in many cases, the nutrient transfer system is not well integrated so that benefits accrue to either the pastoralists or to sedentary farmers often at the expense of soil fertility in part of the agroecosystem (de Leeuw et al., 1993). The sustainability of these range-cropland nutrient transfer systems is determined by whether the rangeland can sustain a sufficient number of animals to meet the manure requirements of the cropland (Turner, 1993; Williams et al., 1995). Attempts to accelerate nutrient availability by increasing the animal stocking rate often result in nutrient loss (Turner, 1993). If the carrying capacity of the rangeland is exceeded, reduced animal productivity and manure quality ultimately diminish the productivity of the cropland. Data from Niger suggest that increasing the intensity of land use is not possible if manure is the primary soil amendment to maintain crop yields (Williams et al., 1995). Forage banks of tree legumes to provide sufficient nitrogen to meet animal requirements often can be beneficial in rangeland-cropland nutrient transfer systems because the N fixation by the legumes helps to maintain soil nitrogen levels. Animal productivity remains high because the legumes provide needed protein to the animals. Fire, deliberate or unintentional, results in significant nutrient loss which may upset the nutrient cycling balance. The modeling exercise by Fernández-Rivera et al. (1993) provides a helpful framework to calculate the number of animals needed to maintain cropland fertility and whether livestock population, livestock location, efficiency of manure collection or availability of feed or land are limiting animal and crop production.

The timing of excretion and manure handling practices affect the efficiency with which nutrients can be recovered. In many parts of Africa, animals are confined in kraals or bomas during the night and graze during the day (Probert et al., 1992). According to results from ILRI (Fernandez-Rivera et al., 1993), approximately 43% of the daily fecal output was excreted between 1800 and 0600 hours when the animals were most likely to be confined. The manure and uneaten crop residues that accumulated in these confinement areas were removed and incorporated into the cropland. The crop responses to the "boma" manure were lower than those observed when manure was collected and stockpiled at a research station where there was less denitrification and leaching loss (Probert et al., 1992). In addition to nutrient losses, dilution of the boma manure with soil was thought to be a factor in the lower crop response.

Animal rearing in confinement using "cut-and-carry" feeding has been recommended to minimize unintended grazing of cropland by unrestrained animals and to reduce disease and parasitism. Total manure collections are possible, but losses due to leaching and denitrification are likely because urine recovery is difficult. Confinement systems have been particularly effective in the humid tropics where feed collection is relatively easy and disease risks are high (Tanner et al., 1993). In Java, more feed than is required by the animals is offered routinely and the uneaten portion is composted with the feces and urine. There are three advantages to this 'overfeeding' system: (1) goats and sheep, both selective feeders, have the opportunity to choose the most nutritious fraction of what they are offered resulting in increased productivity. (2) The overfeeding strategy increases the likelihood that diverse plant species will be available providing a cafeteria feeding situation more like that of a free-ranging animal. This minimizes the negative effects of anti-nutritional compounds. (3) The amount and quality of the manure-compost is improved. Station-based research in upland Java determined the most economical *overfeeding* strategy and showed that the manure-composts resulted in higher maize production than either vegetable compost or commercial fertilizer (Tanner et al., 1993).

2.2.2 Animal Feeding Strategies and Choice of Animal Species

When confinement systems are used to recover manure, the time animals spend grazing and in confinement must be balanced. The amount of grazing time needed for an animal to meet its requirements varies greatly depending on the digestive capacity of the animal, animal condition, and nutrient availability (Dove, 1996; Laca and Demment, 1996). When plant density is low, more bites are needed for the animal to ingest sufficient feed to meet its requirements prolonging the grazing time needed (Ungar, 1996). Animal species plays an important role here: the intake rate of a goat is much slower than that of a water buffalo (Murray and Illius, 1996). Much of the literature on ASP systems focuses on the suitability of different plant species in diverse environments but far less attention is paid to which animal species are included. There are scores of papers on the relative benefits of *Leucaena, Calliandra, Gliricidia* and *Erythrina* but little discussion of the relative merits of sheep and goats. Animal species have very different feeding strategies enabling them to occupy varied ecological niches (Van Soest, 1994). Hofmann's classification system (1989) proposed three classes of herbivores: concentrate selectors (deer, duikers), intermediate feeders (goats, sheep) and roughage eaters (buffalo, cattle). These groups can be further divided into grazers and browsers (Van Soest, 1994). The concentrate selectors are incapable of digesting large amounts of fiber and must rely on their ability to select the most nutritious portions of the plant. Each animal species has different strategies that permit effective occupancy of its nutritional niche. Goats, for example, have a specialized lip structure, the philtrum, which enables them to grab nutrient-rich leafy material from thorny shrubs. As browsers, goats and deer often eat high-tannin legumes. To minimize the toxic effects of the tannins, they secrete proline-rich salivary proteins (Asquith et al., 1987; Robbins et al., 1987). Functionally similar proteins containing high levels of histidine with high tannin affinity have been isolated from human saliva but not yet from other species (Yan and Bennick, 1995). These salivary proteins preferentially bind tannins protecting more valuable dietary protein and digestive enzymes from inclusion in tannin-protein complexes. Proline-rich proteins have not been isolated from the saliva of cattle and sheep whose predominantly grass diets typically contain little tannin.

As ASP systems are developed, attention must be paid to the choices of animals that are used. If trees and shrubs with high levels of secondary compounds are to form a significant portion of an animal's diet, it is wise to use animals that can tolerate these potentially toxic compounds. Goats with their proline-rich salivary proteins and their ability to eat selectively are likely to thrive better than sheep or cattle on browse. The roughage eaters with considerable fiber-digestion capacity use a very different feeding strategy. They eat unselectively and their long ruminal retention times of ingested feed permit extensive fiber digestion. If rice straw or corn stover forms the basis of a diet, roughage eaters such as cattle or buffalo will fare better than selective feeders like goats.

With animals as with plants, "monocultures" lower the efficiency with which resources are used. In the Serengeti, more than 10 species of ungulates may be found in close proximity with several species in mixed foraging groups (Murray and Illius, 1996). The feed preferences of different animal species tend to balance each other preventing over-grazing of some plant species while other plants remain uneaten. This is not to imply that rangeland pasture is a balanced static ecosystem: it is the opposite, a dynamic environment affected by many environmental, animal, and human factors. When external constraints such as restricted migration are not imposed, multi-species systems often result in efficient use of resources. A balance of selective feeders and roughage eaters is recommended because during the dry season livestock spend 50 to 80% of their time grazing crop residues, but browse can constitute 50% of cattle diets and up to 80% of the diets of small ruminants during droughts (Somda et al., 1993). Complementarity and increased productivity often has been observed when sheep and cattle are grazed together (Tainton et al., 1996). Mixed animal systems also provide insurance to risk averse small landholders.

Soil compaction and establishment of trees and shrubs are two issues that are related to choice of animal. Data from Canada suggest that compaction from treading by cattle affects bulk density of the soil (Bezkorowajnyj et al., 1993) which in turn affects water infiltration of the soil and nitrogen uptake by plants. Compaction is more likely with larger animals. The importance of compaction depends on soil type, water availability, level of animal traffic and vegetation.

Establishment of trees and shrubs is one of the biggest problems in ASP systems. During the several years between planting and the first harvest, often no economic benefit is derived from the land, a disincentive to the establishment of ASP systems. Equally problematic is keeping livestock, especially goats, from eating young seedlings. Several commercial and locally available repellents including those based on dog urine extracts, sheep dung, and several different chemicals (Haines et al., 1994; Eason et al., 1996) have been applied to trees with mixed short-term results. Physical barriers surrounding the young trees often are of limited effectiveness. Perimeter fences of unpalatable species like *Crotalaria* and *Tephrosia* surrounding the plantation deter animals but they lengthen the period before the first harvest. Failure to protect young seedlings is a major obstacle to wider adoption of ASP systems.

2.2.3 Year-round Provision of Feed

The biggest challenge in many ASP systems is feeding animals during the dry season. In many parts of the semi-arid tropics, dry seasons lasting up to 8 months are common. Animals rely on crop residues and mature grasses with supplementation from leguminous trees and shrubs. During the dry season, animal feed is likely to be both in short supply and of low nutritional quality. The dry matter digestibility of mature grasses may fall below 60% and crude protein levels may be less than 7%, the level required to maintain feed intake (Van Soest, 1994). Leguminous trees especially have an important role in providing feed during the dry season because they have deep roots that permit water uptake when surface water levels are low, reducing leaf-drop during the dry season. The edible foliage contains 12 to 30% crude protein and 0.4 to 4.7% phosphorus, levels that usually exceed the maintenance requirements of most animals (Norton, 1994a). Protein and carbohydrate levels of leaves from trees and shrubs, unlike grasses and herbaceous legumes, change little during the year. Unfortunately, these legumes frequently have high concentrations of anti-nutritional factors. These compounds restrict the amounts of these legumes that can be fed. Prunings from young plants have more soluble phenolics than older cuttings (Mafongoya et al., 1997). Plant-breeders have tried to develop varieties with low levels of these inhibitory compounds but the new varieties have been susceptible to disease and over-grazing by domestic and wild animals and insects.

An alternative to legumes with their problematical inhibitory compounds is non-leguminous species. Several species that are attractive forages with high levels of N such as *Morus* sp. have high nutrient requirements. Balanced nutrient cycling is difficult without the contribution of biological nitrogen fixation. High-yielding legumes can fix as much as 500 kg N ha^{-1}, although farm results are likely to be lower (Gutteridge and Shelton, 1994). In the savannahs of Latin America, eight different forage legumes fixed from 51 to 237 kg N ha^{-1} on fertilized plots compared to 23 to 101 kg N ha^{-1} on plots that received no supplemental P or K (Thomas and Lascano, 1993). If legumes comprised at least 20% of the aerial dry matter, the N balance in these pastures was balanced or positive. Without the biologically-fixed N, the nitrogen balance sheet would be as imbalanced as that for phosphorus. Working around the problems posed by inhibitory compounds seems more feasible than reliance on external sources of N.

2.3 Anti-Nutritional Factors in Leguminous Trees

2.3.1 Nutritional Effects of Anti-Nutritional Factors

Secondary metabolites in plants such as alkaloids, glycosides, flavonoids, saponins, terpenoids, lignin, and isoflavones (coumarin) protect plants from herbivory by animals and insects, and from viral and bacterial diseases, environmental stress (Salunkhe et al., 1990) and ultraviolet light

(Harborne, 1997). The chemical structures and thus the negative effects of these compounds are varied. In many cases, plants have more than one anti-nutritional factor present at inhibitory concentrations. For example, *Acacia aneura* (mulga) contains tannins and oxalate while *Leucaena leucocephala* contains mimosine, condensed tannins and flavonol glycosides (Norton, 1994b). Whether the effects of the anti-nutritional factors are additive or whether the effect of one can offset another is open to question. The inhibitory effects of tannins and saponins were additive when they were studied *in vitro* (Makkar, 1995), but other research suggests that alkaloids and tannins may form complexes that are not absorbed (Freeland, 1974; Powell, 1994) so that the biological activity of both is reduced.

Anti-nutritional factors may reduce the palatability of a plant, or they may be toxic either to gastrointestinal microbes or to the host animal. Their effects are not limited to nutrition; estrogenic compounds can impair reproductive performance. Because of the diversity of these compounds, it is not possible to consider all of them here and our focus will be on tannins. For a more extensive treatment of the chemistry and toxicity of other compounds, the book by Cheeke (1985) provides a good overview.

Tannins are polyphenolic compounds that form complexes with proteins and other molecules. They have been used commercially for centuries in leather-making and their astringent flavors are relished in fruits, red wine, tea, and beer. Tannins frequently are classified in two groups: condensed tannins or flavonoids that are more correctly called proanthocyanidins, and hydrolyzable tannins, which are esters of a polyol and gallic acid (Haslam, 1989). This grouping is simplistic as hybrid flavonol gallates have been isolated (Porter, 1989) and a third group, the phlorotannins, are found in green algae (Stern et al., 1996). Differences in the chemical structure of tannins reflect their different synthetic pathways and biological activities.

The antinutritional effects of tannins include (Reed, 1995): (1) lowered intake due to decreased palatability, (2) decreased digestibility of carbohydrate and protein due to both substrate and enzyme binding, (3) reduced availability of minerals, especially iron (Mila et al., 1996), (4) decreased nitrogen absorption with higher fecal nitrogen output, and (5) damage to the mucosal lining of the gastrointestinal tract. Long-term feeding caused liver damage in rats (Adachi et al., 1987).

Not all effects of dietary tannins are negative because when low levels are provided in the diet, benefits are evident. Although inclusion of tannin at 2 to 3% of diet dry matter has been recommended to maximize nutritive value (Waghorn et al., 1990), the optimal level of tannin inclusion is dependent on the amounts and forms of protein in the ration. The positive effects of tannins in ruminants include more efficient use of protein and greater N retention, bloat prevention and inhibition of pathogenic bacteria and ectoparasites (Schragle and Muller, 1990; Launchbaugh, 1996). Several mechanisms may be involved in the improved use of protein when moderate amounts of tannin are fed: (1) decreased ruminal protein degradability which increases the intestinal amino acid supply through either ruminal protection due to tannin-protein complex formation or to faster passage rate due to increased salivation, (2) more efficient urea recycling, or (3) increased microbial efficiency (Reed, 1995). The benefits are greatest when levels of soluble true protein are high, a situation found only in high quality tropical pastures. If tannins comprised more than 5.5% of diet dry matter, microbial activity and feed intake were depressed consistently (Waghorn et al., 1990). As Chesson points out in his comparison of ruminal digestion and soil organic matter decomposition (Chesson, 1997), tannins exert similar effects in both ecosystems. Nitrogen release is slowed when plants contain high levels of soluble phenolics (Palm and Sanchez, 1990, 1991). Initial levels of N also are important in prediction of ruminal release or soil mineralization (Constantinides and Fownes, 1994).

2.3.2 Animal Adaptations to Anti-Nutritional Factors

Because many of the trees and shrubs in tropical ASP systems contain tannins at levels that are harmful rather than beneficial, strategies to minimize their negative impacts must be devised. Fortunately, animals can adapt to diets high in anti-nutritional factors using a variety of mechanisms including feed avoidance, and detoxification at both the gut and tissue levels. Feed avoidance in animals as in humans can be either innate or learned. Mammalian herbivores usually instinctively avoid feeds with strong odors or tastes (Provenza et al., 1988) and can learn to avoid toxic plants that have negative gastrointestinal consequences to the animal (Launchbaugh, 1996).

Detoxification at the gut level is an important adaptation mechanism especially in ruminants with their diverse microbial populations. The effectiveness of microbial adaptation was underscored in recent experiments in Ethiopia with two groups of sheep that were fed *Acacia angustissima*. The first group was supplemented with 300 g d^{-1} of *A. angustissima* without adaptation. Although they ate less than 100 g d^{-1} of the acacia, two of the sheep had died within nine days and the third died on day 21 of the trial. When three other sheep were incrementally adapted to the supplement, none of the animals died although they consumed 200 g d^{-1} of the supplement (Odenyo et al., 1997). When ruminal fluid from the adapted animals was added to the rumens of animals that had not previously received *A. angustissima*, the unadapted animals were able to tolerate a challenge of 200 g d^{-1} without adverse effects (Odenyo et al., 1997). When the forage was fermented *in vitro*, levels of gas and volatile fatty acid production were lower than those of the *Sesbania sesban* control. When *in vitro* cultured inoculum enriched for the ability to degrade mulga (*A. aneura*) was administered to Australian sheep, their nitrogen retention improved and their weight loss was less compared to the uninoculated controls (Miller et al., 1997) during the first 57 days of the experiment. After 83 days, the control animals adapted to the mulga diet and there were no differences between the two groups. These two examples as well as the classic *Leucaena*-mimosine story (Allison et al., 1990) clearly indicate that ruminal microbes play an important role in the detoxification of anti-nutritional compounds.

Other mechanisms at the gut level are involved in detoxification including modification of the gut environment to reduce absorption of toxic compounds and tannin binding by salivary proteins or by addition of compounds such as polyethylene glycol (Silanikove et al., 1994) which form complexes with tannins (Launchbaugh, 1996). Increased mucus production protects the gut lining (Bernays et al., 1989). If the anti-nutritional factor is absorbed from the gut, they usually are metabolized by enzyme systems most of which are in the liver but which are located in other tissues as well. These post-absorptive detoxification mechanisms are well described by Launchbaugh (1996) and Bernays et al. (1989).

Animal management plays an important role in minimizing the impact of anti-nutritional factors. Browsing animals seldom eat a large quantity of a single plant species preferring to eat small amounts of many species. When animals are forced to consume larger amounts of less desirable plants because of overstocking or drought, they eat more toxic compounds and are more likely to suffer the consequences. The risk of forcing animals to eat undesirable plants is especially high in cut-and-carry systems unless attention is paid to the diversity of the plants that are harvested. Forage preservation through drying or ensiling can reduce the ability of tannins to bind protein (Terrill et al., 1990; Mafongoya et al., 1997) but oxidation during these processes forms high molecular weight compounds with nutritional effects similar to lignin.

2.4 Adoption of ASP Systems

Long-term adoption of recommended practices in ASP systems has been lower than hoped in many situations. The practices required to integrate livestock and cropping systems require additional labor, a commodity often in short supply (Turner, 1993). Confining animals requires more effort and resources than free-ranging livestock systems because fences must be built and water provided. Water itself is an important consideration, especially in semi-arid areas during the dry season. The range that livestock can use and thus the areas that receive manure is often determined by water availability.

There are many socio-economic considerations in ASP systems: how does crop-livestock integration mesh with land tenure practices? In Botswana, cattle traditionally are kept at cattle posts that are distant from the crop-growing areas or 'lands'. Do livestock have roles beyond their economic contribution? In many parts of Africa, cattle ownership confers status well beyond the economic value of the animal. What are the hidden costs of the system? Does animal ownership keep young boys out of school to tend livestock? Do the new approaches disrupt old markets and exchanges? If sale of manure for cooking fuel provides household income, it is likely to be difficult to convince people to manure their cropland. Do people have sufficient knowledge to effectively use an integrated ASP system? Can people see the benefits of the new methods? Dramatic improvements in crop yields and animal productivity are much less likely than the stemming of a cycle of degradation. Often it is difficult for people to appreciate that their efforts have only stopped things from becoming worse.

The questions in the preceding paragraph are difficult to answer, but without satisfactory answers to these issues, people are unlikely to modify their farming practices even if the P balance sheets show improvement. Some of the essential components of a research/extension program to resolve some of biological and social problems that limit adoption of ASP systems include: (1) farmer identification of problems, (2) some station-based research for preliminary evaluation of new technologies without imposing risk on farmers, and (3) on-farm research with strong farmer involvement to determine obstacles to adoption and to obtain realistic production responses.

3. CONCLUSIONS

Chemical analyses of tropical grasses and legumes show that only 46% of the grasses and legumes assembled in a large data set analyzed by Van Soest (17) meet the maintenance requirements of most ruminants for both energy and protein. About 43% of the grasses were deficient in protein while all of the legumes provided adequate protein. Sixty percent of the legumes failed to provide adequate energy. Problems of forage quality and quantity are most evident during the dry season. The best strategy to meet, or, even better, to exceed, the maintenance requirements is to develop an integrated grass-legume system that can provide forages of adequate quality and quantity throughout both the rainy and dry seasons. As we address future research on ASP systems we must ask what limits the productivity and adoption of the system. Recent data from East Africa (74) show that cumulative forage yield was increased when grass-legume combinations were compared to planting of grasses or legumes alone.

When we look at the ASP system from a livestock perspective, the nutrition 'bottleneck' clearly is the major constraint. Lack of needed energy, protein, and minerals retards growth, limits production, impairs reproduction, and makes animals more susceptible to disease. Legumes can help to resolve this problem but secondary compounds limit the extent to which they can be included in the diet. If the effects of these toxic compounds can be minimized, legumes may play a larger role in animal diets. Selection and management of forages also plays an important role in determining forage quality. There are many biophysical questions that need to be addressed, yet solving these questions alone is not going to change the productivity of ASP systems. If management changes are needed, people must be willing to alter their current practices and use new ideas. Thus, an integrated approach must be adopted including all components of the ASP systems: animals, plants, soil, and people.

REFERENCES

Adachi, H., Konishi, K., Torizuka, K., and Horikoshi, I., 1987, "The *in vitro* effects of tannic acid on rat liver mitochondrial respiration and oxidative phosphorylation," *Chem. Pharm. Bull.* 35, 1176.

Agboola, A. and Kintomo, A. A., 1993, "Nutrient recycling in pastures, rangeland, fallow and cut-and-carry systems in sub-Saharan Africa," in *Livestock and Sustainable Nutrient Cycling in Mixed Farming Systems of Sub-Saharan Africa,* Powell, J. M., Fernández-Rivera, S., Williams, T. O., and Renard, C. Eds., ILCA, Addis Ababa, 247.

Allison, M. J., Hammond, A. C., and Jones, R. J., 1990, "Detection of ruminal bacteria that degrade toxic dihydroxypyridine compounds produced from mimosine," *Appl. Environ. Microbiol.* 56, 590.

Asquith, T. N., Uhlig, J., Mehansho, H., Putnam, L., Carlson, D. M., and Butler, L. G., 1987, "Binding of condensed tannins to salivary proline-rich glycoproteins: The role of carbohydrate," *J. Agric. Food Chem.* 35, 331.

Barnett, G. M., 1994, "Phosphorus forms in animal manure," *Bioresource Technol.* 48, 139.

Bernays, E. A., Cooper Driver, G., and Bilgener, M., 1989, "Herbivores and plant tannins," *Adv. Ecol. Res.* 19, 263.

Bezkorowajnyj, P. G., Gordon, A. M., and McBride, R. A., 1993, "The effect of cattle foot traffic on soil compaction in a silvo-pastoral system," *Agrofor. Sys.* 21, 1.

Brewbaker, J. L., 1986, *Leguminous trees and shrubs for Southeast Asia and the South Pacific*, Blair, G. J., Ivory, D. A., and Evans, T. R., Eds., Australian Centre for International Agricultural Research, Canberra, 43.

Brower, J. and Powell, J. M., 1993, "Soil aspects of nutrient cycling in a manure application experiment in Niger," in *Livestock and Sustainable Nutrient Cycling in Mixed Farming Systems of Sub-Saharan Africa,* Powell, J. M., Fernández-Rivera, S., Williams, T. O., and Renard, C. Eds., ILCA, Addis Ababa, 211.

Cheeke, P. R. and Shull, L. R., 1985, *Natural Toxicants in Feeds and Poisonous Plants*, AVI Publishing, Westport, CT.

Chesson, A., 1997, "Plant degradation by ruminants: Parallels with litter decomposition in soils," in *Driven by Nature: Plant Litter Quality and Decomposition,* Cadisch, G. and Giller, K. E. Eds., CAB International, Wallingford, U.K., 47.

Constantinides, M. and Fownes, J. H., 1994, "Nitrogen mineralization from leaves and litter of tropical plants: relationship to nitrogen, lignin and soluble polyphenol concentrations," *Soil Biol. Biochem.* 26, 49.

Cooper, P. J. M., Leakey, R. R. B., Rao, M. R., and Reynolds, L., 1996, "Agroforestry and the mitigation of land degradation in the humid and sub-humid tropics of Africa," *Expl. Agric.* 32, 235.

de Leeuw, P. N., Reynolds, L., and Rey, B., 1993, "Nutrient transfers from livestock in West African agricultural systems," in *Livestock and Sustainable Nutrient Cycling in Mixed Farming Systems of Sub-Saharan Africa,* Powell, J. M., Fernández-Rivera, S., Williams, T. O., and Renard, C. Eds., ILCA, Addis Ababa, 371.

Dove, H., 1996, "The ruminant, the rumen and the pasture resource: Nutrient interactions in the grazing animal," in *The Ecology and Management of Grazing Systems,* Hodgson, J. and Illius, A. W. Eds., CAB International, Wallingford, U.K., 219.

Eason, W. R., Gill, E. K., and Roberts, J. E., 1996, "Evaluation of anti-sheep tree-stem-protection products in silvopastoral agroforestry," *Agrofor. Sys.* 34, 259.

Fox, D. G., Sniffen, C. J., O'Connor, J. D., Russell, J. B., and Van Soest, P. J., 1992, "A net carbohydrate and protein system for evaluating cattle diets: III. Cattle diets and diet adequacy," *J. Anim. Sci.*, 70, 3578.

Fox, D. G., Juarez, F., Pell, A. N., and Lanna, D. P., 1997, "Predicción de requerimientos nutricionales de los bovinos en condiciones tropicales (Predicting nutrient requirements of cattle in tropical conditions)," in the *Proceedings of the XXXIII Reunión Nacional de Investigación Peucaria*, Veracruz, Mexico, 49.

Fernández-Rivera, S., Williams, T. O., Hiernaux, P., and Powell, J. M., 1993, "Faecal excretion by ruminants and manure availability for crop production in semi-arid West Africa," in *Livestock and Sustainable Nutrient Cycling in Mixed Farming Systems of Sub-Saharan Africa,* Powell, J. M., Fernández-Rivera, S., Williams, T. O., and Renard, C. Eds., ILCA, Addis Ababa, 149.

Freeland, W. J. and Janzen, D. H., 1974, "Strategies in herbivory by mammals: The role of plant secondary compounds," *Am. Na.* 108, 269.

Gressel, N. and McColl, J. G., 1997, "Phosphorus mineralization and organic matter decomposition: A critical review," in *Driven by Nature: Plant Litter Quality and Decomposition,* Cadisch, G. and Giller, K. E. Eds., CAB International, Wallingford, U.K., 297.

Gutteridge, R. C. and Shelton, H. M., 1994, "The role of forage tree legumes in cropping and grazing systems," in *Forage Tree Legumes in Tropical Agriculture,* Gutteridge, R. C. and Shelton, H. M. Eds., CAB International, Wallingford, U.K., 3.

Haines, P. J., Bell, A. B., and Thatcher, L. P., 1994, "Evaluation of some factors involved in reducing browsing damage to eucalypt trees by sheep," *Aust. J. Exp. Agric.* 34, 601.

Harborne, J. B., 1997, "Role of phenolic secondary metabolites in plants and their degradation in nature," in *Driven by Nature: Plant Litter Quality and Decomposition,* Cadisch, G. and Giller, K. E. Eds., CAB International, Wallingford, U.K., 67.

Haslam, E., 1989, *Plant Polyphenols*, Cambridge University Press, Cambridge, U.K.

Hofmann, R. R., 1989, "Evolutionary steps of ecophysiological adaptation and diversification of ruminants: a comparative view of their digestive system," *Oecologia* 78, 443.

Laca, E. A. and Demment, M. W., 1996, "Foraging strategies of grazing animals," in *The Ecology and Management of Grazing Systems,* Hodgson, J. and Illius, A. W. Eds., CAB International, Wallingford, U.K., 137.

Lal, R., 1987, *Tropical Ecology and Physical Edaphology,* John Wiley & Sons, Chichester, U.K.

Launchbaugh, K. L., 1996, "Biochemical aspects of grazing behavior," in *The Ecology and Management of Grazing Systems,* Hodgson, J. and Illius, A. W. Eds., CAB International, Wallingford, U.K., 159.

Mafongoya, P. L., Dzowela, B. H., and Nair, P. K., 1997, "Effect of multipurpose trees, age of cutting and drying method on pruning quality," in *Driven by Nature: Plant Litter Quality and Decomposition,* Cadisch, G. and Giller, K. E. Eds., CAB International, Wallingford, U.K., 167.

Makkar, H. P. S., Blümmel, M., and Becker, K., 1995, "*In vitro* effects of and interactions between tannins and saponins and fate of tannins in the rumen. *J. Sci. Food Agric.*," 69, 481.

McDowell, L. R., Conrad, J. H., and Ellis, G. L., 1984, "Mineral deficiencies and imbalances, and their diagnosis," in *Herbivore Nutrition in the Subtropics and Tropics,* Gilchrist, F. M. C. and Mackie, R. I. Eds., Science Press, Craighall, South Africa, 67.

Mila, I., Scalbert, A., and Expert, D., 1996, "Iron withholding by plant polyphenols and resistance to pathogens and rots," *Phytochemistry* 42, 1551.

Miller, S. M., Klieve, A. V., Plumb, J. J., Aisthorpe, R., and Blackall, L. L., 1997, "An *in vitro* cultured rumen inoculum improves nitrogen digestion in mulga-fed sheep." *Aust. J. Agric. Res.* 48, 403.

Murray, M. G. and Illius, A. W., 1996, "Multispecies grazing in the Sergeneti," in *The Ecology and Management of Grazing Systems,* Hodgson, J. and Illius, A. W. Eds., CAB International, Wallingford, U.K., 247.

Murwira, H. and Kirchmann, H., 1993, "Carbon and nitrogen mineralization of cattle manures, subjected to different treatments, in Zimbabwean and Swedish soils," in *Soil Organic Matter Dynamics and Sustainability of Tropical Agriculture,* Mulongoy, K. and Merckx, R., Eds., Wiley-Sayce, Chichester, U.K., 189.

National Academy of Sciences, 1979. *Tropical Legumes: Resources for the Future,* National Academy Press, Washington, DC.

National Research Council *Sustainable Agriculture and the Environment in the Humid Tropics*, National Academy Press, 1993, Washington, DC.

Norton, B. W., 1994a, "The nutritive value of tree legumes," in *Forage Tree Legumes in Tropical Agriculture,* Gutteridge, R. C. and Shelton, H. M. Eds., CAB International, Wallingford, U.K., 177.

Norton, B. W., 1994b, "Anti-nutritive and toxic factors in forage tree legumes," in *Forage Tree Legumes in Tropical Agriculture,* Gutteridge, R. C. and Shelton, H. M. Eds., CAB International, Wallingford, U.K., 202.

Odenyo, A. A., Osuji, P. O., Karanfil, O., and Adinew, K., 1997, "Microbiological evaluation of *Acacia angustissima* as a protein supplement for sheep." *Anim. Feed Sci. Technol.* 65, 99.

Palm, C. A. and Sanchez, P. A., 1990, "Decomposition and nutrient release patterns of the leaves of three tropical legumes," in *Biotropica* 22, 330. (Abstract)

Palm, C. A. and Sanchez, P. A., 1991, "Nitrogen release from the leaves of some tropical legumes as affected by their lignin and polyphenolic contents," in *Soil Biol. Biochem.* 23, 83. (Abstract)

Porter, L. J., 1989, "Tannins," *Meth. Plant Biochem.* 1, 389.

Powell, J. M., Fernández-Rivera, S., and Höfs, S., 1994, "Effects of sheep diet on nutrient cycling in mixed farming systems of semi-arid West Africa," *Agric. Ecosyst. Environ.* 48, 263.

Power, J. F., 1994, "Understanding the basics: Understanding the nutrient cycling process," *J. Soil Water Conserv.* 49 (Suppl), 16.

Probert, M. E., Okalebo, J. R., Simpson, J. R., and Jones, R. K., 1992, "The role of boma manure for improving soil fertility," *Proc. ACIAR* 41, 63.

Provenza, F. D., Balph, D. F., Olsen, J. D., Dwyer, D. D., Ralphs, M. H., and Pfister, J. A., 1988, "Toward understanding the behavioral aspects of livestock to poisonous plants," in *The Ecology and Economic Impact of Poisonous Plants on Livestock Production,* James, L. F., Ralphs, M. H., and Nielson, D. B. Eds., Westview Press, Boulder, CO, 407.

Reed, J. D., 1995, "Nutritional toxicology of tannins and related polyphenols in forage legumes," *J. Anim. Sci.* 73, 1516.

Reed, J. D. and Bert, J., 1993, "The role of livestock in sustainable agriculture and natural resource management," in *Livestock and Sustainable Nutrient Cycling in Mixed Farming Systems of Sub-Saharan Africa,* Powell, J. M., Fernández-Rivera, S., Williams, T. O., and Renard, C. Eds., ILCA, Addis Ababa, 461.

Robbins, C. T., Mole, S., Hagerman, A. E., and Hanley, T. A., 1987, "Role of tannins in defending plants against ruminants: Reduction in dry matter digestion." *Ecology* 68, 1606.

Roothaert, R. L. and Paterson, R. T., 1997, "Recent work on the production and utilization of tree fodder in East Africa," *Anim. Feed Sci. Technol.*, 69, 39.

Russelle, M. P., 1992, "Nitrogen cycling in pasture and range," *J. Prod. Agric.* 5, 13.

Salunkhe, D. K., Chavan, J. K., and Kadam, S. S., 1990, *Dietary Tannins: Consequences and Remedies, CRC Press*, Boca Raton, FL.

Power, J. F., 1994, "Understanding the basics: Understanding the nutrient cycling process," *J. Soil Water Conserv.*, 49(Suppl), 16.

Schrägle, R. and Müller, W., 1990, "The influence of selected tannin-containing plant species on the tenacity of pathogenic bacteria in an *in vitro* rumen system," *J. Vet. Med. B* 37, 181.

Sharpley, A. N., 1996, "Availability of residual phosphorus in manured soils," *Soil Sci. Soc. Am. J.* 60, 1459.

Silanikove, N., Nitsan, Z., and Perevolotsky, A., 1994, "Effect of a daily supplementation of polyethylene glycol on intake and digestion of tannin-containing leaves (*Ceratonia siliqua*) by sheep," *J. Agric. Food Chem.* 42, 2844.

Somda, Z. C., Powell, J. M., Fernández-Rivera, S., and Reed, J. D., 1993, "Feed factors affecting nutrient excretion by ruminants and the fate of nutrients when applied to soil," in *Livestock and Sustainable Nutrient Cycling in Mixed Farming Systems of Sub-Saharan Africa,* Powell, J. M., Fernández-Rivera, S., Williams, T. O., and Renard, C. Eds., ILCA, Addis Ababa, 227.

Stangel, P. J., 1993, "Nutrient cycling and its importance in sustaining crop-livestock systems in sub-Saharan Africa: An overview," in *Livestock and Sustainable Nutrient Cycling in Mixed Farming Systems of sub-Saharan Africa,* Powell, J. M., Fernández-Rivera, S., Williams, T. O., and Renard, C. Eds., ILCA, Addis Ababa, 37.

Stern, J. L., Hagerman, A. E., Steinberg, P. D., and Mason, P. K., 1996, "Phlorotannin-protein interactions," *J. Chem. Ecol.* 22, 1877.

Strauch, D., 1987, "Hygiene of animal waste management," in *Animal Production and Environmental Health,* Strauch, D. Ed., Elsevier Science, Amsterdam, 155.

Tainton, N. M., Morris, C. D., and Hardy, M. B., 1996, "Complexity and stability in grazing systems," in *The Ecology and Management of Grazing Systems,* Hodgson, J. and Illius, A. W. Eds., CAB International, Wallingford, U.K., 275.

Tanner, J., Holden, S. J., Winugroho, M., Owen, E., and Gill, M., 1993, "Feeding livestock for compost production: A strategy for sustainable upland agriculture on Java," in *Livestock and Sustainable Nutrient Cycling in Mixed Farming Systems of sub-Saharan Africa,* Powell, J. M., Fernández-Rivera, S., Williams, T. O., and Renard, C. Eds., ILCA, Addis Ababa, 115.

Terrill, T. H., Windham, W. R., Evans, J. J., and Hoveland, C. S., 1990, "Condensed tannin concentration in Sericea lespedeza as influenced by preservation method," *Crop Sci.* 30, 219.

Thomas, R. J. and Lascano, C. E., 1993, "The benefits of forage legumes for livestock production and nutrient cycling in pasture and agropastoral systems of acid-soil savannahs of Latin America," in *Livestock and Sustainable Nutrient Cycling in Mixed Farming Systems of Sub-Saharan Africa,* Powell, J. M., Fernández-Rivera, S., Williams, T. O., and Renard, C. Eds., ILCA, Addis Ababa, 277.

Turner, M., 1993, "The sustainability of rangeland to cropland nutrient transfer in semi-arid West Africa: Ecological and social dimensions neglected in the debate," in *Livestock and Sustainable Nutrient Cycling in Mixed Farming Systems of Sub-Saharan Africa,* Powell, J. M., Fernández-Rivera, S., Williams, T. O., and Renard, C. Eds., ILCA, Addis Ababa, 435.

Ungar, E. D., 1996, "Ingestive behavior," in *The Ecology and Management of Grazing Systems,* Hodgson, J. and Illius, A. W. Eds., CAB International, Wallingford, U.K., 185.

Van Soest, P. J., 1994, *Nutritional Ecology of the Ruminant*, 2nd Ed., Cornell University Press, Ithaca, NY.

Waghorn, G. C., Jones, W. T., Shelton, I. D., and McNabb, W. C., 1990, "Condensed tannins and the nutritive value of herbage." *Proc. New Zealand Grassl. Assoc.* 51, 171.

Waterman, P. G. and Mole, S., 1994, *Analysis of Phenolic Plant Metabolites,* Blackwell Sci. Pub., London.

Williams, T. O., Powell, J. M., and Fernández-Rivera, S., 1995, "Soil fertility maintenance and food crop production in semi-arid West Africa: Is reliance on manure a sustainable strategy?" *Outl. Agric.*, 24, 43.

Yan, Q. and Bennick, A., 1995, "Identification of histatins as tannin-binding proteins in human saliva," *Biochem. J.* 311, 341.

3 Water Management with Hedgerow Agroforestry Systems

Susan J. Riha and Beverly D. McIntyre

1. INTRODUCTION

This chapter addresses the question "How do agroforestry systems influence the movement of water, and water use by plants, in comparison to more conventional agricultural systems?" Movement of water in agricultural systems is important not only because water can limit plant growth, but also because water movement can have significant impacts on soil erosion and nutrient cycling. Our objective in focusing on this question is to provide a framework with which to summarize experimental work that has addressed this issue; to assess some of the climate, soil and management conditions influencing water movement; and to suggest areas for further research.

An outline of the methods used to integrate annual crop production with woody perennials, based on the spatial pattern in which trees are grown, is presented by Lal (1989). These methods range from random scattering of trees across a field to alternate strips of trees and annual crops. This latter system has been referred to as alley cropping, avenue cropping, or hedgerow cropping (Lal, 1989). The agroforestry examples we will use to address the question posed above will be drawn mainly from hedgerow cropping experiments conducted primarily in tropical environments. However, many of the concepts used to analyze tropical hedgerow systems could be validly applied

1-56670-294-1/99/$0.00+$.50

to parkland and tree-pasture agroforestry (silvopastoral) systems, as well as to hedgerow systems in temperate environments.

Some researchers have proposed that agricultural systems that contain multiple plant species can exploit more resources through temporal or spatial niche differentiation than can monocultural systems. Thus multiple plant systems can potentially complement each other with respect to resource use and therefore be more productive (Berendse, 1979; Anderson and Sinclair, 1993). Others have criticized the application of niche theory, originally developed to describe resource use in animal communities, to plant communities, because plants compete for the same resources — light, water and nutrients (Welden and Slauson, 1986; Loomis and Connor, 1992). Species in multiple plant systems may not complement each other with respect to the use of these resources, but rather compete with each other, thus using the same resources as each would if they were grown in monoculture. For example, in a review of annual intercropping systems, Morris and Garrity (1993) found that there was generally no increase in the total water use in an intercrop system compared to monoculture. More recently, Cannell et al. (1996) have argued that, from a biophysical perspective, the "benefits of growing trees with crops will occur only when the trees are able to acquire resources of water, light and nutrients that the crop would not otherwise acquire."

To address the question of whether and the degree to which agroforestry systems alter water fluxes relative to other agricultural systems, including the amount of water transpired by plants per given area of land per year, the major pathways of water inputs and losses in the soil-plant-atmosphere system must be considered. Water is lost from the soil system as evaporation from soil and plant surfaces, as transpiration, as deep percolation, and as surface runoff. Increases in the amount of water used by plants in the system is therefore accompanied by, and in some cases contingent upon, decreases in water lost as evaporation, deep percolation and/or surface runoff. The importance of these various pathways for water loss will vary with climate, soil type and the landscape, as well as with the type and management of the cropping system. In addition, water can enter some plant-soil systems not only as precipitation infiltrating the soil surface, but also through the lateral movement of water beneath the surface. Some cropping systems may be able to access more of this water than others, thus increasing the amount of water available to plants.

2. SURFACE RUNOFF/INFILTRATION

Agroforestry systems have been recommended for decreasing soil erosion caused by surface runoff. Surface runoff can occur when the rate of precipitation exceeds the rate of infiltration of water into the soil surface. It can be reduced by the interception of water by plant canopies and plant litter, which can decrease the amount, the intensity and the spatial distribution of the precipitation reaching the soil surface. Interception of rainfall by plants and litter protects the soil surface from the direct impact of raindrops which can cause breakdown of the soil structure, leading to sealing of the surface and large reductions in surface infiltration rates (Marshall et al., 1996). However, runoff can also occur when the soil becomes saturated above a shallow, impermeable layer or when groundwater reaches the surface. In such cases, surface infiltration rates would have little impact on runoff.

The infiltration rate of the soil is influenced by factors that contribute to and maintain soil structure, including soil organic matter, roots and soil biota. Mulching the soil surface not only protects it from the direct impact of raindrops, but also provides organic matter that can be consumed by earthworms and other soil fauna. Thus, mulching can indirectly affect soil aggregate formation and stability. Previous studies have demonstrated that crop residue mulches can effectively reduce surface runoff and increase infiltration. Studies discussed in this paper focused on the impact of agroforestry mulches on surface runoff and infiltration.

Water that does not immediately infiltrate the soil can move across the soil surface and be collected in depressions or ditches that can serve as temporary storage basins for ponded water. Hedgerows contribute to this process by stabilizing the barriers that create the storage areas (Sheng,

1989; Wiersum, 1991; Dano and Siapno, 1992). However, Wiersum (1984; 1991) considers that the main contribution of agroforestry systems to erosion control, and by implication to surface runoff, lies in the the capacity of these systems to establish and maintain a ground cover of litter, rather than in their role in maintaining water storage basins.

2.1 Studies of Surface Runoff

The impact of hedgerow agroforestry systems on soil runoff can be affected by variations in management practices such as mulching, tillage and weeding, as well as variations in the amount of time the soil surface is covered by the hedgerow and the effectiveness of trees in maintaining soil conservation structures (Sheng, 1989). Climate, experimental design and *in situ* factors may complicate the interpretation of runoff results. For example, frequency and intensity of precipitation, the slope, and soil type and depth have large effects on surface runoff. Experimental results depend on the time over which measurements are made and plot size. All these variables make it very difficult to compare the results from different field experiments, since not all variables can be controlled or are even reported. For these reasons, Bregman (1993) has concluded that it is usually not possible to quantify the impact of agroforestry systems on soil erosion and thus accurately assess a system's potential for decreasing soil loss. A similar conclusion could be reached for assessing surface runoff in agroforestry systems, with the consequence that the results of surface runoff studies presented below cannot necessarily be generalized across different agroforestry sites and systems. As an alternative to quantitative assessment, Bregman (1993) proposes a method to estimate the relative erosion control potential of various agroforestry systems, based on their protective functions (cover, barrier and soil reinforcement). This method assigns values to specific features of agroforestry systems (for example, age and arrangement of woody perennials, history of site, and practices influencing soil conditions). Such an approach could be extended to include estimates of relative impact of agroforestry systems on surface runoff as well as soil erosion. The studies discussed below serve as an introduction to the range of environments and management techniques under which surface runoff in agroforestry systems has been measured. These studies could, along with further research, serve as the basis for a method to estimate the relative potential of agroforestry systems to control surface runoff.

Several studies have compared runoff from systems where agroforestry mulches have or have not been applied. Rao et al. (1991) studied the impact of *Leucaena leucocephala* hedgerows in contrast to mulch on a low-sloping (1–2%), shallow Alfisol in semi-arid India. Unmulched hedgerows were effective in reducing annual runoff by up to 55%, with hedgerows at 3-m spacing appearing to reduce runoff more than hedgerows at 5.4-m spacing. In contrast, mulches of hedgerow prunings, either applied to the soil surface or incorporated into the soil, reduced runoff by 86% in sole annual crops. Surface runoff in unmulched sole annual crops averaged only about 10% of the average annual rainfall of 550 mm during the study period. This was considered a low rate of runoff because in an earlier study runoff had comprised 26% of average annual rainfall. Runoff may have been low because the annual rainfall during the study period ranged from 477 to 591 mm year^{-1}, whereas the long term mean annual rainfall is 765 mm.

Ong (1995) discusses the results of an experiment conducted on 14% slope at a semi-arid site in Machakos, Kenya, using *Cassia siamea*. Runoff from plots planted with *Cassia* contour hedgerows at 4-m intervals and mulched with prunings was compared to runoff from plots planted to sole cowpea that were either not mulched or mulched with 3 t ha^{-1} of *Cassia* prunings. For one 40-mm rainfall event, runoff from the unmulched sole cowpea was 8 mm, compared to 2 mm from the mulched sole cowpea and less than 1 mm from the mulched hedgerow-cowpea system.

In a one-year study in Machakos, Kenya, on deep, well-drained Alfisols with 15% slopes, runoff losses from sole maize plots mulched with *Cassia siamea, Gliricidia sepium* and *Grevillea robusta* were 28, 48 and 58% lower, respectively, than from unmulched control plots (Omoro and Nair, 1993). In the unmulched controls, runoff totaled 19.5 mm of water (3.75% of the study period's total rainfall of 519 mm) while losses from treatments mulched at 2.24 and 4.48 t ha^{-1} ranged from 5 to 11 mm. Runoff was significantly less in the long rainy season from the treatment mulched at the higher rate (4.48 t ha^{-1}). Ground covered by the mulch decreased from 80% to less than 20% in 60 days during the long rainy season.

Alegre and Rao (1996) evaluated the combined effect of contour hedgerows and mulching using *Inga edulis* in a six-year plot-scale experiment conducted in the lowland humid tropics (2200 mm annual rainfall) of Peru where the topography is undulating with gentle to medium slopes. The slash-and-burn agricultural system (1 to 2 years of cropping followed by 5 to 15 years of fallow), that is dominant in the region, is considered the major cause of soil degradation and deforestation. The researchers wanted to evaluate whether hedgerow intercropping could control surface runoff and soil erosion and enable continuous cropping on moderately sloping hillsides. The experiment was conducted on a fine, loamy Paleudult, 15 to 20% slope previously covered by a 12-year-old secondary forest. All the prunings from tree hedges and crop residues were placed on the soil surface as mulch in the contour hedgerow plots, while crop residues were placed on the surface in the sole crop plots, also planted on the contour, of rice rotated with cowpea. Plots were not cultivated between seasons, but were hand-weeded during the cropping season. They found that the hedgerow intercrop systems, spaced at 4-m intervals and covering 22% of the land area, annually conserved 287-mm of water compared to the sole crop plots. This conservation represented 83% of the water lost from sole crop plots.

A five-year experiment with *Leucaena leucocephala* and *Gliricidia sepium* contour hedgerows at 2- and 4-m intervals was conducted in Ibadan, Nigeria (forest-savanna transition zone) where average annual precipitation ranges from 1100 to 1300 mm (Lal, 1989). The hedges were planted on an Oxic Paleustalf with approximately 7% slope. This soil is course-textured near the surface but underlain at 20 to 30 cm by a horizon that is 20 to 50% gravel (by weight) embedded in a matrix that is 60 to 70% clay. This layer restricts root growth of annual crops whereas tree roots are able to penetrate it. Surface runoff from the agroforestry treatments was compared to plow-till and no-till systems of maize and cowpea (Lal, 1989). Within the agroforestry systems, runoff was least in the plots planted at 2-m intervals. At both 2-m and 4-m spacings, runoff was less under *Gliricidia* than *Leucaena*. However, in all the systems, runoff was least in the maize/cowpea no-till. In this experiment, trees were pruned an average of three times a year and the prunings were rotated into the soil between the hedges. Lal (1989) notes the relatively greater runoff in the agroforestry systems compared to the no-till treatments, and attributes this difference to the disturbance caused by the incorporation of prunings into the surface soil of the agroforestry systems.

Of course, other cropping and tillage systems have also been considered useful in decreasing surface runoff. Lal (1996) compared a number of restorative farming systems implemented in watershed-scale studies in southwestern Nigeria where annual rainfall averaged 1145 mm during the study period. The objectives of this study included evaluating the effects of several farming systems on reducing surface runoff and erosion on eroded and degraded soils. The restorative farming systems studied were contour alley cropping with *Leucaena leucocephala* at 4-m intervals, a *Mucuna utilis* fallow, and alley farming. Annual runoff, in the three years reported, varied greatly with year and ranged from 8 to 76 mm of water, which represented 1 to 5% of the annual rainfall. The various farming systems studied yielded no significant differences in runoff.

2.2 Infiltration Rates

Direct measurements of surface runoff, even on the small plot-scale level, can be difficult and costly to obtain, as well as difficult to extrapolate to other sites and systems. Researchers have also investigated the impact of agroforestry systems on the infiltration rates of soil. Generally, as ponded water is applied to a dry soil, the infiltration rate will decline quickly as the soil is recharged, approaching an equilibrium value with time that is equivalent to the saturated hydraulic conductivity of the soil. Therefore, some studies report the equilibrium infiltration rate and others report the saturated hydraulic conductivity of the soil in order to assess the effect of agroforestry systems on the soil infiltration rate. For example, Alegre and Rao (1996) found that the saturated hydraulic conductivity after 14 consecutive seasons was 50 cm hr^{-1} in their hedgerow intercrop system vs. 18.5 cm hr^{-1} in their sole crop treatment. These values cannot be used directly to estimate changes in surface runoff, but they can suggest whether decreases in surface runoff might be expected. For example, with infiltration rates as high as 50 cm hr^{-1} any surface runoff that might occur would likely be due to other causes, such as soil saturation above a shallow, impermeable layer. Infiltration values can also be used in conjunction with simulation models to estimate runoff. However, such values may be improved by using additional methods to measure infiltration, such as rainfall simulators, that more realistically mimic soil wetting by rainfall in comparison to the more commonly used infiltrometer methods.

After 5 years of continuous cultivation of an Oxic Paleustalf (formerly in secondary forest) in the Ibadan, Nigeria, study discussed above, Lal (1989) found that equilibrium infiltration rates were greater in *Leucaena leucocephala* and *Gliricidia sepium* plots and the plow-till plots than in the no-till controls (19, 24 and 21 cm hr^{-1} vs. 8 cm hr^{-1}). Thus, infiltration was greatest in the *Gliricidia* plots, whereas surface runoff rates were lowest in the maize-cowpea no-till plots. In all cases, infiltration rates declined over the study period.

At the end of the study discussed above by Omoro and Nair (1993), infiltration rates between mulched and unmulched treatments did not differ, despite differences in surface runoff. This similarity might have been due in part to the exponential decline (from 80 to 20% or less) in ground cover over 60 days of the cropping season. The rate of decline was dependent on the type of mulch, with the decline in ground cover occurring more rapidly with the *Gliricidia* mulch than with the *Grevillea* and *Cassia* mulches. *Gliricidia* mulches also decomposed more rapidly than *Flemingia* and *Cassia* mulches in a study conducted in Ibadan, Nigeria (Yamoah et al., 1986).

As mentioned previously, mulching may have a direct impact on surface infiltration by decreasing the impact of raindrops. Rainfall can cause breakdown of the soil structure, leading to sealing of the surface and large reductions in surface infiltration rates. The impact of mulches on surface runoff is a function of the relation of the length of the rainy season to the length of time a mulch remains effective before decomposing. Budelman (1989) proposed a method to characterize the impact and effective lifetime of mulches with respect to their influence on soil temperature and moisture (see Section 3.2). Such an approach might be extended to include their influence on surface infiltration. The combined results of several studies indicate that the rate of decline of leaf mulches is greatest with *Leucaena leucocephala*, followed by *Gliricidia sepium* and *Cassia siamea*, with *Flemingia macrophylla* and *Grevillea robusta* declining at the slowest rate (Yamoah et al., 1986; Budelman, 1989; Omoro and Nair, 1993). Budelman (1989) suggests that these variations in rates of decline may be due in part to the size of the leaves, with smaller leaves or leaflets decomposing more rapidly than larger leaves.

Studies of soil water infiltration in temperate regions have shown that there is significant temporal, spatial and tillage-induced variability in this parameter. Intrinsic variability (i.e., variability due to measurement error, inadequate sampling volume and randomness) is also high (van Es, 1993). Thus, increased sampling may be needed to clarify variations among different agroforestry systems in terms of their impact on soil water infiltration. Macropores may make important contributions to soil water infiltration in agroforestry systems. Their importance can be evaluated by measuring infiltration using two methods in combination. In the first method, water is applied at a small positive pressure and consequently, macropores, as well as micropores, conduct water. In the second method, water is applied at a small negative pressure and pores greater than 1 mm in diameter remain empty (Marshall et al., 1996). Another approach to measuring infiltration under non-ponded conditions is to use a rainfall simulator. The studies reviewed above suggest that until soil infiltration is better characterized in agroforestry systems, care must be exercised in using measurements of soil water infiltration to assess the impact of different agroforestry systems on surface runoff.

2.3 Biological Activity

The application of mulch can increase soil organic matter and alter soil temperature and water regimes. These changes in turn provoke changes in biotic populations and root density. The activities of biota and roots may enhance or maintain surface infiltration rates by promoting the formation of biochannels and improving soil structure (Lal, 1989). In southwestern Nigeria, Kang et al. (1994) found that the rate of surface casting of earthworms varied among different agroforestry tree species, ranging from 18 to 26 t ha^{-1} per year. Tian et al. (1993) compared earthworm populations under *Acioa barteri*, *Gliricidia sepium* and *Leucaena leucocephala* prunings, maize stover and rice straw and found that the population increased over 41% under any of these types of residues compared to unmulched controls. Termite populations were higher under all mulched plots, but ant populations were only higher under *Leucaena* and *Gliricidia* prunings. Schroth et al. (1992), in an experiment conducted in central Togo, found that mulching with leaves and branches of *Cajanus cajan* increased the number of roots, worms, and other soil invertebrates in soil under the mulch.

2.4 Negative Impacts

Phytotoxicity or increased pest and disease risks may offset the advantage of decreased runoff associated with mulching. Tian and Kang (1994) investigated possible phytotoxic effects of *Gliricidia sepium* prunings. In the laboratory, maize seedling growth was significantly reduced by leachate from the prunings. In the field, leaf chlorosis was observed in maize and cowpea seedlings mulched with *Gliricidia* prunings. The number of chlorotic leaves increased with mulch rate, though growth in the field was not reduced. Applying mulch one week before planting eliminated the phytotoxic effect on maize. Lal (1989) reported better crop establishment when cowpea was sown two to three weeks after fresh prunings of *Leucaena leucocephala* and *Gliricidia sepium* were turned into the soil. Increased weevil (*Cosmopolites sordidus*) populations under mulches have been reported by Price (1993) and Rukazambaga (1996). Rukazambuga noted three- to four-fold increases in weevil populations under mulch treatments, but did not observe a concomitant decrease in crop yield. Preliminary data by Speijer (personal communication) suggest changes in nematode populations (both density and species) under mulch are related to changes in the soil temperature regime.

Fodder is often a higher priority use for leaf and stem prunings than is mulch (Ong et al., 1991). In an analysis of economic returns from a number of hedgerow agroforestry treatments, Rao et al. (1991) reported negative returns on mulched plots in semi-arid India because mulching did not improve crop yields, fodder value was lost and mulching required additional labor. However, Alegre and Rao (1996) argue that in the humid tropics hedgerow agroforestry systems, which provide on-site mulch in the form of hedge prunings, are more viable than mulching systems that require off-site production and transportation.

3. SOIL EVAPORATION

When the soil surface is wet and plant cover is absent, water can evaporate from the soil at rates greater than 5 mm d^{-1}, depending primarily on climatic conditions. This is called the first stage of drying. In a freely draining soil, as the soil surface continues to dry, the evaporation rate will decrease rapidly until a very low rate of evaporation is established. Though the top few centimeters of the soil can become very dry in the absence of plant water uptake, only a few centimeters below the surface the soil may be at field capacity. The depth of drying will generally be greater in fine-textured than coarse-textured soils. This phenomena is the basis of fallow systems in which water that infiltrates below the top few centimeters of soil will be stored for later use by plants. However, in a soil in which there is a water table, the depth of the water table and the hydraulic conductivity of the soil will determine the minimum evaporation rates. The closer the water table is to the surface and the greater the hydraulic conductivity, the higher the minimum evaporation rate will be (Gardner and Fireman, 1958). Under these conditions, water will continue to be lost from soil in the absence of plants, and the soil water content below the surface may be less than field capacity.

In contrast, in the presence of a plant canopy, at least some of the energy that would have been used to drive soil evaporation is absorbed by the crop, thus decreasing the rate of soil evaporation and increasing the rate of evaporation of water intercepted by the plant canopy or transpired through plant leaves. Some of the improvements in water use efficiency of annual cropping systems have resulted from increasing the rate of crop canopy closure, which in turn decreases the amount of water that is lost as soil evaporation. Increases in the rate of crop canopy closure have been obtained by utilizing crop species with rapid canopy growth and/or by increasing plant density.

Continuous perennial covers (such as pastures, orchards, and forests) are land management systems that minimize soil evaporation. Hedgerow agroforestry systems can provide perennial cover, but only over a limited area (ICRAF, 1995). Mulching the soil surface provides additional cover and encourages surface rooting; both the cover and the surface rooting act to lower soil evaporation (Loomis and Connor, 1992). Decaying or dead roots and other macropores, such as those created by rodents, termites and earthworms, minimize soil evaporation by providing channels for rapid subsurface infiltration of water.

3.1 Canopy Closure

In regions with seasonal rainfall, water is frequently lost to soil evaporation at the onset of the rainy season, before crop canopies are well established. Hedgerows are generally severely pruned immediately before this time in order to decrease competition with annual crops, which reduces their potential to decrease soil evaporation. For example, during two cropping seasons in Machakos, Kenya, soil evaporation was responsible for an estimated 42 to 48% of evapotranspiration losses from monocultures of cowpea, maize, and multi-stemmed *Senna spectabilis* hedges, and from intercrops of *Senna* and maize or cowpea (McIntyre et al., 1996). At the beginning of each growing season, hedges were pruned to a height of 0.5 m, but no mulch was applied to these systems. The rate of canopy closure in the agroforestry plots compared to the sole crop plots and multi-stem hedge plots was similar. Much of the rainfall occurred while both annual and hedgerow canopies were beginning to expand and were intercepting less than 20% of the light. When monoculture *Senna* hedges were pruned to single stems, canopy expansion was slower than the other treatments and, consequently, in these systems 52 to 58% of total seasonal evapotranspiration was estimated to be lost through soil evaporation.

If crops reach physiological maturity before the end of the rainy season, hedgerows may take up some superficial soil water that might otherwise be lost to soil evaporation at the end of the annual cropping season. Additionally, hedgerows may utilize water from intermittent rains during the dry season that might otherwise be lost to soil evaporation. The importance of hedgerows in utilizing such water is highly dependent on the annual distribution of rainfall. Moreover, in many agronomic systems, weeds use precipitation occurring between cropping seasons.

3.2 Mulching

Mulches can reduce soil evaporation by reducing the amount of radiation absorbed by the soil and by decreasing air turbulence at the soil surface. Budelman (1989) evaluated the impact of *Leucaena leucocephala*, *Gliricidia sepium* and *Flemingia macrophylla* mulches on surface soil temperature reduction (5-cm depth at 15.00 h) and moisture conservation (percent weight at 0 to 5-cm depth) near Abidjan, Ivory Coast, on a soil that was 85% sand. Leaves of all species were applied at a rate of 5 t ha^{-1} during a 60-day dry period, when rainfall averaged 59.5 mm and the average maximum air temperature was 31°C. The mulches were held in place by light bamboo frames or tree branches. Budelman described the effect of the mulches using two parameters: the initial impact (as measured 10 days after the mulches were applied) and the effective lifetime of the mulches. The initial impact ranged from a reduction in soil temperature of 5.6°C for *Leucaena* to 9.8°C for *Flemingia* and an increase in soil moisture from 4.0% for *Leucaena* to 5.6% for *Flemingia*. The effective lifetime was the time at which the impact was no greater than the least significant difference between the mulched and unmulched treatments. For moisture, this ranged from 44 days for *Leucaena* to more than 90 days for *Flemingia*.

Mulches also may enhance rooting close to the soil surface (Schroth et al., 1992), thus increasing the probability that plants can recover water that might otherwise be lost through evaporation. Also, mulches may enhance the funneling of precipitation entering the soil surface, as described in the next section.

In sandy soils in semi-arid environments, there is evidence that the first stage of drying may last for less than a day; i.e., that the soil surface dries rapidly and evaporation begins to decline in a few hours after rainfall (Pilbeam et al., 1995). Under these conditions, the soils are said to be "self-mulching" and mulches and shading would be expected to have little effect on soil evaporation (Affolder, 1995; Pilbeam et al., 1995). Hanks and Ashcroft (1980) note that it is easier to reduce evaporation during the first, constant stage of evaporation, than during the second, falling rate stage. Mulches are thus more likely to reduce soil evaporation when rainfall events are frequent and on finer textured soils.

3.3 Water Funneling/Deep Infiltration

Soil evaporation also can be reduced by any technique that reduces the amount of water retained near the surface (Marshall et al., 1996) and encourages deep infiltration. Water entering dry, hydrophobic soil is particularly likely to infiltrate unevenly, with certain areas wetting to several centimeters while other areas are still dry at the surface. Hedgerows may tend to funnel intercepted precipitation to their base. This preferential channeling may result in an uneven distribution over the surface and thus to deeper penetration of water into the soil underneath the hedgerow (Lal, 1989; Monteith et al., 1991; Huxley et al., 1994). Huxley et al. (1994) suggested increased maize yields in rows on the windward side of *Grevillea robusta* hedgerows may have been due to increased infiltration of precipitation on this side of the hedgerow, as well as to a wind shelter effect. In the study previously discussed by Lal (1989) in Ibadan, Nigeria, higher soil water contents at 0- to 5-cm depth were noted in the vicinity of the hedgerows. This increase in surface soil moisture was attributed to shading and the channeling of water runoff by the hedge barrier. Some mulches might also promote uneven and thus potentially deeper penetration of water into the soil.

3.4 Surface Rooting

The impact of hedgerows in reducing soil evaporation will depend on the lateral extension of roots and how close roots are to the surface. Studies of root systems in agroforestry systems have been limited and factors controlling the distribution and dynamics of roots in these systems are still not well understood. Vertical and horizontal distribution of roots can be influenced by genetic factors, by the physical and biological environment of the roots, by the phenological stage of the plant and by management. Studies which compare the root system of several tree species grown at the same site give an indication of genetic differences, though these differences may vary in expression at other dissimilar sites. Comparisons of the same species from studies conducted at dissimilar sites ideally should be indicative of the response of the species to different physical environments, but it is often difficult to account for, or control the influence of, seasonal dynamics, management practices and biological factors (e.g., weeds and mycorrhizae). Also, root systems can be measured in different ways, such as weight per volume of soil, root length per volume of soil, or root count per area of soil, and the relationship among these different measures can vary with species and environmental factors. Of these root indices, root length per unit volume of soil is probably most useful in estimating water uptake.

Several factors can affect lateral root extension. For example, the distance roots extend laterally from the base of a tree can vary considerably among agroforestry tree species. In a study of 12 six-year-old tree species growing at 5-m by 5-m spacing in an arid environment in northwestern India, Toky and Bisht (1992) found that the lateral extension of roots varied from 1.4 m to 5.8 m and was less than the crown spread in many of the species. The presence of other plants may impact lateral root extension, either through allelopathic effects or through limiting the above ground growth of trees, and consequently resource allocation to roots. Eason et al. (1992), in a study of the root extension of young ash trees (*Fraxinus excelsior*), reported that roots did not proliferate as greatly in soil planted with ryegrass as compared to clover or bare soil. In a study of roots in a 5-year-old maize-*Grevillea robusta* agroforestry system on an Alfisol in Machakos, Kenya, Huxley et al. (1994) found that in the agroforestry plots *Grevillea* roots extended 4 m laterally, while in plots without maize, *Grevillea* roots extended at least 12 m. Orientation of the *Grevilla* hedgerows had no impact on rooting patterns, which were similar on both sides of the row.

Many agroforestry species appear to have their highest root densities in the surface 0 to 15 cm or 15 to 30 cm of soil. Of the 12 species studied by Toky and Bisht (1992), eight had their greatest percent of total root biomass in the 0 to 15-cm layer and three species had their greatest percent of total root biomass in the 15 to 30-cm layer. The percent of total number of roots of all diameters, as well as the percent of total number of roots in the 0 to 2-mm size class, was greatest in the 0 to 15-cm depth for six species and in the 15 to 30-cm depth for five species. In a study conducted in Tanzania comparing fine root distribution of five two-year-old agroforestry species growing on a sandy loam soil, all but one of the species had the greatest fine root biomass in the 0 to 20-cm soil layer (Jonsson et al., 1988).

Several agroforestry studies with *Leucaena leucocephala* report higher root densities at soil depths below 15 cm. Root length density during the dry season of *Leucaena leucocephala* in a hedgerow experiment conducted on an Alfisol in Hyderabad, India, was approximately 0.1 cm cm^{-3} in the surface 15 cm; this was less than *Leucaena* root length density from 15 to 30 cm (Ong et al., 1991). Govindarajan et al. (1996), in a study of *Leucaena leucocephala* hedgerows in Machakos, Kenya, reported similar values for root length density of *Leucaena* with depth at the end of the short rainy season. As did Ong et al. (1991), Govindarajan et al. (1996) noted root length density appeared to be maximal at the 12 to 25-cm depth. The peak in root distribution at this depth was maintained after pruning. In the study of Toky and Bisht (1992), the percent of total root numbers for *Leucaena leucocephala* was greater at the 15 to 30-cm depth than the 0 to 15-cm depth, though the percent of total biomass was slightly greater in the 0 to 15-cm depth compared to the 15- to 30-cm depth. Jonsson et al. (1988) found fine root biomass of two-year-old *Leucaena leucocephala* from 20 to 40-cm soil depth similar to or greater than that from 0 to 20-cm depth.

In the Toky and Bisht study (1992), the percent of total root biomass in the upper 30 cm of the soil varied among species, ranging from 42 to 84%. There is a possibility that this pattern of root distribution may change with age in some species. For example, in root distribution studies of different agroforestry species in Tanzania, at one year of age over 70% of *Senna siamea* root weight was 0 to 20 cm below the soil surface, but at nine years of age less than 30% of the root weight was in this part of the profile (Kijoti and Chamshama, 1990). However, in the Jonsson et al. study (1988), the percent of fine root biomass in the upper 0 to 20 cm or 0 to 40 cm of soil did not appear to differ greatly between two-year-old and six-year-old *Leucaena leucocephala*.

Other plants, such as annual crops or weeds, may impact tree root distribution patterns with depth. Huxley et al. (1994) present diagrams showing the vertical and horizontal root distribution of sole *Grevillea robusta* and *Grevillea* grown with maize after the onset of the rainy season. As mentioned previously, *Grevillea* lateral root extension was suppressed in a *Grevillea*-maize system. From visual examination of the diagrams, the presence of maize also impacted vertical root distribution of *Grevillea*. When grown with maize, *Grevillea* root density near the soil surface was higher below the hedges and contiguous to the hedges than when *Grevillea* was grown alone. The nearness of other trees also may influence tree root distribution with depth. A study was conducted by Eastham et al. (1990) near Brisbane, Australia, to evaluate water use in *Eucalyptus grandis* — pasture systems with *Eucalyptus* planted at various densities. Although there was an exponential decrease in roots with soil depth in both the high density spacing (2150 stems ha^{-1}) and the low density spacing (82 stems ha^{-1}), trees at the high density were more deeply rooted than at the low density (6 vs. 4 m). A greater proportion of the total root density of the low density trees was in the first meter of soil (68.4 vs. 61%).

Schroth and Zech (1995), in a study at Oume, Ivory Coast, on an Alfisol and an Oxic Tropept planted with *Gliricidia sepium* intercropped with rice followed by maize, report that root length density of *Gliricidia* roots during the cropping season at 0 to 10-cm depth was 0.6 and 0.21 cm cm^{-3} at distances of 1 m and 2.5 m respectively from the hedge. *Gliricidia* root length density was greater in the 0 to 10-cm depth than at 10 to 30 cm. In contrast, the root length density in the agroforestry plots at the 0 to 10-cm depth at 1 m and 2.5 m from the hedge was 13 and 21.7 cm cm^{-3} for rice and 3.5 and 2.3 cm cm^{-3} for maize. Even in the dry season, weeds in the agroforestry plots contributed more of the fine root length density than *Gliricidia* roots to a depth of 50 cm. Schroth and Zech conclude that the root length density of the *Gliricidia* was too low to compete for resources with the annual crops; whether it would have been higher if grown without the annual crops and if weeded is not clear.

Pruning also may impact root distribution. Repeated pruning limits photosynthate production and can therefore decrease total root weight and root length. Schroth and Zech (1995) suggested pruning was responsible for the decline in root length density of *Gliricidia sepium* in the study cited above. In a study of five agroforestry species, Noordwijk and Purnomosidhi (1995) found that pruning to a shorter stem height was associated with a larger number of superficial roots of smaller diameter in all species except *Gliricidia*, which produced thick storage roots. They suggest that pruning may enhance root turnover and lead, through the death of root apices, to the formation of new adventitious roots, thus favoring root proliferation near the soil surface. There is some evidence to suggest that pruning of trees may influence root density as a function of lateral distance from the tree base. Following pruning of *Leucaena leucocephala* hedgerows at the beginning of the rainy season in Machakos, Kenya, new root growth was more pronounced as distance from the hedgerow (2.7 m vs. 0.9 m) increased (Govindarajan et al., 1996). In contrast, 15 weeks after pruning, Nygren and Campos (1995) found a large decrease in the fine root biomass density of *Erythrina poeppigiana*, especially in the surface 25 cm of soil. However, at the onset of the wet season (16 weeks after pruning), fine root growth at this depth appeared to be increasing.

As Noordwijk and Purnomosidhi (1995) suggest, high surface root densities generally may be characteristic of the fast-growing trees used in agroforestry systems. Repeated pruning and wide spacing may enhance this trait. In terms of competition for water uptake with agronomic crops, such a trait is usually considered undesirable. However, a root system that can extend great distances laterally, and in which root density near the surface is high, would in theory be advantageous in capturing water that might otherwise be evaporated from the soil surface or transpired by weeds.

4. DEEP PERCOLATION

There has been considerable interest in the use of agroforestry systems to recover water and nutrients that are lost as deep percolation in other cropping systems. There are several aspects of deep percolation (water moving below the cropping system root zone) that deserve further discussion. Taking up more water at depth in the soil profile is often identified with decreasing water lost to deep percolation, but this is not always the case. In the absence of runoff and water uptake from ground water or laterally transferred water beneath the soil surface, deep percolation can be viewed as the difference between precipitation and evapotranspiration. Under a given rainfall regime, reducing deep percolation requires increasing evapotranspiration. Where the actual rate of evapotranspiration is at the maximum rate (potential evapotranspiration), deep percolation cannot be significantly reduced and thus some soil water will inevitably be lost to drainage. Potential evapotranspiration is controlled primarily by the climate. However, vegetation has some influence on potential evapotranspiration rates. Potential evapotranspiration rates from surfaces covered with agronomic crops or pasture may differ from surfaces covered with trees (Oke, 1987). Evapotranspiration from tree surfaces may be greater than from crop surfaces when the canopy is frequently wetted by small storms and the surface area of the tree canopy is greater than that of the crop canopy. In this case, evaporation of water intercepted by plant canopies can be a significant component of the water budget. In contrast, when canopies are dry, but transpiration is not limited by soil water availability, transpiration from crop surfaces may be greater than from tree surfaces due to higher leaf temperatures and, possibly, lower canopy resistance to vapor transfer.

4.1 Humid Environments

Actual evapotranspiration may be less than potential evapotranspiration when canopy closure is incomplete or when leaf area is reduced as a result of drought or temperature stress, pest damage and/or disease. As mentioned in the previous section, without additional precipitation, a newly wet soil surface evaporates at the potential rate for, at most, only a few days before the rate of evaporation exponentially declines. In comparison, transpiration rates of crops growing in moist soils can continue at the potential rate for several weeks. However, eventually the rate of actual transpiration will fall below the potential rate as water is depleted from the soil. The transpiration rate transitions from climate-controlled to soil-controlled as it falls below potential. Crops or cropping systems that can access more water or are growing in soils that have higher water-holding capacities can, in theory, continue at or near the potential rate for longer periods of time.

In humid climates, monthly precipitation generally equals or exceeds potential evapotranspiration. This means that soil water is not often depleted below the frequently recharged surface horizons. Opportunities to decrease deep percolation will most likely be found in minimizing the time that it takes annual crops to attain a closed canopy and utilizing cropping systems that maintain a constantly transpiring cover. For these climates, increasing rooting depth per se will likely have little impact on deep percolation.

4.2 Semi-Arid Environments

In semi-arid climates, long periods occur in which potential evapotranspiration rates exceed precipitation rates. At first glance it would seem that cropping systems that can access water deep in the soil profile would be able to transpire more water in these climates than more shallow-rooted cropping systems. However, the manner in which water moves deep into the soil must be considered. Significant downward movement of water in soil generally occurs only when the overlying soil is above field capacity. For example, for water to move below a depth of 1.5 m in a deep soil profile, the soil must be recharged so that the water content of the soil profile above 1.5 m exceeds field capacity. If 200 mm of water is required to recharge a soil depleted of plant-available soil water to field capacity to a depth of 1.5 m, it would seem at first glance that even in semi-arid regions, with annual rainfall of 500 mm or more, movement below 1.5 m could occur. However, on warm, sunny days, evaporation from the soil surface combined with transpiration can remove greater than 5 mm of water per day. This means that the opportunities for recharge to occur at depth are limited to times when rainfall events are of high intensity and/or long duration or when potential evapotranspiration rates are low. Such a model of soil water movement predicts that, under the same climatic regime, soils with low water holding capacities, such as sands, will recharge to greater depths than soils with large water holding capacities. However, this model does not account for water movement in some highly structured soils, in which ponded water may enter cracks or root channels and move rapidly through the pore, thus bypassing the surface soil (Scanlon and Goldsmith 1997). Also, in semi-arid regions recharge at depth may occur locally where water collects after a rainstorm (Scanlon and Goldsmith, 1997).

If annual cropping systems are already using water at a depth to which recharge usually occurs, then deep percolation is at a minimum and increasing the depth of rooting in the system will not increase the total transpiration in the system. For example, in a study of *Senna spectabilis* agroforestry systems in Machakos, Kenya, McIntyre et al. (in press) found that during the period 1993 to 1994 soil water recharge below a depth of 0.45 m occurred only once during the cowpea cropping season (short rainy season) and once during the maize cropping season (long rainy season) under all treatments. During their respective growing seasons, maize and cowpea monocultures were able to utilize all the water at this depth. Stored soil water was greater at the beginning of the short rainy season in the sole annual plots compared to all plots where trees were present. This greater soil moisture most likely resulted from the failure of maize during the previous cropping season and the uptake of stored soil moisture by trees during the subsequent dry season. Govindajaran et al. (1996), for the same year and location, found similar results in *Leucaena leucocephala* hedgerow systems. McIntyre et al. (in press) found that soil water at the end of the short rainy season was similar in all plots, indicating that evapotranspiration during this season was greatest for the sole cowpea. In other words, the soil water accessed by the trees in the preceding dry season could have been used effectively by the sole cowpea once the cropping season began. In a 1992 Machakos study, Howard et al. (1995) also found that plots with *Leucaena leucocephala* hedgerows entered the rainy season with 40 mm less water in the top 2 m of the soil. They refer to the continued depletion of water during the dry season by the *Leucaena* as interseasonal competition.

4.3 Sub-Humid/Savanna Environments

Some research suggests sub-humid environments may offer the greatest opportunity for increasing transpiration and decreasing deep percolation by recovering more water at depth (Ong et al., 1991; Hartemink et al., 1996; McIntyre et al., 1996). That is, sub-humid environments are more likely to receive enough precipitation during the rainy season to recharge the soil at depth. By the time soil water at depth is being accessed by roots, agronomic crops may be senescing or transpiring less due to drought stress, while tree crops may still be transpiring and therefore capable of utilizing water that is annually recharged but unavailable to annual crops. Hartemink et al. (1996) compared water and nitrogen contents with depth and over time of soils left in a bare fallow, weed fallow, cropped to maize or planted with *Sesbania sesban* at two sites in western Kenya: Ochinga (mean annual rainfall of 1800 mm) and Muange (mean annual rainfall of 900 mm). At both sites, the soil water content below 1 m of the *Sesbania* treatment was significantly lower than the maize treatment at the end of the first cropping season (September through January). At the end of the subsequent fallow period, there was a heavy rain at the Muange site, so water was relatively similar at depth for all treatments (Buresh, personal communication). At the Ochinga site, soil water at the one to two-m depth was fully recharged during the beginning of the second cropping season (April through August) for the maize and weed fallow treatments, but not the *Sesbania* treatment (ICRAF, 1995). However, during the third cropping season (September through January), soil water at the 1 to 2-m depth was fully recharged in all treatments. This study suggests that soil water recharge below the depth of annual crop uptake may occur regularly and be utilized by deeper rooted tree species.

The woodland-pasture systems of Australia offer another example of trees utilizing water that might otherwise be lost as deep percolation. In a study near Brisbane, Australia (mean annual precipitation 1099 mm), Eastham et al. (1988) report increasing rates of evapotranspiration and decreasing rates of deep percolation as tree density increased in a *Eucalyptus grandis*-pasture system. For the two years of the study, rainfall was below normal. The pasture (predominantly *Setaria sphacelata*) is shallow-rooted, with its roots apparently confined to the upper 50 cm of soil (1988). *Eucalyptus grandis*, in contrast, can root to depths of 4 to 6 m. The rate of evapotranspiration from pasture species growing in lysimeters among the trees fell farthest below the total tree-pasture evapotranspiration rate (Eastham and Rose, 1988) during the wettest months of the year (mostly in the summer), when potential evapotranspiration rates were also the highest. Presumably, deep percolation occurred during these wet periods in the lower density tree treatments. In both years of the study (1985 and 1986), evapotranspiration in the high tree density treatment greatly exceeded precipitation, while in the low density treatment precipitation and evapotranspiration were similar. Differences in evapotranspiration are attributed to the enhanced ability of the more densely planted trees to extract water at depth. Assuming that annual precipitation is sufficient to periodically recharge water to the 5 m depth, the ability to take water up at depth would be an important drought avoidance strategy for the *Eucalyptus* (Loomis and Connor, 1992). However, even without the deep-rooted trees, improved pastures that transpired at greater rates during the rainy season might also result in a significant decline in deep percolation.

Other studies have found that evapotranspiration can be considerably higher in tree-grass systems compared to pure grass systems. In a study conducted on a Vertisol in Guadeloupe, Tournebize (1995) found that during three different time periods actual evapotranspiration from a *Gliricidia sepium-Dichanthium aristatum* (natural savanna grass) system ranged from 75 to 90% of potential evapotranspiration, while actual evapotranspiration from the pure grass ranged from only 30 to 45% of potential evapotranspiration. The leaf area index of the grass stand ranged from 0.7 to 1.3 during the periods studied, with the leaf area index of the pure stand of grass similar to that of the grass in the agroforestry systems. The shrub leaf area index ranged from 1.7 to 2.6, thus much of the difference in actual evapotranspiration can be attributed to the low leaf area index of the grass. It was not clear whether the 50% greater water consumption in the agroforestry system could be sustained. Joffre and Rambal (1993) determined the water balance of two components of Mediterranean rangelands (annual rainfall of 600-800 mm): annual grasses and oak (*Quercus ilex* and *Q. suber*)-perennial grass associations. Mean annual evapotranspiration was 400 mm for the annual grass component and 590 mm for the tree-grass component. In the annual grass component, excess water was lost both as surface runoff and deep drainage, while little, if any, surface runoff or deep drainage occurred in the tree-grass component when annual precipitation was less than 570 mm. The largest differences in evapotranspiration between the two systems occurred during the rainy season, when actual evapotranspiration in the grass component was low and poorly correlated with rainfall. Joffre and Rambal (1993) attribute the lower evapotranspiration from the grass to a poorly developed root system, but do not report the leaf area index of the annual grass. The leaf area indices of the trees were greater than or equal to 3.0, but the actual evapotranspiration from the tree component in the dry, summer months was small, ranging from 30 to 50 mm.

When the results of Tournebize (1995) are compared to the results of Joffre and Rambal (1993) and Eastham et al. (1988), there is some basis to suppose that the low actual evapotranspiration rates in the grasses may be due as much to low leaf area index, and/or an inability to respond to recharge of the soil profile, as to shallow rooting depth. Low leaf area in the grasses may be due to overgrazing, inappropriate pasture species, and/or low soil fertility. Prebble and Stirk (1988) found no effect on evapotranspiration when trees were killed and improved pasture established on an open grassy woodland in Queensland, Australia, because grass growth was enhanced in areas formerly under tree canopies. In a multifaceted study of pasture management in northeastern Queensland, Australia, McIvor and Gardener (1995) found that the response of pastures to tree killing depended on season and site. They suggest that when the wet season is not interrupted by dry spells, competition between trees and grasses will be minimal and response of pasture to increased water will be small. In contrast, when the wet season is broken by several dry spells, competition for water between trees and grasses will be more severe and the response of pasture to tree clearing could be large. However, soil fertility also plays a role; when dry spells are interspersed through the wet season, pasture response to tree clearing may not be great, because the pasture could be nutrient-limited, rather than water-limited.

4.4 Restricted Soils

Trees may be able to access water unavailable to annual crops in deep soils with very low water-holding capacities, or where a relatively shallow soil overlays a subsoil that can be exploited by tree roots, but not annual crop roots. Rao et al. (1991) conducted agroforestry experiments on an Alfisol where annual crop root growth was restricted by an ironstone layer at 0.3 to 0.4 m. However, it was not clear whether evapotranspiration increased under the agroforestry system. In a study of a *Pinus radiata* pasture system near Melbourne, Australia, Connor et al. (1989) reported that tree roots increase infiltration of water into the high clay subsoil.

4.5 Measurement

Decreases in deep percolation in agroforestry systems in comparison to annual cropping systems have not been widely documented, presumably due to difficulties in quantifying deep percolation. The net flux of water below some specified lower boundary of the plant-soil system can be estimated from water content data and integrated over time (Eastham et al., 1988; Joffre and Rambal, 1993). As mentioned previously, under many conditions deep percolation can be estimated as the difference between precipitation and evapotranspiration. Precipitation is easily measured, but measuring evapotranspiration in small agroforestry plots is difficult, though progress has been made recently through the use of heat balance (Howard et al., 1995; Tournebize, 1995; Walsh et al., 1995) or heat pulse methods (ICRAF, 1995). Evapotranspiration may be estimated with a simulation model if the requisite climate, soil and crop data is available (McIntyre et al., 1996), but this approach requires knowledge of soil properties, the dynamics of the plant canopies and some assumptions about the crop's response to drought.

Comparing the soil water status at one or a few points in time to estimate differences in evapotranspiration among various cropping systems presents some problems. For example, in the study by McIntyre et al. (1997), at the end of the dry season there was less water in the agroforestry plots compared to the sole crop plots, suggesting greater water use in the agroforestry system. However, by the end of the rainy season, soil water status was similar among plots, as a result of higher transpiration rates during the rainy season in the sole crop plots than in plots with hedgerows. This difference in seasonal soil water indicates that the time of comparison of profile soil water measurements is important. Moreover, greater biomass production in a cropping system does not necessarily mean that transpiration in that system has increased (and consequently deep percolation has decreased), because species may differ in water use efficiency (Chaturvedi et al., 1988).

5. GROUNDWATER/DRAINAGE WATER

In many parts of the world, destruction of native forests, overgrazing of pasture land and introduction of irrigation systems have resulted in large areas of land being subjected to rising groundwater, which is frequently saline (Greenwood, 1986; Bell et al., 1990; Bari and Schofield, 1991; Bodia et al., 1994; O'Leary and Glenn, 1994). Researchers have identified two biological methods to reduce groundwater levels. The first method involves the removal of superficial water by plants whose roots do not reach the water table (non-phreatophytes). As discussed in Section 4.3, silvo-pastoral systems can be quite effective at reducing or halting deep drainage through the uptake of superficial water. In the second method, plants whose roots grow freely into a water table (phreato-phytes) take up water directly from the saturated zone (Greenwood, 1986). Researchers in Western Australia (Bari and Schofield, 1991) evaluated silvopastoral systems of mixed pine species (*Pinus radiata* and *P. pinaster*), pure *Eucalyptus camaldulensis* and mixed eucalyptus species (*Eucalyptus sargentii, E. wandoo, E. camaldulensis* and *E. calophylla*) at densities that covered about 58% of the ground area. In comparison to nearby pasture, the water table declined over a 10-year period by 1-m and salinity declined 9% in the mixed pine and *E. camaldulensis* site and over an 8-year period the water table declined 2 m and salinity decreased 6% in the mixed eucalypt site. In a study of the hydrological effect of *Casuarina glauca* established on a salt pan in central Queensland, Australia (Walsh et al., 1995), the water table adjacent to the tree was depressed 130 mm relative to the water Table 10 m from the tree. The authors concluded that the rate of water use by the trees was sufficient to inhibit the flow of water to the soil surface and thus halt the concentration of salts by evaporation at the soil surface. This decline in salinity and increase in surface soil moisture would also presumably promote the establishment of grasses between the trees.

Agroforestry systems have also been irrigated with poor quality, subsurface drainage water (Jorgensen et al., 1992; Tanji and Karajeh, 1993; Miyamoto et al., 1994) in order to reduce the volume of such water. The long term management and viability of these systems, however, remain unclear (Tanji and Karajeh, 1993).

6. CONCLUSIONS

Theory suggests, and there is evidence to support the contention, that hedgerow agroforestry systems can reduce surface water runoff relative to some types of agricultural systems. Whether and to what degree agroforestry might have advantages over other conservation systems in decreasing runoff is more open to question. This lack of clarity is in part due to the number of variables that impact surface runoff and the complexity of their interactions. But it is also due in part to an incomplete understanding of how factors that are unique to agroforestry might impact processes influencing surface runoff. For example, mulching is likely to have a considerable beneficial effect on creating and maintaining good surface infiltration, especially on soils that seal easily and in areas where rainfall intensity can be high. Does mulching with tree leaf and stems convey any advantages not present with other mulches? To what degree do tree roots, particularly in degraded or shallow soils, influence surface infiltration? Is stabilizing conservation structures that pond water moving down slope the only impact of tree hedges on surface runoff, when tree litter is not used for mulch? Answers to these questions could be useful in developing a method to estimate the relative potential of agroforestry and other conservation systems to control surface runoff, as well as have practical significance for tree and residue management in agroforestry systems aimed at increasing soil water infiltration and decreasing surface runoff.

Compared to the interest in the utilization of deep water by trees in agroforestry systems, relatively little attention has been paid to minimizing soil evaporation in agroforestry systems. Further experiments could be conducted to quantify the impact of mulches and intercrops on tree surface rooting and the extent to which mulching of the soil surface with hedgerow prunings can reduce water lost to soil evaporation in these agroforestry systems. Manipulation of surface rooting patterns and hedgerow canopy cover by species selection, density of tree planting and pruning present opportunities to decrease soil evaporation. Such opportunities may be greatest in sub-humid or savanna regions where rains occur at the end of the growing season or occur intermittently during the dry season. However, some strategies that lead to reduction in soil evaporation also may result in the tree competing more aggressively for water with the annual crop.

There has been considerable interest in increasing the total water uptake in agroforestry systems compared to annual cropping systems through selecting trees that will root deeply and access water at depth in the soil profile — water that otherwise would not be used by annual crops. Whether enhanced utilization of water at depth in the soil profile may occur in agroforestry systems will depend on several factors. Many important agroforestry tree species do not appear to be especially deep rooted when contrasted to annual crops. Furthermore, the management of these tree species (wide spacing, frequent pruning and interplanting with crops) may promote proliferation of roots near the surface. This suggests that the trees are more likely to compete with crops for water being recharged at the surface than to take water up at depth. In some cases, more complete and vigorous crop or pasture canopies could decrease the amount of water moving below the soil surface horizons, thus limiting the amount of water available for uptake at depth. Frequent, large rain events will enhance the movement of water below the soil surface horizons, especially in soils with low water holding capacity (coarse textured soils and Oxisols) and with poorly developed crop canopies (low fertility soils). If these conditions occur in regions with a sustained dry season, deeply rooted plants may then access water at depth in the soil that has been recharged during the rainy season. Research aimed at increasing water uptake at depth should focus on these areas and on identifying trees that will root deeply in agroforestry systems.

When uptake of water by trees from water tables is considered, the image that frequently comes to mind is deep rooted trees accessing water tables at great depths in the soil. Since some tree species have the ability to grow roots into a water table and/or survive under conditions of fluctuating water tables, they can increase system water uptake by utilizing water in shallow water tables that would not be used by many annual crops and pastures. This can have a positive environmental effect of lowering water tables that are close to the soil surface and might otherwise contribute to soil salinization. In areas where soil and water salinization has already occurred, agroforestry systems might also prove valuable, as some trees are tolerant of saline conditions. Design and testing of agroforestry systems for these conditions has been limited and should be further pursued.

From the studies presented in this paper, we can conclude that the influence of an agroforestry system on water movement and plant water use depends on many factors. Some biophysical factors that affect surface runoff, infiltration and soil evaporation include litter decomposition rates, ground and canopy cover, and channels formed by soil biota and roots. Cultural practices such as tillage, pruning, species selection, weeding and plant spacing also affect surface runoff, infiltration and soil evaporation. In addition, water loss below the root zone can be reduced by deep-rooted species, or by species capable of increased superficial uptake. Advantages of agroforestry systems over conventional cropping systems to improve water use and movement will ultimately depend on the system, climate, soil type and landscape.

REFERENCES

Affholder, F., 1995, "Effect of organic matter input on the water balance and yield of millet under tropical dryland condition," *Field Crops Res.*, 41: 109-121.

Alegre, J. C. and Rao, M. R., 1996, "Soil and water conservation by contour hedging in the humid tropics of Peru," *Agri. Ecosyst. Env.*, 57: 17-25.

Anderson, L. S. and Sinclair, F. L., 1993, "Ecological interactions in agroforestry systems," *Agrofor. Abs.*, 6: 57-91.

Bari, M. A. and Schofield, N. J., 1991, "Effects of agroforestry-pasture associations on groundwater level and salinity," *Agro. Syst.*, 16: 13-31.

Bell, R. W., Schofield, N. J., Loh, I. C., and Bari, M. A., 1990, "Groundwater response to reforestation in the Darling Range of Western Australia," *J. Hydrol. (Amsterdam)*, 115: 297-317

Berendse, F., 1979, "Competition between plant populations with different rooting depths," *Oecologia*, 43: 19-26.

Bodla, M. A., Baig, M. S., and Shamsi, S. R. A., 1994, "Forage production from salt-affected and water-logged areas in Punjab, Pakistan," in *Halophytes as a Resource for Livestock and for Rehabilitation of Degraded Lands*, Squires, V.R. and Ayoub, A.T., Eds., Kluwer Academic Publishers, 231–234.

Bregman, L., 1993, "Comparison of the erosion control potential of agroforestry systems in the Himalayan region," *Agro. Syst.*, 21: 101-116.

Budelman, A., 1989, "The performance of selected leaf mulches in temperature reduction and moisture conservation in the upper soil stratum," *Agro. Syst.*, 8: 53-66.

Cannell, M. G. R., van Noordwijk, M. V., and Ong, C. K., 1996, "The central agroforestry hypothesis: The trees must acquire resources that the crop would not otherwise acquire," *Agro. Syst.*, 34: 27-31.

Chaturvedi, A. N., Sharma, S. C., and Ramji, S., 1988, "Water consumption and biomass production of some forest tree species," *Int. Tree Crops J.*, 5,: 71–76.

Connor, D. J., Sands, R., and Strandgard, M., 1989, "Competition for water, light and nutrients in agroforestry associations of *Pinus radiata* and pasture," in *Meteorology and Agroforestry*, Reifsnyder, W. and Darnhofer, T., Eds., International Council for Research in Agroforestry, Nairobi, 451–462.

Dano, A. M. and Siapno, F. E., 1992, "The effectiveness of soil conservation structures in steep cultivated mountain regions of the Philippines," in *Erosion, Debris Flow and Environment in Mountain Regions*, Walling, D., Davis, T., and Hasholt, B., Eds., International Association of Hydrological Sciences Pub. No. 209, 399–405.

Eason, W. R., Tomlinson, H. F., and Hainsworth, C., 1992, "Effect of ground vegetation on root distribution of ash trees," *Asp. A Biol.*, 29: 225-231.

Eastham, J. and Rose, C. W., 1988, "Pasture evaporation under varying tree planting density in an agroforestry experiment," *Ag. Water Mgt.*, 15: 87–105.

Eastham, J., Rose, C. W., Cameron, D. M., Rance, S. J., and Talsma, T., 1988, "The effect of tree spacing on evaporation from an agroforestry experiment," *Ag. For. Meteorol.*, 42: 355-368.

Eastham, J., Rose, C. W., Cameron, D. M., Rance, S. J., Talsma, T., and Charles, E. D. A., 1990, "Tree/pasture interactions at a range of tree densities in an agroforestry experiment. III. Water uptake in relation to soil hydraulic conductivity and rooting patterns," *Aus. J. Ag. Res.*, 41: 709-718.

Gardner, W.R. and Fireman, M., 1958, "Laboratory studies of evaporation from soil columns in the presence of a water table," *Soil Sci.*, 85:244-249.

Govindarajan, M., Rao, M. R., Mathuva, M. N., and Nair, P. K. R., 1996, "Soil-water and root dynamics under hedgerow intercropping in semiarid Kenya," *Agron. J.*, 88: 513-520.

Greenwood, E. A. N., 1986, "Water use by trees and shrubs for lowering saline groundwater," *Rec. Revege. Res.*, 5: 423-434.

Hanks, R. J. and Ashcroft, G. L., 1980, *Applied Soil Physics*, Springer-Verlag, New York.

Hartemink, A. E., Buresh, R. J., Jama, B., and Janssen, B. H., 1996, "Soil nitrate and water dynamics in sesbania fallows, weed fallows, and maize." *Soil Sci. Soc. Am. J.*, 60: 568-574.

Howard, S. B., C. K. Ong, Rao, M. R., Mathuva, M., and Black, C. R., 1995, "The partitioning of light and water in Leucaena-maize agroforestry systems," in *Ecophysiol. Trop. Intercrop.*, Sinoquet, H. and Cruz, P., Eds., Institut National de la Recherche Agronomique, Paris, 123–135.

Huxley, P. A., Pinney, A., Akunda, E., and Muraya, P., 1994, "A tree/crop interface orientation experiment with a Grevillea robusta hedgerow and maize," *Agro. Syst.*, 26: 23-45.

ICRAF Annual Report, ICRAF, 1995

Joffre, R. and Rambal, S., 1993, "How tree cover influences the water balance of Mediterranean rangelands," *Ecol.*, 74: 570-582.

Jonsson, K., Fidjeland, L., J. A. Maghembe, and P. Hogberg, 1988, "The vertical distribution of fine roots of five tree species and maize in Morogoro, Tanzania," *Agro. Syst.*, 6: 63-69.

Jorgensen, G. S., Solomon, K. H., and Cervinka, V., 1992, "Agroforestry systems for on farm drain water management," in *Drainage and water table control: Proc. Sixth International Drainage*, American Society of Agricultural Engineers, 484–490.

Kang, B. T., Akinnifesi, F. K., and Pleysier, J. L., 1994, "Effect of agroforestry woody species on earthworm activity and physicochemical properties of worm casts," *Biol. Fert. Soils*, 18: 193–199.

Kijoti, W. E. and Chamshama, S. A. O., 1990, *Root development and plantation performance of some tree species at Dodoma, Tanzania*, Morogoro, Tanzania, Faculty of Forestry, Sokoine University of Forestry.

Lal, R., 1989, "Agroforestry systems and soil surface management of a tropical alfisol: I. Soil moisture and crop yields," *Agro. Syst.*, 8: 7–29.

Lal, R., 1989, "Agroforestry systems and soil surface management of a tropical alfisol: II. Water runoff, soil erosion and nutrient loss," *Agro. Syst.*, 8: 97–111.

Lal, R., 1989, "Agroforestry systems and soil surface management of a tropical alfisol: V. Water infiltrability, transmissivity and soil water sorptivity," *Agro. Syst.*, 8: 217–238.

Lal, R., 1996, "Deforestation and land-use effects on soil degradation and rehabilitation in western Nigeria. III. Runoff, soil erosion and nutrient loss," *Land Deg. Dev.*, 7: 99–119.

Loomis, R. S. and Connor, D. J., 1992, *Crop ecology: productivity and management in Ag. Syst.*, Cambridge University Press, Cambridge, U.K.

Marshall, T. J., Holmes, J. W., and Rose, C. W., 1996, *Soil Physics.*, Cambridge University Press, Cambridge, U.K.

McIntyre, B. D., Riha, S. J., and Ong, C. K., 1996, "Light interception and evapotranspiration in hedgerow agroforestry systems," *Ag. For. Meteorol.*, 81: 31-40.

McIntyre, B. D., Riha, S. J., and Ong, C.K., 1997, "Competition for water in a hedge-intercrop system," *Field Crops Res.* 52: 151-160.

McIvor, J. G. and Gardener, C. J., 1995, "Pasture management in semi-arid tropical woodlands: effects on herbage yields and botanical composition," *Aus. J. Exp. Ag.*, 35: 705-715.

Miyamoto, S., Glenn, E. P., and Singh, N. T., 1994, "Utilization of halophytic plants for fodder production with brackish water in subtropic deserts," in *Halophytes as a Resource for Livestock and for Rehabilitation of Degraded Lands*, Squires, V.R. and Ayoub, A.T., Eds., Kluwer Academic Publishers, 43–75.

Monteith, J. L., Ong, C. K., and Corlett, J. E., 1991, "Microclimatic interactions in agroforestry systems," *For. Ecol. Mgt.*, 45: 31–44.

Morris, R. A. and Garrity, D. P., 1993, "Resource capture and utilization in intercropping: Water," *Field Crops Res.*, 34: 303–317.

Noordwijk, M. V. and Purnomosidhi, P., 1995, "Root architecture in relation to tree-soil-crop interactions and shoot pruning in agroforestry," *Agro. Syst.*, 30: 161–173.

Nygren, P. and Campos, A., 1995, "Effect of foliage pruning on fine root biomass of *Erythrina poeppigiana* (Fabaceae)," in *Ecophysiol. Trop. Intercrop.* Sinoquet, H. and Cruz, P., Eds., Institut National de la Recherche Agronomique, Paris, 295–302.

Oke, T. R., 1987, *Boundary Layer Climates*, Methuen, London.

O'Leary, J. W. and Glenn, E. P., 1994, "Global distribution and potential for halophytes," in *Halophytes as a Resource for Livestock and for Rehabilitation of Degraded Lands*, Squires, V.R. and Ayoub, A.T., Eds., Kluwer Academic Publishers, 7–17.

Omoro, L. M. A. and Nair, P. K. R., 1993, "Effects of mulching with multipurpose-tree prunings on soil and water run-off under semi-arid conditions in Kenya," *Agro. Syst.*, 22: 225–239.

Ong, C. K., 1995, "The 'dark side' of intercropping: manipulation of soil resources," *Ecophysiol. Trop. Intercrop.*, 45-65.

Ong, C. K., Corlett, J. E., Singh, R. P., and Black, C. R., 1991, "Above and below ground interactions in agroforestry systems," *For. Ecol. Mgt.*, 45: 45–57.

Ong, C. K., Odongo, J. C. W., Marshall, F., and Black, C. R., 1991, "Water use by trees and crops: five hypotheses," *Agrofor. Today* 3: 7–10.

Pilbeam, C. J., Daamen, C. C., and Simmonds, L. P., 1995, "Analysis of water budgets in semi-arid lands from soil water records," *Exp. Agri.*, 31: 131–149.

Prebble, R. E. and Stirk, G. B., 1988, "Hydrological effects of land use change on small catchments at the Narayen Research Station, Queensland," *Aus. J. Soil Res.*, 26: 231-242.

Price, N. S., 1993, *Preliminary weevil trapping studies in Cameroon. Biological and Integrated control of highland banana and plantain pests and diseases*, Cotonou, Benin, IITA.

Rao, M. R., Ong, C. K., Pathak, P., and Sharma, M. M., 1991, "Productivity of annual cropping and agroforestry systems on a shallow Alfisol in semi-arid India," *Agro. Syst.*, 15: 51–63.

Rukazambuga, N. D. T. M., 1996, *The effects of banana weevil (Cosmopolites sordidus) on the growth and productivity of bananas (Musa AAA EA) and the influence of host vigor on attack*, University of Reading, Reading, U.K.

Scanlon, B. R. and Goldsmith, R. S., 1997, "Field study of spatial variability in unsaturated flow beneath and adjacent to playas," *Water Resou. Res.*, 33: 2239–2252.

Schroth, G. and Zech, W., 1995, "Root length dynamics in agroforestry with Gliricidia sepium as compared to sole cropping in the semi-deciduous rainforest zone of West Africa," *Plant and Soil*, 170: 297–306.

Schroth, G., Zech, W., and Heiman, G., 1992, "Mulch decomposition under agroforestry conditions in a subhumid tropical savanna: processes and influence of perennial plants," *Plant and Soil*, 147: 1–11.

Sheng, T. C., 1989, *Soil conservation for small farmers in the humid tropics*, Food and Agricultural Organization of the United Nations, Rome.

Tanji, K. K. and Karajeh, F. F., 1993, "Saline drain water reuse in agroforestry systems," *J. Irrig. Drain. Eng.*, 119: 170–180.

Tian, G., Brussaard, L., and Kang, B. T., 1993, "Biological effects of plant residues with contrasting chemical compositions under humid tropical conditions: effects on soil fauna," *Soil Biol. Biochem.*, 25: 731–737.

Tian, G. and Kang, B. T., 1994, "Evaluation of phytotoxic effects of *Gliricidia sepium* (Jacq.) Walp. prunings on maize and cowpea seedlings. *Agro. Syst.*," 26: 249–254.

Toky, O. P. and Bisht, R. P., 1992, "Observations on the rooting patterns of some agroforestry trees in an arid region of north-western India," *Agro. Syst.,* 18: 245–263.

Tournebize, R., 1995, "Water relations in a shrub-grass intercrop and comparison with a pure stand," in *Ecophysiology of tropical intercropping,* Sinoquet, H. and Cruz, P., Eds., Institut National de la Recherche Agronomique, Paris, 189-196.

van Es, H. M., 1993, "Evaluation of temporal, spatial, and tillage-induced variability for parameterization of soil infiltration," *Geoderma,* 60: 187-199.

Walsh, K. B., Gale, M. J., and Hoy, N. T., 1995, "Revegetation of a scalded saline discharge zone in Central Queensland. 2. Water use by vegetation and watertable drawdown," *Aus. J. Exp. Ag.,* 35: 1131–1139.

Welden, C. W. and Slauson, W. L., 1986, "The intensity of competition vs. its importance: An overlooked distinction and some implications," *Quat. Rev. Biol.,* 61: 23–44.

Wiersum, K. F., 1991, "Soil erosion and conservation in agroforestry systems," in *Biophysical Research for Asian Agroforestry,* Avery, M. E., Cannell M. G. R., and Ong, C. K., Eds., Winrock International, New Delhi, 209-230.

Wiersum, K. F., 1984, "Surface erosion under various tropical agroforestry systems," in *Symposium on Effects of Forest Land Use on Erosion and Slope Stability,* O'Laughlin, C. L. and Pearce, A. J., Eds., University of Hawaii, 231–239.

Yamoah, C. F., Agboola, A. A., Wilson, G.F., and Mulongoy, K., 1986, "Soil properties as affected by the use of leguminous shrubs for alley cropping with maize," *Agri. Ecosyst. Env.,* 18: 167–177.

4 Managing Ground Cover Heterogeneity in Coffee (*Coffea arabica* L.) Under Managed Tree Shade: From Replicated Plots to Farmer Practice

Charles Staver

1. INTRODUCTION — COFFEE IN AGROFORESTRY SYSTEMS

Rural households throughout the low and mid-altitude tropics grow coffee in a diversity of agroforestry systems. In an individual country in Latin America or even within a single locality, coffee

1-56670-294-1/99/$0.00+$.50

can be found prospering in multi-strata mixed orchards, as a monocrop under mixed and single species shade, as well as in sun-grown monocrop (Figure 1). These highly contrasting systems are the product, on the one hand, of local knowledge based on many decades of grower tinkering and experimentation and, on the other hand, of more recent green revolution-type technology generation and transfer which has also undergone modification by growers. Traditional coffee production is low cost, low risk, and resource conserving, but also low yielding. With the use of improved varieties, fertilizers, and pesticides, coffee yields have increased several fold for a small segment of households, but costs, risks, and resource degradation have also increased. This situation represents a clear challenge to scientists and practitioners in coffee and agroforestry. Can we develop tools and procedures which go beyond the strengths of both models for the benefit of the broad majority of coffee-growing households often located in environmentally fragile and heterogeneous upper watersheds?

In this paper we draw on research advances and development experience from weed management in coffee in Nicaragua and Central America to propose and illustrate an approach which combines replicated small plots, larger systems studies, and participatory farmer training and experimentation. The proposed outcome of this work is needed not only for improved weed control and soil conservation in coffee agroforestry, but also for piloting methods to achieve strengthened farmer capacity in ecologically-based decision-making for the management of ground cover heterogeneity in perennial crops under trees.

2. THE CASE OF WEED CONTROL IN COFFEE

In traditional systems for coffee production, weeds had only minor significance. Their management derived primarily from the design of the production system itself. Except in certain limited conditions predominated by cloudiness, shade trees either naturally occurring or planted were essential to coffee productivity. Shade levels of 40–80%, the accumulated leaf litter layer, and leaf prunings from shade management combined to minimize weed problems which were easily managed with relatively low-cost, rapid, and infrequent slashing with machete. Weed levels were also minimal since coffee was usually planted in forested lands with a minimum of weedy vegetation and limited weed seed banks.

The "modernization" of coffee production technology has been based on the planting of new compact varieties at higher densities. For these varieties to achieve their higher production potential, shade levels have been reduced or eliminated and fertilizer inputs greatly increased. These changes in the production technology as well as the globalization of coffee pests have contributed to increased pest levels in modern coffee (Staver, 1994). The expanded pest complex has been managed on a pest by pest basis, usually with pesticides. Weeds, for example, which grow more aggressively under reduced shade and the minimal leaf litter layer which characterize modern coffee fields, are controlled with herbicides. With the repeated use of herbicide mixtures, especially pre-emergents, modern coffee growers are able to maintain their fields free of weeds for up to 12 to 16 weeks, although more commonly the weed-free period of bare soil is 4–6 weeks. (See Njoroge, 1994 for the herbicide perspective in weed control in coffee.)

In practice, only a small sector of growers systematically use a complete herbicide routine, although among most growers the integrated weed management typifying the traditional system has also been eroded. Weed management among the diverse types of coffee growers has a number of common features.

First, weed biomass in coffee fields varies from excessive immediately before weed control to bare soil for two–six or more weeks right after weed control, a clear indication of grower efforts to control weeds, while keeping costs down. This pattern results simultaneously in weed competition with coffee plants, abundant weed seed production, poor control and a subsequent build-up of the most-difficult-to-control weeds, as well as periods of bare soil.

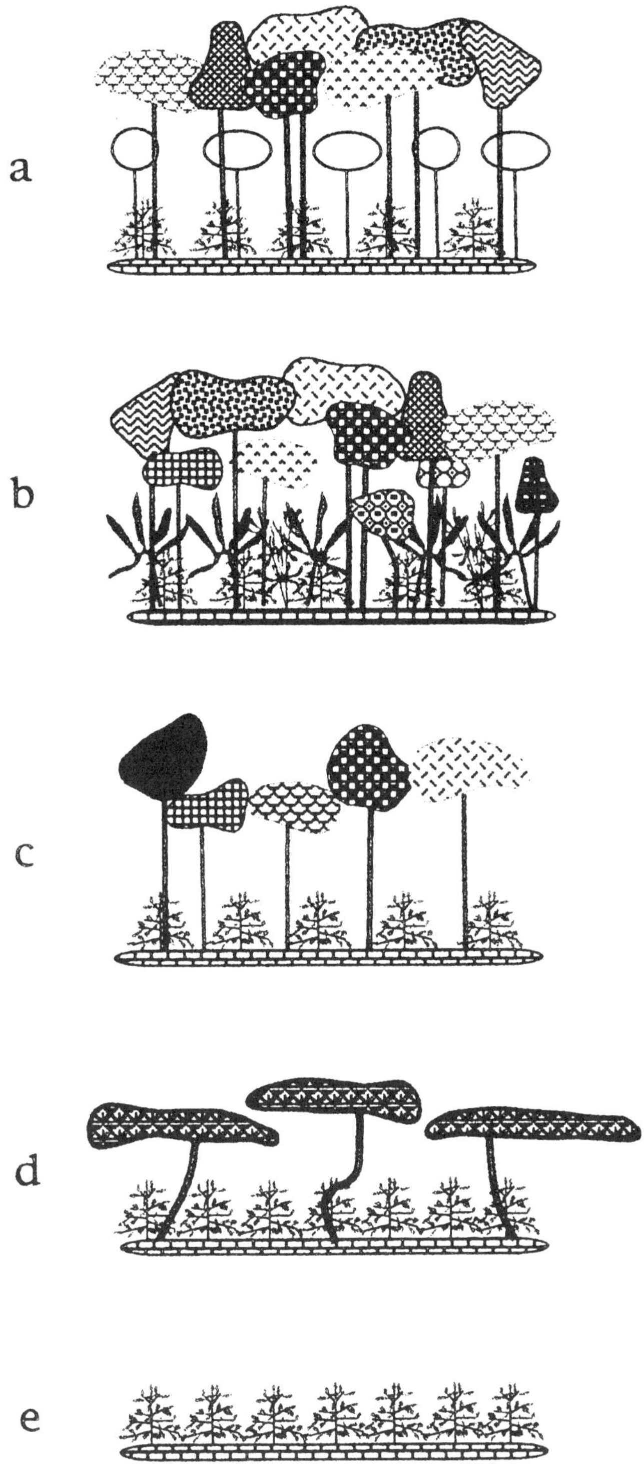

FIGURE 1. Diversity of types and levels of shade found in coffee fields. a= thinned forest shade; b= mixed species planted shade including banana, fruit trees, and timber trees, similar to dooryard garden; c= simplified mixed species planted shade including fruit, timber, and shade trees; d= single species planted shade: e= open sun coffee without shade. Fertilizer and pesticide use increases from a to e. (after Moguel and Toledo, 1995)

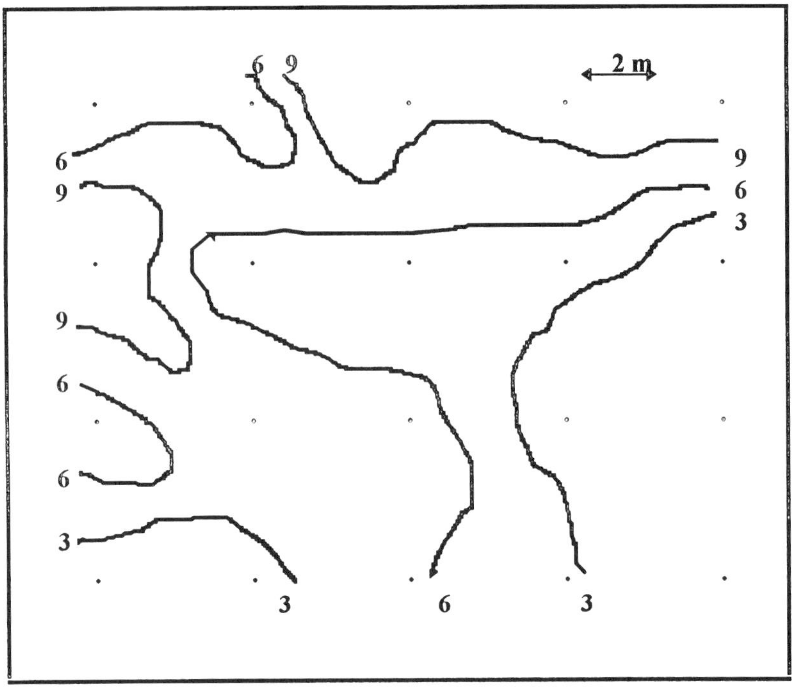

FIGURE 2. Spatial variability of shade trees in a Mexican coffee field as indicated by distribution of photosynthesis-active radiation (PAR ($10^6\mu E/m^2/day$)) at ground level. Canopy cover varied from 20% (PAR=3) to 80% (PAR=9). (with permission from Goldberg and Kigel (1986)).

Second, weed control is generally carried out independently from other crop management practices such as soil conservation or shade management. Similarly, these other practices are not carried out for their possible weed control effect. This is understandable among regular users of herbicides from which a highly specific and short-term result is expected. The sporadic use of purchased synthetic inputs such as herbicides which characterizes many coffee growers has also led to decreased emphasis on agronomic practices which have multiple and cumulative effects on the crop environment including weed growth, crop nutritional status, and soil productivity.

Third, weed control is generally applied uniformly across fields and farms without much regard for different types of weeds, even though the major tools for weed control are the machete and backpack sprayer, both of which are hand operated.

Fourth, coffee fields are characterized by within-field heterogeneity. This is a product of several factors. The sloped and irregular topography which typifies most coffee growing regions is associated with highly variable soils. Variable growth and vigor of coffee bushes and unevenly distributed shade trees result in part from the irregular topography and soils, but also from untimely management and grower inattention. The variability in shade levels can be more than three fold over short distances as shown in Figure 2. Weed types and biomass can also be highly variable, a product of the variability in soil and shade levels as well as the inherent clustered nature of both vegetatively-propagated as well as seed-bearing weeds.

3. COFFEE MANAGEMENT PARADIGM: UNIFORMITY OR HETEROGENEITY-BASED

In the past several decades the agenda guiding the development of agricultural production technology has broadened. Initially increased yields were the sole objective. Now we are also concerned with off-site impacts of chemical inputs, soil and water quality, the level of local input use, and biodiversity. In coffee, modernization based on purchased inputs was achieved to the greatest extent in Costa Rica and El Salvador which still have among the highest country average yields per hectare in the world (Rice and Ward, 1996). However, soil erosion, ground water contamination (Reynolds, 1991), the loss of bird habitat and biodiversity (Perfecto et al., 1996), and the price squeeze between increasing input costs and coffee price fluctuation have motivated doubts about modern production techniques (Fernandez, 1990; Rice and Ward, 1996; Staver, 1994).

The implicit coffee management paradigm derived from replicated plot research and the use of purchased chemical inputs is based on uniformity as a fundamental assumption. Inputs are applied uniformly to a supposedly uniform crop/soil condition with an expected uniform yield response. Some fine tuning may be achieved through the definition of recommendation domains in which farmers are grouped by farm size, soil type, and climate. In the low-input version of the paradigm, extremely robust and widely applicable plant varieties or production techniques such as tree species or cover crops are developed for ready transfer to farmers over broad areas. In this paradigm based on the uniform use of either high or low levels of purchased inputs, technology development drives progress and the transfer of technology is simply a matter of communication from extensionist to farmer.

This assumption of uniformity, however, has proven to be of limited use for improving either the productivity or sustainability of coffee production among a vast majority of coffee producers who still have low yields. We have already mentioned the heterogeneity which characterizes coffee fields and growers. Within-field and between-field variability, differences among growers, and year-to-year and season-to-season variability in prices, pests, yields or weather factors are the rule rather than the exception. The results have been both increased economic and production inefficiency and high risk. This poses a serious question of how to adjust the management paradigm to incorporate heterogeneity.

According to Borlaug and Dowswell (1994), the main thrust of agricultural development for low-income countries should be more inputs and improved varieties, especially for less literate farmers who they suggest may not have the ability to employ management-intensive approaches as described in this chapter. Studies on farmer input use in Asia suggest that conventional technology transfer of input packages was more successful with farmers with more formal education (Jamison and Lau, 1982). On the other hand, follow-up studies on farmer field schools designed to improve farmer decision-making in integrated pest management of rice showed that farmer learning was equally effective for all levels of schooling and farm sizes (Useem et al., 1992). The important issue for greater efficiency and productivity through management of heterogeneity would appear to be neither farmer education level nor input availability, but rather the development of effective approaches to farmer learning.

A management paradigm based on the ecological principles of variability and heterogeneity draws on experiences from many different fields. These include integrated pest management (Kenmore, 1991; Staver et al., 1996; Vandermeer, 1996), management-intensive grazing (Voisin, 1988), range management (Savory, 1988; Westoby et al., 1989), and participatory community resource management (Bimbao et al., 1995).

This paradigm, in which the farmer, farm household, and community hold the pivotal role as decision makers, has a number of important concepts. First, an awareness of the patterns and potential in crop performance, soils, weather and climate, farmer knowledge and experience, and other local conditions at different scales is the starting point for change. Second, the ability of different members of rural households and communities to observe and analyze patterns and differences and relate them to possible causal factors must be strengthened. The scientist or extensionist has a supporting role in developing and demonstrating methods to broaden observational and analytical skills. Third, farmer observation of specific local conditions and available resources must be related to an appropriate understanding of ecological principles for the identification of opportunities for improved management. Again, the scientist or extensionist serves as a

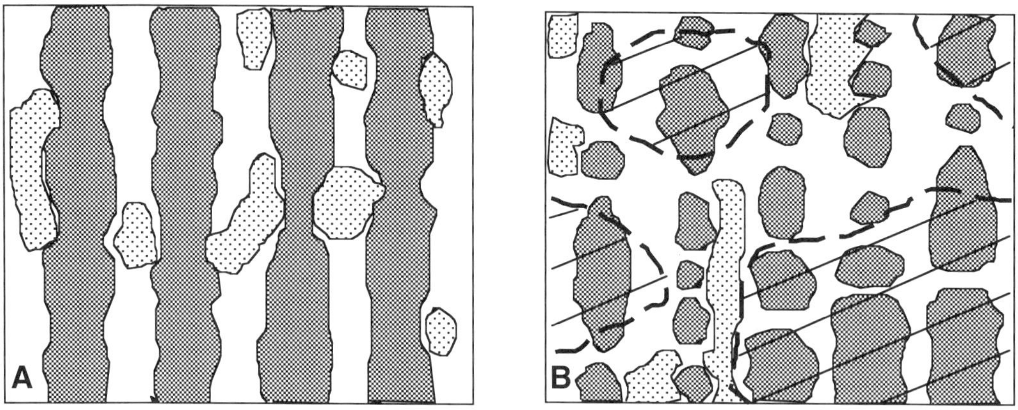

FIGURE 3. Distribution of coffee plants ( ), difficult weeds ( ), and shade ( ) is highly variable from one coffee field to another. A= sun-grown uniform coffee hedges with patches of difficult-to-control weeds. B= field with high degree of heterogeneity in shade distribution, vigor and distribution of coffee bushes, and weed patches.

source of ecological background knowledge as well as possible improved practices which may be tried by growers in an experimental approach to the validation, modification, and widespread use of technologies. Fourth, the process itself of observation and analysis of heterogeneity and the parallel design and modification of management plans within a field, between fields, and from year to year, depending on the variability in soil resources, crop and pest biology, and interactions with climate, weather, farmer resources, and economic factors, systematically strengthens the abilities of farmers, households, and communities to use ecological principles for decision-making. Similarly, scientist and extensionist involvement with farmer groups enriches the research and extension agendas.

The development of this paradigm is still incipient. Nonetheless, we foresee that heterogeneity might be managed either to move towards greater uniformity or to maintain heterogeneity. In the first case the management of heterogeneity operates as a gradual transition to more uniform crop and resource conditions as the optimum practices are identified and fine-tuned in each local environment. In the second case the management of heterogencity and variability is on-going with an emphasis on diversity, cycles, niches, and the emergence of new patterns and opportunities. In this case the basic resource and production matrix continues to be heterogeneous.

What are the expected outcomes from a management paradigm based on heterogeneity? First and foremost, natural human abilities for observation, experimentation, and learning are broadened and strengthened. Second, farm household decision making routines incorporate more systematic observation and information, broadened analytical skills, and a more reasoned consideration of possible outcomes and risks. Third, farm household members have increased capacity to share learning and experiences with neighbors and family. Fourth, land, labor, and on-farm and off-farm resources are used more efficiently for a more stable and diversified income. Fifth, the farm and community resource base, principally soils and water, is conserved and improved.

The contrast between uniformity and heterogeneity as the basis for management can be illustrated in the case of weeds and ground cover in coffee. The range in the degree of heterogeneity in a coffee field can be wide (Figure 3). In example A, the sun-grown coffee growing on flat land is highly uniform and weeds are minimal, although patches of more difficult-to-control weeds are common throughout the field. In example B, both the coffee and the shade are uneven in their distribution and the ground is covered not only with patches of different weeds, but also with a variegated pattern of shade and leaf litter.

With uniformity as the paradigm, field-wide practices such as machete weeding and/or herbicides are imposed on variable conditions to eliminate weeds. The major decision by growers is what technique to use and when. Even if weed control and soil conservation are combined within a single practice like the use of planted ground covers, the practice is applied over the entire field. In either of the two examples shown in Figure 3, the application of uniform practices frequently produces a mismatch between the practice applied and the prevailing conditions both in terms of the weeds and the crop, shade, and ground covers.

With heterogeneity as the basis for management, two situations are possible. Heterogeneity can be taken into account to gradually achieve uniformity. For example, modern coffee growers may use spot application of herbicides to eliminate residual weeds after earlier blanket applications. The combined effect leaves the soil completely bare. The heterogeneity in example B also might be taken into account to shift to more uniform coffee and shade tree distribution through the elimination of undesired coffee plants and trees and replanting of single species at regular spacings. With a more uniform crop-shade tree matrix, a uniform weed-free condition could also be more easily achieved.

Heterogeneity can also be managed and maintained. The grower observes the within-field variability in coffee bush distribution, shade species and levels, and weeds to formulate a management plan. On a given small patch of the field, they may find it easier to accumulate leaf prunings, select for low-growing non-competitive weeds, do nothing since shade and natural leaf fall minimize weed presence, or clean-weed if the area is at the base of a coffee bush. Viewed as a whole the ground of a single field would be covered by a mosaic of different patches. In each field the mosaic would be different. By managing heterogeneity for continued heterogeneity, a grower builds from their existing conditions to an improved situation through a diversified approach which is low cost, conserves the soil resource, and continues to be flexible and improvable in time.

Fundamental to management either for increasing uniformity or continued heterogeneity is an iterative grower decision-making process based on goal-setting, observation, analysis, action, and evaluation which leads back to observation.

4. THE BEGINNINGS OF A MODEL: REPLICATED PLOTS TO FARMER PRACTICE

Since 1989 in the coffee regions of Nicaragua, Central America, a multi-disciplinary team from CATIE (Central American Institute for Research and Teaching in Tropical Agriculture and Natural Resources) with financing from the Norwegian and Swedish development agencies has been addressing the question of how to improve pest management among small coffee-growing households. The work of the team, in collaboration with many universities, grower associations, individual growers, and non-governmental projects, covers insect pests, nematodes, and diseases as well as weeds and ground cover which are reported on here. During more than six years of work in the field, the team has undertaken a wide range of activities focused on how to make integrated pest management (IPM) effective in the hands of farmers. These have included small plot research, short courses for researchers and field staff, participatory replicated research plots, integrated management plots, field schools, and participatory training workshops. More importantly, the team has begun to identify how all these activities can contribute to a heterogeneity-based management paradigm for greater effectiveness. The activities, instead of focusing singularly on the development of techniques for pest management and their transfer, have also been designed to identify underlying ecological principles, sampling and observational methods, and causal factors for variability and heterogeneity which are important in putting the decision-making paradigm into practice with growers, albeit in a preliminary form.

In retrospect, we can identify five distinctive activities which form a rough chronological process to the work. These are not linear stages, but rather groups of activities which overlap, feedback, and modify each other as the work advances. In the remainder of this section, we briefly describe these activities.

Many of the teams' diverse initial activities contributed to our familiarity with the different coffee systems and heterogeneity and variability on-farm. How do coffee fields vary from one field to another and what are the explanatory factors? How uniform are coffee fields internally? How do farm households make decisions? This was a long and somewhat inefficient learning period which might have been consolidated had we realized the utility of a heterogeneity-based decision making framework. In retrospect we among the team were working with implicit and piecemeal analysis. An explicit team exercise to assemble a first approximation of the ecological basis for pest variability and its relationship to the management of the coffee ecosystem, based on existing knowledge and initial fieldwork, would have made our later work more efficient and productive.

The second set of activities, carried out in small replicated plots, examines the biological feasibility of alternative management approaches and the range of variability in response and seeks options to specific crop or pest problems, while at the same time generating ecological information and principles for use in interpreting on-farm pest and crop dynamics and variability. For organisms which can be studied in small plots, interactions can also be analyzed. In this stage, uniformity of conditions, rather than heterogeneity, is important for clearer results.

Third, the problem we address is developing methods for farmers to improve their quantification and observational skills. Conventionally in IPM scouting is associated with thresholds for pest control. Simple, standardized observation methods are also useful to visualize within-field variability, to promote discussions among farmers about why their fields differ one from the other, to compare different practices which are being tried, and to track changes over time. The sampling problem here is quite different from replicated plots. Farmers need simple methods to sample key minimal data from large, heterogeneous fields for analysis and decision-making, while in the previous stage, plots were smaller, more homogeneous and subject to more intensive data collection by scientists.

Once the results from small replicated plots begin to come in, larger-plot, longer-term systems studies become more feasible to establish either on-farm or on the research station. These studies are an opportunity to address numerous objectives. The most promising options from small plot studies can be subjected to practical and economic evaluation more easily in large rather than small plots. Larger plots maintained for longer periods are more appropriate to study the interaction of certain management variables like shade with pests or with soil quality factors. At the same time we can also monitor whether the same ecological principles we observed in small plots continue to operate in large plots. These systems studies should probably be limited to a few, highly contrasting treatments as the general goal is careful observation of interactions under relatively uniform conditions rather than the fine-tuning of specific management options under uniform conditions.

Fifth, operating in parallel to the previous groups of activities, is field experimentation and widespread multiplication of heterogeneity-based decision-making with groups of growers and households in coffee. Starting with the first approximation of the ecological basis for pest variability and adding methods for observation and counting, scientists and selected extensionists work with pilot groups of growers. As the management paradigm is consolidated and enriched with field experience, more extensionists and groups of growers are incorporated. The framework is participatory with cumulative and mutual learning directed towards strengthening the abilities of the farm household to make decisions in the face of incomplete and uncertain information. In the case of coffee IPM in Nicaragua, groups of growers meet with extensionists at regular intervals corresponding to critical phenological and decision-making moments in the annual coffee cycle. In each meeting farmers practice observational skills, discuss the data collected, and relate the data, the crop stage, weather factors, and farmer resources to possible management options. At the end of each meeting growers propose the activities they will complete in their own fields before the next meeting. These may include data taking or alternative management practices. The group also may designate certain plots for experimental comparison of agreed-upon practices (see SIMAS, 1996 for a more complete description of working methods).

For the scientist this set of activities combines long and short-term perspectives, work on management practices and on general ecological mechanisms, and attention to replication and predictability as well as to heterogeneity and uncertainty. The extensionist gains an ecological framework for organizing extension activities and methods to go beyond simply giving technical recommendations. The set of activities equips growers both with alternative management options as well as with a better understanding of crop-pest-resource interactions and a strengthened ability for observation and analysis, crucial to decision-making for timely use and modification of cropping practices.

5. MANAGING WEEDS AND GROUND COVER

The multi-disciplinary working environment of the CATIE IPM project from its first years provided an important context for the characterization of weed problems in different coffee systems and for the definition of approaches to weed management. In initial surveys during 1990-91 of cross sections of coffee fields varying from traditional to modern, the team observed that pest levels (diseases, insects, pests, and weeds) were highly variable by region and by type of grower, but also among fields with relatively similar management. These studies also demonstrated the fundamental role of shade management in pest levels both within fields and between fields (Monterroso, 1993). In 1992 the project team began work with 8 groups of coffee growers to develop and validate IPM options under field conditions (Siman and Staver, 1992). These meetings every two months with growers and extensionists over a period of two years provided the first understanding of how growers make decisions, indicated the need for training oriented towards decision-making, and demonstrated the importance of the practical integration of the different pest management disciplines in the hands of growers.

During this early period the working model for weed control was modified several times. The initial concept of *uniform field-wide weed control* shifted to the *management of weed heterogeneity*, then to *heterogeneous ground cover management* (Staver et al., 1993), and finally to *heterogeneous ground cover as habitat management* (Staver, 1996 following Hogue and Neilsen, 1987). The initial objective of weed control was to minimize competition between the crop, coffee in this case, and weeds. A highly contrasting objective was added in ground cover management to protect the soil from rainfall impact and soil erosion. Finally, the effect of ground covers on other pest problems also needed to be evaluated. At a minimum, the ground covers should not make other pest problems worse, but also should be managed to reduce the severity of other pest problems. For coffee grown under shade, a number of different mechanisms potentially meet these multiple objectives: certain naturally occurring ground cover weed species, the overhead tree canopy and its leaf litter, leaves and small branches pruned from trees, and planted ground covers.

In the following sections, the progressive development of this framework from research to farmer practice will be described in greater detail. First, we discuss replicated small plot research on the biological feasibility of different ground cover options. Then, we review the results from a large-plot systems study. Finally, after we present sampling methods for grower observation and quantification of ground cover, we describe the procedures for working with groups of growers on the heterogeneity-based management paradigm.

5.1 Small Plot Research on Ground Cover Management Options

The variability within and among coffee fields that was identified in early studies suggested several approaches to the management of weeds and ground cover. Initially we focused just on the weeds, their possible competitive effects, and how to change the floristic composition. Quickly we realized that the tree component could also be managed for weed control and soil protection. The small plot research reported in the following sections was done to determine the biological feasibility of certain methods, to examine the variability of response of different treatments, and to begin to observe weed population dynamics in coffee under more controlled conditions.

5.1.1 Selective Management of Weeds

The idea to manage floristic composition of weeds in coffee first arose as a preliminary step to research on the competitive effect of different weed types on coffee productivity. We needed to manage weed populations to create and maintain plots with weed stands of a single type such as grasses, annual broadleafs, or herbaceous perennials to study their competitive effects on coffee. The results of the first experiments demonstrated that selective weed management had great practical value in reducing weed competition and also maintaining ground cover to reduce soil erosion.

In a coffee field under partial elevated shade with a mixed weed population of grasses and perennial and annual broadleafs, single interrows of 1 meter were subjected to different manual and herbicide treatments simulating the twice per season frequency of farmer weed control. Within two years the floristic composition of the plots had diverged by treatment (Figure 4). In the plots with only machete weeding, perennial broadleafs and grasses which resprout readily after machete slashing had increased. The plots in which 2,4-D, a selective broadleaf herbicide, was used, were dominated by grasses. In the plots which received the conventional mix of herbicides used by growers to create weed-free soil conditions, annual broadleaf and grass weeds were most common. These weeds germinated from the seed bank once the effect of the pre-emergent herbicide had been lost. In a fourth treatment, weed control practices were flexible, varying as the floristic composition changed. In this treatment selective herbicides were used to control the more damaging weeds and to leave uncontrolled the species considered non-competitive with coffee. Weeds with extremely shallow root systems and a low creeping growth habit such as *Oplismenus burmanii* (Retz) P. Beauv. were conserved. By the end of the second year these ground cover weeds made up more than 50% of the weed biomass in the plots with flexible management practices, while in the other treatments ground cover weeds made up only 10–20%.

This experiment verified mechanisms in applied plant ecology and demonstrated a number of practical management practices, both of which proved useful later in the development of the management paradigm for work with groups of growers. First, the 27 weed species found in the plots were readily grouped by their reproductive cycle, growth habit, and differential response to management practices like slashing and selective herbicides into four–seven groupings for statistical analysis and for practical observation. Second, the relative proportions of weed types differed according to the management practices that were used. This suggests that the future weed populations are a product of the current weed and land management practices with problem weeds being a product of weed management itself. Third, the flexible management treatment demonstrated that the floristic composition can be changed with deliberate objectives in mind. Although in the experiment the time period was short and the plots extremely small ($10m^2$), a strategy in which weeding practices change as the weed floristic composition shifts was effective. Fourth, rather than generally applicable weed control practices which seek to reduce weed biomass across the board, weed management practices in coffee should be focused on specific weed groups. Practices should be directed at the weeds which are most difficult to control and which are most likely to compete with the coffee crop. These practices, however, should be designed and carried out to affect minimally the ground cover weeds which are present in the field. In this particular field the most common damaging weeds were *Alternanthera*, *Amaranthus*, *Mirabilis*, *Digitaria*, and *Euphorbia*, while *Oplismenus burmanii* was the predominant ground cover weed. In other fields with other weed species, the general strategy is still applicable, although the specific practices would vary with the floristic composition.

In this theme of selective management of weed types at least three topics warrant additional study. First, the competitive effects of different weed types on coffee, including different coffee species and varieties, within and between rows, have still not been quantified. Our current recommendations and training materials are based on extrapolations from differences in weed growth and rooting habits. The recommendation to maintain the area within the coffee bush canopy free of weeds derives from studies of the coffee root system which is concentrated in the upper 30 cm

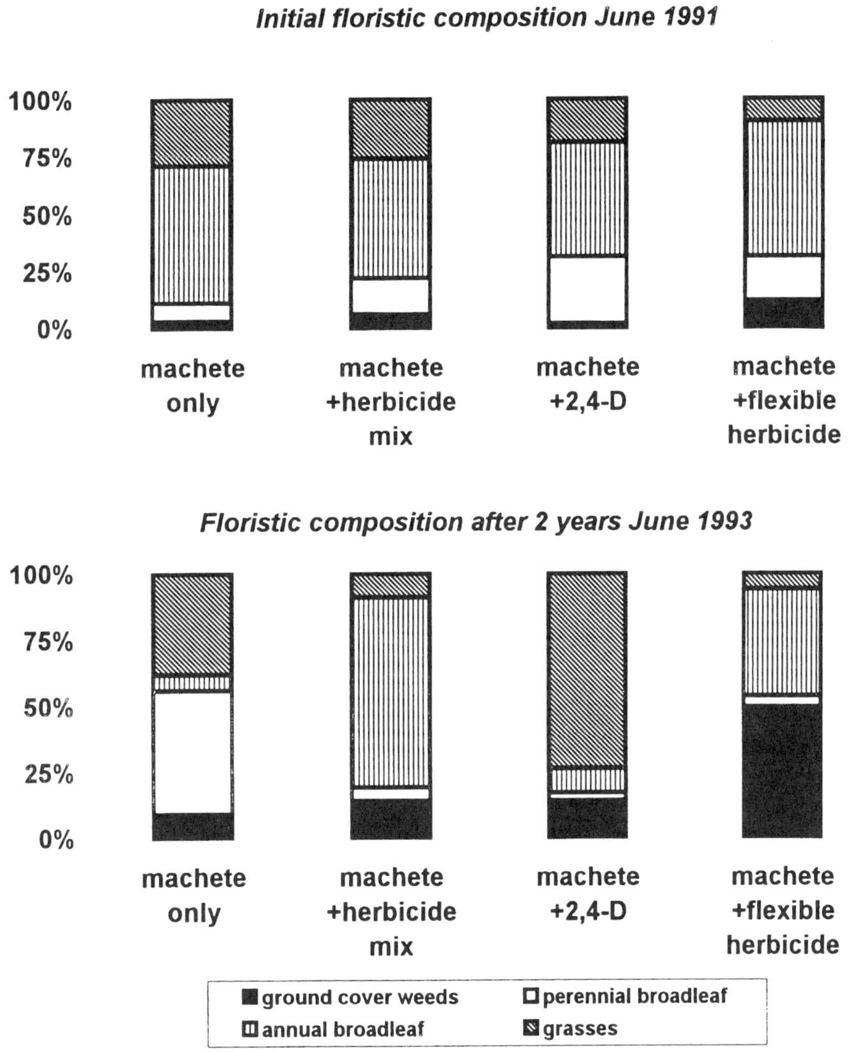

FIGURE 4. Divergence in floristic composition of coffee weed complex in small plots due to different weed control treatments from 1991 to 1993, Southern Watershed, Managua, Nicaragua.

as well as from other pest management recommendations. For example, a clean soil surface facilitates the gleaning of fallen coffee fruits which are the overwintering sites for the coffee berry borer (*Hypoyhenemus hampei* Ferrari). Second, ground cover weeds take up and retain nutrients which might otherwise be lost through leaching or erosion. Aguilar and Staver (1997) found that in three successive years 8.8–10, 1.1–1.2, and 11–12 kg/ha of N, P, and K, respectively, were taken up and held by *O. burmanii* as ground cover during the rainy season. In the same plots 12–14, 1.1–3.8, and 24–68 kg/ha of N, P, and K were taken up and returned to the orchard floor for reabsorption, the result of five–seven selective weed slashings and weed regrowth. Are these nutrients being withheld from the coffee? What would be the fate of these nutrients if there were no ground cover weeds? Are nutrients recycled within the weed component or is there transfer between the weeds and coffee? Third, there may be allelopathic effects from certain ground cover weeds (Ramos et al., 1983). A fourth research topic is the effect of weed distribution on the practicality of selective weed management practices. Informal observation suggests that fields vary in the degree of local clumping of weed types. On the one hand, two or more weed types may be highly intermingled in areas smaller than 100 cm^2, while in other cases clumps of a single weed occupy 1–4 m^2. The steps to selective management may depend both on the weed types and the degree of clumping. In situations where the weeds are highly intermingled, practices to promote greater clumping or to simplify the weed population may need to be prioritized. Our understanding of the spatial and temporal dimensions of weed patchiness in perennial crops and agroforestry in general is limited and merits greater study.

5.1.2 Trees in Ground Cover Management in Coffee

Trees are planted with coffee primarily to regulate the sunlight available for photosynthesis by the coffee plant and to stabilize yields over a period of several productive years (Kimemia and Njoroge, 1988; Carvajal, 1984). At the same time both the shade and the natural leaf litter alter the growing conditions for weeds on the orchard floor (Kimemia and Njoroge, 1988; Goldberg and Kigel, 1986; Nestel and Altieri, 1992), although no formal documentation of any differential effects by tree species is available. The natural growth of most tree species results in greater than optimum shade levels and an uneven distribution. The regulation of shade levels through pruning as frequently as once or twice annually improves both the percent and distribution of shade, and also produces additional leaf litter which has potential utility for localized weed control and ground cover.

Two small plot studies, one without shade (Rivas, 1994) and the other under approximately 50% shade (Vado, 1996), were carried out to document the effect of fresh leaves from different common shade tree species for weed control. Not surprisingly, depending on mulch thickness and species, weed numbers were reduced compared with the unmulched control for up to 60 days (Table 1). In the shaded plots mulch controlled both weeds originating from seed (94%) and weeds resprouting vegetatively (88%), in contrast with Budelman (1988), working in open sun, who found that leaf mulches did not control resprouting weeds. Mulch was less effective in reducing total weed biomass, with no effect in the open sun plots by 30 days, except in the most decomposition resistant species at double thickness. In both the open sun plots and in the shaded plots where differences in weed biomass were still found at 60 days, individual weed weights were greater in the presence of mulch. In multiple regression analysis of the two studies which were conducted in different fields in different years, the three most significant explanatory variables of weed numbers were mulch species, mulch thickness, and species by thickness. In open sun a species like *Gliricidia sepium* (Jacq.) Walp. even at double and triple thicknesses, decomposed rapidly with a 65% loss in biomass in 63 days, while *Inga paterno* Harms, *Simarouba glauca* D.C., and *Clusia rosea* Jacq. decomposed much more slowly with only a 15% loss in biomass in the same period. Under more humid, shaded conditions, all species had decomposed to small leaf pieces by 65 days, although visually *Clusia* appeared to be more intact than *Gliricidia*. Barradas and Fanjul (1986) found that open sun coffee was characterized by higher temperatures, greater fluctuations in temperature, and lower air humidity than shaded coffee. This suggests that coffee with lower shade levels is a more variable microclimate in terms of wetting and drying and less conducive to leaf decomposition.

This work with leaf prunings confirmed the practical utility of several ecological mechanisms in weed management. First, the reduction of plant growth resources, in this case sunlight, through the use of mulch is applicable in coffee under shade, although tree species differences in leaf decomposition rate, a readily observable trait, must be taken into account. Second, the use of mulch will be more effective for reducing seed germination than for suppressing sprouting. Such weeds as shade-tolerant perennial vines which are especially problematic in coffee will be less easily controlled with mulching. However, mulch should be highly effective against annual weeds in less shaded patches of a coffee field. Last, weed plasticity means that weeds which escape through leaf mulch layers have importance far beyond their small numbers both in terms of their increased biomass and seed production which was not measured in the studies cited above. However, the

TABLE 1
Effect of tree species and thickness of leaf mulch on weed number and biomass in southwestern Nicaragua — rainy season 1992–1993

	Mulch Treatments: Species X Thickness											
	No leaf mulch	Gs[a]	Gd	Gt	Ss	Sd	St	Cs	Cd	Ct	Id	It
Open Sun Plots												
27 days after mulch weed numbers/m²	1641	—	1090	400	—	306	180	—	206	69	329	183
33 days after mulch weed biomass/m²	381	—	551	239	—	316	213	—	249	88	350	141
39 days after mulch all weeds eliminated												
63 days after mulch weed numbers/m²	655	—	581	382	—	74	82	—	57	118	74	71
Shaded plots												
17 days after mulch weeds from seed/m²	129	18	0	0	18	3	3	4	0	0	—	—
weed sprouts/m²	108	21	1	1	35	15	8	17	6	2		
45 days after mulch weeds from seed/m²	95	14	11	6	11	5	3	2	3	1	—	—
weed sprouts/m²	136	29	17	8	25	31	10	18	14	4		
65 days after mulch weed biomass/m²	254	93	51	37	79	71	44	62	40	16	—	—

[a] G=*Gliricidia*; S=*Simarouba*; C=*Clusia*; I=*Inga*; s=single layer (0.4 kg dry matter/m²); d=double layer (0.8 kg kg dry matter/m²); t=triple layer (1.2 kg dry matter/m²).

cost of slashing a few isolated large weeds escaping through leaf mulch may be less than slashing the thick low weed growth in the absence of mulch.

A number of aspects of leaf mulches as ground cover warrant additional study. First, while leaf litter is effective against raindrop impact (Wiersum, 1984; Young, 1989), how effective is it against the other components of soil erosion, laminar flow and concentrated flow? Is this an important element in the extensive use of mulch for ground cover in coffee on sloped lands? Second, what is the potential production of leafy material through pruning of different tree species? To achieve a layer of 0.3–1.2 kg dry matter/m² on 50% of the orchard floor which were the levels used in the experiments in Table 1, 3–25 tons/ha of fresh leaf matter must be produced, about 25–154 kg/tree planted at 100–277 trees/ha. What are the actual production levels of leaf matter with different pruning regimes, and what is the potential for planting trees at a higher density with a more severe pruning regime to increase the leaf mulch production rate? Third, there also may be an allelopathic effect of specific leaf mulches which could be exploited as part of a weed control program (Alan and Barrantes, 1988).

There are also a number of other themes in weed and ground cover management which could contribute to the practical design of the tree component of coffee fields. Different tree species have contrasting leaf size and arrangement, canopy structure, and seasonality in leaf emergence and fall which alter the quantity and quality of light at the soil surface (Figure 5). Is there a difference in weed species and growth under different shade trees with similar light interception? Does the combination of canopy structure and natural leaf litter make some trees species more effective in weed control than others? Naturally occurring ground cover species like *Oplismenus burmanii* and *Commelina diffusa* Burm. f. appear to be more common in shade. However, we do not know if they are found more commonly under certain tree species than others. Are the more shade-tolerant problem weeds associated with certain tree species? Informal observation suggests that *Syngonium podophyllum*, a creeping, climbing aroid which is difficult to control, for example, may be more common under *Inga*.

For initial studies with the interactions among weeds and ground cover with tree species, multiple criteria for tree selection should be used to define a preliminary list of most probable tree species. These criteria would include income potential from trees, tree species as wildlife habitat (Greenberg et al., 1997), and tree effects on coffee productivity, on other pest problems like diseases or nematodes, and on soil fertility and structure.

FIGURE 5. Comparison of leaf size and arrangement and branching pattern for *Inga paterno* (A) and *Gliricidia sepium* (B). *Gliricida* has lower levels of light interception due to more erect branches and smaller leaves and suffers complete leaf loss for several months in the dry season.

TABLE 2
Within and between row coffee root densities under different planted leguminous covers and weed management at the end of the dry season in southwestern Nicaragua — May 1992

	Roots at the coffee bush dripline		Roots between coffee rows	
	cm roots/cm³ soil			
	0–30 cm	30–50 cm	0–30 cm	30–50 cm
perennial *Arachis*	1.7	0.7	1.3	0.4
perennial *Desmodium*	2.9	0.9	0.6	0.4
weedy control (annual weeds senescing at end of rainy season)	4.5	0.9	0.8	0.5
grower management (weed free from late rainy to end of dry season)	6.4	3.9	1.3	0.4
Statistical significance	0.05	0.05	not significant	not significant

5.1.3 Planted Ground Covers

The concept of replacing weedy vegetation in tropical plantation crops, including tea, rubber, and oil palm, with a planted leguminous cover crop is not new (Sampson, 1928; Broughton, 1977). Planted ground covers in coffee have been studied for many years in Africa (Bouharmont, 1978, 1979) and Latin America, particularly Colombia (A. Gomez, Cenicafe, Chinchina, Colombia, personal communication). Recently there has been renewed interest in Nicaragua where coffee was renovated under open sun in wide rows of up to 3.5 m (Rice 1990). Increasing labor costs and concern for soil productivity and erosion in Costa Rica where herbicides are used extensively in coffee have also given rise to on-farm testing of perennial cover crops (Garcia et al., 1996).

In Nicaragua the initial choice of species was guided by the definition of a series of characteristics — perennial legume; low-growing with dense, creeping growth habit; and planting material readily available and easy to multiply. Many species which are twining in their growth habit (*Centrosema* sp, *Macroptylium* sp), non-creeping herbaceous (*Stylosanthes* sp), annual (*Canavalia ensiformis* (L.) D.C., *Mucuna pruriens* (L.) D.C., *Dolichos lablab* L.), difficult to multiply (*Indigofera spicata, Desmodium heterophyllum, Zornia* spp), or shrubby (*Cajanus cajan* (L.) Millsp., *Crotalaria* spp) were set aside as inappropriate for initial feasibility testing of the concept of a low-cost, planted ground cover system. Two species, *Arachis pintoi* Pinto and *Desmodium ovalifolium* (Prain) Ohashi were chosen for a preliminary look at the biological feasibility of planted legume covers in coffee. Many accessions of these species have been collected by CIAT. In a long term research program, both more accessions and many species mentioned previously may warrant further consideration (Firth and Wilson 1995).

In a three-year stand of sun-grown coffee in a region with a five-month dry season seed, from *Desmodium* and vegetative stems from *Arachis* were planted and given initial intensive weed control to ensure good stand establishment (Bradshaw, 1993; Bradshaw and Lanini, 1995). Once the legumes were established, they provided excellent control of annual weeds in the interrow area in four sampling dates (Figure 6), although manual weeding was used to keep the coffee row itself free of both weeds and planted covers. In other plots *Arachis pintoi* was observed to be less effective against perennial weeds. No differences were found in coffee growth and yield in the coffee plots with legume covers compared with a weedy control or grower management, although data were highly variable. Coffee plants in the legume plots and the weedy control had greater leaf water stress in the last month of the dry season (-35 MPa) compared with grower management which was free of weeds during the dry season (-29 MPa). Gravimetric measurements of soil water content showed that by the end of the dry season soil water levels were lowest in the plots with grower

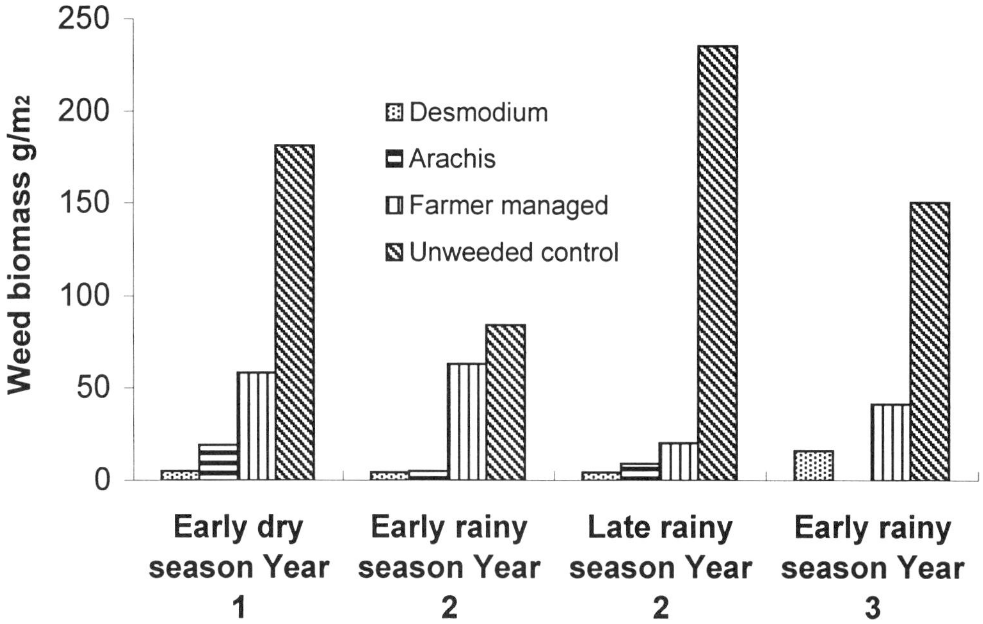

FIGURE 6. Weed biomass on four sampling dates for different leguminous covers and weed management in coffee, 1991–1992, San Marcos, Nicaragua (with permission from Bradshaw and Lanini, 1995).

management, 23%, compared with 27% in the other treatments. Coffee roots at the edge of the coffee plant canopy were found to be most abundant in the grower management plots (Table 2), suggesting that the roots from the legume cover crops or weeds had reduced coffee root expansion and increased plant water stress as seen earlier. When interrow weed biomass was low in the end of the growing season and into the dry season under typical grower management, the more extensive coffee root system more effectively extracted soil moisture with less water stress in the plant.

In the north of Nicaragua *Arachis* and *Canavalia* were compared in recently established sun-grown coffee (H. Mendoza, UNICAFE/Pasolac, personal communication). Coffee with *Arachis* had reduced growth compared to coffee with *Canavalia*. Differences were attributed to possible competition from *Arachis* (water, see above, or nitrogen, (Dominguez and de la Cruz, 1991)) which grew freely to the base of the coffee plants, a wind break effect of taller-growing *Canavalia*, and a possible nitrogen contribution from the prunings of *Canavalia* with a twining growth habit which needed to be manually controlled. Plots with *Arachis* had lower weed levels and soil erosion once established, but required extensive weeding and soil disturbance during the establishment year when soil erosion levels were high.

The studies of planted legumes as ground cover in perennial crops demonstrated several simple ecological points. First, herbaceous plant biomass associated with coffee can have positive or negative effects on the crop, depending on plant characteristics in terms of competition, soil protection, seasonal biomass production, and wind protection. While this is not a sophisticated point, it reconfirms the utility of selective management and on-site observation for decision-making. Second, both the seasonality and spatial distribution of living ground cover needs to be taken into account in selective ground cover management. The area at the base of the coffee plant should be free of all living vegetation. For both the width and the amount of biomass in the interrow area a number of critical moments need to be taken into account — periods with a greater frequency of erosive rainfalls, often at the beginning of the rainy season when living ground cover is scarce; periods when resource competition is more likely, towards the end of the rainy season; and abnormally droughty periods when rainfall is low.

A number of themes merit further research. First, when and where are coffee roots growing, a crucial factor for managing the row-interrow interface, and what is the effect of ground cover vegetation with different types of root systems? Second, can planted legumes be managed through slashing or pruning for multiple objectives — to protect soil, minimize competition with the coffee, and also produce nitrogen-rich mulching material? Third, how much nitrogen is fixed by legumes as the levels of shade increase? Do species and accessions vary for this trait?

5.1.4 Ground Cover Interactions with Pests

Ground cover plant species influence the levels of crop pests in a number of ways (Prokopy, 1994; Bugg and Waddington, 1994). They may be specific alternate hosts for pests. They may be preferred habitat for generalist pests or specialist or generalist natural enemies. Different ground cover species may create different microclimates which alter pest or natural enemy levels. Ground cover species also improve or deteriorate the physiological status of the coffee plants which affects their susceptibility to pest problems. The principal coffee pests including leaf diseases such as rust (*Hemileia vastatrix* Berk and Br), brown spot (*Cercospora coffeicola* Berk and Cook), and *Colletotrichum* spp, and insects such as coffee berry borer do not have alternate hosts. However, certain ground cover species are alternate hosts for pests such as nematodes which attack coffee roots. Different ground cover species also may alter the fauna of a coffee field as, for example, the presence of a rust spore-eating *Diptera* larva which feeds on spores from rusts infecting several weeds as well as from coffee rust. The incidence of brown spot disease is influenced by coffee plant nutritional status which may be affected by competition from weedy ground cover species or shade levels. Depending on the mobility and dispersal mechanisms of the pest organisms or natural enemies, only certain of these interactions can be studied in small plots.

The three principal naturally-occurring ground cover species in Nicaragua, *Oplismenus burmanii*, *Panicum trichoides* Sw., and *Commelina diffusa*, were found to not be hosts of the parasitic coffee nematode *Meloidogyne incognita* (Rofoid and White) Chitwood (Jorge Jarquin, Central American University, Managua, Nicaragua, unpublished data).

In several other studies the effects of the planted ground covers *Arachis* and *Desmodium* on levels of the nematodes *Meloidogyne* and *Rotylenchulus* were measured (Bradshaw and Calderon, 1993; Herrera, 1995). While both *Arachis* and *Desmodium* sometimes reduced nematode populations and galls on coffee roots, results did not show a consistent effect of cover crop species on different nematode species or strains. In small field plots of coffee with *Arachis* or *Desmodium*, *M. incognita* populations were lower in coffee with cover crops than in coffee alone. However, *Rotylenchulus* levels were lower in coffee with *Arachis*, but higher in coffee with *Desmodium* (Herrera, 1995). Much remains to be studied on the effect of leguminous and non-leguminous ground cover species on the soil environment and pest and beneficial soil fauna, not only between isolated single plant and animal species, but also between mixed plant and animal communities where both spatial arrangements and differing time periods of interaction may be important.

Coffee berry borer reproduces exclusively in coffee berries, although they may take momentary refuge in other species. Differences in ground cover may affect the microclimate around fallen berries where the pest overwinters and multiplies. Differences in ground cover also may make certain recommended management practices more difficult, such as gleaning of fallen berries. No differences were found in berry borer population dynamics under four different ground covers — leaf litter, tall weeds, low weeds, and bare soil (Mendez, 1992). However, more than 80% of fallen berries were found under the coffee plants and only 20% were found between rows where ground covers would be present. An even lower percentage of the overwintering berry borers, 5–13%, were found in the interrow area, since a greater percentage of the fallen fruits under the coffee plants were infested. The recommendation for both ground cover and berry borer management is to keep the area under the coffee plants free of weeds and ground cover, especially as fruits begin to ripen, and to glean fallen berries from the same area. The role of ground cover diversity as possible habitat for natural enemies of borer such as ants remains to be studied in greater detail, although the greater habitat diversity in shaded coffee compared with open sun coffee has been measured to have greater arthropod diversity in general (Perfecto et al., 1996). The effect of ground covers on more mobile insects such as the *Cephalanomia stephanoderis* Betrem wasp which parasitizes the coffee berry borer may need larger plots than were used in the studies cited here.

Small plots studies in agroforestry systems cannot address certain important interactions. Notable here is the relationship between ground covers and coffee leaf diseases. Small plot studies are also not appropriate for looking at shade tree-crop-ground cover relations. Again of great relevance are the role of microclimate and coffee plant nutritional status in coffee diseases such as *Cercospora*, rust, and *Colletotrichum*, as well as the role of trees and ground cover in nutrient cycling.

Small plot studies in the area of weeds and ground cover in coffee agroforestry have proven useful to sort through the biological feasibility of possible management options and to begin to understand ecological mechanisms for weed and crop response. Small plots are also useful to achieve within-plot uniformity for highly variable and diverse weed populations. With a background of on-farm surveys and small plot studies, we were also able to simplify the possible treatments for larger plot studies of alternative weed and ground cover management systems.

5.2 Systems for Weed and Ground Cover Management

With the preliminary feasibility of selective weed management and planted ground covers provided in small plot studies, we moved to address questions which could not be examined on a small scale nor in existing coffee fields which in Nicaragua are often characterized by heterogeneity in shade and coffee plant productivity. Are selective management of ground cover weeds or the planting of leguminous ground covers practical and how much do they cost? What is the effect of these ground covers on coffee growth and yield? We also hoped to look at the effect of ground cover management on soil quality factors and on other pest problems. Some of these studies are underway, others are not possible due to certain site factors which limit instrumentation. In this section we will focus primarily on the ground cover–coffee interactions and costs.

The conventional herbicide program for weed control in coffee was compared in replicated plots of 240 coffee plants to four selective ground cover management alternatives with and without herbicides and with and without planted *Arachis pintoi*. In the conventional program, machete slashing of weeds twice during the rainy season in the interrow space was followed once weeds had begun to regrow by the application of the herbicide tank mix of 2,4-D, paraquat and a pre-emergent. In the selective chemical–mechanical treatment, herbicides and machete slashing were used selectively to promote ground cover weeds and reduce potentially competitive weeds in the interrow; in the chemical–mechanical treatment with *Arachis*, *Arachis* was first planted in the interrow where weeds predominated, and then herbicides and machete slashing were used selectively to promote ground cover weeds and *Arachis* and to reduce competitive weeds. The remaining two treatments (mechanical only) with and without *Arachis* were similar with the exception that only machete slashing and not herbicides were used to promote ground cover weeds and reduce competitive weeds. Several replantings of *Arachis* were needed in the first and second years of the study. The area within the coffee row was managed uniformly free of vegetation in all plots, as was the application of fertilizer, coffee pulp, and fungicides for leaf diseases. The study was managed under a uniform combination of temporary (*Ricinus communis* and *Cajanus cajan*) and permanent shade (*Gliricidia sepium*, *Simarouba glauca*, *Inga* sp, and *Clusia rosea*). This is not a common practice among growers who have abundant practical experience and strong and often divergent opinions about their preferred shade species. The strategy in the study area was to use mixed species at higher densities with possible pruning or thinning at a later stage.

After five years, a number of results and lessons are prominent (Aguilar et al., 1997). First, the shift in the floristic composition of the weed cover in response to the four selective management alternatives was rapid (Figure 7). By the second year the percentage of ground cover weeds had doubled and even tripled in some cases. The changes were less dramatic in succeeding years with continued decline in broadleafs and grasses and moderate increases in ground cover and *Arachis*. Second, the selective management approach and the increase in ground cover species in combination led to greater soil protection. During the entire rainy season from 25–95% of soil surface was covered with living annual shallow-rooted cover in contrast with the conventional management in which the soil had 0% cover in three–six weeks after weed control to 50–100% competitive weed cover right before weed control. During the prolonged dry season the ground in selective plots was protected by a mulch of dried out annual cover weeds, while the soil was often bare in the conventional plots. Third, the higher levels of weed and ground cover biomass in the *Arachis* plots and the machete only selective plots reduced first year coffee yields in the third year of the study (Figure 8). While the conventional and selective chemical/mechanical plots had less than 1000 g of fresh biomass/m^2, the other three treatments exceeded 2400 g/m^2. For each of the following two years and for the total over three years yields were not different among treatments, although on average the conventional and selective chemical/mechanical treatments with 230 and 219 kg fresh berries/plot had higher cumulative yields than the other three treatments with 200, 200, and 185 kg/plot. Fourth, herbicide use was greatest in the conventional treatment, while for selective management herbicide use was lower and labor use was higher (Table 3). Labor use was highest in the *Arachis* plots due to the planting costs, although these costs might be lowered with practical studies on alternative planting methods. Fifth, the growth of the coffee plants and the shade trees over the five year period led to a number of changes in ground cover. By year five grasses made up only a minimal part of the weed complex; weed and ground vegetation in general declined with all treatments in the range of 9–456 g/m^2; and tree leaf litter increased to nearly 50% coverage. By years four and five the ground cover condition, while variable among treatments, was highly favorable in all cases. Competitive weed biomass was minimal, while the soil was protected by a combination of shade tree canopy, leaf litter, ground cover weeds, or *Arachis*. The relative success of all the treatments we attributed not only to the shade and coffee development, but also to the timely execution of agronomic practices, which is often not the case among growers, as we will see later in this section. In shaded coffee the use of *Arachis* appeared to be redundant both in terms of soil conservation and organic matter contributions. In practical terms, the extra cost for establishment, possible dry season competition for water, and rodent infestations also reduced the attractiveness of this planted ground cover.

This large plot study generated a number of additional questions, particularly for the management of weeds and ground covers in recently planted coffee. How much vegetation biomass can be permitted in the interrow zone and how wide should the vegetation-free band be in the coffee row, a study which could be carried out in smaller plots? Shade tree-crop-ground cover interactions were also not clarified to the extent planned by the study. The rodent *Orthogeomys* sp (Hilje, 1992) preferentially felled *Gliricidia* shade trees in or adjacent to *Arachis* plots which created patchiness in shade levels and confounded any interpretation of data on coffee leaf disease levels by plot. The interactions among shade levels and tree species, coffee varieties, pest problems, ground cover, and input levels and sources need both multiple localities and multiple year periods to begin to provide the basis for improved organic and low-input agroforestry coffee production.

The promising results from the experiment also posed the question whether similar results could be achieved on-farm and in coffee areas with wetter, higher altitude conditions and differing weed flora. In half hectare plots of recently established coffee on five large coffee farms in northern Nicaragua we compared selective weed management side by side with the conventional management used on each farm. Fields varied in their weed complexity (Table 4), although grasses and vines were often present in significant levels and naturally occurring ground cover weeds were, if at all, present in very low levels. Shade levels were low and patchy within all fields, made up primarily of volunteer species. Shade trees had also recently been planted in two of the fields. In bi-monthly visits during two years we surveyed weed cover and discussed possible management options with

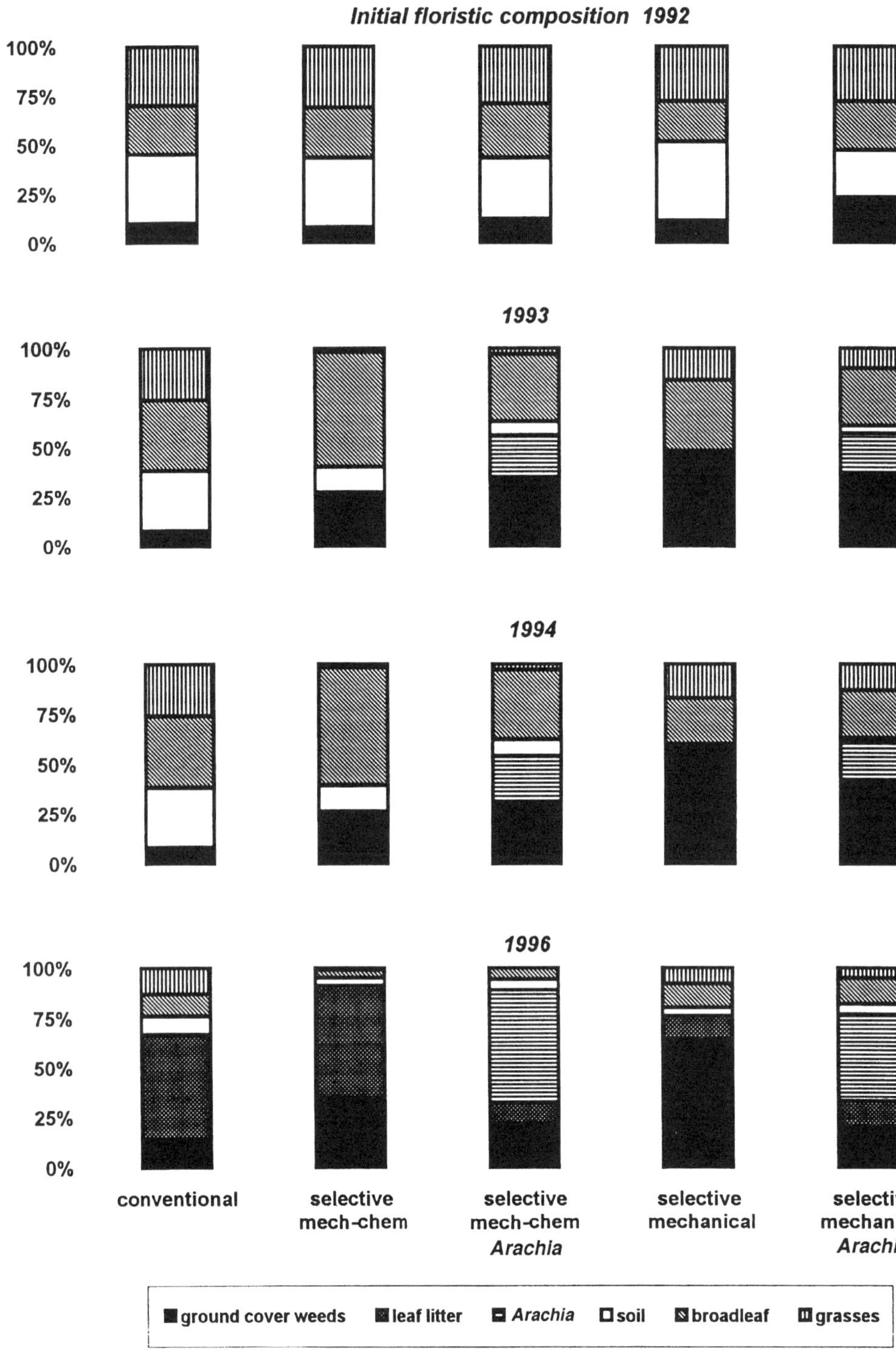

FIGURE 7. Shifts in floristic composition of ground cover vegetation in shaded coffee under five management systems, 1992–1995, Masatepe, Nicaragua.

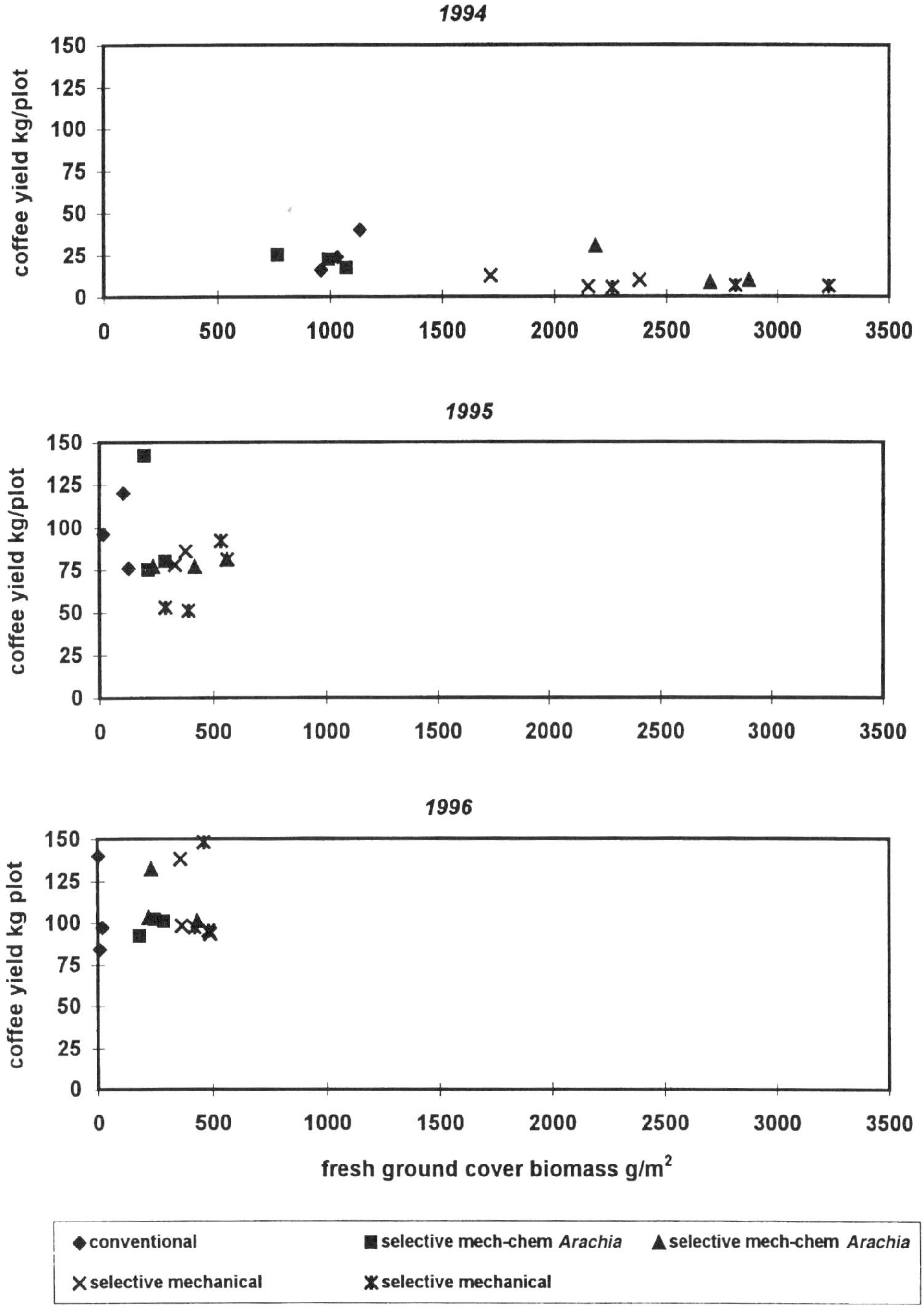

FIGURE 8. Variability in total ground cover biomass and coffee yields by individual plots for five ground cover management systems, 1994–1996, Masatepe, Nicaragua.

TABLE 3
Use of labor and herbicide in the establishment and management of alternative weed and ground cover management systems 1992–1995, Masatepe, Nicaragua. Amounts indexed with conventional control in 1993 = 100.

		conventional grower weed control with herbicides and machete	selective management with herbicides and machete	selective management with herbicides, machete, and *Arachis*	selective management with machete	selective management with machete and *Arachis*
1992	Labor	47	47	517	152	571
	Herbicides	72	48	31	0	0
1993	Labor	100	119	418	191	691
	Herbicides	100	75	55	0	0
1994	Labor	101	174	228	279	366
	Herbicides	230	99	75	0	0
1995	Labor	126	138	157	239	192
	Herbicides	0	43	18	0	0
1996	Labor	33	45	56	95	67
	Herbicides	72	37	25	0	0
Total	Labor	407	523	1376	956	1887
92–96	Herbicides	474	302	204	0	0

TABLE 4
Variability in percent weed cover by growth habit in five coffee fields in Northern Nicaragua — June 1995.

	% weed cover by coffee field				
	La Suana	El Carmen	El Hular	La Estrella	La Fundadora
perennial broadleafs	7	26	16	2	19
annual broadleafs	1	2	—	12	7
vines	11	25	37	20	10
grasses	47	12	14	25	31
sedges	9	2	—	6	1
broadleaf cover weeds	1	6	2	1	9
grass cover weeds	1	1	—	1	3
leaf litter	17	26	31	29	1
dead weed litter	3	—	—	—	19
bare soil	3	—	—	4	—

the technical advisor and field supervisor from each farm. Options were restricted by the economic and logistic constraints of each farm which supplied inputs and labor.

The results, while largely based on personal observation, contrast with the study of alternative ground cover systems. First, we found that floristic composition in the selective management plots has made only minor shifts. Grasses and vines have declined in some cases, although ground cover weeds have generally not increased. We observed that selective management plots regularly experienced excessive weed seed production, frequent periods of weed competition, and little improvement in shade or leaf litter levels. Second, the soil in both conventional and selective management has been well protected with a heavy cover of weed biomass, often the result of delayed weed control. Soil erosion never has been even remotely probable in any of the plots, although weed competition has been frequent. Third, management of the coffee fields in general has been characterized by neglected practices, poor or incomplete practices, or untimeliness. These were attributed on differing occasions to credit and cash flow problems, problems with input purchase, staff turnover or the prioritization of other competing activities. We interpreted these factors to represent more a problem with decision-making than with technology. Carrying out weed or ground cover management through untimely or misdirected practices, inappropriate herbicides or inattention to the crop-shade tree matrix we understood not to be a problem of cost or even the availability of technology, but rather of management ability to translate available resources into timely cropping practices.

5.3 Grower Decision Making for the Management Of Heterogeneity

In the initial pages of this chapter we characterized coffee fields as heterogeneous both within and between fields in their soils and slope, in the size, shape, and distribution of shade trees and coffee plants, and in their weed complex. Coffee growers themselves vary in their available resources, their experience with the crop and in agriculture, and in their management skills. These factors in and of themselves suggest that better weed management will not come simply through new technology generation and transfer. In addition we have proposed that better weed management is not only the minimization of weed competition with the crop, but also ground cover management for soil conservation and the possible reduction of other crop pest problems. Other multiple objectives of weed management might also include minimization of costs, of negative health and environmental effects from herbicide use, and of damage to the coffee plants from the use of herbicides or machetes. Growers currently use short-term practices to momentarily suppress all weed growth. In this section we describe key components in how we have attempted to move towards multiple objective decision-making by growers faced with field heterogeneity in weed management.

5.3.1 Participatory Training in Coffee Pest Management

Most growers who spend time in their coffee fields readily recognize differences in the growth habit of weeds. They know which weeds are perennials or annuals. They have observed how roots grow and how different weeds resprout after weeding. They also have practical experience that different weeds cause different damage to crops. Even though this knowledge seldom plays a role in current weed control decisions, it provides an important basis for working with growers on the reorientation of weed control to ground cover management.

The participatory training on ground cover management forms part of an integrated approach to better farmer decision-making in coffee pest management. This integrated process has a number of characteristics. First, groups of farmers come together to share their experiences, what they have observed and what they have tried, to analyze their current situation and prioritize their problems, and to propose changes to be tried. Second, meetings are held in the field where growers feel more comfortable and where live study materials, weeds of different types, insects, damaged coffee leaves or fruits, abound. Third, the dialogue with growers incorporates biological and ecological concepts which draw on what they have observed and can observe and which strengthen their ability to analyze why pest problems vary within fields, from field to field, and from year to year. The field technician acts as facilitator of the dialogue incorporating participatory field exercises and data which growers bring from their own fields to fill in gaps in the group's knowledge. Fourth, the dialogue is organized to link what growers know and have observed to the decisions they make and the practices they use in pest management. Making decisions and carrying them out in a timely fashion is emphasized. At the end of each session growers agree to take data in their fields and try different practices which they report on in the following session. The field technician also may describe alternative practices with which the group is not familiar. Fifth, training sessions form a sequence in time which permit the group to follow-up their own discussions systematically with each session coinciding with phenological and critical decision-making moments in the crop cycle.

The goal of participatory training in weed control is to assist growers in organizing their practical observations and experience with different weed types into an experimental ground cover management plan for their coffee fields. Two components of the training will be discussed here, methods for observing and quantifying weeds and the organization of a dialogue to connect observations to decision-making.

5.3.2 Methods to Observe and Count Weeds and Ground Cover

An important first step in the management of weed heterogeneity is observing which and how many weeds are present in coffee fields. On the short term the decision is whether weeds need to be controlled based on an assessment of their likely current competition with the crop. Also on the short term, but on the medium term as well, is an assessment of the proportions of different weed types by growth habit. This information is useful for a discussion about how each group of weeds may potentially compete with the coffee plants, for an analysis of how current control methods correspond to the problem weed groups in the field, for the identification of new possible practices to favor certain groups and reduce other weed types, and for monitoring changes in the ground cover in response to changes in practices.

Weed sampling methods for routine use by growers are relatively uncommon, especially for perennial crops which may have a relatively continuous ground cover with as many as 20–30 species within a few hectares. Sampling methods for grassland and range vegetation provided some initial guidelines (Evans and Love, 1957; Johnston, 1957) which were then modified and field tested in a formal study (Staver, 1993) and have since been used with numerous groups of growers and field technicians in Nicaragua and other countries of Central America.

5.3.2.1 Shoe tip monitoring of weed species and ground cover type

In this method the grower walks a 1–3 hectare coffee field during 30–60 minutes observing the type of weed or ground cover in an imaginary point the size of a pencil point adjacent to a predefined spot on their shoe tip. Upwards of 150–200 points are observed, distributed throughout the field. This can be achieved by walking every 3–10 interrows and observing the ground cover approximately every 3–10 footsteps. At each observation the grower identifies the ground cover present in the imaginary point. Seven groups of weeds by growth habit and life cycle are commonly designated: annual broadleafs, perennial broadleafs, grasses, sedges, vines, broadleaf ground cover weeds, grass ground cover weeds. This grouping correlates both with the likely competitive effect of the weeds on the coffee plant as well as with weed response to different management practices including herbicides. However, the groups can be modified to suit the needs of a particular zone or group of growers. In addition to seven weed groups, the grower also may observe bare soil, tree leaf litter, or dead weed litter. A mark is made on the sampling format in the appropriate category. Before moving to the next observation spot, the grower also observes the coffee plant on one side or the other to determine whether a vine weed is in close proximity to or on the plant and also marks the appropriate category. At the end of the walk the observations are summed up by category and divided by the total observations to calculate the percentage of each weed or ground cover type in the field as well as the percentage of coffee plants with vine weeds (Figure 9). Experience has shown that this observation method is easiest to do when weeds are from 5–10 cm in height.

5.3.2.2 Matrix of cover type by height

While the shoe tip method provides an assessment of the proportions of weed types and ground cover, it is not suitable for short-term decisions to determine how soon weeds need to be controlled. As a complement to the first method, a faster and simpler second method was developed using a matrix (Figure 10) which quantifies ground cover in broad categories as well as by actual height.

Field name: La Luna	**Observer:** Juana Venado	**Date:** 13 June 95	
Weed groups or ground cover observed	**Mark here each time weed or ground cover observed**		%
perennial broadleafs	///////// /////	15	9
annual broadleafs	///// ///// ///// ///// //	22	14
vines	///// ///// ///// //	17	11
grasses	///// ///// ///// ///// ///// ///// ///// ////	39	25
sedges	///// /	6	4
broadleaf cover weeds	//	2	1
grass cover weeds	///// ////	9	6
tree leaf litter	///// ///// ///// ///// ///// ///// ///// //	37	24
dead weed litter			
soil	///// ////	9	6
Total		156	100
vines near			
or on coffee bushes	///// ///// //	12	8

FIGURE 9. The data tally sheet for the shoe tip weed and ground cover monitoring method illustrating how observations might be noted and tabulated.

The grower walks their field during 20–30 minutes observing a circle 20–30 cm at their shoe tip. Approximately 50–100 circles should be sufficient to evaluate the current state of weeds and ground cover in the field. This number of points can distributed throughout the field by walking every 5–15 rows and observing the ground cover every 5–15 steps. At each observation spot the grower determines the average state of the ground cover by five categories — ground cover weeds, leaf litter, bare soil, normal weeds, and severe weeds — and also quickly determines the height with reference to their height (Figure 10). Determining the average ground cover for the circle is generally favored by the tendency of weeds to aggregate. When several types of cover occur in a single circle the grower decides whether to emphasize the positive or negative. After marking the appropriate box on the matrix, the grower also observes whether there are vine weeds growing on the coffee plant to their right or left and makes a mark in the box. At the end of the walk the state of the field can be analyzed visually. If most of the marks are in the lower left corner, the field is in excellent shape both in terms of soil protection and minimal actual and potential weed competition. If most of the marks are in the lower right, weeds are still small, but are likely to need control within a few weeks. If the marks are primarily in the upper right, immediate weed control is urgent. If the marks are spread throughout the matrix, then the field is highly heterogeneous in its weeds and ground cover, indicating that different practices should be directed at different parts of the field. This method can be used at any stage of weed growth in the field.

In practice both methods have found acceptance with field technicians and have been used in training sessions with growers. Growers have also practiced the shoe tip method in their own fields. The methods are not easily employed by illiterate growers. There have also been some difficulties with both methods in maintaining the size of the sampling point and in the basic arithmetic for calculating percentages. A number of studies would be useful to improve the applicability of the ground cover sampling methods in the participatory training process. Do growers have their own methods to observe weeds? Do they form categories? How do they decide when to control weeds? These studies also may need to take into account field workers and field supervisors in the case of larger farms. In addition, simple and reliable sampling methods are needed for tree shade measurement which estimate not only average shade levels, but also the heterogeneity in its distribution.

<table>
<tr><td>waist</td><td></td><td></td><td></td><td>///// ////</td><td></td></tr>
<tr><td>knee</td><td></td><td></td><td></td><td></td><td>///</td></tr>
<tr><td>calf</td><td></td><td></td><td></td><td></td><td>/////</td></tr>
<tr><td>ankle</td><td>///</td><td></td><td></td><td>///// ///// ///</td><td>///// ///</td></tr>
<tr><td>shoe sole</td><td>///// ////</td><td>///// ///// //</td><td>///// ///</td><td></td><td></td></tr>
<tr><td>WEED HEIGHT</td><td>ground cover weeds</td><td>leaf litter</td><td>bare soil</td><td>normal weeds</td><td>severe weeds</td></tr>
<tr><td colspan="6">TYPE OF GROUND COVER</td></tr>
</table>

FIGURE 10. Matrix of cover type by height for weed and ground cover sampling. Each observation is located by both the type of cover which is on the horizontal axis as well as by height which is on the vertical axis. The pattern of observations in the matrix shows a very heterogeneous coffee field with patches of problem weeds which need immediate control as well as patches of ground cover which protect the soil without competing with the coffee.

Similarly, a better understanding of how growers estimate tree shade levels would also improve the participatory training process.

5.3.3 Growers Discuss, Study, and Try Ground Cover Management

Based on the characteristics of participatory training in coffee pest management described in Section 5.3.1 and our accumulated understanding of ground cover management through field surveys and small and large plots studies, we formulated key elements for participatory training in weed control. These included: (1) draw on grower experience with weed types and formulate agreed-upon groups of weeds by growth habit; (2) emphasize not only weed competition, but also soil conservation, and their relationship to weed and ground cover categories; (3) stimulate an interest in weed sampling and train in simple methods described above; (4) analyze how weed control methods affect the different weed types; and (5) link the discussions on weeds and ground cover to management decisions and possible testing of alternative approaches to weed control. These elements were translated into exercises to be carried out with groups of small coffee growers which will be described in the following sections.

5.3.3.1 Growers analyze good and bad weeds in a matrix

A dialogue on weed control is not usually the most important theme for groups of coffee growers in Nicaragua who are often more interested in leaf diseases and coffee berry borer. However, once the group has expressed interest in weed control, the field technician arranges a workshop to coincide with the early rainy season when weed control is one of the priority tasks. To set the tone for the session and create a common starting point, the group first discusses what the word weeds means to them and draws up a list of other words with similar meanings. They also explain the different problems they have with weed control which provides the field technician the opportunity to make reference to their problems throughout the rest of the dialogue.

From there the group moves to a nearby coffee field where growers are asked to roam the field and bring back three different weeds, one very damaging to coffee, one not so damaging, and one which does not affect the coffee plant. The extensionist facilitates a discussion of the characteristics of each weed and how the different weeds might be grouped according to their growth habit and potential damage to coffee plants. The weeds are placed in piles on a large white sheet which has been laid out with a matrix. The columns correspond to weed types and three rows are used to illustrate each weed type in terms of its growth and root system, how it reproduces (seeds and/or sprouts), and its advantages and disadvantages (competition, soil protection, etc.). Symbols like seeds or plant parts or red and green buttons for advantages and disadvantages are used instead of lettering to fill out the matrix. The weed groups mentioned in Section 5.3.2.1 are commonly formed, although growers may decide to form other groups. Certain weed groups also may be absent from the field. Once the weed groups have been formed, growers mention other types of ground cover in their coffee fields like leaf litter and dead weed litter which are also incorporated into the matrix.

The group steps back from the matrix to reflect on the information and draw the conclusion that there are several types of weeds and ground covers and that it should be possible to promote those weeds and ground covers which offer advantages like soil protection without the disadvantages of competition with the coffee plants.

5.3.3.2 Growers quantify weeds and ground cover

The extensionist then proposes to the group that although they know more about the weeds and ground cover in the field, they don't really know how much of each type there is and proposes an exercise in sampling using the shoe tip method. Groups of two–three growers walk a part of the field observing the weed and ground cover in 20–40 points. Each group adds up and then presents their results in a large group session. The facilitator uses the opportunity to discuss the variability within the field and the importance of a well distributed sample. The group also reviews the arithmetic of each small group, calculates the percentages of the different categories for the field, and then reviews the results. Which weed group covers most? How much of the field has the soil protected without weed competition? Which is the weed group most difficult to control? Which weed groups should be reduced in presence?

5.3.3.3 Growers analyze current weed practices and propose changes

To analyze why the field has the weeds it does and to identify the possible short and medium term changes in field management to improve the ground cover, the group facilitator uses a set of cards on stiff paper with drawings of the different weed groups. These cards help the group visualize and discuss how certain practices favor certain weeds and disfavor other weed types. Cards can be partially covered to illustrate slashing to different stubble heights or turned over to symbolize elimination by hand weeding or herbicides. Cards with weeds in flower are used to emphasize the importance of seed production in future weed presence. Cards can be moved laterally to illustrate one weed type increasing at the expense of another. In this way the current practices can be analyzed with the group to predict how the weed complex might be changing. Are the worst weed groups likely to increase or decrease? Are the practices favoring or disfavoring ground cover weeds? Similarly, the cards can be used to generate a list of alternative practices which reduce the spread of or eliminate the undesirable weed types and promote the spread of cover weeds.

This structured brainstorming sets the stage for the final group discussion. Who will try weed sampling in their own fields? How many of the alternative practices for ground cover management will be tested in the action-research plot where the group does its field training and tries practices not only with ground cover, but also with coffee berry borer and leaf diseases? Who will try an alternative ground cover practice in their field? When will the next meeting be?

5.3.3.4 Growers continue to analyze ground cover management

Follow-up to the starting dialogue described above is essential and should be an on-going theme with each group of coffee producers. The potential for and the nature of the follow-up depend on the interest and commitment shown by different participants in the farmer group. Field experience in Nicaragua has shown that groups are very variable both in data taking and the testing of alternative practices. Follow-up should draw on the learning experiences of the group itself in such a way that those most dedicated growers continue to advance, that those who are just beginning to understand or become interested have the opportunity to reinforce their advances, and that the group stays motivated to meet again and improve their coffee pest management decision-making.

6. CONCLUSIONS — MULTIPLYING AND IMPROVING THE HETEROGENEITY-BASED PARADIGM

Will the heterogeneity-based paradigm which we have described here as being built on ecological concepts prove useful for the world's coffee growing households? In closing, we suggest that it should be evaluated in two ways.

First, how easily can it be made accessible to an ever increasing number of farm households? In Nicaragua several hundred extensionists and perhaps several thousand growers have been improving their observation skills of coffee pests and abilities in ecological reasoning to understand pest variability and to improve management. Our participatory training methods and ecological information base are still rudimentary, yet we can see the potential for each participant to become a more effective multiplier of the approach both within formal training and extension programs and through informal rural communication networks. There is a clear need for a better understanding of how different types of information flows in rural communities from farmer to farmer, within farm households, and among all members of rural communities. This understanding could provide the foundations for farm household- and rural community-based multiplication and strengthening of the heterogeneity-based paradigm.

Second, will the paradigm serve scientists, extensionists, and growers in developing a better understanding of the ecological bases and interactions in coffee and perennial crops in agroforestry? Based on our limited beginning in pest management, we perceive great potential to develop positive feedback in the coffee ecosystem, among coffee, multi-use trees, the soil, and the wide diversity of associated macro- and micro-flora and fauna, for higher, more efficient, and sustainable production. This potential can begin to be realized through the interaction of multiple and overlapping learning approaches (Liebman et al., 1998), which may be applicable in other plantation crops and agroforestry systems. These include, as described in this chapter, (1) groups of farmers, extension, and a multi-disciplinary research team in a co-learning process in the time and space of farmers' planning and decision-making; (2) simple methods to observe, measure, and understand the ecological basis of the heterogeneity of system components such as weeds in farmers' fields; and (3) small plot and long term systems studies under more controlled conditions to study ecological relationships and their expression in practical management options.

REFERENCES

Aguilar, A., Aguilar, V., Somarriba, S., and Staver, C., 1997, *Selective weed management for soil conservation in young coffee: an evaluation of mechanical-chemical and mechanical systems with and without Arachis pintoi* (in Spanish), Proceedings of 18th Latin American Coffee Symposium, San Jose, Costa Rica, 85–92.

Aguilar, V. and Staver, C. 1997, *The uptake and return of nutrients in weed biomass under three ground cover management systems in coffee in the Pacific Zone of Nicaragua* (in Spanish), Proceedings of 18th Latin American Coffee Symposium, San Jose, Costa Rica. 115–124.

Alan, E. and Barrantes, U., 1988. *Allelopathic effect of Gliricidia sepium on the germination and initial growth of some tropical weeds* (in Spanish), Turrialba, 38: 271–278.

Barradas, V. and Fanjul, L., 1986, "Microclimatic characterization of shaded and open-grown coffee (*Coffea arabica* L.) plantations in Mexico," *Ag. For. Meteor.* 38: 101–112.

Bimbao, M., Lopez, T., and Lightfoot, C., 1995, "Learning about sustainability," *ILEIA Newsletter*, 11(2):28-30.

Borlaug, N. and Dowswell, C., 1994, "Feeding a human population that increasingly crowds a fragile planet," *Supplement of Transactions of 15th World Congress of Soil Science*, International Society of Soil Science, Acapulco, Mexico, 3–4.

Bouharmont, P., 1978. *The use of cover crops in robusta coffee in Cameroon* (in French), Cafe, Cacao, *The*, 22: 113–138.

Bouharmont, P., 1979. *The use of cover crops and mulch in arabica coffee in Cameroon* (in French), Cafe, Cacao, The, 23:75-102.

Bradshaw, L., 1993, *Perennial Coffee Crops in Nicaraguan Coffee Orchards: Mechanisms and Manipulations of Plant Competition*, Ph.D. dissertation, University of California, Davis.

Bradshaw, L. and Calderon, M., 1993, *Preliminary tests on the control of Meloidogyne incognita in young coffee plants with two leguminous cover crops* (in Spanish), 33rd Meeting of American Plant Pathology Society-Caribbean Division, San Salvador, El Salvador, Sept. 26–29.

Bradshaw, L. and Lanini, T., 1995, "Use of perennial cover crops to suppress weeds in Nicaraguan coffee orchards," *International. J. Pest Mgt.*, 41(4):185-194.

Broughton, W. 1977, "Effect of various covers on soil fertility under *Hevea brasiliensis* Muell. Arg. and on growth of the tree," *Ag. Ecosystm. Env.*, 3(2):147-170.

Budelman, A., 1988, "The performance of the leaf mulches of *Leucaena leucocephala*, *Flemingia macrophylla* and *Gliricidia sepium* in weed control," *Agrofor. Sys.* 6:137-145.

Bugg, R. and Waddington, C., 1994, "Using cover crops to manage arthropod pests of orchards: a review," *Agri. Ecosys. Env.*, 50:11-28.

Carvajal, J., 1994, *Cafeto: cultivo y fertilizacion*, International Potassium Institute, Switzerland, 76.

Dominguez, J. and de`la Cruz, R., 1991, "Nutrient competition by *Arachis pintoi* Pinto as a cover crop during the establishment of peach palm *Bactris gasipaes* H.B.K." (in Spanish), *Revista de Manejo Integrado de Plagas*, 18: 1–7.

Evans, R. and Love, R., 1957, "A step-point method of sampling — a practical tool in range research," *J. Range Mgt.*, 10: 208–212.

Fernandez, C, "Sustainable coffee production" (in Spanish). 1990, *Boletin Promecafe-IICA*, April–July.

Firth, D. J. and Wilson, G. P., 1995, "Preliminary evaluation of species for use as permanent ground cover in orchards on the north coast of New South Wales," *Tropical Grasslands*, 29: 18–27.

Garcia, B., Cheves, L., Valverde, B., and Garita, I., 1996, "The establishment of six living covers in a young coffee (*Coffea arabica*) plantation in Juan Vinas, Costa Rica," *Memorias VI Congreso Internacional de Manejo Integrado de Plagas*, Mexico, 102.

Goldberg, A. D. and Kigel, J., 1986, "Dynamics of the weed community in coffee plantations grown under shade trees: effects of clearing," *Is. J. Bot.*, 35:121-131.

Greenberg, R., Bichier, P., Cruz Angon, A., and Reitsma, R. 1997, "Bird populations in shade and sun coffee plantations in central Guatemala," *Cons. Biol.*, 11(2): 448-459.

Herrera, I., 1995. *The effect of leguminous cover crops for the control of plant parasitic nematodes in coffee* (in Spanish), M.Sc. thesis, CATIE, Turrialba, Costa Rica.

Hilje, L., 1992, "The biology and ecology of rodent pests of Costa Rica" (in Spanish), *Revista Manejo Integrado de Plagas*, 23: 17–25.

Hogue, E. and Neilsen, G., 1987, "Orchard floor management," *Hort. Rev.*, 9:377-430.

Jamison, D. and Lau, L., 1982. *Farmer Education and Farmer Efficiency*, Johns Hopkins Press, Baltimore, MD.

Johnston, A., 1957, "A comparison of the line intercept, vertical point quadrat, and loop methods as used in measuring basal area of grassland vegetation," *Can. J. Plant Sci.*, 37: 34–42.

Kenmore, P., 1991, *Indonesia's IPM-A Model for Asia.* FAO, Manila.

Kimemia, J. K. and Njoroge, J. M., 1988, "Effect of shade on coffee — A review," *Kenya Coffee*, 53: 387–391.

Liebman, M., Mohler, C., and Staver, C., 1998, "Knowledge, science, and practice in ecological weed management: research-extension-farmer interactions," in *Ecological Management of Agricultural Weeds*, Cambridge University Press, Cambridge, U.K.

Mendez, E., 1992, *Effect of different ground covers on the survival of coffee berry borer Hypoyhenemus hampei after coffee harvest* (in Spanish), MSc thesis, CATIE, Turrialba, Costa Rica.

Moguel, P. and Toledo, V., 1996, "Coffee in Mexico — sustainability and native and peasant resistance" (in Spanish), *La Jornada del Campo*, Sept. 26.

Monterroso, D. 1993, "A proposal for the low-input management of the primary leaf diseases of coffee" (in Spanish), in *Resumenes de la Semana Cientifica*, December 8–10, CATIE, Turrialba, Costa Rica.

Nestel, D. and Altieri, M., 1992, "The weed community of Mexican coffee agroecosystems: effect of management upon plant biomass and species composition," *Acta Ecologica*, 13: 715–726.

Njoroge, J. M., 1994, "Weeds and weed control in coffee," *Exp. Agri.*, 30: 421–429.

Perfecto, I., Rice, R. Greenberg, R., and Van der Voort, M., 1996, "Shade coffee: A disappearing refuge for biodiversity," *Biosci.* 46:598-608.

Prokopy, R., 1994, "Integration in orchard pest and habitat management: A review," *Agri. Ecosys. Env.*, 50: 1–10.

Ramos, L., Anaya, A., and Nieto de Pascual, J., 1983, "Evaluation of allelopathic potential of dominant herbaceous species in a coffee plantation," *J. Chem. Ecol.*, 9: 1079–1097.

Reynolds, J., 1991, *Soil nitrogen dynamics in relation to groundwater contamination in the Valle Central, Costa Rica*, Ph.D. dissertation, University of Michigan, Ann Arbor, MI.

Rice, R., 1990, *Transforming agriculture: The case of coffee leaf rust and coffee renovation in southern Nicaragua*, Ph.D. Dissertation, Geography Department, University of California, Berkeley.

Rice, R. and Ward, J., 1996, *Coffee, Conservation, and Commerce in the Western Hemisphere*, Smithsonian Migratory Bird Center, National Resources Defense Fund, Washington, DC.

Rivas, C., 1994, *Mulches from pruned leaves of shade trees for weed control in coffee* (in Spanish), undergraduate thesis, National Agrarian University, Managua, Nicaragua.

Sampson, H., 1928, "Cover crops in tropical plantations," *Tropical Horticulturalist*, 71: 153–170.

Savory, A., 1988. *Holistic Resource Management*, Island Press, Washington, DC.

Siman, J. and Staver, C., 1992, *Working with limited resource farmers in the generation and validation of coffee pest management options in Nicaragua* (in Spanish), 15th Symposium on Latin American Coffee Production, Xalapa, Mexico.

SIMAS (Mesoamerican Information Service for Sustainable Agriculture), 1996, "Nicaragua: Experiences with IPM in the CATIE-INTA/IPM Project," in *New Partnerships for Sustainable Agriculture*, ed., L. Thrupp, World Resources Institute, Washington, DC, 75–84.

Staver, C., 1993, *Weed scouting in coffee for growers: methods for floristic composition and height/cover* (in Spanish), 16th Latin American Coffee Symposium, Managua, Nicaragua.

Staver, C. 1994, *Post-modern coffee production: Instead of convincing producers to change the cropping practices, provide them with elements to improve their decision-making* (in Spanish), SIMAS, Managua.

Staver, C., Guharay, F., Jimenez, C., Monterrey, J., Monterroso, D., Calderon, M., Mendoza, R. Aguilar, A., and Mendez, E. 1996, *Dialogue with coffee producers about the biology and ecology of diseases, berry borer, weeds and dry season pests in coffee: methods and procedures for ecological pest management* (in Spanish), Proceedings VI International IPM Congress, Acapulco, Mexico, Sept. 30–Oct. 4, 101.

Staver, C., Bradshaw, L., and Somarriba, S., 1993, *Ground cover management in Central American shaded coffee: selective weeding, mulch, and perennial living covers*, Agronomy Abstracts–American Agronomy Society Annual Meetings, November 7–12, 61.

Staver, C., 1996, "*Arachis pintoi* as a cover crop in coffee-research results and farmer experience in Nicaragua during 1990–95" (in Spanish), in *Experiencias Regionales con Arachis pintoi y Planes Futuros de Investigacion y Promocion de la Especie en Mexico, Centroamerica y el Caribe*, ed. P. Argel, CIAT, Cali, 150–170.

Useem, M., Setti, L., and Pincus, J., 1992, "The science of Javanese management: Organizational alignment in an Indonesian development program," *Public Administration and Development*, 12:447-471.

Vado, N., 1996, *The effect of leaf mulches from three shade trees for weed control in coffee* (in Spanish), undergraduate thesis, National Agrarian University, Managua, Nicaragua.

Vandermeer, J., 1996, "Ecological understanding and complexity for integrated pest management in the post-modern world" (in Spanish), *Revista de Manejo Integrado de Plagas*, 41: 37–44.

Voisin, A., 1988. *Grass Productivity*, Island Press, Washington, DC, 177.

Westoby, M., Walker, B., and Noy-Meir, I., 1989, "Opportunistic management of rangelands not at equilibrium," *J. Range Mgt.*, 42:266-274.

Wiersum, K., 1984, "Surface erosion under various tropical agroforestry systems," in *Symposium on effects of forest land use on erosion and slope stability*, Eds. C. O'Loughlin and A. Pearce, The East-West Center, Honolulu, HI, 231–239.

Young, A., 1989, *Agroforestry for Soil Conservation*, Science and Practice of Agroforestry #4, CAB International/ICRAF, Wallingford, U.K.

5 The Science and Practice of Black Walnut Agroforestry in Missouri, U.S.A.: A Temperate Zone Assessment

Harold E. Garrett and Larry S. Harper

1. INTRODUCTION

The Missouri agroforestry program represents a successful merging of traditionally conflicting cultural practices — those associated with conventional agriculture and forestry. Traditionally, forested lands in Missouri have been cleared to accommodate row cropping or pasture establishment. In particular, vast areas were converted from forest to pasture during the 1960s and early 1970s. With the proper financial incentive, however, many landowners are willing to reestablish trees on lands that were once forested. In recent years, a movement has been observed in Missouri to reforest land that was previously held in agriculture. Approximately 120,000 new forested hectares were established between 1972 and 1989 (Hahn, 1991). Most of the conversion back to woodlands has occurred without government support and is an indication of the strength of a movement of the people to reforest lands perceived as best serving Missouri's needs in a forested state. While the reasons for reforestation vary from landowner to landowner, all have a resource, economic, environment or social underpinning.

As a consolidated practice, agroforestry is not new but is relatively recent and unique to the state of Missouri and the midwestern United States in general. The Missouri agroforestry program, which requires that tree rows be widely spaced and horticultural or agronomic crops be planted in

1-56670-294-1/99/$0.00+$.50

the alleyways, provides owners who are interested in placing trees back on the land, a financially viable alternative to conventional cropping practices. The system is enhanced in its flexibility when nut trees (e.g., walnut (*Juglans nigra* L.) or pecan (*Carya Illinoensis* (Wanyenh.) K. Koch)), are used due to the value of the nuts and the advantages of creating an earlier cash flow from the tree than is possible when employing species that have timber value only.

2. HISTORY

Alley cropping in Missouri, as a recognized form of agroforestry, began during the early 1970s. It was at that time that Hammons Products Co. of Stockton, MO initiated plantings of black walnut at spacings sufficiently wide to accommodate the physiological needs of companion crops grown in alleyways between the rows.

Hammons Products Co. is the single largest processor of eastern black walnut nuts in the United States. But in spite of their long history, they have and continue to be totally reliant upon nut collection from the wild population of walnut distributed throughout the eastern half of the United States (Jones et al., 1989). Demand for their products, both nutmeat and industrial shell, has increased several fold since the company's inception. With this increase has come shortfalls in the availability of the nut resource. Within a wild population not under management, yields from nut trees can be expected to fluctuate due to the normal variation in yearly weather patterns and the alternate-bearing (i.e., heavy nut crops occur at intervals of two, three or more years) characteristic of nut trees. Such has been observed to be the case for nut production in black walnut. Jones (1975), in a study of nuts purchased by Hammons Products in 17 southwestern Missouri counties from 1960 to 1974, found a high correlation between the number of nuts available for purchase, the previous year's nut crop, and rainfall patterns from June through August of the preceding year. During many years, the availability of the wild nut crop was substantially less than market demand. It was because of this lack of reliability of the native crop that a decision was made to develop practices that would encourage landowners to plant and manage black walnut. Furthermore, it was reasoned that the practices developed had to provide a cash flow during the early years of plantation development while the landowner was awaiting nut or wood production from the trees. Logic suggested that the greatest probability for success would come from planting tree rows at wide spacings and growing conventional agricultural crops (i.e., corn (*Zea mays* L.), soybeans (*Glycine max* L.) etc.), already familiar to landowners, between the rows.

In what has become known as the "Missouri System of Agroforestry," tree species are planted in geometric patterns that permit the successful growing of various forms of cash crops within alleyways.

The Missouri program emphasizes the use of alley cropping for both production and conservation benefits. As is the case with all U.S. agroforestry practices, design is accomplished with emphasis on four key "i" words. Trees and crops are *intentionally* managed as a whole unit. The practices are *intensively* managed to maintain their productive and protective functions. The biological and physical *interactions* between the tree and crop components are manipulated to enhance the production of more than one harvestable component at a time. And, finally, the tree and crop components are structurally and functionally combined into a single *integrated* management unit (Slusher et al., 1997).

3. BIOLOGICAL ASPECTS

3.1 Tree Crop

3.1.1 Tree Selection

To accomplish specific goals in alley cropping requires many important decisions to be made regarding the trees. However, none is more important than the selection of the proper species. Many tree species of high economic value are available for use in alley-cropping practices in Missouri. However, selection of the "best" species must be made with knowledge of the site to be planted and the alley-cropping practice to be implemented. The physiological requirements of the tree species must be met by the site, and the companion crop's physiological requirements must be accommodated by the tree.

Because of its potential value for nuts and wood, and its adaptability to a wide range of soil types, black walnut has become the preferred species for planting in Missouri. However, pecan is emphasized on soils along rivers and streams that are subject to occasional short-term overflows. Oaks, especially the high value lumber species, white (*Quercus alba* L.) and red (*Q. rubra* L.), along with a number of other conventional lumber species such as white ash (*Fraxinus americana* L.), are planted on well-drained soils that do not satisfy the needs of walnut and pecan, but still have the capability to support good tree growth. On soils that are more poorly drained in bottomlands, several fast-growing hardwood species such as cottonwood (*Populus deltoides* Bartr.) and silver maple (*Acer Saccharinum* L.) are grown and marketed for lumber and pulp. Futuristically, these species may take on even greater value in the oriented strand board (OSB) market (Spelter, 1996) and for energy production (Hall et al., 1993).

Because an early decision was made to use eastern black walnut as the model species in Missouri alley cropping, and, therefore, the greatest amount of research has been conducted on this valuable species, black walnut will be emphasized in illustrating Missouri agroforestry.

3.1.1.1 Black Walnut's Adaptability to Alley Cropping

Assuming that black walnut's physiological needs are satisfied by the site and walnut and the desired companion species are compatible, walnut is an excellent choice for an alley cropping program and perhaps most closely represents the "ideal" alley cropping tree species for Missouri. It is the single most valuable timber species in the United States and produces a marketable nut. Therefore, it possesses great potential for producing income. Walnut has growth and shade characteristics that are compatible with the growth of many companion crop species. It leafs out late in the spring and drops its foliage early in the fall. Even with its maximum leaf surface area, walnut admits sufficient light to satisfy the photosynthetic needs of many potential companion species (Hupe, 1980).

Black walnut's root system is also ideally suited for agroforestry. While walnut grows best on deep, well-drained soils with a neutral pH, root characteristics are similar over a fairly broad range of soil conditions (Pham et al., 1978; Yen et al., 1978). Walnut typically produces a deep taproot that may penetrate more than 2 m in the absence of physical barriers (Figure 1). While large lateral roots stay close to the surface, most of the smaller roots turn down sharply into the soil. Feeder roots (roots that are most active in water and nutrient uptake), normally occupy a zone 10–20 cm below the surface. This provides a shallow zone near the soil surface within which companion crops can compete for water and nutrients.

Black walnut offers incentives that few other species can match. The potential for high returns on wood from individual trees, combined with regular income from annual or periodic nut crops makes it an attractive investment species. When combined with the proper companion crop, practices can be created that are financially competitive with conventional agriculture, especially on marginal agricultural lands.

Once agreement has been reached on the species to be planted (based upon site compatibility and the landowner's needs), spacings are determined. Actual spacings between rows vary from a few m (4.5–6.0) to in excess of 30 m. Between-row spacings are dictated by the short- and long-term goals of the landowner, (e.g., emphasis placed on production vs. conservation benefits), the companion crop(s) of choice (shade tolerant or intolerant), and, the duration of each projected cropping regime. Even the width of available farm equipment must be considered if the efficiency of the program is to be maximized. Alleyway widths, ideally, should be a multiple of the width of the primary equipment to be used (e.g., combines).

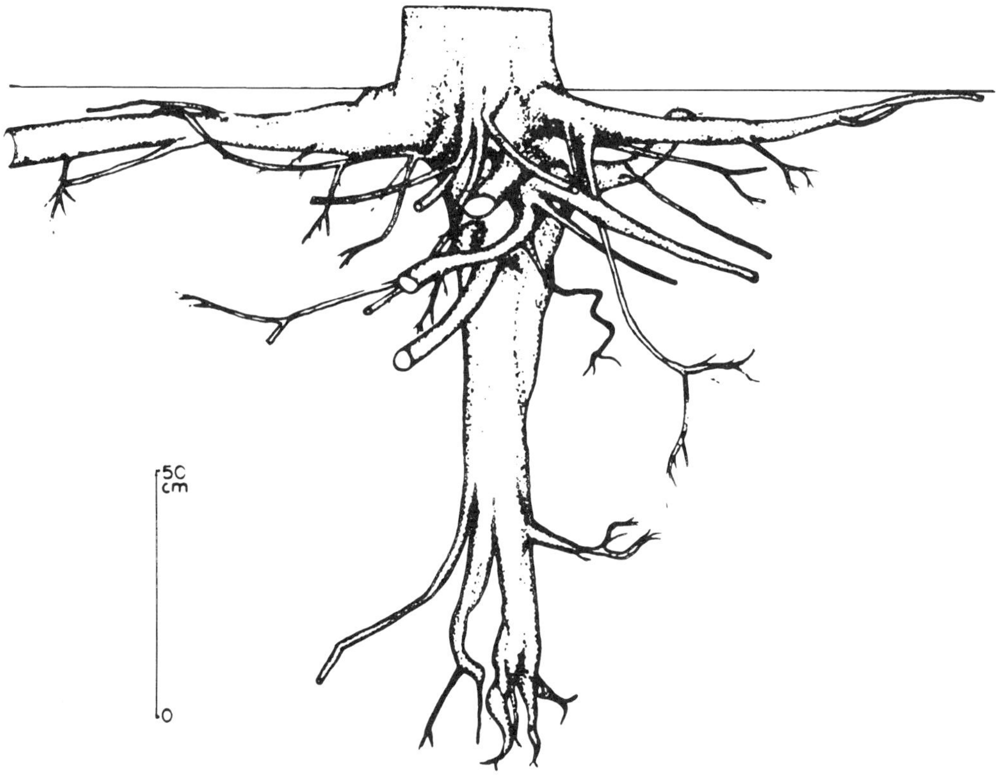

FIGURE 1. Sketch of a typical root system for a 30-year-old eastern black walnut grown in a deep fertile soil without physical obstructions.

3.1.2 Tree Management

Management of trees will vary greatly between landowners. However, black walnut, like most tree species, performs best when intensively managed. Research strongly suggests that weed control is a prerequisite to good growth and nut production in black walnut. While both mechanical and chemical control have proven to be beneficial (Garrett et al., 1992), because of the ease of application, chemical control is most commonly used in Missouri. Following planting, herbicides are applied in a circle around each seedling creating an opening 1 m or more in diameter or along both sides of a row of trees. The width of the control zone along the row will vary but should be a minimum of 1 m on both sides. As the tree crowns expand, widths of vegetation-free zones increase correspondingly. Because of its effectiveness on difficult-to-kill species like tall fescue (*Festuca arundinacea* Schreb.), Roundup™ in combination with one of several pre-emergent herbicides is used extensively throughout the state.

While alley-cropping fertilization prescriptions still are being researched, current recommendations for black walnut are similar to those for English walnut (*Juglans regia* L.) in California (Ramos, 1985). In English walnut, nitrogen has been reported to be the most limiting nutrient. Ramos (1985) reported that 185 kg of actual N must be added annually to each hectare to support satisfactory nut production and growth. Similarly, Wienbaum et al., (1990) suggested that to replace nitrogen removed from a hectare of English walnut orchard, between 138 and 185 kg must be added annually.

Significant increases in black walnut nut yields, however, have been observed in Missouri with the application of only a small percentage of the fertilizer recommended for English walnut by altering the timing of application. In a study by Jones et al. (1995), nut yields were 49% greater for trees fertilized in August with only 0.18 kg of NPK (13-13-13) per cm of tree diameter than for trees receiving the same fertilizer treatment in the early spring. In a follow-up study (Gray, 1997), nuts filled better and yielded a higher percentage kernel with late-summer fertilization than with spring fertilization.

While the biological reasons for the increased quantity of nut production with late summer fertilization is not fully understood, it may in part relate to the timing of pistillate flower development in black walnut. Schaffer et al. (1996), through light microscopic observations, found pistillate floral primordia in black walnut during the winter. Although it remains to be determined how soon after anthesis in the previous growing season these primordia appear, the organo-genetic sequence of pistillate flower formation appears to be very similar to that of the English walnut. It is known that the pistillate floral primordia in English walnut is produced in late-summer of the year prior to nut production. If this is also true for black walnut, late-summer fertilization might serve both to enhance the number of pistillate floral initials produced and nut retention by improving tree vigor (Gray, 1997).

Since it is a fundamental goal of Missouri alley cropping to provide a reasonable investment return for the landowner, both pruning and thinning become important considerations. The timing and intensity of thinnings vary with site and the alley-cropping practice (current and projected). Trees grow faster on good than on intermediate-quality sites, and assuming equal spacings between trees, thinnings will be required sooner on the good sites. Since walnut is somewhat shade intolerant, any shading can inhibit growth and reduce the financial returns. Therefore, thinnings are conducted on an ongoing basis. The final objective is to reduce, through thinning, the number of original trees to just enough to satisfy the environmental requirements of the projected alley crop while optimizing the financial gain from the tree component. A greater number of trees are retained where wood production is the priority than where nut production is the priority. If the projected companion crop requires shade (i.e., ginseng (*Panax quinquefolium* L.)) more trees are reserved than would be the case if alleyways are to be planted with light-demanding crops (i.e., forages, small grains etc.). Regardless of the timing of the thinnings in walnut alley cropping, they are designed to select trees for removal based upon stem quality (straightness, diameter, height and apical dominance) and nut production characteristics (precociousness, bearing regularity, percentage crackability, quality and quantity).

Contrary to the conventional approach of maximizing the clear length of a walnut log, agroforestry management in Missouri advocates shorter boles with larger crowns. Numerous financial analyses (Kurtz et al., 1991; Garrett and Kurtz, 1983) have demonstrated that internal rate of return and present net worth are maximized through the sacrifice of clear-log length in favor of greater crown area for nut production. Therefore, in Missouri alley cropping, walnut trees are pruned to a height of only 2.4 to 3 m on most upland sites (site index 16–21 m) and 3.7 to 4.9 m on bottom land, bench and higher quality upland sites (site index > 21 m). Where nut production is not a consideration, pruning heights range from 5 m to as much as 8 m.

3.2 Companion Crops

Alley cropping can be designed to accommodate the biological requirements of most crops from those requiring a deep shade to those requiring full light. The creation of the proper microenvironment and the timing of its creation are products of selecting the correct tree species and spacing. Therefore, with proper planning, companion crop selection is not dictated by the alleyway microenvironment, as that is under the control of the individual designing the practice. It instead becomes a function of complex socio-economic factors, land quality, personal interests, and the ability of the landowner to respond to constantly changing markets.

3.2.1 Row Crops

In Missouri, there are many examples of crops that have been planted in alleyways including Christmas trees, energy species, botanicals, raspberry (*Rubus idaeus* L.), blueberries (*Vaccinium* s), landscaping ornamentals, etc. However, the most common alleycrop plants are conventional row crops (i.e., corn, soybeans, winter wheat (*Triticum aestivum* L.)) and forages (grasses combined with legumes). In early research performed by Garrett and Kurtz (1983), upland and bottom-land sites were planted with walnut and the alleyways "dual" cropped with soybeans and winter wheat. Trees were spaced 3 m apart within rows and 12 m between rows providing 267 trees ha^{-1}. Soybeans averaged 1505 and 2007 kg per ha^{-1} on the upland and bottom-land sites, respectively, during the first five years. Wheat yields averaged 2571 kg ha^{-1} on both sites during the same time frame. While wheat yields changed little between years five and 10, soybean yields decreased by nearly 20%. Due to the percentage change in yields, soybean production was found not to be profitable after the tenth year under the conditions of the early work. While the percentage decrease in yield attributable to increased shade, competition for water and nutrients or allelochemical inhibition is unknown, controlling the competition through the pruning of tree roots or widening the spacing between tree rows extends the life of the cropping regime. Studies conducted in Indiana demonstrated a 62% yield increase in corn planted in the alleyways of eight-year-old black walnut from the severing of lateral roots growing into the alleys (Jose et al., 1995). Through the management of below-ground competition by root pruning beginning at the time of tree establishment, it may be possible to train roots in the plow layer to extend deeper into the profile (Kang, 1993). If so, this would greatly reduce the competition between the companion crop and the tree and allow the tree to accommodate its water and nutrient needs from depths deeper in the profile where water may be more abundant during summer droughts. If started at the time trees are planted, no permanent damage occurs to the trees and, ultimately, the trees benefit due to the improved vertical distribution of their root systems.

3.2.1.1 *Cool-Season Row Crops*

Cool-season row crops are in many ways better suited to Missouri alley cropping than warm-season crops. During early establishment, cool-season crops (i.e., winter wheat, barley (*Hordeum vulgare* L.) etc.), can be planted to within one meter of a tree row (Figure 2). This allows the landowner to minimize the hectareage removed from row crop production during the early years. With a spacing of 12 m between tree rows (probably the most common width of alleyways in Missouri), 87% of the land during the first few years following establishment is planted to the cool-season crop, only 13% is in trees. As the trees grow, the distance between tree and crop must be extended due to branches growing into the alleyways (Figure 3). However, during the initial years, cash flow is maximized by minimizing the land taken out of production. Furthermore, cool-season crops like winter wheat will normally compete less with trees for water than warm-season crops and create fewer management problems. The exception to this may occur during years of low spring rain. One study in Canada found significant reductions in black walnut growth when intercropped with small cool-season grains as compared to soybeans and corn. Growth differences were attributed to lower soil moisture availability in the late spring with cool-season grains than with the warm-season crops (Gordon and Williams, 1991). It is reasonable to assume, however, that if sufficient spring rains occur to recharge the soil profile, this would not be a problem.

3.2.1.2 *Warm-Season Row Crops*

Warm-season row crops must be established at distances from the trees that are two or more times greater than for cool-season crops (Figure 4). Since most tree species (black walnut included) do not compete well for moisture and nutrients, it is critical to the early growth and establishment of the tree component in any alley-cropping program to maintain a competition-free zone or, at the least, a zone of minimal competition near the trees. Since most weed species are also warm-season plants, sufficient space must be left between the trees and crop to permit weed control. While the

FIGURE 2. Winter wheat cropped to within one meter of two-year-old black walnut.

FIGURE 3. Winter wheat being harvested in 12 m alleyways between four-year-old black walnut. The space between trees and crop has been extended from one meter to a meter and a half to avoid branches becoming entangled in the harvesting equipment.

spacing may vary depending upon the method of control employed (i.e., chemical vs. mechanical), a minimal distance of 2.1 m is recommended under Missouri conditions. With tree rows spaced at 12 m, this leaves only 62% of the land for row-crop planting which can significantly alter income during the early years prior to the beginning of commercial nut production. Some savings in space may be possible by using a commercial "weed barrier" mulch and planting the tree seedlings directly into it.

Competition for water and nutrients is inevitable when trees and companion crops are planted together. However, such competition is especially compelling when both species are vying for a

FIGURE 4. Six-year-old black walnut planted on a 12 × 3 m spacing and alley-cropped with soybeans. A spacing of 2.3 m has been maintained between trees and crop since establishment to enable weed control.

resource that is in short supply. Missouri is noted for its late July-August droughts during which time moisture can be severely limiting. Black walnut, due to its deep rooting pattern and its ability to reduce water loss by dropping its leaves, is well adapted to Missouri's drought conditions. However, it is logical to assume that even black walnut's problems are exacerbated under drought conditions when planted with a companion crop.

Although there is a shallow zone near the soil's surface that contains a reduced number of black walnut roots, most row crops have root systems that plunge deeper into the profile and are in direct competition with the walnut. Differential responses, however, can occur. During the summer of 1980, a severe drought year in Missouri, milo (*Sorghum bicolor* (L.) Moench), planted in the alleyways of five-year-old black walnut (3 m × 12 m) revealed effects that obviously were related to competition for moisture. Plants in a circle formation around each tree were one week longer in heading out and the heads required 10 additional days to mature to the bronze color stage than for plants located in the center of the alleyway. In contrast, soybeans grown in the alleyways of walnut of the same age yielded 10% more beans than soybeans grown in the open (H. E. Garrett, unpublished data). Differences in responses between the milo and soybeans may have been due to inherently different root systems or differences in shade percentages of the two sites. The greater shade associated with the soybean site may have led to the soybeans being under less heat and moisture stress than the milo. Irrespective of the problems associated with planting warm-season row crops, many Missourians use them with trees as a transition between a monoculture of row crops and a combination of trees and forages.

3.2.2 Forage Crops

Missouri is the second leading cow-calf state in the United States behind Texas. Because of this, there is considerable potential for alley cropping with forages or converting alley cropping practices to silvopastoral management. Alley cropping plantings can, and often do, take on a livestock dimension in Missouri. In the early years of an agroforestry program, a landowner might be inclined to row crop or cater to a local or regional specialty market by growing landscaping species, small fruits or some other commodity with trees. Such practices are well within the definition of alley cropping. However, if the species grown are shade intolerant, at some point in time, unless the initial spacing between rows is quite wide, conversion to more shade tolerant crop species is required. At this juncture, many Missouri landowners make the decision to switch to a forage cover crop. If the landowner hays the alleyways, as with growing any other agronomic or horticultural crop, the practice is alley cropping. However, if the alleyways are grazed, regardless of the animal type used, an entirely new agricultural dimension has been added and the landowner has converted an alley-cropping program to silvopastoral management, another of the five widely recognized agroforestry practices in North America (i.e., alley cropping, silvopastoral management, windbreaks, riparian buffers and forest farming). Similarly, if trees are established in an alley-cropping configuration, forages planted and grazed from the beginning, the practice is silvopastoral management even though the geometric configuration of the planting is that of alley cropping. This distinction is necessary due to the vast differences between the management requirements of livestock and plant crops and because all occur in Missouri agroforestry.

The prominence of Missouri in cow-calf operations is not happenstance. Missouri has an abundance of land that is well-suited to forage production, nearly 5.2 million ha of nonfederal pasture and rangeland combined. More than 2.8 million of these hectares would readily lend themselves to alley cropping or silvopastoral management (Garrett et al., 1994). Adding trees to pastures has the potential to yield additional income from wood and/or nuts while enhancing wildlife, biodiversity and the aesthetic value of the land. Furthermore, Missouri has nearly 2.4 million ha of cropland with an erodibility index (EI) greater than 8, much of which would better serve the state and landowner in grass. Approximately half the hectares are recommended for forestry plantings and could be placed in trees and grass in support of the cow/calf industry.

Since most forage species are adapted to open fields and full sunlight, landowners interested in converting pure pastures to alley cropping or silvopastoral management must do so with an eye to the future. During the early years following the addition of trees, shade will not be a consideration. However, as the trees grow, new microenvironments are established and forage species that are shade intolerant will soon disappear. Therefore, shade tolerance is one of the most important characteristics to consider in selecting a forage for either alley cropping or silvopastoral management. Unfortunately, since forages are normally managed under open conditions in Missouri, reliable data on the shade tolerance of many forage species are lacking. Tall fescue is, however, known to be shade tolerant and Missouri boasts of having an estimated 2.4 million hectares planted to this species. Allard et al., (1991), demonstrated that tall fescue shoot growth was similar at full sun or 40% shade. Orchardgrass (*Dactylis glomerata* L.), perhaps the second most widely planted grass species in Missouri pastures, also performs well under trees, as its name implies. Blake et al. (1966), reported that neither yield nor persistence of orchardgrass were affected when grown at 1/3 incident sunlight. Among the legumes commonly grown in Missouri, red clover (*Trifolium pratense* L.) is rated as being shade tolerant and has been shown to saturate at relatively low light intensities (Bula, 1960).

Even within these forages, however, a complete knowledge of their shade tolerance is lacking. Little, for example, is known about how drought and soil fertility levels influence shade tolerance. Moreover, for many of the other forage species recommended for planting in Missouri, even less is known. Because of their value as cash crops, many landowners are interested in combining plantings of alfalfa (*Medicago sativa* L.) or timothy (*Phleum pratense* L.) with their trees, but little or no guidance is available due to the dearth of shade tolerance research. High-quality alfalfa hay is currently selling in the U.S. for \$82 to \$198 per Mkg and timothy hay for \$99 to \$135 per Mkg, depending on location (Anonymous, 1997), and could serve as excellent cash crops in agroforestry.

Assuming that the forage is shade tolerant, reasonable to good hay yields can be expected from an alley-cropping planting. An alley-cropping program consisting of walnut trees planted 12.2 × 3.0 m on a good site in Missouri and a red clover/orchardgrass mix in the alleyway, averaged producing 6.3 Mkg ha^{-1} yr^{-1} of hay between years 15 and 20 (H.E. Garrett, University of Missouri — data unpublished).

In an attempt to provide much needed information, the University of Missouri Center for Agroforestry (Columbia, Missouri) initiated a major shade tolerance screening program on forages

TABLE 1.
Shade tolerance of some cool-season grasses and legumes with potential for Missouri agroforestry practices. Trials carried out under full irrigation.

	Low	Moderately Shade Tolerant	Very Shade Tolerant	High
Grasses				
Kentucky Bluegrass *Poa pratensis* L.		■		
Orchardgrass (Benchmark) *Dactylis glomerata* L.			■	
Smooth Bromegrass *Bromus inermis* L.		■	■	■
Tall Fescue (KY 31) *Festuca arundinacea* Schreb				■
Timothy *Phileum pratense* L.	■			
Red Top *Agrostis gigantea* Roth	■	■	■	
Reed Canarygrass *Phalaris arundinacea* L.			■	
Legumes				
Alfalfa (cody) *Medicago sativa* L.		■		
Berseem Clover *Trifolium alexandrinum* L.		■		
Ladino Clover *Trifolium repens* L.		■		
Red Clover *Trifolium pratense* L.		■		

in 1995. Grasses and legumes are grown in 8.8-foot pots under artificial shade cloth (three replications) with drip irrigation. Shade treatments are full sun, 50% shade and 80% shade. It has been the goal of this program to identify forage species suitable for growth in Missouri that are shade tolerant. The effect of drought on shade tolerance has not been considered. While, in general, warm-season grasses that are native to Missouri have been found to be intolerant of the shade levels tested, several cool-season grasses and legumes have been identified in the early trials that are shade tolerant (Table 1, Lin et al., 1995).

Of the grasses, orchardgrass is most often used in tree plantings in part because it is commonly thought to be shade tolerant and is fairly shallow rooted, which minimizes its competition for water with the trees. Red clover is usually the legume of choice. While tall fescue is also very shade tolerant, it is not widely planted in Missouri agroforestry due to its deep and prolific rooting characteristics and reputation for being a fierce competitor. Timothy, because of the high quality hay it produces, could play an important future role in agroforestry. In spring–early summer screenings, timothy yields were reduced by 0.33% at 50% shade. However, even with this reduction, dry matter yields were comparable to that of many of the other cool-season grasses when grown under full sun. Moreover, in late summer–fall trials, yields at 50% shade were the same as those in full sun (Lin et al., 1995).

Of the other grasses showing shade tolerance (Table 1), smooth-brome (*Bromus inermis* Leyss.) is especially prominent in northern Missouri and could offer good opportunities for landowners interested in introducing trees into their pastures. Reed canarygrass (*Phalaris arundinacea* Roth), which prefers wetter sites, may offer opportunities for agroforestry in flood plain areas, especially in association with pecan plantings or fast growing soft hardwoods.

Great interest has been expressed in growing alfalfa as an alleycrop with black walnut and pecan. In Missouri's early trials, alfalfa has performed well under partial shade conditions and would appear to be a viable candidate for planting with trees. However, it and most of the other grass and legume species discussed have not been sufficiently tested as a companion crop with walnut or other tree species to merit a recommendation. Only their shade tolerance has been evaluated. Alfalfa produces an extensive root system which is more competitive for soil water than many other forage species such as orchardgrass (Chamblee, 1958). Planting alfalfa with walnut or pecan could reduce nut yields due to competition for water. Since the generation of income is an important consideration in designing any agroforestry practice, the financial "trade-offs" of planting alfalfa or any other forage species with nut trees in a Missouri agroforestry practice must be assessed before recommendations can be made. With the current knowledge of forage/tree interactions, only a combination of orchardgrass and red clover has been sufficiently field-tested to be recommended.

The success of any tree/forage practice is directly correlated with the shade and drought tolerance of the forage or forages selected for planting, forage value, the tree species planted, the geometric pattern and density of the trees, and the age of the trees. In designing practices, all factors must be considered. Vast differences exist in crown and root characteristics of trees and the density of planting will directly influence light availability. Success depends upon selecting a tree species that will accommodate the biological requirements of the forage while yielding valuable wood and/or nut products and planting the trees using appropriate densities and geometric patterns. Since forages vary in shade tolerance and value, financial gain will, in part, depend upon the ability of the landowner to match the forage species with the correct light regime. Alley cropping is dynamic and microenvironments change. Such changes often dictate the need to change the companion species.

4. FINANCIAL ASPECTS

The Missouri agroforestry program adds significant diversity to existing or planned agricultural systems. This added diversity is, in general, financially advantageous. It provides a landowner the opportunity to develop short- and long-term investments thus allowing risk spreading through diversification. In addition, for owners with land rated as marginal for row crop production, it provides a means by which such lands can be removed from row crops over an extended period of time without financial loss and with great social benefits (Kurtz et al. 1996; Garrett et al., 1994).

The production economics of Missouri agroforestry is representative of the multiple output model where several fixed and variable inputs are combined to produce at least two products. An important characteristic of this model is that yields from all outputs can be varied through deliberate management decisions. This provides the opportunity for the landowner to optimize production efficiency through the creation of various combinations of inputs and outputs that best serve the achievement of production objectives (Kurtz et al. 1996; Garrett et al., 1994).

Various economic analyses have been conducted on Missouri agroforestry practices (Kurtz et al., 1991; Garrett et al., 1994; Kurtz et al., 1996; Kurtz et al., 1984). Depending upon the combination of crops studied, internal rates of return (IRR) have been found to range from 4 to 11% (Kurtz et al., 1991; Garrett et al., 1994). In general, returns tend to increase with management complexity and site quality. However, obvious factors such as market value of crops grown, cash-flow relationships and even risk-taking influence profitability. In particular, early returns from the planted intercrops (grain, forage etc.) coupled with nut production are important factors in creating high financial yields. Of these, nut revenues appear to be most important (Kurtz et al., 1984).

There are few reports available on actual walnut nut production under managed conditions (Jones et al., 1989; Garrett et al., 1995; Jones et al., 1994). Because of this, it is difficult to project nut yields for the purpose of obtaining a satisfactory financial analysis of Missouri walnut agroforestry. However, one may assume that yields of selected cultivars will exceed those of trees growing in a wild and unmanaged condition. Slusher (1985) reported nut yield data for six open-grown, unmanaged walnut trees. The trees ranged in age from 70 to 100 years and were growing on a deep, well-drained silty clay loam. During a 17-year period, the average yield tree^{-1} year^{-1} was 44 kg's with a range of 21 to 73 kg's. If one assumes 74 trees per hectare (this approximates what a "mature," Missouri walnut alley cropping practice would contain), the total nut yield per ha would equal 3256 kg's. At a price of \$0.44 kg^{-1}, one ha would have the potential to produce nuts with a value of \$1,435.00 per year. While the nut yields of these wild mature trees appear to be sufficient to justify the planting of black walnut in agroforestry, yields from the establishment of agroforestry plantings using seedlings from a wild population do not.

In Missouri agroforestry plantings made in 1975 and 1976, using wild, genetically unknown walnut seedlings, early nut production has been found to be erratic and, in general, poor. An evaluation of 934 trees from ages 7 through 15, revealed that 80% of the trees studied produced an average of 100 or fewer nuts tree^{-1} year^{-1}, while only 2% produced more than 200. Forty-one trees had borne no fruit by age 15, while only three averaged producing more than 300 nuts per year (Garrett et al., 1995; Jones et al., 1994). The monitoring of nut production on more than 13,000 trees each year since age 15 has revealed only minimal improvement. Due to the value of nut production in providing a cash-flow for the agroforestry adopter, and its ultimate effect on increasing the internal rate of return, it is critical that trees selected for use in agroforestry produce fruit early, regularly and heavily. While old "wild" trees may be good producers, trees of unproven genetics from a wild population, in general, do not come into production early enough to allow maximization of the financial benefits of a Missouri alley cropping or silvopastoral management program. To achieve the financial success that is desirable in the Missouri program will require the use of grafted trees. Only minor areas of grafted Sparrow, Emma K and Kwik Krop (three of the best overall yielding walnut varieties for Missouri) are currently found in Missouri and little data are available on nut production potential. However, from the limited data that is available, yields of from 900 to 1100 kg's ha^{-1}, or more, appear to be achievable between years 15 and 20. Early yields within this range in combination with the value of companion crops provides an excellent financial incentive for the adoption of alley-cropping and silvopastoral management in Missouri.

As an additional financial incentive to adopt, Missouri passed legislation in 1990 which was amended in 1993 entitled, "The Missouri Economic Diversification and Afforestation Act of 1990." It is the purpose of this legislation to complement and extend the federal Conservation Reserve Program (CRP) which pays landowners to remove highly erodible land from agricultural production. This act has greatly stimulated interest in Missouri agroforestry by providing: (1) a funding base for incentive payments, and (2) a structure of cooperating organizations for implementation.

The legislation provides:

1. annual incentive payments for up to 10 years which, when added to any gross cash or in-kind income produced by crops raised on the land, are substantially equal to the amount previously paid while the land was under CRP, or which would have been paid to the landowner if the land had been in CRP, and
2. financial assistance to share the cost (up to 75%) of establishing the trees and/or shrubs to be used in an agroforestry management program.

5. CONCLUSIONS

Black walnut alley cropping in Missouri has evolved from a little-known research effort initiated in the mid 1970s to a much discussed and well-received practice during the 1990s. While many tree species are used in Missouri alley cropping, black walnut, more than any other, epitomizes the ideal species. Prized for its wood which typically commands high prices, its flavorful nut meat, and its nut shell which is marketed as an abrasive, it is unequaled in value in Missouri and in the United States as a whole.

Its root and crown morphology are ideally adapted to alley cropping. This species is one of the last to leaf out during the spring and one of the first to defoliate in the fall. A short growing season, that is often less than 135 days, lends itself well to the growth of companion crops. A black walnut's root system extends deep into the soil. A minimum of feeder roots are found near the soil's surface providing a shallow zone within which companion crops can effectively compete for moisture and nutrients. Because of its unique morphological characteristics, many alley cropping practices that incorporate row crops, cover crops and specialty crops with black walnut are being adopted for both conservation and production benefits.

REFERENCES

Allard, G., Nelson, C. J., and Pallardy, S. G., 1991, "Shade effects on growth of tall fescue: I. Leaf anatomy and dry matter partitioning," *Crop Sci.*, 31, 163.

"Hay market update," 1997, *Hay and Forage Grower*, Vol. 12, No. 2, 76.

Blake, C. T., Chamblee, D. S., and Woodhouse, W. W., Jr., 1966,"Influence of some environmental and management factors on the persistence of ladino clover in association with orchardgrass," *Agron. J.*, 58, 487.

Bula, R J., 1960 "Vegetative and floral development in red clover as affected by duration and intensity of illumination," *Agron. J.*, 52, 74.,

Chamblee, D. S., 1958, "The relative removal of soil moisture by alfalfa and orchardgrass," *Agron. J.*, 50, 587.

Garrett, H. E., Buck, L. E., Gold, M. A., Hardesty, L. H., Kurtz, W. B., Lassoie, J. P., Pearson, H. A., and Slusher, J. P., 1994, *Agroforestry: An integrated land-use management system for production and farmland conservation*, Resource Conservation Act Appraisal Document, USDA, SCS, Washington, DC, 58.

Garrett, H. E., Loewenstein, E. F, Jones, J. E., and Rink, G., 1995, "The importance of selection gain in walnut agroforestry management," in *Proceedings, 8th Biennial Southern Silvicultural Research Conference*, Edwards, M. B., Ed., USDA, Forest Svc. General Tech. Rpt. SRS-1, 364.

Garrett, H. E. and Kurtz, W. B., 1983, "Silvicultural and economic relationships of integrated forestry farming with black walnut," *J. Agro. Sys.*, 1, 245.

Garrett, H. E., Kurtz, W. B., and Slusher, J. P., 1992, *Walnut agroforestry*, University of Missouri, MU Guidesheet G5020, 4.

Gordon, A. M. and Williams, P. A., 1991, "Intercropping valuable hardwood tree species and agricultural crops in southern Ontario," *The Forestry Chronicle*, 67, 200.

Gray, D., 1997, *Black walnut nut yield components: Effects of N form and timing of application*, unpublished Masters thesis, School of Natural Resources, University of Missouri, 133 pp.

Hahn, J. T., 1991, *Timber resources of Missouri*, USDA, Forest Svc. Resource Bulletin, NC-119.

Hall, D. O., Rosillo-Calle, F., Williams, R. H., and Woods, J., 1993, "Biomass for energy: Supply prospects," in *Renewables for fuels and electricity*, Johnson, Kelly, Reddy and Williams, Eds., Island Press, Washington, DC, Ch. 14.

Hupe, S. C., 1980, *The effects of pole-size black walnut trees on six forage species*. MS Thesis. University of Missouri, Columbia, MO.

Jones, J. E., 1975, *Factors which influence regional nut production of eastern black walnut*, 66th Annual Report of the Northern Nut Growers Association, 108.

Jones, J. E., Garrett, H. E., and Haines, J.K., 1989, "Production, harvesting, utilization and marketing of eastern black walnut nuts — an industrial viewpoint," in *Proceedings, Black Walnut Symposium*, July 30–Aug. 2, Carbondale, IL, 242.

Jones, J. E., Garrett, H. E., Haines, J., and Loewenstein, E. E., 1995, "Genetic selection and fertilization provide increased nut production under walnut-agroforestry management," *Agro. Syst.*, 29, 265.

Jones, J. E., Haines, J., Garrett, H. E., and Loewenstein, E. F., 1994, "Genetic selection and fertilization provide increased nut production under walnut-agroforestry management," in *Proceedings 3rd North American Agroforestry Conference*, Schultz, R. C. and Colletti, J. P., Eds., Aug. 15–18, 1993, Iowa State University, Ames, IA, 39.

Jose, S., Gillespie, A. R., and Seifert, J. R., July 1995, "The microenvironmental and physiological basis for temporal reductions in crop production in an Indiana alley cropping system," in *Proceedings, 4th North American Agroforestry Conference*, Ehrenreich, J. H., Ehrenreich, D. L., and Lee, H. W., Eds., Boise, ID, 54.

Kang, B. T., 1993, "Alley cropping: past achievements and future directions," *Agro. Syst.*, 23, 141.

Kurtz, W. B., Garrett, H. E., and Kincaid, W. H., 1984, "Investment alternatives for black walnut management," *J. Forestry*, 82, 604.

Kurtz, W. B., Garrett, H. E., Slusher, J. P., and Osburn, D. B., 1996, *Economics of agroforestry*, University of Missouri, MU Guidesheet G5021.

Kurtz, W. R., Thurman, S. E., Monson, M. J., and Garrett, H. E., 1991, "The use of agroforestry to control erosion — financial aspects," *Forestry Chronicle*, 67, 254.

Lin, C. H., McGraw, R. L. George, M. F., Garrett, H. E., Piotter, B. J., and Alley, J. L., 1995, "Effects of shade on forage crops that have potential use in agroforestry," *Agron. Abstracts*, 53.

Pham, C. H., Yen, C. P., Garrett, H. E., and Cox, G.S., 1978, "Slope position, soil-water storage capacity, and black walnut root development," in *Proceedings, Soil moisture — site productivity symposium,* Palmer, W E., Ed., Myrtle Beach, S.C. Nov. 1-3, 1977, 326.

Ramos, D. E., 1985, *Walnut orchard management*, Cooperative Extension, University of California, Publication #21410, 178.

Schaffer, K. L., George, M. F., Peleg, M., Garrett, H. E., and Cecich, R.A., 1996, "Pistillate flower development in eastern black walnut," *Can. J. Forest Res.*, 26, 1514.

Slusher, J. P., 1985, *A Missouri case study of black walnut nut production*, 76th Annual Report of the Northern Nut Growers Association, 20.

Slusher, J. P., Garrett, H. E., and Kurtz, W. B., 1997, "Agroforestry in Missouri," *The Forest Landowners Manual*, 56, 52.

Spelter, H., 1996, *Capacity, production and manufacture of wood-based panels in the United States and Canada*, USDA Forest Service, FPL-GTR-90, 17 pp.

Weinbaum, S.A., Murooka, T. T., Gatlin, P. B., and Kelley, K., 1990, "Utilization of fertilizer N by walnut trees," in *Walnut Research Reports* (The Walnut Marketing Board of California), 234.

Yen, C. P., Pham, C. H., Cox, G. S., and Garrett, H. E., 1978, "Soil depth and root development patterns of Missouri black walnut and certain Taiwan hardwoods," in *Proceedings, Root form of planted trees symposium*, Eerden, E. V. and Kinghorn, J. M., Eds., Victoria, B.C., Canada, May 16–19, 1978.

6 Silvopastoralism: Competition and Facilitation Between Trees, Livestock, and Improved Grass-Clover Pastures on Temperate Rainfed Lands

Steven H. Sharrow

1. INTRODUCTION

Livestock grazing in forests is common throughout temperate areas of the world. It is largely an opportunistic activity that makes beneficial use of forest herbs and shrubs which would otherwise go unused by humans. Mature humid forests typically produce little grazeable understory vegetation once the tree canopy has coalesced into a continuous overhead layer. Forest openings are created by natural disturbances such as wildfire, wind-throw, and insect attack, and through human managed events such as clearcut tree harvesting or prescribed burning. Openings often produce considerable amounts of grazeable understory vegetation until trees reoccupy the site and canopy closure reoccurs. Regenerating forest stands are sufficiently common and important as a forage resource that range managers call them "transitory rangelands" (Spreitzer, 1985). They are temporarily available for livestock grazing between the stand opening event and subsequent tree canopy closure.

1-56670-294-1/99/$0.00+$.50

FIGURE 1. Open canopied forests, such as this ponderosa pine forest in eastern Oregon, support considerable amounts of grazeable forage.

The rate of tree growth and canopy closure is largely dependent on soil and climatic factors. Canopy closure occurs more quickly on deep fertile soils under humid to subhumid climates than it does on shallow or infertile soils in semi-arid climates. In fact, some semi-arid woodlands never attain canopy closure. These open canopy forests produce grazeable vegetation throughout their life and are commonly managed as forested rangelands. Together, open canopy forests and regenerating closed canopy forests contribute substantially to world forage resources. In the United States, for example, approximately one quarter of all forest land is grazed by livestock (USDA, 1982). This contributes 70 million hectares, or 13% of the total grazed land in the United States (USDA, 1986). By comparison, pasture contributes 53 million hectares of grazeable land, while grazed cropland provides only 26 million hectares (Figure 1).

Although forest grazing combines the tree, forage, and livestock components of a structural silvopastoral agroforestry system, it frequently lacks the system perspective in managing interactions among components required of true agroforestry systems. This is especially true for government owned or large industrial forest tracts where local livestock owners are permitted to graze, but have little influence on forest management decisions. Recent interest in prescription livestock grazing as a tool to facilitate coniferous forest regeneration provides an impetus for silviculturalists, forest wildlife biologists, and livestock producers to work together as a management team. Livestock can be a safe, cost effective, and socially acceptable means of managing understory vegetation in young tree stands (Doescher et al., 1987; Sharrow, 1994). However, this approach is most successful when it is carefully integrated with other stand practices such as site preparation prior to tree planting, planting of forage species, tree regeneration method (seedlings planted vs. natural regeneration from seeds or sprouts) and wildlife management practices. The resulting management strategies represent an intentional combination of trees and livestock with a clear intent to manage the interactions between them, that is, silvopastoral agroforestry. (See Chapter 5.)

Indigenous forest grazing practices, together with the more intensive management typical of improved pastures form the basis for silvopastures. Silvopastoral systems are a form of structural agroforestry in which tree, forage, and animal components all share the same hectare of land at the same time. The forage component is typically an improved pasture of introduced grasses and/or legumes which is managed under normal agronomic principles. Silvopastures may be simplistically described as "trees in pastures." They are a subcategory of silvopastoralism which differs from forest prescription grazing primarily in being a very intensive system as opposed to open forest grazing which is a more extensive approach to resource use (Merwin, 1997).

Silvopastoralism is by far the most common form of agroforestry practiced in developed countries. Within the temperate zone, silvopastures are most common in the coniferous forest regions of the Pacific Northwest (Sharrow and Fletcher, 1995) and Southeastern United States (Dangerfield and Harwell, 1990), New Zealand (Knowles, 1991), and Southwestern Australia (Mead, 1995). These commercial silvopastures share several common features: (1) they are most often based upon growing of conifers, eucalyptus, or other evergreen trees which contain antiherbivory chemicals rather than more palatable deciduous hardwood trees; (2) they have been mainly promoted for and adopted by non-industrial foresters, farmers, and livestock producers rather than for large commercial forests or government owned properties; (3) they are predominately taungya-type systems, intended to make agricultural use of young forest plantations, but to produce a full commercial stand of trees when mature; and (4) the tree, pasture, and livestock components are, predominately, those already widely used by local people.

Considerable potential exists to expand conifer-based silvopastoral systems within the traditional grazing and timber growing regions of the world. New Zealand had more than 60,000 ha of silvopastoral forest plantations in 1986, out of an estimated 168,000 ha of suitable forest (Knowles, 1991). Great potential also exists for silvopasture development on a portion of New Zealand's 18 million ha of farm land. In the United States, similar potential exists in the traditional livestock and timber producing areas of the Southeastern and Pacific Northwestern states. Together they contain approximately 40 million hectares of forest and woodland suitable for silvopasture management (Garrett, 1994). As well as the established areas already mentioned, silvopastures are being evaluated in South America (Penaloza et al., 1985; Garrison, 1992), Europe (Dupraz, 1994), and Great Britain (Campbell et al., 1994).

Silvopastures offer environmental advantages, such as reducing erosion and reversing salt intrusion of agricultural soils (Bari and Schofield, 1991; Malajczuk et al., 1996). However, their main attractiveness to landowners is primarily economic (Merwin, 1997). A properly designed and well managed combination of trees and livestock can be more profitable than traditional agriculture or forestry options (Clason, 1995; Sharrow and Fletcher, 1995; Malajczuk et al., 1996). Because silvopastures are a multiproduct system, they reduce the risk of loss due to unfavorable markets, weather, or political decisions. Combining long-term income from timber sales with yearly income from livestock or hay sales, provides landowners with greater income continuity over the life of the tree plantation than would be obtainable from traditional forest management (Dangerfield and Harwell, 1990; Sharrow, 1994) (Figure 2).

Silvopastures are difficult systems to design and to manage. Polycultures, such as agroforestry, are by their nature knowledge intensive undertakings. The complex web of interactions between system components as well as with outside actors such as naturally occurring mammals, birds, insects, and plants must be understood if management is to be proactive rather than reactive. The pasture component of silvopastures is often itself a polyculture of planted and volunteer grasses, legumes, and other broad leaved herbs. It is, therefore, a herbaceous polyculture within the forest polyculture. Discussion of within pasture dynamics is beyond the scope of this chapter. For further information on pasture ecology and management, consult Langer (1990). The high level of management intensity and knowledge needed to design and operate silvopastures, and lack of adequate technical support to provide needed information are major disincentives for adoption of silvopasture technology (Lawrence and Hardesty, 1992; Merwin, 1997).

FIGURE 2. A grass/clover/Douglas-fir silvopasture near Corvallis, OR offers early income from grazing, longer-term income from log sales, and higher total productivity than pasture or forest alone.

Silvopastoralists manage agroecosystem processes primarily through manipulating species composition and spatial structure. Successful integration of forests and pastures relies upon understanding the impacts of species composition and spatial structure on ecosystem processes such as succession, facilitation, competition, and herbivory at both silvopasture and individual plant scales. These four processes can operate together producing a net increase or decrease in the productivity of individual components. For example, Callaway et al. (1991) observed that some individual oak trees (*Quercus douglasi*) increased pasture production beneath their canopy while others nearby reduced pasture yield. Further investigation suggested that oaks facilitated nutrient availability beneath their canopy, but that trees with strong fibrous roots near the soil surface competed with understory grasses for access to soil resources. The net result was increased grass growth beneath deeper rooted trees and reduced grass growth associated with shallow rooted trees.

2. SILVOPASTURE INTERACTIONS

2.1 Succession

Newly established silvopastures go through a successional process which mimics that of natural forests. Vegetation typically progresses from short to taller stature and from short to longer lived plants, passing from annual grass/forb > perennial grass/forb > shrub > hardwood tree > conifer stages. Depending upon such factors as climate, soil, presence of plant propagules, fire, grazing, or other perturbations, stages may be skipped or the process may become stable (end) at an intermediate stage. Plants from different successional stages co-occur mainly during the transitions from one stage to another. These are times of high biological diversity but relatively low stability. It is reasonable to expect that silvopastures which attempt to combine elements from different successional stages will be difficult to maintain. This has proven to be the case with annual clover/perennial grass pastures, for instance, where the early successional clover requires carefully applied grazing in order to reduce competition from the later successional perennial grasses (Sharrow et al., 1981; Langer, 1990).

The successional process can be manipulated through initial floristics, the planting of desired trees and pasture plants. Once trees are established, they increasingly capture site resources eventually excluding pasture plants. The pasture understory is a valued economic and ecological component of silvopastures. Integration of pasture with trees requires that silvopastoralists act as a stabilizing force to moderate the natural momentum of the system to move quickly from the grass through the shrub to a closed canopy forest. Balancing the needs of tree, pasture, and livestock components requires an understanding of how silvopastoral components may interfere with or facilitate each other's success. The predominant forms of interference in silvopastures are competition and herbivory.

2.2 Competition

Plants in polycultures compete for scarce resources. Light, soil moisture, and soil nutrients are the most common limiting resources in plant communities (Harper, 1977). Trees and ground vegetation compete for both above ground (light) and below ground (soil moisture and nutrients) site resources. However, beneficial as well as competitive effects accrue from the interactions between plants as they attempt to use the resources available to them (Hunter, 1988).

2.2.1 Above Ground Factors

Competition for light in established silvopastures is largely one-sided (Cannell and Grace, 1969). That is, the tallest plants get first access to resources while smaller stature plants use what is left. Young trees may be overtopped and shaded by rapidly growing herbaceous or shrubby vegetation. Physical crushing of small trees when herbaceous vegetation is lodged under snow can flatten small trees in areas which receive considerable amounts of snow or ice, keeping them prostrate below the herbaceous canopy. Once tree canopies have emerged above the ground vegetation, understory plants must compete with trees for light while radiation interception by overstory trees is not affected by the understory.

Tree canopies impact both the quantity and quality of light reaching the pasture. In order for plants to grow, daily photosynthesis must exceed daily respiration. For most herbaceous plants approximately 10% of full sunlight is required to reach this compensation point (Gardner et al., 1985). Photosynthesis rapidly increases beyond the compensation point until the leaves become light saturated. Typical light saturation levels for cool season plants (C3 pathway) and warm season (C4 pathway) plants are approximately 50 and 85% of full sunlight, respectively (Gardner et al., 1985). Low light under trees may be somewhat compensated for by cooler temperatures in the shade, which increases the efficiency of light use by C3 plants (Brown, 1982) during warm days. Thus, light reduction under tree canopies is probably more important for warm season than it is for cool season pastures. It is unlikely to impact cool season pasture production until trees intercept over half of the total incoming solar radiation. This is consistent with results reported for shade tolerant temperate conifers (Mitchell and Arnott, 1995) which required light reductions of at least 60% before they began to show either growth reductions or significant morphological reactions. Understory forage production has been observed to increase substantially when nitrogen and phosphorous fertilizer were applied to a thinned mid-rotation *Pinus pinaster* agroforest where 30% sunlight reached the pasture (Braziotis and Papanastasis, 1995), but failed to respond to fertilization with only 11% sunlight. Hart et al. (1970) reported similar results for mid-rotation warm season perennial grass — *Pinus elliottii* silvopastures. Trees had no effect upon forage production until they were seven years old and intercepted 84% of sunlight. Pasture yields under these conditions were still 56 to 79% of open pasture yields and were approximately doubled by fertilizer application. The following year, although pastures under trees received only 7% of incident radiation, forage yields were 19 to 38% of open pasture yields and again could be approximately doubled by fertilizer application. Clearly, pastures are not entirely limited by light even under relatively dense tree shade.

Silvopastures are generally planted at a lower tree density than are commercial forests. On a stand level, young open canopied silvopastures seldom exceed 50% interception of sunlight by trees. Anderson (1987) reported light interception of seven-year-old (>8 m tall) *Pinus radiata* was only 31% for high density silvopastures. Harris and Sharrow (Unpublished) found little effects of Douglas-fir (*Pseudotsuga menziesii*) on light available to pasture within a six-year-old silvopasture. Light is unevenly distributed about individual trees, however. Yunusa et al. (1995) reported that a stand of three-year-old *Pinus radiata* (2.7–2.9 m tall) only intercepted 40–50% of incoming solar radiation during the spring. However, grass growing within 1 meter on the shady side of a tree received only 25% sunlight while grass on the sun side of the tree was in full sun. Strikingly similar patterns of light interception occur for Douglas-fir silvopastures (Harris and Sharrow, Unpublished).

Overhead canopies selectively absorb the photosynthetically active red and blue portions of the spectrum, leaving their shade proportionately higher in orange, yellow, green, and infrared wavelengths (Krueger, 1981). Although light quality is seldom studied in agroforestry systems, its central importance in regulation of plant growth processes clearly warrants consideration. The phytochrome system of plants interacts with red and infrared wavelengths of light to influence the amount of growth regulating hormones, indole-3-acetic acid, abscisic acid and gibberellins (Baraldi et al., 1995). Tillering in grasses (Davis and Simmons, 1994; Frank and Hofmann, 1994), stem production in clovers (Robin et al., 1994), seed development (Felker et al., 1995), and other basic plant growth processes (Moe and Heins, 1990) all respond to the phytochrome system. Lack of adequate red light reaching the growing crown of grasses tends to reduce tiller numbers, increase individual tiller size, and increase the proportional allocation of plant resources from roots to shoots (Motazedian and Sharrow, 1987). Only a brief exposure to red light (i.e., a flash) is required to activate phytochrome mediated plant responses. Although pasture growing under a tree canopy may be shaded much of the day, sufficient solar radiation to activate the phytochrome system may be received during the morning or afternoon when light arrives at sufficiently low angles of incidence to penetrate the tree canopy.

It appears that direct competition for light probably has little stand level influence on tree/pasture relations in early to mid rotation, open canopied silvopastures. Light can contribute to the patterns of forage production around individual trees, but these effects are minor compared to other factors, such as competition for soil moisture. Light availability becomes limiting in silvopasture forage production when trees are sufficiently large that their canopy produces dense shade over a large proportion of the pasture. Pruning lower branches of trees in mid to late rotation silvopastures reduces tree leaf area, thus increasing light reaching the pasture and reducing water use by trees. However, the main reason for pruning is generally to enhance tree value by increasing the volume of knot-free wood in the stem. Increased pasture production is a welcome by-product of pruning.

Forage quality of understory plants may be affected by the low light and lower temperatures associated with tree canopies. Warmer temperatures during growth increase the structural portion of plant cells relative to the more easily digested cell contents in both cool season and warm season grasses, thus decreasing their potential digestibility by animals (deSilva et al., 1966; Henderson and Robinson, 1982). Plants growing under a tree canopy often exhibit delayed flowering and remain green longer into the summer than open grown plants (Krueger, 1981). Higher protein content of both grasses (Hart et al., 1970; Walters and Stafford, 1989) and shrubs (Blair et al., 1983; Happe et al., 1990) have been observed in conifer forests compared to open areas. Digestibility of forest understory plants may be lower than that of plants at the same stage of growth in the open (Blair et al., 1983), perhaps due to competition for soil moisture and nutrients near trees. However, digestibility on a specific date may be increased by reduced inhibitory compounds such as tannins

(Happe et al., 1990) in shrubs or by delaying the decline in forage quality associated with maturity in herbaceous plants (Krueger, 1981). (See Chapter 2.)

As Cannell and Grace (1969) point out, plants rarely compete for light without concurrently competing for nutrients and water. As a general rule, both tree root density and canopy effects are greatest near the stem and decrease with increasing distance from the tree. For this reason, it can be very difficult to separate canopy and root effects by simple examination of spatial understory patterns. In addition, both water soluble macronutrients (Marschner et al., 1991) and micronutrients (Xu et al., 1996) are obtained by plants through mass flow of water containing the dissolved nutrients. The water potential gradient which provides the force to support water movement from soil to root to shoot is very sensitive to canopy temperature and humidity, which in turn, is greatly influenced by the amount of incoming solar radiation absorbed by leaves. This basic physical linkage between plant moisture uptake, soil nutrient extraction, and evapotranspiration from leaves makes separation of competition for nutrients, water, and solar radiation very difficult.

2.2.2 Below Ground Factors

Competition for below ground factors is two-sided (Cannell and Grace, 1969). Silvopasture trees and pasture plants compete with each other for available soil moisture and nutrients. Production of both tree and pasture components may suffer from this competition. Forage plants are often fierce competitors for soil resources within the top portion of the soil profile because their dense fibrous root systems develop quickly during the growing season. Grasses and forbs have been observed to reduce both establishment and subsequent growth of young conifers (Nambiar and Zed, 1980; Eissenstat and Mitchell, 1983; Miller et al., 1991) especially on dry sites with shallow soils (Cole and Newton, 1986; McDonald, 1986). Suppression of competing vegetation for one of two years after tree planting is a common recommendation for establishment of silvopastures (Sharrow and Fletcher, 1995). This is most often accomplished by spraying a 1–2 m circle or strip with herbicide around trees to provide a weed-free zone for tree establishment. Mowing (Lewis, 1985) and grazing (Doescher et al., 1987; Sharrow et al., 1992) have also proven effective for controlling competition between pastures and recently planted conifer trees, but introduce some risk of mechanical or browsing damage to trees. Once tree roots have extended beneath the pasture root zone, tree growth is less affected by moisture use of pasture plants. Uncontrolled herbaceous plants may, however, continue to reduce conifer growth until canopy closure (Cole and Newton, 1986). Nitrogen, sulphur, and other nutrients associated with soil organic matter are most plentiful near the soil surface. It is likely, therefore, that competition between established trees and pasture for nutrients may be more important than competition for moisture in the upper soil layers.

Presumably, herbaceous production under dense tree canopies is limited by light, while that of younger or more open canopied forest is reduced by competition with trees for soil resources. Work by Young and Smith (1982; 1983) showed that light or water could be the most limiting factor depending on the environmental characteristics during the year. Although water is generally assumed to be the primary limiting factor for recently established trees, Woods et al. (1992) noted that nitrogen fertilization completely offset the negative impacts of herbaceous weed competition on two–three-year-old *Pinus radiata* seedlings, suggesting that competition was mainly for soil nitrogen. Several authors concluded that competition between large conifers and ground vegetation is primarily for soil resources (Krueger, 1981; McCune, 1986; Riegel et al., 1991). McCune (1986) observed that trenching increased forb canopy cover from an initial 7% to 55% five years later in a closed canopy *Abies grandis* forest, suggesting understory plants may be limited by tree withdrawal of soil resources even in low light, humid understory conditions. As mentioned previously, understory forage plants often respond strongly to nitrogen fertilization, even under dense tree canopies (Hart et al., 1970; Braziotis and Papanastasis, 1995). Below ground competition may vary seasonally with soil nutrients being most important in the spring, while competition for soil moisture is dominant in the summer (Riegel et al., 1991; Sharrow, 1995). Overall, competition for below ground resources appears to influence both tree and pasture growth more than does competition for light until pasture is excluded by tree canopy closure. A similar conclusion was reached for agrosilvicultural intercrops in the semiarid tropics (Ong et al., 1991).

2.3 Facilitation

Facilitation occurs when one silvopasture component supports the productivity of another component. Hunter and Aarssen (1988) list nine ways in which plants help other plants. Of these, modification of the microclimate, increases in nutrient availability, and competitive exclusion are common ways that plants and animals benefit each other in silvopastures.

2.3.1 Microclimate Modification

Conifers in pastures reduce wind speed, increase grass minimum daily temperatures, and reduce soil temperatures (Hawke and Wedderburn, 1994), all of which may benefit cool season forages or cereal crops. Although crop increases attributable to close association with conifer trees are rarely reported, Frost and McDougald (1989) reported increased annual grassland production under *Pinus sabiniana* in California. They attributed increased production to soil moisture conservation due to shading by pine canopies. Both cereal crop (Akbar et al., 1990) and pasture yields (McClaran and Bartolome, 1989; Ratliff et al., 1991) have been observed to increase under hardwood trees. A review of shelter belt literature by Korte (1988) includes seven references to increased hay production, with an average increase of 42% due to windbreaks. Animals may benefit from the shelter provided by trees during inclement weather (McArthur, 1991). Tree shelter belts are sometimes planted for this purpose (Wilson et al., Undated).

2.3.2 Nutrient Availability

Plants take soil nutrients up through their root systems, use them to build tissue, then return them to the necromass pool as roots and leaves senesce. Although roots of woody plants often extend well beyond their canopy and can reach soil depths greater than 10 m (Eis, 1987; Stone and Kalisz, 1991; Kuiper and Coutts, 1992), deposition of both leaf litter and fine roots is concentrated under the canopy. Birds, native ungulates, and livestock are attracted to the habitat provided by trees. Animals may redistribute nutrients and organic matter from pasture toward the vicinity of trees in dung and urine excreted as they loaf near trees. Both of these mechanisms tend to favor increased soil organic matter and nutrient content under tree canopies. This "biological transport" has been proposed as the mechanism responsible for the "fertile islands" often seen associated with woody vegetation in arid climates (Garner, 1989). For example, Jackson et al. (1990) reported greater soil carbon, nitrogen, cation exchange capacity, and phosphate P in soils under oak (*Quercus douglasii*) canopy than in grassland between oaks in grazed savannah. Changes in soil organic matter and soil structure beneath tree canopies can also alter their hydrologic characteristics, increasing water captured and stored in soils under tree canopies compared to grassland openings between scattered trees (Joffre and Rambal, 1993).

Care must be taken to differentiate between nutrient enrichment and nutrient redistribution when stand level processes are inferred from nutrient distribution about individual trees. Biological transport does not add additional nutrients, it merely relocates them and their associated plant production within the silvopasture. Such redistribution may produce noticeable patterns of forage production surrounding trees without changing overall pasture production appreciably. Silvopastures are enriched when one or more components adds additional nutrients, or increases the availability of nutrients already present. Trees used in temperate silvopastoral systems are rarely nitrogen fixers. Therefore, they are unlikely to increase the amount of nutrients available for plant

FIGURE 3. Subclover planted under southern pines provides high quality cool season forage for livestock while enriching soils through nitrogen fixation.

use on a stand level except through more efficient retention of existing nutrients against losses by leaching or soil erosion. Their primary impacts are probably through nutrient redistribution (Figure 3).

Pasture legumes, on the other hand, can biologically capture substantial amounts of atmospheric nitrogen during their growing season. Annual nitrogen fixation rates in rainfed grass/clover pastures exceeding 350 kg nitrogen/ha have been reported (Hoglund et al., 1979). Goh et al. (1996) observed that more than 90% of the nitrogen used for spring growth of clovers in clover-perennial grass silvopastures comes from biological fixation of atmospheric nitrogen. Subclover (*Trifolium subterraneum*) in this trial fixed 70–210 kg nitrogen per ha annually. Although direct transfer of biologically fixed nitrogen from legumes to associated plants does occur (Catchpoole and Blair, 1990), the quantity is relatively small compared to that derived from the soil nitrogen pool. Most legume nitrogen becomes available when plant parts die and decompose. This can be a slow process which requires several years of clover growth before the soil nitrogen cycle is sufficiently enriched to impact growth of associated trees (Pearson et al., 1994). Waring and Snowdon (1985), for example, reported that while subclover competition tended to depress newly planted *Pinus radiata* growth during the first three years, soil nitrogen levels were 36% greater and tree diameters 14% larger compared to controls seven years after planting. Livestock grazing can speed up the cycling of forage nutrients through digestion and redeposition of nutrients in urine and dung (Dawson and McGuire, 1972; Parsons et al., 1991). (See Chapter 2.) They serve as a primary transfer mechanism in silvopastures to move nutrients from legumes to associated grasses and trees. Dawson and Mcguire (1972) suggest that this biological transfer mechanism is so efficient that the same nutrient molecule may be taken up and used for plant growth repeatedly during a single growing season, thereby reducing the amount of total nutrients required to support a high level of forage production.

2.3.3 Competitive Exclusion

Competitive exclusion occurs when competition between two plants excludes a third, potentially more aggressive competitor. This is the rationale for planting forage plants for weed suppression in young forest plantations (McDonald, 1986). Forage plants initially may compete with young trees, but some trees will eventually benefit if more competitive herbaceous and woody plants are prevented from establishing. A good example of this principle was reported for a young Douglas-fir plantation by Sharrow et al. (1992). The plantation was seeded with orchardgrass (*Dactylis glomerata*) and heavily grazed with sheep twice each year for three years. At the end of the grazing period, Douglas-fir trees were slightly smaller as a result of grass competition and heavy browsing by sheep. The combination of grass and grazing, however, delayed the establishment of red alder (*Alnus rubra*), a fast-growing competitive hardwood tree. Ten years after the trial began, trees from the grazed plantation were 22% greater in diameter than those from the ungrazed plantation, largely due to reduced red alder competition. Total tree basal area per hectare was the same for grazed and ungrazed plots. But, without grazing, red alder contributed half of the total tree basal area, compared to only 20% on grazed plots.

2.4 Herbivory

Livestock grazing as a tool to manage understory vegetation in temperate conifer forests and silvopastures was recently reviewed by Sharrow (1994). Although many hardwood trees are readily consumed by livestock, palatability of conifers is often less than associated herbaceous vegetation (Leininger and Sharrow, 1987; Newsome et al., 1995). This provides an opportunity to harvest forage in conifer plantations without incurring unacceptable defoliation levels of crop trees. Conifer palatability to livestock and native ungulates is generally highest just after bud break when young twigs are 2–4 cm long, foliage is not yet fully expanded, and their complement of antiherbivory compounds (phenolics and tannins) is not yet fully established (Leininger and Sharrow, 1987, 1989; Newsome et al., 1995). Mature conifer foliage is relatively unattractive to livestock when other green herbaceous feed is available. Since livestock like some variety in their diet, they will consume small amounts of conifer foliage each day during the grazing period, much as we consume condiments (Sharrow et al., 1992). This condiment effect is more pronounced in silvopastures, where the only woody vegetation is crop trees, than it is in native forests where a more varied array of vegetation is on offer. Once livestock perceive alternative feed to be in short supply, they will begin to actively feed upon tree foliage and young branches. Considerable defoliation of young trees can occur quickly when livestock density is high. This makes close monitoring of livestock in young silvopastures essential.

Herbivory has both direct and indirect effects upon plants. Defoliation directly removes photosynthetic surface and reduces the nutrient capital of the plant proportional to the amount of nutrients present in the foliage removed. Potential for future growth may be reduced if buds are consumed. Plant resources may be redirected for defensive chemicals (Walters and Stafford, 1989) or physical defenses such as thorns (Bazely et al., 1991), reducing the amount available for growth and to maintain helpful symbionts such as mycorrhizal fungi (Gehring and Whitham, 1991). Grazing of grasses and herbs often hinders maturation and stimulates regrowth, extending the green feed period for herbivores (Rhodes and Sharrow, 1990). Grazed forest plantations green-up earlier in the spring, further extending the green feed period (Rhodes and Sharrow, 1990). High producing improved pasture plants require periodic defoliation to control a tendency to self shade (Motazedian and Sharrow, 1987; Langer, 1990; Frank and Hofmann, 1994). Undergrazing of such pastures is as damaging as overgrazing in maintaining a healthy pasture and excluding invasion by weeds (Sharrow et al., 1981).

Young conifer trees are surprisingly tolerant of defoliation provided that the terminal leader and more than 50% of current year's foliage remains (Lewis, 1980; Osman and Sharrow, 1993).

For example, Markkola (1996) observed that repeated 25% defoliation of Scots pine (*Pinus sylvestris*) had little impact on tree biomass. Repeatedly removing 50 or 75% of seedling foliage reduced both root and shoot biomass by approximately 40%. This maintained a fairly constant root/shoot ratio across all defoliation treatments. Ectomycorrhizae were reduced by defoliation proportionally to reductions in root biomass. Since ectomycorrhizae fungi are important contributors to phosphorous uptake in host plants, their reduction has serious implications for nutrient uptake by defoliated seedlings. The tendency of plants to maintain a balance between photosynthetic shoot surface and absorptive fine root surface also occurs in herbaceous plants (Briske, 1991). Grazed plants typically reestablish a favorable root/shoot ratio by reducing the rate of fine root production compared to root senescence. This makes them less competitive for soil resources. Grazing may, therefore, indirectly benefit plants by shifting the balance of competition to favor ungrazed plants or those with the ability to rapidly regrow following defoliation. Karl and Doescher (1993), for instance, observed that grazing orchardgrass in a young Douglas-fir/Ponderosa pine (*Pinus ponderosa*) plantation reduced subsequent height and root production of grasses. Tree water potential and growth both increased as a result of reduced moisture competition from grazed grasses (Doescher et al., 1989). Interestingly, cattle grazing was as effective in promoting tree growth as total vegetation removal. Even when livestock browse young trees, the accompanying reduction in competing vegetation may result in a net benefit, particularly when trees are competing with other woody plants (Sharrow et al., 1989, 1992; Sharrow, 1994). An analysis of protection techniques to reduce browsing of young conifers by deer, for instance, concluded that weed control was more effective in promoting tree growth than was protection from animal damage (Gourley et al., 1990).

Timing, intensity, and duration of grazing are used to manipulate pasture forage dynamics in order to maintain a significant herbaceous legume component (Wilson, 1978; Langer, 1990). Silvopasture grasses and trees ultimately benefit from soil nitrogen enrichment by nitrogen-fixing legumes. The relative balance of negative direct effects and positive indirect effects of herbivory ultimately determine if individual plant components are advantaged or disadvantaged. Experience in silvicultural prescription grazing in commercial forests has generally been positive with growth increases of several conifer species ranging from 5–44% in height and 8–61% in diameter reported as a result of grazing (Sharrow, 1994).

3. SILVOPASTORAL SYSTEM RESPONSES

3.1 Net Effects

3.1.1 Stand Level Scale Responses

Uncontrolled competition between established pasture plants and newly planted trees interferes with conifer establishment in existing pastures. Newly planted pastures are less competitive with trees for soil resources, and vegetation control the first growing season may not be required. Although forage yields during the first few years after tree planting are similar to those of open pasture, grazing use must be restricted to reduce potential browsing damage to trees. In New Zealand silvopastures, for example (Knowles, 1991), only 20%, 40%, and 80% of normal pasture grazing may be obtained during the first three years, respectively. The lax grazing regime during this period favors grasses over clovers (Knowles, 1991). However, clovers will quickly recover from only one or two years of lax grazing once normal grazing resumes (Knowles, 1991). Once trees are well established and the top whorl of foliage is above the reach of livestock, normal grazing can resume provided that forage is always adequate. Livestock damage to trees is most likely to occur in summer when pastures are dry and other green forage is scarce or just after bud break when new succulent growth is present (Leininger and Sharrow, 1989; Sharrow et al., 1992). Removal of competing vegetation by skillfully applied grazing in silvopastures helps reduce moisture stress on young conifers during hot dry periods (Doescher et al., 1989; Carlson et al., 1994) and may increase conifer growth (Jaindl and Sharrow, 1988; Doescher et al., 1989).

As trees grow, so does their influence on silvopasture productivity. Silvopasture trees are typically planted at lower density and into a fertile pasture environment where competition with ground vegetation is controlled. Their growth rates often meet or exceed those obtained from commercial forests on similar sites. However, competition generally exceeds facilitation for the forage component once trees become established. Stand level investigations of tree/understory relationships in mid to late rotation forested rangelands have shown a general reduction of forage production with increasing conifer tree basal area (Clary et al., 1975; Tapia et al., 1990) or canopy (Clary. 1979; Johnson et al., 1986; Joyce and Mitchell, 1989). Although forage overstory/understory models are often presented as linear relationships (Joyce and Mitchell, 1989; Mitchell and Bartling, 1991), the true relationship over the life of a silvopasture is most likely curvilinear with little effect until tree canopy exceeds 30–50% coverage (Krueger, 1981; Joyce and Mitchell, 1989), followed by a rapid decline in understory production as tree canopies coalesce. Total live tree canopy length has proven useful as a predictor of understory pasture yield for pruned conifers (Knowles, 1991; Sibbald et al., 1994) in established silvopastures.

The same general trend of reduced understory pasture and livestock production with increasing tree leaf area, resulting from changes in tree density and/or growth has been reported from silvopastures (Johnson et al., 1986; Anderson, 1987; Hawke, 1991; Knowles, 1991; Sharrow et al., 1996). Clover content of pastures tends to decline with increasing tree canopy (Johnson et al., 1986). Subclover is relatively tolerant of tree overstories and has been observed to yield 1500–4500 kg/ha of dry matter under 50% canopy cover (Johnson et al., 1986). Although subclover is considered to be relatively shade tolerant, the ability to coexist with trees may be explained by its ability to prosper on dry sites. Hawke (1991) noted that forage production of white clover (*Trifolium repens*), which requires a somewhat less xeric site than subclover, declined relative to perennial ryegrass in mid-rotation silvopastures, while understory in closed canopy silvopastures was largely annual grasses.

As one might expect, reducing the number of trees initially planted, early thinning of existing stands, and canopy removal by pruning trees helps to maintain pasture production longer into the timber rotation (Anderson, 1987; Hawke, 1991; Knowles, 1991). Pruning and non-commercial thinning can generate considerable amounts of debris which cover pasture and provide safe sites for rodents and weeds. Anderson (1987) reported that such debris covered approximately 8–35% of land area within a six–nine-year-old *Pinus radiata* silvopasture. Although not widely researched, pattern of tree planting can influence both competition between trees and overstory-understory interactions (Huxley, 1985; Buck, 1986). Aggregating trees together into rows or clusters reduces the tree:crop interface (Huxley, 1985) along which trees and pasture interact compared to traditional rectangular forest grid plantings. The wide pasture interspaces between single or double rows of trees support higher levels of pasture production and facilitate livestock movement and other agricultural operations (Sharrow, 1991). Tree pattern is relatively unimportant at low tree density, but becomes increasingly influential as tree density increases and as trees become larger (Sharrow, 1991). Conceptually, aggregation of trees fosters earlier competition between them for site resources. There is some evidence that this does indeed occur (Hawke, 1991), but the issue is readily dealt with by planting fewer trees initially, or by earlier thinning and pruning (Figure 4).

3.1.2 Individual Tree Scale Responses

Net interactions between trees and ground vegetation often result from a complex set of competitive and facilitative processes whose relative importance varies spatially around trees (Scanlon, 1982; Sharrow, 1991). Scanlan's (1982) model of woody-herbaceous plant relationships separates plant interactions into competitive and stimulatory elements which both inde-

FIGURE 4. Early pruning of KMX hybrid pines (foreground) and Douglas-fir (background) in western Oregon reduces tree impacts on pasture production while increasing saw log quality. (Photo Courtesy of Rick Fletcher)

pendently tend to decrease with distance from the tree. Depending upon the relative rates of decrease, the interplay of competition and facilitation can produce patterns of crop response similar to those reported by Khan (1975) in which *Dalbergia sissoo* trees reduced wheat grain yields within 6 m of the trees, but increased yield 6–13 m from trees. Data collected from Douglas-fir/subclover silvopastures by Sharrow (1991) suggest that, while forage production decreases near trees, a zone of net facilitation in which pasture production exceeds that of open pasture exists about three canopy diameters out from trees. Many authors (Woods et al., 1982; Joyce and Mitchell, 1989; Cameron et al., 1991) have noted a pattern of detrimental effects which lessen with increasing distance from the tree. Understory plant response to these effects varies by tree and forage species. Distinct compositional zones of native forage plants have been described around both juniper (*Juniperus monosperma*) (Arnold, 1964) and *Pinus monophylla* (Everett et al., 1984) trees. These zones may be asymmetric with the north side of trees supporting more mesic plant communities compared to the more xeric south side (Everett et al., 1984). Pieper (1990) found that production of cool season grasses increased under canopy cover of *Pinus edulis* and juniper trees while that of warm season grasses declined. Data presented for several cool season grass/clover silvopastures by Goh et al. (1996) suggest that while total annual pasture production was highest in the center between rows of trees, production was higher on the sunny north edge of rows in early spring and lower in late summer compared to row centers. A reverse trend, higher forage production in late summer/lower in spring was apparent for more shady south edges of rows. Overyielding was also evident with ryegrass/clover silvopastures producing more forage during summer, particularly on the north side of tree rows, than was produced in open pastures without trees (Yunusa et al., 1995) (Figure 5).

FIGURE 5. Pasture under widely spaced southern pines provides a park-like setting for combined livestock and saw log production. (Photo courtesy of Terry Clason)

3.2 Overyielding

A well designed system is more than the sum of its parts. Resource sharing in time and space (Buck, 1986) is fundamental to achieving synergy in silvopastures. Resource sharing includes interactions between components drawing from the same resource pool (true sharing), as well as resource segregation in which components draw from resource pools separated in either time or space. To the extent that facilitation exceeds competition for shared resources and resource use can be segregated, silvopastures should be able to outproduce pasture or forest monocultures. Resource sharing is inherent in the tree/pasture interactions previously discussed. Resource segregation in time may occur when pasture and tree growth periods do not entirely overlap. In Oregon Douglas-fir/perennial ryegrass/subclover pastures, for example, pastures produce more than 95% of their annual yield during the warm moist spring (March–June), a period when Douglas-fir trees produce approximately 65% of their total annual diameter growth. Since rainfall is frequent during most of this period, competition between plants for soil moisture is limited to the end of the pasture growing period. In addition, 95% of the root systems of the pasture are contained in the top 16 cm of soil (Motazedian and Sharrow, 1987), leaving the lower soil profile to support summer growth of Douglas-fir. This resource segregation in both time and space is a favorable situation for over-yielding. Forage production of these silvopastures equalled that of open pastures during the initial five years, while tree diameter and height growth were 6% and 14% greater, respectively, than forest plantations on the same site. Thus, a hectare of silvopasture produced as much forage as a hectare of pasture plus more tree growth than a hectare of forest.

The system productivity of polycultures may be compared to that of their component monocultures by calculating the Land Equivalency Ratio (LER). The LER (Vandermeer, 1981) is the amount of land in monoculture required to equal the combined output of a polyculture. Hawke's (1991) observation that pasture production under a low density *Pinus radiata* stand exceeded that of open pasture suggests an opportunity for overyielding in low density silvopastures. Sharrow et al. (1996) reported a LER of 1.6, indicating that 1.6 hectares of pasture and forest plantation would be required to equal the forage and tree output of one hectare of young tall fescue (*Festuca arundinacea*)/subclover/Douglas-fir silvopasture in western Oregon. Calculations using data presented in Anderson's (1987) review of pine/subclover based silvopastures in Australia suggest LER ranging from 1.18–1.35 for these systems. The LER of young *Pinus radiata* silvopasture in New Zealand was 1.71 for ryegrass/clover and 1.95 for ryegrass understories (Yunusa, 1995). The LERs reported for temperate cool season pasture/conifer silvopastures are within the top 5% of the 506 forage polycultures reviewed by Hiebsch and McCollum (1987). The relatively high LER reported for silvopastures probably reflects synergy obtained from both efficient resource sharing and resource segregation between plant components differing in physical stature and life cycles. Although warm season pasture/conifer silvopastures are becoming common in summer rainfall areas of the world, little has been reported about their potential for overyielding.

Conceptually, agroforests go through three stages: (1) a pasture dominated stage immediately after tree establishment; (2) an intermediate stage where pasture, trees, and livestock share resources; and (3) a forest stage after canopy closure. Production of stages 1 and 3 are dominated by a single pasture or tree component to an extent typical of monocultures. The LER of these stages should be close to 1. Prospects for overyielding are present predominately in the 2nd stage and should reach a maximum LER value somewhere near the middle of the period. Agroforests are frequently designed and managed to maintain period 2 as long as possible. Management actions which increase the length of this period, such as early tree thinning and pruning, should greatly increase the opportunities for overyielding. Silvopastures managed as open canopied forests may never enter period 3. Although seldom discussed, it appears that fertilization may overcome some of the competition effects of trees during the latter portion of period 2. If this proves to be true, it offers hope to increase LER well beyond the high values already reported. Residual effects of pasture production and fertilization on soil structure and nutrients also may support increased growth of silvopasture trees during period 3.

4. CONCLUSIONS — THE FUTURE

Silvopastures are complex ecosystems. They are purposefully designed to serve human needs, not to mimic nature. Nevertheless, they must follow the same chemical, physical, and biological laws which operate in unmanaged "natural" systems. The very complexity of both natural and human created ecosystems makes a predictive understanding of their operations difficult. Silvopasture design is currently based upon knowledge and principles from the existing fields of animal science, agronomy, forestry, and ecology. All knowledge exists within a context, however. Most of our current paradigms are drawn from contexts which differ from those of silvopastures. Great care must be exercised when transferring agricultural or forestry information to more multidimensional undertakings such as agroforestry. Much of our current understanding is empirical, derived from trial and error. Silvopastoral research is mostly less than 20 years old. In this short period, production systems have been developed which rival the biological and economic efficiency of the best agronomic polycultures. The main advances in silvopasture design are still to be realized. Understanding the interactions between plant, animal, and environmental components of silvopastoral systems promises to provide a firm base for future silvopasture design and to open a new window into how organisms interrelate in natural systems.

REFERENCES

Akbar, G., Ahmad, M., Refique, S., and Babar, K. N., 1990, "Effect of tree on the yield of wheat crop," *Agro. Syst.*, 11, 1.

Anderson, G. W., 1987, "Integrated livestock and timber production: beneficial systems in Australia," *Wa. State U., Agroforestry Consortium Tech. Bull. #4*, 11.

Arnold, Journal, 1094, "Zonation of understory vegetation around a juniper tree," *J. Range Mgt.*, 17, 41.

Baraldi, R., Bertazza, G., Bogino, J., Luna, V., and Bottini, R., 1995, "The effect of light quality on *Prunus cerasus* II. Changes in hormone levels in plants grown under different light conditions." *Photochemistry–photobiology*, 62, 800.

Bari, M. A. and Schofield, N. J., 1991, "Effects of agroforestry-pasture associations on groundwater level and salinity," *Agro. Syst.*, 16, 13.

Bazely, D. R., Myers, J. H., and deSilva, K. B., 1991, "The response of numbers of bramble prickles to herbivory and depressed resource availability," *Oikos*, 61, 327.

Blair, R. M., Alcaniz, R., and Harrell, A., 1983, "Shade intensity influences the nutrient quality and digestibility of southern deer browse leaves," *J. Range Mgt.*, 36, 257.

Braziotis, D. C. and Papanastasis, V. P., 1995, "Seasonal changes of understory herbage yield in relation to light intensity and soil moisture content in a *Pinus pinaster* plantation," *Agro. Syst.*, 29, 91.

Briske, D. D., 1991, "Developmental morphology and physiology of grasses," in *Grazing management: an ecological perspective*, Heitschmidt, R. K., and Stuth, J. W., Eds., Timber Press, Portland, OR, 85.

Brown, R. H., 1982, "Response of terrestrial plants to light quality, light intensity, temperature, CO_2, and O_2 carbon dioxide and oxygen," in CRC-Handbook-Biosol-Resources, CRC Press, Boca Raton, FL, 1, 185.

Buck, M. G., 1986, "Concepts of resource sharing in agroforestry systems," *Agro. Syst.*, 4, 191.

Callaway, R. M., Nadkarni, N. M., and Mahall, B. E., 1991, Facilitation and interference of *Quercus douglasii* on understory productivity in central California, *Ecology*, 72, 1484.

Cameron, D., Rance, S., Jones, R., and Charles-Edwards, D., 1991, "Trees and pasture: a study on the effects of spacing," *Agrofor. Today*, January–March, 8.

Campbell, C. D., Atkinson, D., Jarvis, P, G., and Newbould, P., 1994, "Effects of nitrogen fertilizer on tree/pasture competition during the establishment phase of a silvopastoral system," *Ann. App. Bot.*, 124, 83.

Cannell, M. G. R. and J. Grace, J., 1969, "Competition for light: detection, measurement, and quantification," *Can. J. For. Res.*, 23.

Carlson, D. H., Sharrow, S. H., Emmingham, W. H., and Lavender, D. P., 1994, "Plant-soil-water relations in forestry and silvopastoral systems in Oregon," *Agro. Syst.*, 25, 1.

Catchpoole, D. W. and Blair, G. J., 1990, "Forage tree legumes. II Investigation of nitrogen transfer to an associated grass using a split-root technique," *Aus. J. Ag. Res.*, 41, 531.

Clary, W. P., 1979, "Grazing and overstory effects on rotationally burned slash pine plantation ranges," *J. Range Mgt.*, 32, 264.

Clary, W. P., Kruse, W. H., and Larson, F. R., 1975, "Cattle grazing and wood production with different basal areas of ponderosa pine," *J. Range Mgt.*, 28, 434.

Clason, T. R., 1995, "Economic implications of silvipastures on southern pine plantations," *Agro. Syst.*, 29, 227.

Cole, E. C. and Newton, M., 1986, "Fifth-year responses of Douglas-fir to crowding and nonconiferous competition," *Can. J. Forest Res.*, 17, 181.

Dangerfield, C. W., Jr., and Harwell, R. L., 1990, "An analysis of a silvopastoral system for the marginal land in the southeast United States," *Agro. Syst.*, 10, 187.

Davis, M. H. and Simmons, S. R., 1994, "Tillering response of barley to shifts in light quality caused by neighboring plants," *Crop Sci.*, 34, 1604.

Dawson, M. D. and McGuire, W. S., 1972, "Recycling nitrogen and sulfur in grass–clover pastures," *Or. Ag. Exp. Sta. Bull. 610*. 12.

deSilva, J. H. S., Johnson, W. L., Burns, J. C., and Anderson, C. E., 1966, "Growth and environment effects on anatomy and quality of temperate and subtropical forage grasses," *Crop Sci.*, 27.

Doescher, P. S., Tesch, S. D., and Alejandro-Castro, M., 1987, "Livestock grazing: a silvicultural tool for plantation establishment," *J. Forest.*, 85, 29.

Doescher, P. S., Tesch, S. D., and Drewien, W. E., 1989, "Water relations and growth of conifer seedlings during three years of cattle grazing on a southwest Oregon plantation," *NW Sci.*, 63, 232.

Dupraz, C., 1994, "Prospects for easing land tenure conflicts with agroforestry in Mediterranean France: a research approach for intercropped timber orchards," *Agro. Syst.*, 25, 181.

Eis, S., 1987, "Root systems of older immature hemlock, cedar, and Douglas-fir," *Can. J. Forest Res.*, 17, 1368.

Eissenstat, D. M. and Mitchell, J. E., 1983, "Effects of seeding grass and clover on growth and water potential of Douglas-fir seedlings," *Forest Sci.*, 29, 166.

Everett, R. L., Sharrow, S. H., and Meeuwig, R. O., 1984, "Pinyon-juniper woodland understory distribution patterns and species associations," *Bull. Torrey Bot. Club,* 110, 454.

Felker, F. C., Doehlert, D. C., and Eskins, K., 1995, "Effects of red and blue light on the composition and morphology of maize kernels grown *in vitro*," in *Plant Cell,-tissue-organ-culture,* Dordrecht, Netherlands: Kluwer Academic Publishers, 42, 147.

Frank, A. B. and Hofmann, L., 1994, "Light quality and stem numbers in cool season forage grasses," *Crop Sci.*, 34, 468.

Frost, W. E. and McDougald, N. K., 1989, "Tree canopy effects on herbaceous production of annual rangeland during drought," *J. Range Mgt.*, 42, 281.

Gardner, F. P., Pearce, B. B., and Mitchell, R. L., 1985, *Physiology of crop plants*, Iowa State University Press, Ames, IA.

Garner, W, and Steinberger, Y., 1989, "A proposed mechanism for the formation of fertile islands in the desert ecosystem," *J. Arid Env.*, 16, 257.

Garrett, H. E., 1994, *Agroforestry: an integrated land-use management system for production and farmland conservation*, USDA, SCS, unnumbered report, 58.

Garrison, M. and Pita, M., 1992, "An evaluation of silvopastoral systems in pine plantations in the central highlands of Ecuador," *Agro. Syst.*, 18, 1.

Gehring, C. A. and Whitham, T. A., 1991, "Herbivore-driven mycorrhizal mutualism in insect-susceptible pinyon pine," *Nature*, 353, 556.

Goh, K. M., Mansur, I., Mead, D. J., and Sweet, G. B., 1996, "Biological nitrogen fixing capacity and biomass production of different understory pastures in a *Pinus radiata* — pasture agroforestry system in New Zealand," *Agro. Syst.*, 34, 33.

Gourley, M., Vomocil, M., and Newton, M., 1990, "Forest weeding reduces the effect of deer-browsing on Douglas-fir," *For. Ecol. Mgt.*, 36, 177.

Happe, P. J., Jenkins, K. J., Starkey, E. E., and Sharrow, S. H., 1990, "Nutritional quality and tannin astringency of browse in clear-cuts and old-growth forests," *J. Wildlife Mgt.*, 54, 557.

Harper, J. L., 1977, *Population biology of plants*, Academic Press, London, 892.

Harris, N. and Sharrow, S. H., *Unpublished data.* Department of Rangeland Resources, Oregon State University, Corvallis, OR.

Hart, R. H., Hughes, R. H., Lewis, C. E., and Monson, W. G., 1970, "Effect of nitrogen and shading on yield and quality of grasses grown under young slash pines," *Agron. J.*, 62, 285.

Hawke, M. F., 1991, "Pasture production and animal performance under pine agroforestry in New Zealand," *For. Ecol. Mgt.*, 45, 109.

Hawke, M. F. and Wedderburn, M.E., 1994, "Microclimate changes under *Pinus radiata* agroforestry regimes in New Zealand," *Ag. For. Meteorol.*, 71, 133.

Henderson, M. S. and Robinson, D. L., 1982, "Environmental influences on yield and *in vitro* true digestibility of warm-season perennial grasses and the relationship to fiber components," *Agron. J.*, 74, 943.

Hiebsch, C. K. and McCollum, R. E., 1987, "The area X equivalency ratio: a method for evaluating the productivity of intercrops," *Agron. J.*, 79, 15.

Hoglund, J. H., Crush, J. R., Brock, J. L., Ball, R., and Carran, R. A., 1979, "Nitrogen fixation in pasture XII. General discussion," *New Zealand J. Exp. Agri.*, 7, 45.

Hunter, A. F. and Anderson, L. W., 1988, "Plants helping plants," *Biosci.*, 38, 34.

Huxley, P. A., 1985, "The tree/crop interface — or simplifying the biological/environmental study of mixed cropping agroforestry systems," *Agro. Syst.*, 3, 251.

Jackson, L. E., Strauss, R. B., Firestone, M. K., and Bartolome, J. W., 1990, "Influence of tree canopies on grassland productivity and nitrogen dynamics in deciduous oak savanna," *Ag. Ecosyst. Env.*, 32, 89.

Jaindl, R. G. and Sharrow, S. H., 1988, "Oak/Douglas-fir/sheep: A three-crop silvopastoral system," *Agro. Syst.*, 6, 147.

Joffre, R. and Rambal, S., 1993, "How tree cover influences the water balance of Mediterranean rangelands," *Ecology*, 74, 570.

Johnson, M. K., Davis, L. G., Ribbeck, K. F., Render, J. H., and Pearson, H. A., 1986, "Management of subterranean clover in pine forested range," *J. Range Mgt.*, 39, 454.

Joyce, L. A. and Mitchell, J. E., 1989, "Understory cover/biomass relationships in Alabama forest types," *Agro. Syst.*, 9, 205.

Karl, M. G. and Doescher, P. S., 1993, "Regulating competition on conifer plantations with prescribed cattle grazing," *Forest Sci.*, 39, 405.

Khan, G. S., 1975, "A study of the shading effects of shisham trees on wheat crop," *Pak. J. Ag. Sci.*, 12, 281.

Knowles, R. L., 1991, "New Zealand experience with silvopastoral systems: A review," *For. Ecol. Mgt.*, 45, 251.

Kort, J., 1988, "Benefits of windbreaks to field and forage crops," *Ag. Ecosyst. Env.*, 22, 165.

Krueger, W. C., 1981, "How a forest affects a forage crop," *Rangelands*, 3, 70.

Kuiper, L. C. and Coutts, M. P., 1992, "Spatial disposition and extension of the structural root system of Douglas-fir," *For. Ecol. Mgt.*, 47, 111.

Langer, R. H. M., 1990, *Pastures, their ecology and management*, Oxford University Press, New York, 499.

Lawrence, J. H. and Hardesty, L. H., 1992, "Mapping the territory: agroforestry awareness among Washington state land managers," *Agro. Syst.*, 19, 27.

Leininger, W. C. and Sharrow, S. H., 1987, "Seasonal diets of herded sheep grazing Douglas-fir plantations," *J. Range Mgt.*, 40, 551.

Leininger, W. C. and Sharrow, S. H., 1989, "Seasonal browsing of Douglas-fir seedlings by sheep," *W. J. App. Forestry*, 4, 73.

Lewis, C. E., 1980, "Simulating cattle injury to defoliated slash pine: Defoliation," *J. Range Mgt.*, 33, 345.

Lewis, C. E., 1985, "Planting slash pine in a dense pasture sod," *Agro. Syst.*, 3, 267.

Malajczuk, G., Moore, R., and Anderson, G., 1996, "The economics of agroforestry with pine and pasture in the 500 to 700 mm annual rainfall zone of western Australia," *Agro. Syst.*, 33, 51.

Markkola, A. M., 1996, "Effect of artificial defoliation on biomass allocation in ectomycorrhizal *Pinus sylvestris* seedlings," *Can. J. Forest Res.*, 26, 899.

Marschner, H., Haussling, and M., George, E., 1991, "Ammonium and nitrate uptake rates and rhizosphere pH in non-mycorrhizal roots of Norway spruce [*Picea abies* (L.) Karst]," *Trees:-Structure- and-Function*, 5, 1.

McArthur, A. J., 1991, "Forestry and shelter for livestock," *For. Ecol. Mgt.*, 45, 93.

McClaran, M. P. and Bartolome, J. W., 1989, "Effects of *Quercus douglasii* Fagaceae on herbaceous understory along a rainfall gradient," *Madrono*, 36, 141.

McCune, B., 1986, "Root competition in a low-elevation grand fir forest in Montana: a trenching experiment," *NW Sci.*, 60, 52.

McDonald, P. M., 1986, "Grasses in young conifer plantations — hindrance and help," *NW Sci.*, 60, 271.

Mead, D. J., 1995, "The role of agroforestry in industrialized nations: The southern hemisphere perspective with special emphasis on Australia and New Zealand," *Agro. Syst.*, 31, 143.

Merwin, M. L., Ed., 1997, *The status, opportunities and needs for agroforestry in the United States*, Association for Temperate Agroforestry, Lexington, KY, 41.

Miller, J. H., Zutter, B. R., Zedaker, S. R., Edwards, M. B., Haywood, J. D., and Newbold, R. A., 1991, "A regional study on the influence of woody and herbaceous competition on early loblolly pine growth," *So. J. App. Forestry*, 15, 169.

Mitchell, A. K. and Arnott, J. T., 1995, "Effects of shade on the morphology and physiology of amabilis fir and western hemlock seedlings," *New-Forests*, 10, 79.

Mitchell, J. E. and Bartling, P. N. S., 1991, "Comparison of linear and non-linear overstory-understory models for ponderosa pine," *For. Ecol. Mgt.*, 42, 195.

Moe, R. and Heins, R., 1990, "Control of plant morphogenesis and flowering by light quality and temperature," *Acta Horticulturae*, 272, 81.

Motazedian, I. and Sharrow, S. H., 1987, "Persistence of a *Lolium perenne–trifolium subterraneum* pasture under differing defoliation treatments," *J. Range Mgt.*, 40, 232.

Nambiar, E. K. S. and Zed, P. G., 1980, "Influence of weeds on water potential, nutrient content and growth of young radiata pine," *Aus. Forest Res.*, 10, 279.

Newsome, T., Wikeem, B., and Sutherland, C., 1995, "Sheep grazing guidelines for managing vegetation on forest plantations in British Columbia," *B.C. Ministry of Forests, Land Management Handbook #34*, 47.

Ong, C. K., Corlett, J. E., Singh, R. P., and Black, C. R., 1991, "Above and below ground interactions in agroforestry systems," *For. Ecol. Mgt.*, 45, 45.

Osman, K. A. and Sharrow, S. H., 1993, "Growth responses of Douglas-fir (*Pseudotsuga menziesii*) to defoliation," *For. Ecol. Mgt.*, 60, 105.

Parsons, A. J., Orr, R. J., Penning, P. D., and Lockyer, D R., 1991, "Uptake, cycling and fate of nitrogen in grass-clover swards continuously grazed by sheep," *J. Ag. Sci.* Cambridge, U.K., 116, 47.

Pearson, H. A., Elliott-Smith, M. L., Martin, Jr., A., and Baldwin, V, C., 1994, "Subterranean clover: nutrients for pines," *Proc. 3rd Annual Southern States Chemical Science Meeting*, Rials, T. G., Ed., U.S. Department of Agriculture., Southern For. Res. Exp. Sta. Gen. Tech. Rep. SO-10, New Orleans, LA, 14.

Penaloza, R., Herve, M., and Sobarzo, L., 1985, "Applied research on multiple land use through silvopastoral system in southern Chile," *Agro. Syst.*, 3, 59.

Pieper, R., 1990, "Overstory-understory relations in pinyon-juniper woodlands in New Mexico," *J. Range Mgt.*, 43, 413.

Ratliff, R. D., Duncan, D. A., and Westfall, S. E., 1991, "California oak–woodland overstory species affect herbage understory: Management implications," *J. Range Mgt.*, 44, 306.

Rhodes, B. D. and Sharrow, S. H., 1990, "Effect of grazing by sheep on the quantity and quality of forage available to big game in Oregon's Coast Range," *J. Range Mgt.*, 43, 235.

Riegel, G. M., Miller, R. F., and Krueger, W. C., 1991, "Understory vegetation response to increasing water and nitrogen levels in a *Pinus ponderosa* forest in Northeastern Oregon," *NW Sci.*, 65, 10.

Riegel, G. M., Miller, R. F., and Krueger, W. C., 1992, "Competition for resources between understory vegetation and overstory *Pinus Ponderosa* in northeastern Oregon," *Eco. App.*, 2, 71.

Robin, C., Hay, M. J. M., Newton, P. C. D., and Greer, D. H., 1994, "Effect of light quality (red:far-red ratio) at the apical bud of the main stolon on morphogenesis of *Trifolium repens* L.," *Ann. Bot.*, 74, 119.

Scanlon, Journal, 1982, "A model of woody–herbaceous biomass relationships in eucalypt and mesquite communities," *J. Range Mgt.*, 45, 75.

Sharrow, S. H., 1991, "Tree planting pattern effects on forage production in a Douglas-fir agroforest," *Agro. Syst.*, 16, 167.

Sharrow, S. H., 1994, "Sheep as a silvicultural management tool in temperate conifer forest," *Sheep Res. J.*, Special issue, 97.

Sharrow, S. H., 1995, "Western Oregon hill land agroforestry: Interactions among components," *Proc., 4th North American agroforestry conference: growing a sustainable future*, Boise, ID, July 23–28, 138.

Sharrow, S. H., Carlson, D. H., Emmingham, W. H., and Lavender, D. P., 1992, "Direct impacts of sheep upon Douglas-fir trees in two agrosilvopastoral systems," *Agro. Syst.*, 19, 223.

Sharrow, S. H., Carlson, D. H., Emmingham, W. H., and Lavender, D., 1996, "Productivity of two Douglas-fir/subclover/sheep agroforests compared to pasture and forest monocultures," *Agro. Syst.*, 34, 305.

Sharrow, S. H. and Fletcher, R. A., 1995, "Trees and pastures: 40 years of agrosilvopastoral experience in Western Oregon," *Proceedings, Agroforestry and sustainable systems symposium,* Fort Collins, CO, Aug. 7–10, 1994, USDA Gen. Tech. Rep. RM-GTR-261. 47.

Sharrow, S. H., Krueger, W. C., and Thetford, Jr., F. O., 1981, "Effects of stocking rate on sheep and hill pasture performance," *J. Anim. Sci.*, 52, 210.

Sharrow, S. H., Leininger, W. C., and Osman, K. A., 1992, "Sheep grazing effects on coastal Douglas-fir growth: A ten-year perspective," *For. Ecol. Mgt.*, 50, 75.

Sharrow, S. H., Leininger, W. C., and Rhodes, D. B., 1989, "Sheep grazing as a silvicultural tool to suppress brush," *J. Range Mgt.*, 42, 2.

Sibbald, A. R., Griffith, J. H., and Elston, D. A., 1994, "Herbage yield in agroforestry systems as a function of easily measured attributes of tree canopy," *For. Ecol. Mgt.*, 65, 194.

Spreitzer, P.N., 1985, "Transitory range: a new frontier," *Rangelands*, 7,33.

Stone, E. L. and Kalisz, P. J., 1991, "On the maximum extent of tree roots," *For. Ecol. Mgt.*, 46, 59.

Tapia, L. A. B., Folliott, P. F., and Guertin, D. P., 1990, "Herbage production–forest overstory relationships in two Arizona ponderosa pine forests," *J. Range Mgt.*, 43, 25.

USDA, 1982, "Land for all reasons," *Farmline*, 3, 4.

USDA, 1986, *Grazing lands and people: a national program statement and guidelines for the cooperative extension service*, United States Department of Agriculture, Extension Service unnumbered report, December 1996, 17.

Vandermeer, Journal, 1981, "The interference production principle: an ecological theory for agriculture," *Biosci.*, 31, 361.

Walters, T. and Stafford, H. A., 1989, "Variability in accumulation of proanthocyanidins (condensed tannins) in needles of Douglas-fir (*Pseudotsuga menziesii*) following long-term budworm defoliation," *J. Chem. Ecol.*, 10, 1469.

Waring, H. D. and Snowdon, P., 1985, "Clover and urea as sources of nitrogen for the establishment of *Pinus radiata*," *Aus. Forest Res.*, 15, 115.

Wilson, J. R., Ed., 1978. *Plant relations in pastures*, CSIRO, Melbourne, Australia, 425.

Wilson, G. J., Kellas, J., and Kirby, G., Undated. *Livestock Havens*, Department of Conservation and Environment, New South Wales, Australia.

Woods, R., Betters, D., and Mogren, E., 1982, "Understory herbage production as a function of Rocky Mountain aspen stand density," *J. Range Mgt.*, 35, 380.

Woods, P. V., Nambiar, E. K. S., and Smethurst, P. J., 1992, "Effect of annual weeds on water and nitrogen availability to *Pinus radiata* trees in a young plantation," *For. Ecol. Mgt.*, 48, 145.

Xu, M. G., Zhang, Y. P., Sun, B. H., Xu, M. G., Zhang, Y. P., and Sun, B. H., 1996, "Mechanisms for the movement of Fe, Mn, Cu and Zn to plant roots in loessal soil and lou soil," *Pedosphere*, 6, 245.

Young D. and Smith, W., 1982, "Simulation studies on the influence of understory location on transpiration and photosynthesis relations of *Arnica cordifolia* on clear days," *Ecology*, 63, 1761.

Young D. and Smith, W., 1983, "Effect of cloud cover on photosynthesis and transpiration in subalpine understory species, *Arnica latifolia*," *Ecology*, 64, 681.

Yunusa, I. A. M., Mead, D. J., Lucas, R. J., and Pollock, K. M., 1995, "Process studies in a *Pinus radiata* — pasture agroforestry system in a subhumid temperature environment. II. Analysis of dry matter yields in the third year," *Agro. Syst.*, 32, 185.

7 Pest Management in Energy- and Labor-Intensive Agroforestry Systems

Mary Ellen Dix, Badege Bishaw, Sarah W. Workman, Michael R. Barnhart, Ned B. Klopfenstein, and Anne M. Dix

1. INTRODUCTION

Pests destroy more than 30% of the food, fiber, feed, and energy produced globally in agricultural systems (James, 1989). To reduce these losses, producers annually expend nearly $20 billion on pesticides and $150 million on other plant protection techniques worldwide (James, 1989). Integrated techniques for managing pests are extensively researched and well documented for agricultural systems

1-56670-294-1/99/$0.00+$.50

especially those consisting of extensive monocultures. However, in agroforestry systems integrated pest management (IPM) is less widely studied and the approach depends on the degree of energy inputs or labor inputs within the system. This chapter provides an overview of current and potential IPM in both high energy (mechanical) and high labor-intensive (indigenous) agroforestry systems. Techniques will be discussed that are applicable to both types of agroforestry systems with examples from diverse geographic locations.

2. INTEGRATED PEST MANAGEMENT

The 20th century agricultural revolution in industrial countries occurred in part because pesticides were developed. During the 1970s, the use of single pest control techniques to solve pest problems lost acceptance in industrial countries because pesticide resistance and subsequent pest resurgence threatened the stability of global agriculture. Technical packages for cash crops in developing economies included pesticides and high input techniques that resulted in social ills, hazards to human health, as well as pesticide resistance. Integrated pest management (IPM) concepts were developed to address the problem of pesticide resistance and ensure sustainable management of environmental resources vital to global economies and human well being (NRI, 1991; NRI, 1992; Hardy et al., 1996; Morse and Buhler, 1997).

IPM is a strategy for maintaining vertebrate and invertebrate pest, weed, and pathogen populations at levels below those that cause economic loss. It provides guidelines and techniques to identify pests and their natural regulators, recognize damaging pest thresholds, and manage pest populations when damaging thresholds are present (Epila, 1986; NRI, 1991; Hall, 1995). Rather than the transfer of specific control technologies, like pesticide application or biological control, IPM integrates all available pest management strategies, guidelines, techniques, and economic activities into the whole farming system. IPM should be regarded as a strategy to reduce losses that integrates methods and disciplines while considering environmental values and socio-economic parameters (Zethner, 1995; Hardy et al., 1996).

Ideally, practitioners can customize pest management strategies by selecting techniques appropriate to their individual situations, thereby minimizing unneeded intervention (NRI, 1991; Zethner, 1995; Hardy et al., 1996). However, a United States Department of Agriculture Forest Service survey of 57 specialists in 27 nations and examination of related agroforestry databases found that pesticides and biological control agents were the predominant pest management techniques on high-energy farms. These primary techniques were seldom integrated with other management techniques. Furthermore, few specialists recognized the contribution of passive management techniques and indigenous practices in pest management (Dix, 1996). The survey also found that implementation of IPM strategies was dependent on their perceived complexity, reliability, effectiveness, and impact on profits (Dix, 1996). Choice, however, is something limited to practitioners due to lack of access or knowledge. Lack of resources and knowledge can limit the IPM strategies and technologies implemented on energy- and labor-intensive agroforestry systems of temperate countries, Africa, Asia, and Central and South America (NRI, 1991; NRI, 1992; Black and Sweetmore, 1994; Mengech et al., 1995; Dix, 1996, Hardy et al., 1996; Morse and Buhler, 1997).

2.1 IPM Agroforestry Systems of Temperate Countries

In industrialized countries of North America, Europe, Asia, and Australia, only a small proportion of the population is directly involved in agricultural production. Most farms consist of large areas devoted to extensive crop or tree monocultures. Energy and resource inputs are generally high, and include large inputs of agrichemicals (Dix, 1996; Hardy et al., 1996). In these systems, trees frequently occur around field edges in windbreaks, shelterbelts, riparian areas, orchards, natural areas, and near homesteads (Forman and Baudry, 1984).

Pest management is an integral part of sustainable production on these farms, and it requires coordinating cultural farming practices with pest-resistant plants, pesticide application, biological control, and other pest management techniques (Schertz, 1991; Higley and Boethel, 1994; Meinke, 1995). For example, no-till, ridge-till, mulch-till, and other conservation tillage practices leave plant residues on the soil surface, thereby reducing soil erosion, decreasing surface runoff of agrichemicals, and saving fuel, time, and money. Surface residues can serve as overwintering sites for some pest species and their natural enemies (Schertz, 1991). Crop rotation is used to revitalize soil and reduce pest abundance. With crop rotation, availability of suitable hosts and varying cultural practices can adversely impact pest life cycles and survival in the field. Effective use of pest-resistant plants can also slow pest development, and reduce population vigor (Higley and Boethel, 1994). Pesticides can be applied at lower rates when combined with pest-resistant plants. Conventional pesticide application is an effective and economical means to reduce insect and disease damage on crops, especially for mid- and late-season pests (Higley and Boethel, 1994). Pesticide application is, however, not problem free. As an example, insect resistance to insecticides is reported on many crops, including soybeans (*Glycine max*) and corn (*Zea mays*) in the southwestern and central Great Plains of the United States (Higley and Boethel, 1994; Meinke, 1997). By minimizing pesticide use, populations of native natural enemies are allowed to build and hold pest populations in check (Higley and Boethel, 1994). Abundance of natural enemies and other beneficials is encouraged by growing cover crops in orchards, establishing hedgerows or windbreaks adjacent to crops, and leaving unmaintained vegetation corridors in and around the crop fields. These areas contain flowers and other vegetation essential to survival of natural enemies and bees (Apidae) (Altieri and Schmidt, 1985; Altieri, 1991). Unmaintained grassy corridors, for example, suppress populations of leaf- or stem-sucking insects in soybeans, while increasing abundance of their predators and parasites (Rodenhouse et al., 1992).

Edge vegetation is recognized as an important source of biodiversity that contributes to sustainability of agroforestry ecosystems by providing food and shelter to crop pests and their natural enemies (Altieri et al., 1983; Forman and Baudry, 1984; Altieri, 1991; Dix et al., 1997). Areas with low diversity are vulnerable to outbreaks of small mammal and arthropod pests or decreased crop pollination. Unlike birds and grasshoppers (Orthoptera), which can migrate, and many insects, which go into diapause, rodents do not possess the physiological and behavioral characteristics for surviving long periods of scarcity. In agroforestry systems, hedgerows, shelterbelts, woody edges, and other ecological enrichments provide food and shelter for rodents and their predators (Leus and Mercetis, 1997). For example, in the Knesha area of Bulgaria, an outbreak of field rodents resulted when vegetative habitat for predatory birds was removed (Dix et al., 1995). In Alberta, Canada, winter damage by small mammals to seedlings and small trees is more prevalent when brush piles and other shelters of their weasel predators are removed (Fiedler, 1986). In Taiwan, rodent densities are highest in areas where crops are intercropped with woody and herbaceous vegetation (Kuo and Liao, 1986). A Eurasian tree sparrow (*Passer montanus*) eradication program in China predisposed crops to increased insect damage to crops (Suyin, 1959). In Great Britain, low bumblebee diversity is attributed to intensive agriculture in areas where marginal lands are drained and hedgerows removed (Williams, 1986). Conversely, edge vegetation and associated debris can harbor overwintering curcullionid, pysllid, and chrysomellid pests, as well as coccinellid predators (Slosser and Boring, 1980; Roach and Thomas, 1991; Dix, 1993; Dix et al., 1993).

Edges can be managed to enhance biodiversity and increase stability of agroforestry systems. Flowering trees, shrubs, grasses, annual flowers, and successional vegetation serve as alternative sources of pollen and nectar for bees, bumblebees, butterflies, parasitic insects, and syrphid flies (Leius, 1967; Ayers and Harman, 1992; Lagerlöf et al., 1992). Trees and tall vegetation can provide vertical structure essential to survival of spiders and birds (Dix et al., 1995). Goldenrod and other vegetation can support aphid populations that serve as alternative food for coccinellid beetles when their primary prey is limited (Altieri and Whitcomb, 1979). In Germany, nest boxes are placed in fruit orchards to increase Eurasian tree sparrows, a predator of fruit tree pests and chrysomelid beetles (Summers-Smith, 1988).

2.2 African IPM Agroforestry Systems

African agriculture is typified by farmers who cultivate small holdings, from less than one hectare to a few hectares, in agroforestry systems composed of annual crops associated with trees (Bishaw et al., 1994; Zethner, 1995). Crops are often cultivated in mixtures of annual and perennial species under scattered trees (e.g., parklands of semiarid areas) or annual species intercropped or alley cropped with woody species (e.g.,humid areas). These traditional agroforestry systems ensure food security of annual and perennial species, optimal use of soil and space, maintenance of soil fertility, erosion control, limited weed damage, and, in some cases, reduced incidence of insect pest attack while maintaining lower pest-control costs (Matteson et al., 1984; Dent, 1991; Zethner, 1995).

Although most African farmers use traditional pest control practices on food crops for home consumption, few attempts have been made to compile details of such practices. In Tanzania, farmers grow maize, beans, and other crops under scattered trees for on-farm consumption. For example, they use both deliberate and incidental practices to manage insect pests and diseases. The September–December season is best suited for bean cultivation, because it is typified by reliable and well distributed rains and is neither cold nor humid to favor fungal disease development (Zenther, 1995). Since early planting reduces insect pest infestation and produces the best yields, planting commences as soon as the rains start. Overall yields of early-sown crops are higher because they receive a full season's rainfall, suffer less weed competition, and benefit from initially high soil nitrate levels (Dent, 1991; Gebre-Amlak et al., 1989; Zethner 1995). In Ethiopia, early-sown maize has fewer maize stalk borers *(Busseola fusca),* a pest that causes 10 to 25% loss, and higher yields than late-sown maize (Gebre-Amlak et al., 1998). In beans, maize, and other home garden crops, mechanical picking and crushing is used to control grasshoppers and caterpillars (Zethner, 1995).

Most African farmers have substantial knowledge of insect and other animal pests, weeds, and to a lesser degree, pathogens, even if not strongly based on scientific backgrounds. They also have developed resistant varieties by always selecting seeds from tolerant and resistant varieties (Dent, 1991; Zethner, 1995). When planning and implementing pest and disease management projects, these farmers provide valuable practical technical expertise and a historical perspective. Traditional agricultural systems and effective indigenous methods should form the starting point for development and implementation of successful IPM strategies (Matteson et al., 1984).

2.2.1 Weed Management

Weeds (e.g., *Gramineae* (annual or perennial grasses), *Cyperaceae*, *Euphorbiaceae*, and *Scrophulariaceae*) often cause greater losses than other pests in home gardens and field crops, with losses among different cropping systems varying between 28 and 100%. Most weed control is accomplished by hand and can require from 20 to 50% of a farmer's time. Herbicides are occasionally used by small farmers who lack money or labor for weeding. In dry areas of Africa, witch weed (*Striga* sp.), a parasitic plant that feeds on roots of maize, sorghum, millet, sugarcane, and rice can cause yield losses of 10 to 30% (Adegoroye et al., 1989). Few reliable control techniques are available for *Striga* sp. In Kenya, crops grown under *Sesbania seban* trees contained less *Striga hermonthica*, and the soil had fewer *S. hermonthica* seeds (NRI, 1992; Anon, 1996; Esilaba and Ransom, 1997; Esilaba et al., 1997). In Gambia, rotational cropping systems that appear to prevent *Striga* sp. increases have been adopted by farmers. Other practices include hand pulling weeds, herbicides, and a combination of close spacing and clean weeding. Although monitoring systems and decision thresholds have been developed for some crops, they usually have not been implemented (NRI, 1992).

Many woody species (e.g., *Lantana camara*, *Prosopis juliflora*, *Leucaena leucocephala*, *Albizia* sp. *Sesbania seban,* and *Chromolaena odorata*) which were introduced into agricultural systems to provide fodder, organic mulch, shade, and other benefits, also may become weeds (Zenther, 1995). For example, *C. odorata*, a perennial semi-woody shrub from South America, has become an invasive weed in vegetable crops and plantations of South Africa. *C. odorata* is controlled by herbicides, applied to the foliage, stump, or soil, mechanical or cultural methods, and by site rehabilitation practices such as over-sowing or natural succession (Goodall and Erasmus, 1996). Mulch and leaves from trees can be used to control weeds. Organic mulch from *Acacia auriculiformis*, *Calliandra calothrsus,* and *L. leucocephala* is used to suppress weeds in coconut plantations (*Liyarage*, 1993).

2.2.2 Plants with Insecticidal Activity

Chinaberry (*Melia azedarach*) and pepper tree (*Schinus molle*) are exotic species, well adapted to Ethiopian conditions, that are planted for fuelwood, landscaping, and windbreaks. Leaves of pepper trees are routinely placed on tables to deter flies (Diptera) from residence areas and preliminary tests show insecticidal activity. Chinaberry, closely related to the neem tree (*Azadirachta indica),* is known for its insecticidal activity against *Aphis fabae*, *Atteva fabriciella,* and other insects. Mature fruits of endod (*Phytolacca dodecandra*), a native herb that grows as a weed in many parts of Ethiopia, are used as detergent for cleaning clothes in many rural communities (Gebre-Amlak et al., 1998).

In Ethiopia, extracts of all three plant species were found to significantly reduce larval abundance and infestation levels of maize stalk borer, and help increase yields. In post treatment sampling of infested maize for living *Busseola fusca* (maize stalk borer) larvae, treatments with leaves and fruits of Chinaberry, or endod and leaf extracts of pepper trees produced significantly fewer *B. fusca* (Gebre-Amlak et al., 1998). Before these plant extracts can be used as a viable IPM alternative to synthetic insecticides, further testing is necessary to determine their mode of insecticidal action and persistence against the maize stalk borer. Because such extracts of local plants are easily and inexpensively prepared, they offer a viable and attractive pest control strategy for subsistence farmers in Africa (Gebre-Amlak et al., 1998).

2.2.3 Coffee and Coffee–Banana Systems

Coffee (*Coffea arabica* and *C robusta)* is grown by small (average plots of 0.5 ha) and large (4 ha to more than 500 ha) landowners in the east African highlands characterized by bimodal rainfall. This system grows coffee as a cash crop and integrates crops, livestock, and trees. Most coffee is grown by small holders and is marketed through farmer owner cooperatives or consumed on the farm. Coffee yields are dependent on canopy management, fertilizers, and insect pest and disease control (Minae, 1988; Hoekstra et al., 1990).

The most damaging coffee diseases are coffee berry disease (*Colletotrichum coffeanum*), which attacks ripening berries, coffee leaf rust (*Hemilea vastatrix*), which causes premature leaf fall and reduction in flower abundance, and bacterial blight (*Pseudomonas syringae*). The most damaging insects are leafminers (*Leucoptera* s), fried egg scale (*Aspiridiotus* sp.), soft green scale (*Coccus alpinus*), *Icerya pattersoni*, antesia bugs (*Antestiopsis* sp.), defoliators, and thrips. Many aphids (Aphidae), scales (Coccidae), and mealybugs (*Pseudococcus* sp.) species are tended and protected by ants in the subfamilies of Myrmecinae, Campotinae, and Dolichoderinae (le Pelley, 1968; Zethner, 1995). In the early 1920s, the coffee mealybug (*Planococcus kenyae*) was accidentally introduced into Kenya from Uganda. It was later controlled by two parasites introduced from

Uganda (Zethner, 1995). The Kenya Coffee Research Foundation has developed an IPM program that includes planting lower yielding coffee varieties that are resistant to coffee berry disease and leaf rust. They recommend pruning trees to reduce humidity and concurrent risk of coffee berry and leaf rust diseases while limiting refugia of antesia bugs in leaf clusters. Mulching is used to prevent thrips from pupating in the soil, but favors pupation by leafminers. The Kenya Coffee Research Foundation has established thresholds for pesticide use. However, their use on large and small farms varies from none to two or three times recommended rates. Such misuse of pesticides can cause secondary pest outbreaks and environmental contamination (Zethner, 1995).

Banana and plantains, primary food crops in Uganda, Kenya, Burundi, Rwanda, and northern Tanzania, are often interplanted with coffee in home gardens and large holdings. These crops are threatened by a weevil (*Cosmopolites sordidus*) and nematodes (*Pratylenchus* sp.). Cultural control is commonly recommended by extension services. It includes using clean plant material, weeding, and removing or chopping harvested pseudo stems and corms during April and October to eliminate potential weevil breeding sites. However, few farmers actually implement these practices because they perceive these practices as too labor demanding, and instead, they opt to apply excessive dosages of insecticides (Oduol and Aluma, 1990; Gold et al., 1993; Gowen, 1993; Rugalema et al., 1994). Both farmers and national programs recognize that this approach is not cost effective, adversely impacts non-target organisms, and promotes the development of pesticide resistance. Pesticide recommendations for coffee and banana need revision and suitable training materials should be prepared for traditional farmers (Gold et al., 1993; Herren, 1993; Zethner, 1995). Biological control of the banana weevil with three egg predators from western Kenya also appears promising (Allard, 1993; Koppenhoffer, 1993).

2.2.4 Cocoa and Coconut Agroforestry Systems

Cocoa *(Theobroma cacao)* originated in the Brazilian Andes, and was introduced into Ghana, Nigeria, and Côte d'Ivoire in the 17th century (Hill and Wallner, 1988). Small-scale farmers cultivate this small tree under coconut, oil palm, and other planted trees and in forest openings. The cocoa is also interplanted with a variety of cash and food crops (NRI, 1992; Zethner, 1995). In Africa, citrus mealy bugs (*Planococcus citre*), a complex of mirid bugs (Miridae), a cossid moth (*Eulophontus myrmeleon*), swollen shoot virus, black pod disease (*Phytophthora* sp.), and rodents are the most damaging and widespread pests and diseases of cocoa. In Southeast Asia, pod borer (*Conopomporpha cramerella*) and stem borers (*Xyloborus* s) are also important (Thurold, 1968; Entwistle, 1985; Hill and Wallner, 1988; NRI, 1992). Weeds are a problem during planting establishment when there is little shade.

In general, large plantations have well defined pest control programs using herbicides in newly established plantations, insecticides and fungicides, phyto-sanitary procedures (i.e., pruning of diseases stems and pods), as well as careful maintenance of shade and canopy. Decision thresholds based on borer and mirid abundance have been developed for implementing pesticide applications in Asia, but similar thresholds have not been established for Africa or the Americas (NRI, 1992).

Crematogaster sp. and *Pheidole* sp. ants tend the mealybugs and spread the swollen root virus and black pod diseases. Other genera of ants, *Oecophylla* sp. and *Macromischoides* sp., protect the trees from these diseases by preying on their arthropod vectors. Majer (1974) proposed an IPM strategy to control these vectors by manipulating the predator ants. He recommended the removal of forest trees and vegetation from in and around the proposed plantation; and replacement with coconut. A periphery planting of coconut prevents invasion of *Crematogaster* sp. ants from neighboring bush, while encouraging *Oecophylla* sp. colonization. The cocoa planting must be shaded to prevent invasion by capsid bugs and protect habitats of beneficial ants. The planting should be monitored periodically to detect nests of pest ant species (Majer, 1974).

In East and West Africa, a plant sucking bug *(Pseudotheraptus* sp.) causes 30 to 65% nut losses of coconuts. Palms inhabited by a predator ant (*Oecophylla* sp.) are almost completely protected from damage by this bug (Way and Khoo, 1992). This ant is absent in most plantations because dominant competing genera have displaced them. In Côte d'Ivoire, an effective IPM program was established that uses insecticides to control the pest and competing ants on trees where *Oecophylla longinoda* is absent. There, *O. longinoda* is introduced on palms, and scrub and ground vegetation that favor the beneficial ant is encouraged. Insecticide treatment is only used on trees that are not colonized by the beneficial ant. Damage to coconuts is insignificant with this management strategy (Fataye and de Taffin, 1989, Zethner, 1995). IPM successes on these plantation crops have been mostly local while wide spread pesticide usage continues in much of Africa. A sustained effort is needed to develop and implement widespread IPM strategies for cocoa and coconut and to maintain quarantines that prevent spread of pests and diseases.

2.2.5 African IPM Programs

Few comprehensive IPM programs can be identified in African agriculture, but work on single IPM components (e.g., resistant varieties, biological, training) occurs continent wide (Kiss and Mierman, 1991). In addition, relevant research results are rare and personnel adequately trained in crop protection and IPM programs are few (Zethner, 1995). An exception is the African Highlands Initiative (AHI), which was developed in 1995 through collaborative efforts of National Agricultural Research Institutions (NARs) and International Agricultural Research Centers (IARCs) in the highlands of eastern and central Africa. The AHI goal is to improve sustainability, while enhancing the productivity in intensive land-use systems. This goal is addressed by working with farmers to evolve policies and develop technologies that increase agricultural productivity without diminishing natural resource quality. AHI has successfully implemented training activities, and developed supporting information. IPM research supports AHI's priority theme of soil productivity as an approach to natural resource management in the region. The IPM work focuses on crop-pest complexes that affect soil fertility of intensive agricultural systems. In 1995, research projects were initiated on several crop pest complexes: striga (*S. hermonthica*) in maize and sorghum; bean stem maggots (*Ophiomyia* sp.) and root rots (*Pythium* sp., *Rhizoctonia solani*, *Fusarium solani*, *Sclerotium rolfsii*) in beans; potato-bacterial wilt (*Pseudomonas dolanacearum*) in potato; and weevils (*Cosmopolites sordidus*, *Temnoschoita nigroplagiata*, *R. basipennis*, and *T. erudiatia*) in banana. These development-oriented research projects involve farmer groups and multiple partner organizations, focus on simple strategies of field sanitation and cultural practices rather than pesticide use, and are associated with soil degradation and declining fertility. Relevant literature reviews and diagnostic reports are available though ICRAF's AHI Technical Report Series (Braun et al., 1996; Okech and Gold, 1996; Berga, 1997; Esilaba and Ransom, 1997; Esilaba et al., 1997; Nderitu et al., 1997; Wang'ati, 1994). Another example is the Kenyan project, Integrated Pest Management of Crop Borers for Resource Poor Farmers in Africa (ICIPE, 1991).

2.3 Asian IPM Farming Systems

In many Asian nations, most people are directly or indirectly dependent on agriculture for their livelihoods (Saxena et al., 1989; van den Beldt et al., 1994). Twenty-five percent of farms in these countries are extensive monocultures, where pesticides are the preferred control method for crop pests. Other low-maintenance techniques are preferred for tree pests (NRI, 1991; Ciesla et al., 1995; Showler, 1995). Many agroforestry research and extension efforts are directed toward these resource-rich farms (Reddy, 1991; James, 1989). In these regions, IPM efforts are fragmented and usually focus on pests of major crops (e.g., rice, cotton, maize, sorghum, vegetables, and wheat), and the leucaena psyllid or cypress aphid that attack high value crops and trees (Greathead, 1988; NRI, 1992; Napometh, 1994; Ciesla et al., 1991; Ciesla et al., 1995; Raheja, 1995; Showler, 1995). Pest survey programs have been established in China, India, Indonesia, Malaysia, Korea, Pakistan, Philippines, and Thailand (Raheja, 1995). The primary pest control methods include pesticides, resistant plants, biological controls, cultural and mechanical practices, and quarantines (NRI, 1992; Raheja, 1995). Because the primary goal is to increase productivity, many Asian governments recommend pesticide application instead of IPM methods and provide subsidies for pesticide purchases (Raheja, 1995). Such policies result in increased pesticide use. In 1990, Asia consumed more than 26% of the world pesticides (GIFAP, 1992).

2.3.1 National IPM Program in Indonesia

Rice, the most important crop, occupies about 60% of the land in Indonesia. In the 1950s, Indonesia imported most of its rice. In the 1960s, the government initiated an eight-year BIMAS (Bimbingan Masal) program to increase agricultural productivity and sustainability of rice cultivation while making Indonesia more self sufficient. However, subsequent failure of this program was attributed to inherent fraud and corruption. In 1971, the government initiated a new program that handled credit and distribution through ordinary market channels. This program introduced farmers to the Green Revolution technologies and provided subsidies for pesticides and herbicides (Morse and Buhler, 1977). Farmers were given subsidies to apply pesticides typically three to four times during the rice growing season to control stem borers and the brown planthopper (BPH), *Nilaparvata lugens*. Consequences of pesticide overuse were evident by 1984. A 1986 outbreak of BHP was attributed to the elimination of natural enemies and development of resistance in BHP (Whalon et al., 1990; Wardhami, 1991). Consequently, contamination by these pesticides virtually eliminated rice–fish systems in many areas (Garrity and Sajise, 1991). Furthermore, this situation was exacerbated by lack of crop rotation and staggered planting thereby providing a continuous food source for BHP. In 1986, Presidential Decree No. 3 banned 57 registered brands of broad-spectrum pesticides and left only a few narrow-spectrum pesticides for use on rice. Pesticide subsidies were decreased from more than 70% in 1986 to 0 in 1989. Five years after the Presidential Decree, rice production had risen 15% over 1986 levels, and pesticide use had fallen 60% (Oka, 1990; Wardhani, 1991; Siswomihardjo, 1991). This National IPM Program is a social movement that links scientific development of ecological concepts with intensive farmer training in sound field management technologies. This program provides an example of second-generation Green Revolution technologies that link environmental conservation, public health, and farmer profitability (Wardhani, 1991). Farmer involvement is one of the cornerstones of this program. Farmer field schools are held for three to five hours every week over the 10 to 12 week crop cycles. At these schools, farmers are trained in basic pest identification and IPM strategies, such as efficient application of fertilizers and water management. Farmers can then adapt these methods to their own farming systems. Research conducted by the International Rice Research Institute (IRRI), government funding for IPM development, and extension efforts have provided Indonesia with one of the best examples of farmer oriented IPM systems in the world. This IPM strategy serves as a prototype for programs in Bangladesh, China, India, Korea, Malaysia, the Philippines, Vietnam, Sri Lanka, and Thailand (Morse and Buhler, 1997).

2.3.2 Upland Rice Agroforestry Systems

Upland rice culture is used in diverse agroforestry settings and highly developed sustainable systems. Farmers routinely exchange their planting seed because they have observed that any variety can develop pest problems when grown continuously on the same land. In the vast areas of Asia under traditional and small-scale cultivation, farm to farm variations in cropping systems provide a temporal, spatial, and genetic diversity that helps limit pest damage. In Northern Luzon, Philippines, the Ifugao people use a farm to grove system where wet rice fields are cultivated within a mosaic of fuel and fruit trees or other grove crops, such as bamboo and rattan (Olofson, 1981). Food web interactions between rice pests and their natural enemies are complex and help provide a low stable population (Matteson et al., 1984). For example, pond cypress (*Taxodium ascendens*) is planted around rice (*Oryza sativa* L.) and wheat (*Triticum aestivum* L.) fields in China to reduce the adverse climatic conditions and provide alternative habitat for spiders that keep leafhoppers in check (Shi and Gao, 1986). In the Philippines, hedgerows of *Senna spectabilis* enhance environmental sustainability in upland rice fields (Van Houten, 1996).

2.3.3 Indigenous Multistory Gardens

In southeastern and southern Asia, farmers intercrop coffee, coconuts, palms, upland rice, bananas, pineapples, taro, tuber crops, fruit trees, and other crops with shade producing trees such as *Gliricidia sepium* in multi-storied gardens. Each tree or crop occupies a different position in the vertical strata (Gordon and Bentley, 1990; Nair, 1993). They protect cypress, *Eucalyptus* sp., *Pinus* s and other flowering trees that wild bees use to produce honey (Sumitri, 1991). *Morus* s trees are managed for their leaves that are fed to silkworms (Sumitri, 1991). IPM strategies for each garden are unique and are adapted to the conditions in that garden. The dispersal of susceptible species in the gardens may minimize insect pest and disease problems and the need for pesticides (Schreiner, 1991; Friday and Westcom, 1996). They build fences around their crops to prevent foraging by wild pigs and rodents, that often severely damage the crops. Owls and other avian predators that live in the diverse systems are important predators of rodents (Lenton, 1978; Lenton, 1983; Fiedler, 1986). They have developed a variety of indigenous techniques for controlling *Imperata cylindrica*, *Saccharum spontaneum* and other weeds that have invaded their garden, slash and burn sites, and hilly sites. These include crushing the weed and replacing it with kudzu (*Pueraria phaseoloides*) or other plants that can out-compete the weeds, using pineapples to out-compete the weeds, and mechanical weeding. *Calliandra calothyrsus*, *Leucaena diversifolia*, and other nitrogen fixing trees produce nutrients and shade out weeds (ATIK, 1992). By maintaining species diversity and varying plant density over time and space, these Asian cropping systems provide greater stability, resilience to pests, economic risk aversion, and production diversity.

2.3.4 *Leucaena*-Crop Systems

Leucaena leucocephala, a leguminous tree, has been widely planted in both high-labor and high-energy agroforestry systems throughout Asian and Pacific regions. In Hawaii, *L. leucocephala* is a weedy pest in sugar plantations, however, it is also used to improve soil fertility, to prevent soil erosion, as cattle fodder, and as honeybee forage. In the Philippines, it occurs as a common hedgerow species, know as ipil-ipil, and in Indonesia, it is planted on 1.2 million ha as shade trees for coffee, vanilla, cocoa, cardamom, and other estate crops. In southeastern Asia, this tree is used for fuel, soil reclamation, green manure, leaf meal, vegetable production and charcoal. Between 1983 and 1992, the leucaena psyllid *(Heteropsylla cubana)*, a pest of *Leucaena* sp. native to Florida, U.S.A., progressively moved through the Pacific Island nations to southern Asia and East Africa, economically, socially, and ecologically impacting the agroforestry systems (Napompeth, 1994).

In Hawaii, classical biological control was the primary pest management technique used against the leucaena psyllid. Several native predators, such as the coccinellids, *(e.g., Curcinus coeruleus* and *Olla v-nigrum)*, feed on the psyllid nymphs and adults (Funaski et al. 1990). Based on work in Trinidad and Tobago, *Psyllaephagus yaseeni*, a eupelmid nymphal parasite, was released and became established (Funaski et al., 1990). *C. coeruleus*, *O. nigrum* and *P. yaseeni* were subsequently introduced into Guam, Samoa, Saipan, Philippines, Indonesia, Papua, New Guinea, Thailand, Tahiti, Tonga, New Caledonia and other Asian-Pacific areas infested with the psyllids (Waterhouse and Norris, 1987; Chazeau et al., 1989). More recently, *P. yaseeni* and another hymenopteran parasite, *Tamaricia leucaeneae* (Eulophidae) were released in East Africa (Van Houten, 1995).

Although a single pest management technique can be effective, an IPM program is feasible. Other opportunities for managing the psyllid include developing entomopathogens as microbial control agents that can be used control psyllid populations under suitable climatic conditions.

Germplasm screening and interspecific hybridization for psyllid tolerance in *Leucaena* spp is underway, but progress has been slow (Van Houten, 1995; Van Houten, 1996). *Leucaena* s can be replaced with alternative, multi-purpose tree species, such as *Gliricidia* sp. and *Calliandra* sp. Chemical pesticides are available but should be applied only to protect young seedlings because of undesirable effects on non-target organisms, high cost, and potential for resistance development by the target insect (Napometh, 1994). The population dynamics, seasonal occurrence and temperature regimes of the psyllid have been extensively studied throughout its range (Napompeth, 1990; Waage, 1990; McClay, 1990; Van Den Belt and Napompeth, 1992).

2.3.5 Rubber Systems

In Indonesia, especially on the islands of Sumatra and Kalimantan, rubber (*Hevea* sp.) is the main source of income for small-scale farmers. More than 70% of the country's total rubber is produced by these smallholder farmers, whose household economies depend on production of rubber in extensive agroforestry systems (Penot, 1995). Most of these farmers use seedlings and unimproved stock gathered from natural forests as planting material, and subsequently obtain low latex yields. These low-input forest rubber systems are established by slash and burn agriculture techniques. After clearing and one to two years of rice production, rubber is interplanted with various nut, fruit, and timber trees to produce a biologically diverse secondary forest (de Foresta 1992). The soil mesofauna and mammal abundance in these secondary forests is similar to primary forests, while plant and bird abundance is 30 and 50% of that in native forests, respectively (Deharveng, 1992; Sibuea and Herdimansyah, 1993; Michen and de Foresta, 1995; Thiollay, 1995). The foliage in these diverse systems screens the rubber trees from other members of the genera, and acts as a barrier to windborne pathogens (Thurston, 1984).

The Rubber Association of Indonesia is working cooperatively with a number of international groups on rubber agroforestry system (RAS) research to intensify and improve production and management of forest rubber systems. Genetically adapted clones and seedlings with improved pest- and disease-resistance are being used in RAS. A variety of complex agroforestry systems, incorporating multipurpose trees with rubber and intercropping annuals (improved upland rice) for 3–4 years are being tested (1992; Noordwijok, 1995). Some of the intensive RAS are being established on alang-alang grasslands dominated by *Imperata cylindrica*, an invasive weed species occurring on degraded lands in Indonesia, in hopes of preventing growth and spread of this weed (Penot, 1995; Noordwijok, 1995). Weed growth in these plantings also is controlled by hoeing, rolling and pressing the weeds, cultivating with animals and tractors and occasionally herbicide application. Sheep reared under rubber trees produce meat and biologically control weeds in the planting (Vergage and Nair, 1985; Tajuldin, 1986). One of the most effective chemical control methods is slashing the alang-alang, allowing the grass to begin regrowth then applying herbicides. Because the alang-alang thrives after fire, community based fire prevention is promoted (Noordwijok, 1995).

In Malaysia, agroforestry systems are being developed as viable and profitable options to monoculture plantations of rubber, oil palm and cocoa. Increasing production costs, labor shortages and wages have eroded profit margins in recent years, and left rubber trees untapped on both estates and small holdings (Mahmud, 1988). Options for integrating timber species into rubber, oil palm, or fruit tree plantations include establishing mosaics within the existing planting system, modifying existing systems, or developing new combinations. Following the well-established plantation model in Malaysia, all tree crops can be planted at the same time, with timber harvest scheduled for an early stage or at the time of agricultural tree felling. Timber species can be planted into existing agricultural tree crop systems, with timber on the more hilly terrain and the rubber, oil palm, or fruit trees on more level terrain. If the area is infested with root disease, timber species can be planted at suitable densities to protect against pathogen spread, (*e.g.,* alternate rows with oil palm) (Mahmud, 1988).

Modified systems include diversified Taurgya systems where vegetables, banana, or maize are grown between *Hevea* rows for up to three years until the rubber canopy develops. An alternative system, the hedge planting system was developed by the Rubber Research Institute of Malaysia in which rubber is planted at wider between row spacing with less distance between individual plants within the rows in a hedges formation (Abdul Ghani et al., 1991). The inter-hedge space is planted with durian fruit trees or timber species, such as teak or sentang *(Azadirachta excelsa)*. Rotational hedges are used to control weeds, and live hedges are also used to deter pigs from cash crops (Noordwijok, 1995). In Malaysia, abundance of barn owls (*Tyto alba*), an important predator of rat pests (e.g., *Rattus tiomanicus*, *R. argentiventer*, and *R. exulans*), can be increased by placing nesting boxes among the oil palms (Lenton, 1978; Lenton, 1983).

In Asia, diseases are the greatest threat to plantation culture of rubber. White root disease *(Rigid-oporus ligosus)* and powdery mildew (*Olpidium heveae*) are the most damaging diseases in West Africa, and Africa and Asia, respectively (Imle, 1979; Thurston, 1984). The first line of defense in preventing the spread of these diseases to uninfected countries is quarantines. For example, quarantines have prevented the spread to Africa and Asia of the South American leaf blight (*Microcyclus ulei*), a disease that has seriously curtailed (practically eliminated) plantation rubber culture in South and Central America. (Rao, 1973; Thurston, 1973). Other methods include sanitation and application of fungicides (Cook, 1981; Thurston, 1984). Low yielding native trees are being screened for disease resistance and highly resistant cultivars are grafted onto highly productive root stock. Crown budding when the trees are six feet tall is also used to protect against mildews (Thurston 1984; Davis, 1997).

2.3.6 Trees with Insecticidal Activity

The roots of *Derris elliotica,* a bushy leguminous vine found near river banks or streams, contain rotenoids, highly potent insecticidal chemicals. Rotenoids, which are used to kill insects in homes, in fields, and on animals, have a short residual toxicity and are relatively non toxic to man (ATIK, 1992).

Neem trees (*Azadirachta indica*), fast growing evergreens that are found in windbreaks, towns and villages from sea level to 670 meters, contain azadirachtin, a potent pesticide. The leaves, twigs, seeds and roots are repellant to 200 species of insects, mites, and nematodes that are household and economic pests. (Benge, 1986; Jacobson, 1988; Ruskin, 1992).

2.4 Central and South American IPM Farming Systems

Seventy-five percent of the agroforesters in these developing countries are traditional farmers who cultivate a few hectares of land, practice shifting cultivation, or use swidden-fallow management. Trees are used for food, fodder, fences, pest control, construction material, dye, and fuel (Beets, 1984; Altieri and Farrell, 1984; Greathead, 1988). Few agrichemicals are applied in these labor-intensive and low-energy systems (Litsinger and Moody, 1976). Resource poor farmers follow practices developed by their ancestors, and they learn from personal experience, informal contact with other agroforesters, pesticide sellers, nongovernment organizations, and foreign assistance programs (den Biggelear, 1991; Chandler, 1991). IPM strategies include selecting pest-resistant crop or tree varieties, plant spacing, tilling, intercropping, water and fertilizer management, manual removal of pests, and use of trees or tree parts to repel pests (Reddy, 1991; Altieri and Farrell, 1984; NRI, 1991; Poswal et al., 1993). However, the most common strategy, practiced for centuries, is the active maintenance or passive allowance of high crop and plant diversity in or around farms. Over time and space, this high diversity helps prevent pest buildup while providing food and habitat for beneficial organisms.

2.4.1 Swidden-Fallow Systems

Shifting cultivation systems in the Amazon and other parts of Central and South America use high ecosystem diversity to keep pest populations low. A parcel of forest is cut and burned, releasing nutrients and reducing weeds. The area is farmed until the soil fertility declines, then abandoned before pest populations increase. An abundance of birds, spiders, ants, and other natural enemies helps prevent pest outbreaks (Altieri and Anderson, 1986). In the Yucatan of Mexico, most resource-poor farmers manage garden, overstory, and edge vegetation to generate plant products for household use and income from non-timber forest products such as honey. Competing growth is removed around 34 melliferous plant species at field edges and in neighboring woodlands. This pruning requires in-depth knowledge of flowering phenology of melliferous and polliniferous species in the area. High diversity within these systems helps keep diseases and arthropod pest populations low (Chemas and Rico Gray, 1991). Rodent and avian pests also may rapidly respond to food abundance within the crop fields. For example, in Haiti, field edges harbor rats, woodpeckers, and weaver birds that destroy up to 40% of the harvested and unharvested grains (Samedy et al., 1986). The diverse edges also provide habitat for their predators.

2.4.2 Coffee Systems

Central America produces almost 14% of the world's coffee, while 20% comes from the rest of Latin America. Most of this production is on small-scale farms (Fernandez, 1984; Perfecto et al., 1996). Coffee is grown unshaded primarily in Brazil, and some parts of Central America, whereas it is grown shaded throughout the rest of Central and South America (le Pelley, 1968). In traditional coffee production more than 40 tree species, including nitrogen-fixing and fruit trees, shade the coffee. In northern South America and Central America, shade coffee plantations contain more than 90 species of bee-pollinated plants (Chazaro, 1982). This diversity offers food and nesting sites for predators, parasites, and various leaf-, fruit-, and nectar-feeding organisms. In addition, ant diversity in the coffee canopy is higher than in natural forests, and spiders and other arthropod assemblages are highly diverse (Torres, 1984; Perfecto et al., 1996). Shaded coffee plantations support large populations of migratory birds, especially forest migrants, and mammals (Gallina et al., 1996; Perfecto et al., 1996). They provide fruit and nectar for birds during the dry season. Pesticide applications to these systems are minimal (Perfecto et al., 1996).

The most important insect pests in coffee plantings are the coffee berry borer, mealybug, scales and leaf miners. Insecticides can control these pests, but excessive amounts may promote outbreaks of leaf miners (Coste, 1992; Ramírez et al., 1992). Shade trees used in coffee also may host pest insects. For example, the flowers and seed pods of *Leucaena glauca* are an alternative host for *Planococcus citri* and *Ferrisia virgata*, two species that also infest coffee. June beetles (*Phyllophara menetriesl*) as adults feed on the leaves of *Erythrina* spp., a tree that shades coffee and other crops, as larvae feed on the coffee and crop roots (Hilje et al., 1993; Westley and Powell, 1993). Infestation can be minimized by removal of this shade tree and substitution with a different species, or removal of its flower clusters and pruning its foliage. In Puerto Rico, *Mymelachista ambigua*, an ant that defoliates coffee, nests in shade trees (*Inga laurina* and *Inga vera*). In Salvador, *Inga* spp host the defoliating caterpillar *Hemiceras* sp. (Coste, 1992).

Coffee rust *(Hemileia vastatrix)* is the most serious disease of coffee in the Americas. Coffee rust, a windborne pathogen, originated on wild coffee in Ethiopia, was introduced to Brazil in 1970 and had spread throughout most of South and Central America by 1982 (Scriber and Zentmyer, 1984). In 1976 and 1977, an attempt to eradicate the disease in Nicaragua by spraying fungicides on the trees prior to their removal and then restocking with more resistant trees was successful until civil war stopped the control efforts (Schieber and Zentmyer, 1984; Javed, 1984). A combination of fungicide applications and thinning and pruning to increase air circulation will control the disease (Javed, 1984; Rodriques, 1984; Fulton, 1984).

In 1978, a regional project, PCOMECAFE, was initiated in Central America, Mexico and the Dominican Republic to improve control of coffee pests and diseases by training technicians, providing advice, and conducting research to improve genetic resistance of coffee to insects and diseases. Other research studies determined the biology and epidemiology of the coffee rust and the biology of the coffee bean borer in Central America, evaluated pesticides and analyzed pesticide residues, developed coffee varieties resistant to the borer and rust, developed technology transfer techniques, and expanded the information network (Fernandez, 1984). Advances in forecasting, discovery of more effective fungicides, improved fungicide application methods, and development of host resistance have enabled growers to more economically control the disease (Coste, 1992).

Sun coffee plantations initially were established to control coffee leaf rust (*Hemileia vastatrix*). Because the farmers believed that high humidity in shade plantations increased the prevalence of coffee leaf rust, they converted shade plantations to sun plantations by removing shade trees (Agrios, 1988; Beer, 1987). These updated plantations are more costly to operate, require large inputs of agrichemicals and other resources, are more prone to water and soil runoff, have low biodiversity, and threaten the sustainability of the system (Rice, 1990). Abundance of spiders, ants, and other predators is low in sun coffee plantations, and the lack of shade trees and food limits the abundance of migratory birds and bats (Perfecto et al., 1996).

2.4.3 Cocoa Systems

Cocoa (*Theobroma cacao*), a woody scrub native to South America, is grown as a cash crop throughout much of Central and South America. Cocoa plantings are in a constant state of flux. Aphids, psyllids, mealybugs (*Planococcus citre*) and Lepidoptera are the predominant pests of young trees, while bugs (Miridae), pod-boring caterpillars (Lepidoptera), ants (Formicidae) and grasshoppers (Orthoptera) are found on mature trees. *Atta* sp. ants defoliate the trees, while the little fire ants, *Wasmannia auropunctata* (Rogers), tend mealybugs. In Trinidad, the mealybug is the primary vector of cocoa virus disease. Other serious diseases are black pod rot, manila pod rot and witches broom disease. In Ecuador and Columbia, manila pod rot can destroy 30 to 90% of the pods (Youdeowei and Service, 1983; Thurston, 1984; NRI, 1992; Delabie et al., 1994). IPM techniques established to control pests and diseases of cocoa, are similar to those used in Africa and Asia. They must be initiated when the tree is established and most continue throughout the tree's existence. Borer infested trees are pruned, harvested and then sprayed with insecticides. Shade trees used to reduce mirid abundance must be monitored for pests because they also can harbor defoliators, mealybugs and pathogens that can later infest cocoa. Sanitation is extremely important because most insects and many diseases are transmitted on dead and decaying vegetation. Pruning and sometimes uprooting is used to control black pod rot, witches broom disease, and other diseases. Fungicides are applied over a three-week period. Although some resistant varieties have been identified, they are not used in cocoa plantings. Quarantines have slowed the spread of diseases among and within the Americas, Africa and Asia (Youdeowei and Service, 1983; Thurston, 1984; NRI, 1992).

2.4.4 Development of IPM

Development of IPM among resource-rich growers in Latin America is closely related to cotton production, excessive use of organopesticides during the 1940s, and the subsequent development of widespread resistance to these pesticides during the 1950s. As a consequence, most Central and South American nations, in conjunction with International Agricultural Research Centers in Mexico and Peru, focused on developing IPM for primary crops such as potatoes and corn (Brader, 1979). Techniques developed include pesticides, biocontrols, and resistant plants. By 1988, biological control agents, such as entomopathogens, were commercially available to producers. In spite of these advances and the potential risks of pesticide resistance, most agroforesters continued to use only pesticides in their IPM programs, and did not integrate chemical, cultural, and biological pest management techniques. Notable exceptions are broccoli, snow peas, and other non-traditional export products, where consumer demand and import country standards forced export companies to minimize pesticide use and incorporate cultural and biological management techniques. Thus, several pest management techniques are used on products for human consumption; whereas, pesticides remain as a predominant method to protect cut flowers, ornamental ferns, and other non-ingested products. In agroforestry systems, IPM is more commonly practiced because resources are scarce. However, this resource scarcity coupled with environmental concerns and consumer interest in organic coffee or similar natural products should support implementation of successful IPM approaches.

3. FUTURE NEEDS

Because many agroforesters lack the equipment and resources for highly technical operations, all farming systems need an array of new and improved pest-management practices for use under various resource conditions. Techniques are needed to monitor pest populations and predict damage levels. Such techniques must be appropriate for both large- and small- scale agroforestry systems, incorporate the local agroforesters perceptions of damage thresholds, and be correctly applied (Ashby, 1990; Chambers, 1990; NRI, 1991; NRI, 1992; Black and Sweetmore, 1994; Mengech et al., 1995; Hardy et al., 1996; Morse and Buhler, 1997). Improved knowledge of and techniques for identification, monitoring, and controlling pest populations are essential.

3.1 Identification and Assessment of Pest and Natural Enemy Populations

Correct identification of pest species and their natural enemies is the first step in an IPM program (Dix, 1996). Taxonomists are developing keys to identify some pest species; however, specialized taxonomists are necessary to identify difficult pest specimens. A critical problem is that the number of taxonomic specialists is declining. Pest identification is too costly for independent support by any single institution or nation. Thus, international and interdisciplinary collaborative taxonomic networks are needed to target organisms for study, develop identification techniques, and coordinate training of taxonomists (Claridge, 1991).

Biosystematic analysis techniques such as protein analysis (e.g., enzyme-linked immunosorbent assays (ELISA)) and DNA analysis (e.g., polymerase chain reaction, DNA sequencing) offer promising alternatives to morphological keys and behavior techniques used by taxonomists in both industrialized and developing regions (Klopfenstein and Kerl, 1995; Hardy et al. 1996). From these techniques, identification kits can be developed to allow quick identification of pests and their natural controls at field sites or regional diagnostic centers. On-line web sites that provide current diagnostic information and user friendly keys can be made available to computer-literate, pest-management specialists and agroforesters.

Currently, taxonomic services primarily benefit agroforesters on high-resource and high-energy farms, but are unavailable to most traditional indigenous agroforesters on labor intensive farms. In contrast, traditional indigenous agroforesters have developed practical systems to identify damaging pest stages and understand pest biologies based on historical practices. This knowledge, which is frequently transmitted through social customs or embedded oral folk stories, is only slowly or not yet incorporated into pest identification systems. Practitioner-friendly keys, guides, computer programs, or tests to identify pests and beneficials must be developed interactively with all partners, including both resource-limited and resource-rich practitioners. Delivery systems must recognize farmer skills and resources, and should be implemented by governments and non-governmental organizations, such as service groups, religious groups, and relief organizations (NRI, 1991; NRI, 1991; Matteson, 1992; Dix, 1996, Morse and Buhler, 1997).

In many agroforestry systems especially in developing countries, scouting, trapping, and other methods are needed to measure pest and natural enemy populations and population trends (Zethner, 1995). In addition, long-term monitoring of pest occurrence, population levels, and overall damage is expensive and time consuming. Cheaper and faster detection and monitoring methods are needed. Molecular diagnostic kits that can identify species and assess potential damage levels are one possibility. Another rapidly advancing field is remote sensing of pests and their damage. Both methods require inputs of national resources and leadership (Dix, 1996; Hardy et al., 1996; Klopfenstein and Kerl, 1995).

3.2 Knowledge of Pest Biology

A serious lack of knowledge about the biologies of important pests hampers implementation of most IPM programs. Collecting such information is an ongoing effort requiring long-term research and administrative commitments. User input on problem identification, pest biology, damage thresholds, and indigenous farming practices is essential to the adoption of IPM practices. Researchers, pest management specialists, agroforesters, and other partners must develop interactive decision-making strategies and guides for established IPM techniques that synthesize indigenous and study-based knowledge (NRI, 1992; Dix, 1996; Hardy et al., 1996).

3.3 Control of Pests

An array of long- and short-term pest-management strategies are available for controlling pests under various resource levels (Litsinger and Moody, 1976; NRI, 1992; Hardy et al. 1996). In recent years, techniques for applying low dosages of pesticides, pesticides that target specific pests, organic farming, genetic engineering, crop rotation, specialized tilling, edge manipulation, modifying planting dates, biological control, and scouting fields for economic thresholds of pest damage were developed to minimize pesticide usage. However, additional knowledge is needed to evaluate the effectiveness of these techniques over time and under various environmental conditions. Application of these techniques is dependent on farmer knowledge and available resources (Litsinger and Moody, 1976). Pesticides or resistant plant varieties alone can individually result in spectacular short-term benefits, but these benefits are frequently lost over the long-term when pests evolve and develop resistance to pesticides and resistant plants (Litsinger and Moody, 1976; Mengech et al., 1995; Hardy et al., 1996).

Transgenic plants are a new strategy that shows considerable potential for incorporation into agroforestry systems. Since 1997, more than 2,000 small-scale field trials were conducted in the United States on more than 44 genetically engineered plants including corn, soybeans, cotton, potato, carrot, melon, rapeseed, turfgrass, strawberry, apple, plum, papaya, walnut, poplar, and spruce (Snow and Palma, 1997). In 1997, *Bt* corn, corn containing a *Bacillus thuringensis* (*Bt*) endotoxin gene, was widely planted in the Midwest and central Great Plains of the United States. Fields planted with *Bt* corn had minimal insect damage and crop yield was up to three-fold higher than fields planted with traditional corn. Furthermore, direct *Bt* effects on non-target natural enemy populations was not apparent (Foster, personal communication). Clones of poplar, a tree commonly used in temperate agroforestry systems, are being developed that are insect-, disease-, or herbicide-resistant, and can tolerant a range of adverse soil conditions (Klopfenstein et al., 1997). Within the next decade, most widely cultivated plants in North America will contain genetically engineered traits. Undesirable consequences must be thoroughly addressed before use of transgenic plants can become widespread (Snow and Palma, 1997).

Conversely, traditional farmers use techniques that have been effective for years and in some cases centuries. Modifying indigenous practices may provide methods that are readily acceptable and have the potential for more longevity (Saxena et al., 1986; Black and Sweetmore, 1994; Dix, 1996; Morse and Buhler, 1997). For example, synchronized planting can reduce pest buildups, reduce labor costs, improve efficiency of available labor, and enhance other aspects of pest management. Community acceptance of IPM techniques is essential and may increase the sustainability of such practices (Saxena et al., 1986; Dix, 1996).

In the developed world, pesticide use is tightly controlled and practices are generally well regulated and relatively safe. In less well developed countries, however, the situation is reversed (Cox, 1994; Mumford and Stonehouse, 1994). Furthermore, as resource-rich nations minimize and restrict pesticide usage, older less efficient pesticides are often shipped to non-industrial nations with less restrictive policies. Although non-industrial nations have not fully integrated pesticides into their agricultural systems, they often use pesticides that are outdated, illegal, or in excess supply in industrialized nations (Poswal et al., 1993; Cox, 1994; Mumford and Stonehouse, 1994; Morse and Buhler, 1997). To achieve short-term profits, many land managers convert their farms to extensive monocultures. They ignore, or are unaware of, the resulting environmental problems such as non-point source water and air pollution, soil erosion, loss of biodiversity, and detrimental effects on non-target organisms (Beets, 1984; Altieri and Farrell, 1984; Greathead, 1988).

Effects of IPM practices on the farmer and environmental health is another concern that is frequently ignored in non-industrialized nations. Pest-management specialists, government organizations, and non-government organizations must collectively establish minimum safety standards and procedures for pesticide use. These standard practices and guidelines must be promptly communicated to all practitioners and farmers through extension services, educational programs, how-to pamphlets, and other publications (Cox, 1994; Mumford and Stonehouse, 1994; Dix, 1996).

4. CONCLUSIONS

Successful IPM systems must incorporate a variety of long- and short-term techniques for managing pests in multiple crops over an entire cropping period, thereby avoiding focus on growth of single crops. Short-term techniques, such as using pesticides or resistant plant varieties, can result in spectacular short-term benefits. However, these benefits are frequently lost over time when pests evolve and develop resistance to pesticides (Litsinger and Moody, 1976; Raheja, 1995; Mengech et al., 1995; Hardy et al., 1996). Introduced natural enemies can provide immediate control of a pest, but long-term benefits depend on the natural enemy's ability to survive in the foreign ecosystem. For this reason, long-term enhancement of native natural enemies through management of edge vegetation and other practices may be more successful in diverse agroecosystems. Furthermore, it is often necessary to conduct pest control activities over several pest generations or tree/crop rotations. Modifying indigenous practices may provide methods that are more readily accepted and applied over the long term.

Successful IPM systems should be sustainable, environmentally safe, and adaptable to the technical, institutional, economic, social, legislative, and educational needs of the agroforesters. Trade-offs between crop production, forestry output, and environmental needs are expected in these dynamic systems (Campanhla et al., 1995; Dix, 1996; Hardy et al., 1996). The challenge in developing sustainable IPM systems for the 21st century is obtaining an adequate understanding of pests and agroforestry systems, developing of effective short- and long-term techniques, gaining acceptance of these practices by the institutions and community, establishing legislative control measures, and adapting techniques for various economic resources (Black and Sweetmore, 1994; Dix, 1996; Hardy et al., 1996). The shortage of qualified specialists to carry out coordinated research and extension activities make this challenge even more difficult (Marshed-Kharusy, 1994). Governments must take actions that will make chemical control less attractive through legislation, registration, taxation, and removal of subsidies, while providing financial support and appropriate incentives for IPM programs (Mengech et al., 1995). Forestry and agricultural organizations at national and international levels should play major roles in developing, and coordinating, and implementing IPM programs at the local, national, and international level.

REFERENCES

Abdul, G. I., Tajuddin, I., and Najib, L. A., 1991, "Potensi sistem penanaman campuran getah dengan buah-buahan untuk pekebun kecil." *Prosiding Simposium Buah Buahan Kebangsaan, Pahang,* 84–87.

Adgoroye, A. S., Akinyemiju, D. A., and Bewaji, F., 1989, "Weeds as a constraint to food production in Africa," in *Pest Management and the African Farmer.* Zethner, O., Ed., ICIPE Science Press, Nairobi, 83–99.

Agrios, G. N., 1988, *Plant Pathology,* Third edition, Academic Press, San Diego, CA.

Alcala, A. C., 1991a, "Soil and Water Conservation (SWC) Technologies and Agroforestry Systems." *Agroforestry Technology Information Kit (ATIK).* Nov. 1, 13, 105–112, 156–161, 162–163.

Allard, G. B., Nankinga, C., and Ogenga-Latigo, M., 1993, "The potential of indigenous fungal pathogens as a component of integrated management of banana weevil, *Cosmoplites* in Uganda: A new research project," in *Biological and Integrated Control of Highland Banana and Plantain Pests and Diseases.* Gold, C. S. and Gemmil, B., Eds. Proceedings of a Research Coordination Meeting, Cotonou, Benin, The Printer, Davis, CA, 118-123.

Altieri, M. A., 1991, Increasing biodiversity to improve insect pest management in agroecosystems," in *The Biodiversity of Microrganisms and Invertebrates: Its Role in Sustainable Agriculture.* Hawksworth, D. L., Ed. CAB International, Wallingford, U.K., 165–181.

Altieri, M. A. and Anderson, M. K., 1986, "An ecological basis for the development of alternative agricultural systems for small world farmers," *Am. J. Alt. Ag.* 1. 30-38.

Altieri, M. A. and Ferrell, J., 1984, "Traditional farming systems of south central Chile, with special emphasis on agroforestry," *Agro. Syst.* 2. 3-18.

Altieri, M. A. and Schmidt, L. L., 1986, "The dynamics of colonizing arthropod communities at the interface of abandoned, organic and commercial apple orchards and adjacent woodland habitats," *Ag. Ecosyst. Env.* 16. 29-43.

Altieri, M. A. and Whitcomb, W. H., 1979, "The potential use of weeds in the manipulation of beneficial insects," *Hort. Sci.* 14. 12-18.

Altieri, M. A., Letourneau, D. K. and Davis, J. B., 1983, "Developing sustainable agroecosystems," *Biosci.*, 33. 45-49.

1991a. *Draft report of the conference on IPM in the Asia–Pacific Region,* Sept. 23–27, 1991, Kuala Lumpur, Malaysia.

1991b, "Spotlight — Pesticide data collection system," *RENPAP Gazette,* UNDP/UNIDO — Quarterly Publication of the Regional Network on Pesticides in Asia and the Pacific. 3, 1–51.

Ashby, J. A., 1990, "Small farmer's participation in the design of technologies," in *Agroecology and Small Farm Development.* Altieri, M. A. and Hecht, S. B., Eds., CRC Press, Boca Raton, FL, 245–253.

Ayers, G. S. and Harman, J. R., 1992, "Bee forage of North America and potential for planting for bees," in *The Hive and the Honey Bee.* Graham, M., Ed. Dadant & Sons, Hamilton, IL., 437–535.

Baxter, J., 1995, "*Chromolaena odorata*: weed for the killing or shrub for the tilling?" *Agrofor. Today,* April–June, 6–8.

Beer, J., 1987, "Advantages, disadvantages and desirable characteristics of shade trees for coffee, cocoa and tea," *Agro. Syst.* 5, 3–13.

Beets, W. C., 1984. *Aspects of traditional farming systems in relation to integrated pest management,* paper presented at the African Seminar on the Use and Handling of Agricultural and Other Pest Control Chemicals, Oct. 30–Nov. 6, 1983, Nairobi, Kenya.

Benge, M. D., 1986, "Neem the cornucopia tree," *S&T/FENR Agro-forestation Technical Series No. 5, Rm. 515-D, Sa- 18.* Agency for International Development, Washington, DC, 90.

Berga, I., 1997, "Integrated control of potato bacterial wilt: Literature review and work plan. 1995–1997," *AHI Technical Report: Series No. 3.*

Bishaw, B., Buckman, R. E., and Brodie, J. D., 1994, "Agroforestry and tree planting potential for the Alemaya Basin, Hararghe Highlands, Ethiopia," in *Opportunities for Agroforestry in the Temperate Zone Worldwide.*

Black, R. and Sweetmore, A., Eds., 1994, "Monograph No. 61, Crop protection in the developing world," in *Proceedings of an International Symposium held at the Brighton Metropole Hotel.* Nov. 21, British Crop Protection Council, Farnham, UK, 21–27.

Braun, A. R., Smaling, E. M. A., Muchugu, E. I., Shepherd, K. D., and Corbett, J. D., Eds., 1996, "Maintenance and improvement of soil productivity in the highlands of Ethiopia, Kenya, Madagascar and Uganda: An inventory of spatial and non-spatial survey and research data on natural resources and land productivity," *AHI Technical Report Series. No. 6.*

Brunt, A. A. and Kenten, R. H., 1971, "Viruses infecting cocoa," *Rev. Plant Pathol.* 50, 591–602.

Campanhola, C., de Moraes, G. J., and de Sá, L. A. N., 1995, "Review of IPM in South America, " *in Integrated Pest Management in the Tropics: Current States and Future Prospects.* Mengech, A. N., Saxena, K. N., and Gopalan, H. N. B., Eds., John Wiley & Sons, New York, 121–152.

Chambers, R., 1990, "Farmer-first: A practical paradigm for the third world agriculture," in *Agroecology and Small Farm Development.* Altieri, M. A. and Hecht, S. B. Eds, CRC Press, Boca Raton, FL, 237–244.

Chandler, P. M., 1991, "The indigenous knowledge of ecological processes among peasants in the Peoples' Republic of China," *Ag. Hum. Val.* 8, 59–66.

Cházaro, M., 1982, "Flora epicola de la zona cafetales de Coatepec, Veracruz," in *Estudios Ecologicas en al Agrosistema Cafetalero, Xalapa (Mexico)*, Jimenez, E., Ed., Instituto Nacional de Investigaciones Sobre Recursos Bióticos. 95–102.

Chazeau, J.S., Bouye, E., and Bonnet de Larbogne, L., 1989, "*Lutte biologique contre le psylle Heteropsylla cubana, ravaguer du faux-mimosa* Leucaena leucocephala *en Nouvelle Caledonie ORSTOM, Centre de Noumia,*" Institut Francais de Recherche Scientifique pour le Development en Cooperation, 82.

Chemas, A. and Rico-Gray, V., 1991, "Agriculture and management of associated vegetation by the Mayan of Tixcacaltuyub, Yucatan, Mexico," *Agro. Syst.* 13, 13–25, 1991,

Ciesla, W. M., Mbugua, D. K., and Ward, J. D., 1995, "Ensuring forest health and productivity: A perspective from Kenya," *J. Forest.*, 93, 36-39.

Ciesla, W. M., Odera, J., and Cock, M. J. W., 1991, "Exotic aphid pests of conifers: A crisis in African forestry," in *Proceedings of a workshop at the Kenya Forestry Institute*, Institute of Food and Agriculture Organizations of the United Nations, Rome.

Claridge, M. F., 1991, "Genetic and biological diversity of insect pests and their natural enemies," in *The biodiversity of microrganisms and invertebrates: Its role in sustainable agriculture*, Hawksworth, D. L., Ed. CAB International, Wallingford, U.K., 183–194.

Cook, J. R., Zentmyer, G. A., and Cowling, E. B., Eds., 1989, *Annual Review of Phytopathology*, 27, 503–531.

Coste, R., 1992, *Coffee: The plant and the product.* Macmillan, London, 328.

Cox, J. R., 1994, "Requirements for the safe and effective management of pesticides in less-developed countries," in *Monograph No. 61*, Nov. 21, 1994 Proceedings of an International Symposium held at the Brighton Metropole Hotel, Nov., 21 1994, Black, R. and Sweetmore, A., Eds. British Crop Protection Council, Farnham, U.K., 21–27.

Datta, S. C. and Banerjee, A. K., 1978, "Useful weeds of West Bengal rice fields," *Econ. Bot.*, 32, 247–310.

Davis, W., 1997, "The rubber industry's biological nightmare," *Fortune*, Aug. 4, 136. 86–98.

de Foresta, H., 1992a, "Botany contribution to the understanding of smallholder rubber plantations in Indonesia: An example from South Sumatra," *Symposium Sumatra Lingkungan dan Perbanhunan*, BIOTROP, Bogor, Indonesia.

de Foresta, H. 1992b, "Complex agroforestry systems and conservation of biological diversity: For a larger use of traditional agroforestry as timber in Indonesia, a link between environmental and economic development. Proceedings of an international conservation of tropical biodiversity," *Malayan Nature Journal.*

Delabie, J. H. C., 1988, "Ocorrencia de *Wasmannia auropunctat* (Roger, 1863) (Hymenoptera, Formicidae, Myrmicinae) em cacuais na Bahia, Brasil," *Revista Theobroma, Brazil*, 18, 29-37.

Dent, D. 1991, *Insect Pest Management,* CAB International, Redwood Press LTD, Wiltshire, U.K., 1–12, 373–396.

Den Biggelaar, C, 1991, "Farming systems development: Synthesizing indigenous and scientific knowledge systems," *Ag. Hum. Val.* 8, 25-36.

Dean, C. G., 1979, "Red ring disease of *Cocus nucifera* L. Caused by *Radinapnelenchus cocophilus* (Cobb, 1919) Goodey, 1960. An annotated bibliography and review," *Commonw. Agric. Bur.*, Farnham Royal, U.K., 70.

Deharveng, L., 1992, "Field report for the soil mesofauna studies," in *Alternatives to slash-and-burn in Indonesia, summary report of phase 1, ASB-Indonesia Report Number 4, 1995.* Noordwijk, M. V., Tomich, T. P., Winahyu, R., Murdiyarso, D., Suyanto, Partoharjono, S. and Fagi, A. M., Eds., Bogor, Indonesia.

Delabie, J. H. C. and Cazorla, M. I., 1991, "Danos de *Planococcus citri* (Homoptera, Pseudococcidae), na producao do cacaueiro," *Agrotropica, Brazil,* 3, 53-57.

Delabie, J. H. C., 1988, "Ocorrencia de *Wasmannia auropunctat* (Roger, 1863) (Hymenoptera, Formicidae, Myrmicinae) em cacuais na Bahia, Brasil," *Revista Theobroma, Brazil,* 18, 29–37.

Dix, M. E., 1993, "Conservation and enhancement of arthropod natural enemies," in *Proceedings of Biological Control of Forest Pests in the Great Plains: Status and Needs — A symposium.* Great Plains Agricultural Council, Forestry Committee, Forest Biological Control Task Force, Bismarck, ND, June 13–16, 1992, 72–75.

Dix, M. E., Johnson, R. J., Harrell, M. O., Case, R. M., Wright, R. J., Hodges, L., Brandle, J. R., Schoeneberger, M. M, Sunderman, N. J., Fitzmaurice, R. L., Young, L. J., and Hubbard, K. G., 1995, "Influence of trees on abundance of natural enemies of insect pests: A review," *Agro. Syst.* 29, 303–311.

Dix, M. E., 1996, "Pest management in agroforestry systems: Worldwide challenges," *J. Forest., 94, 8-12.*

Dix, M. E., Klopfenstein, N. B., Zhang, J. W., Workman, S. W., and Kim, M.-S., 1997, "Potential use of *Populus* for phytoremediation of environmental pollution in riparian zones," in *Micropropagation, Genetic Engineering and Molecular Biology of Populus,* Klopfenstein, N. B., Chun, Y. W., Kim, M.-S., and Ahuja, M. R., Eds. Gen. Tech. Rep. RM-GTR-297, U.S. Department of Agriculture, Forest Service, Rocky Mountain Research Station, Fort Collins, CO, 206–211.

Deharveng, L., 1995, "Field report for the soil mesofauna studies," in *Alternatives to slash-and-burn in Indonesia, Summary Report of Phase 1. ASB-Indonesia Report Number 4, 1995,* Noordwijk, M. V., Tomich, T. P., Winahyu, R., Murdiyarso, D., Suyanto, Partoharjono, S., and Fagi, A. M., Eds. Bogor, Indonesia,

Du Pleissis, C. and Lea, H. A. F., 1943, "The maize stalk borer, *Calamistisfusca* (Hmpn)," *Bull. Dept. Ag., South Africa.* 1–51.

Entwistle, P. F., 1985, "Insects and Cocoa," in *Cocoa.* Wood, G. A. R. and Lass, R. A., Eds. 366–443.

Entwistle, P. F., 1972, "Pests of cocoa," *Tropical Science Series.* Longman, London.

Epila, J. S. O., 1986, "The case of insect pest management in agroforestry research," *Ag. Syst.* 19, 37–54.

Esilaba, A. O. and Ransom, J. K., 1997, "Striga in the Eastern and Central African countries: A literature review," *AHI Technical Report Series No.1.*

Esilaba, A. O., Reda, F., Mulatu, T., Ransom, J. K., Woldewahid, G., and Fitwy, I., Eds., 1997, "A diagnostic survey of striga in the Northern Ethiopian Highlands," *AHI Technical Report Series No. 5, ICRAF.* Nairobi, Kenya.

Eskes, A. B., 1982, "The effect of light intensity on incomplete resistance of coffee to *Hemileia vastatrix.*" *Neths. J. Plant Pathol.* 88, 192–202.

Eskes, A. B., 1982, "The use of leaf disk inoculations in assessing resistance to coffee leaf rust (*Hemileia vastatrix*)," *Neths, J. of Plant Pathol.* 88, 127–141.

Fataye, A. and de Taffin, G., 1989, "Lutte integree contre *Pseudotteroptus devostans* et especes voisines," *Olexgineus.* 44, 525–530.

Fernandez, C. E., 1984, "Central America coffee rust project," in *Coffee Rust in the Americas.* Fulton, R. H., Ed. The American Phytopathological Society, St. Paul, MN, 84–92.

Fiedler, L. A., 1986, Rodent problems and control in agroforest systems, in *Proceedings for the Short Course in Agroforestry.* Fort Collins, CO, Aug. 10–23, 86–103.

Flemming, N., Guinand, Y., and Okorio, J., 1995, "Farmer participatory diagnostic research of the Lakeshore banana–coffee land–use system of Uganda (with emphasis on Mukono District, Uganda)," a publication of the *Agroforestry Research Network for Eastern and Central Africa (AFRENA-ECA),* in collaboration with the International Center for Research in Agroforestry ((ICRAF) and the Forestry Research Institute of Uganda (FORI), 105.

Forman, R. T. T. and Baudry, J., 1984, "Hedgerows and hedgerow networks in landscape ecology," *Env. Mgt.* 8, 495-510.

Frischknecht, P. M., Dufek, J. U., and Baumann, T. W., 1986, "Purine alkaloid formation in buds and developing leaflets of *Coffea arabica,*" *Phytochemistry.* 5, 613–617.

Fulton, R. H., Ed., 1984, *Coffee Rust in the Americas,* Symposium Book No. 2, Annual Meeting of the American Phytopathological Society, Aug. 9, 1982, Salt Lake City, UT, 120.

Fulton, R. H., 1984, "Chemical control of leaf rust in central America," in *Coffee Rust in the Americas.* The American Phytopathological Society. St. Paul, MN, 75–83.

Funaski, G. Y., Lai, P. Y., and Nakahara, L., 1990, "Status of natural enemies of *Heteropsylla cubana* Crawford (Homoptera: Psyllidae) in Hawaii," in *Leucaena psyllid: Problems and Management*, Winrock International/IDRC/NFTA/F-FRED. Napompeth, B. and MacDicken, K. G., Eds. Bangkok, Thailand, 153–158.

Galliag, S., Mandujana, S., and Gonzalez-Romero, A., 1996, "Conservation of mammalian biodiversity in coffee plantations of Central Veracruz, Mexico," *Agro. Syst.* 33, 13–27.

Garrity, D. P. and Sajise, P. E., 1991, "Sustainable land use systems research in Southern Asia," in *Sustainable Land Use Systems Research, Proceedings of an International Workshop.* Feb. 12-16, 1970, New Delhi, India, Oxford & IBH Pub. Co., New Delhi, 59–76.

Gebre-Amiak, A., Sigvald, R., and Pettersson, J., 1989, "The relationship between sowing data, infestation and damage by the maize stalk borer, *Busseola fusca* in Awassa, Southern Ethiopia," *Trop. Pest Mgt.*, 35, 143-145.

Gebre-Amiak, A. and Azerefegne, F., 1998, "Insecticidal activity of chinaberry, endod and pepper tree against maize stalk borer (Lepidoptera: Noctuidae) in Southern Ethiopia," *International. J. Pest Mgt.*

GIFAP, 1992, *Objectives and Structure — Asia Working Group*, International Group of National Associations of Manufactures of Agrochemical Products, Brussels.

Gold, C. S., Ogena-Latigo, M. W., Kashaija, I., and Tushemereirwe, W., 1993, "Farmer perceptions of banana pest constraints in Uganda: result from a rapid rural appraisal," in *Biological and Intergrated Control of Highland Banana and Plant Pests and Diseases.* Gold, C. S. and Gemmil, B., Eds. Proceedings of a Research Coordination Meeting, Cotonou, Benin. Davis, CA. 3–24.

Goodall, J. M. and Erasmus, D. J., 1996, "Review of the status and integrated control of the invasive alien weed, *Chromolaena odorata*, in South Africa," *Ag. Ecosyst. Env.* 56, 151–164.

Gordon. C. J. and Bentley, W. R., 1990, *A handbook on the management of agroforestry research.* Winrock International and Oxford & ISH Pub., New Delhi, 72

Gowen, S. R., 1993, "Yield losses by nematodes to different banana varieties and some management techniques appropriate to African farmers," in *Biological and Integrated Control of Highland Banana and Plantain Pests and Diseases*, Gold, C. S. and Gemmil, M., Eds., Proceedings of a Research Coordination Meeting. Cotonou, Benin, Davis, CA, 199–208.

Greathead, D. J., 1988, "Crop protection without chemicals: Pest control in the third world," *Asp. Biol.*, 17, 19–28.

Greenberg, R., 1977, "Criteria working group thought paper," in *Proceedings of the First Sustainable Coffee Congress.* Rice, R. A., Ashley, M. H., and Jennifer, M., Eds., September 1996, Smithsonian Migratory Bird Center, Washington, DC, 403–409.

Guillermo, R. O. 1984, "An effort to control and possibly eradicate coffee rust in Nicaragua," in *Coffee Rust in the Americas,* Fulton, R. H., Ed., The American Phytopathological Society, St. Paul, MN, 93–104.

Hall, R., 1995, "Challenges and prospects of integrated pest management," in *Novel Approaches to Integrated Pest Management.* Reuveni, R. Ed., Lewis Publishers, Boca Raton, FL, 1–19.

Hardy, R. W., Beachy, R. N., Browning, H., Cauldea, J. D., Charudattan, R., Faulkner, P., Gould, F., Hinkle, M. K., Jaffee, B. A., Knudson, M. E., Lewis, W. J., Loper, J. E., Mahr, D. L., and Van Alfen, N. K., 1996, *Ecologically Based Pest Management, New Solutions for a New Century,* Board on Agriculture National Research Council, National Academy Press, Washington DC.

Herren Gowen, S. R., 1993, "Yield Losses by nematodes to different banana varieties and some management techniques appropriate to African farmers," in *Biological and Integrated Control of Highland Banana and Plantain Pests and Diseases.* Gold, C. S., and Gemmil, M., Eds. Proceedings of a Research Coordination Meeting. Cotonou, Benin, Davis, CA, 199-208.

Higley, L. G. and Boethel, D. J., Eds., 1994, *Handbook of Insect Pests of Soybean.* Entomological Society of America. Hyattsville, MD, 136.

Hilje, L., Shannon, P. J., and Coto, D., 1993, "Insects associated with *Erythrina* species in Costa Rica," in *Erythrina in the New and Old Worlds*, Westley, S. B. and Powell, M. H., Eds., 330–339.

Hill, D. S. and Waller, J. M., 1988, *Pests and diseases of tropical crops. Volume 2, Handbook of Pests and Diseases*, Longman, New York.

Hoekstra, D., Torqubequ, E., and Bishaw, B., Eds. 1990, *Agroforestry: Potential and Research Needs for the Ethiopian Highlands*, ICRAF Report No. 21. Nairobi, Kenya, 115.

Horwley, B., 1977, "Central America gears to fight coffee rust," *Foreign Agric.*, June 27, 11–14.

ICIPE, 1991, *Integrated Pest Management of Crop Borers for Resource–Poor Farmers in Africa. Achievement and Highlights. 1980 –1990*, ICIPE. Science Press, Nairobi.

Imle, E. P., 1979, "Hevea rubber — past and future," *Eco. Bot.* 3, July–September 1978, 32, 264–277.

Jacobson, M., Ed., 1992, "1988 Focus on Phytochemical Pesticides Volume 1 The Neem Tree," in *Selected Neem Reprints, S&T/FENR Agro-forestation Appendix to: Tech. Series No.5, Rm. 515-E, SA-18.* Agency for International Development, Washington, D C, 77–101.

James, C., 1989, *Implementation of integrated pest management (IPM) in the third world — institutional aspects*, International Committee on Integrated Pest Management, Mexico City.

Javed, Z. U. R., 1984, "Leaf rust in Africa and what it means to American programs," in *Coffee Rust in the Americas.* Fulton, R. H., Ed., The American Phytopathological Society, St. Paul, MN, 15–34.

Kang, B. T., Akobundu, I. O., Manyong, V. M., Carsky, R. J., Sanginga, N., and Kueneman, E. A., Eds., 1995, "Moist Savannas of Africa: Potentials and constraints for crop production, " in *Proceedings of an IITA/FAO Workshop.* Sept. 19–23, 1994, Cotonou, Republic of Benin, 382.

Kenten, R. H. and Legg, J. T., 1971, "Varietal resistance of cocoa to swollen shoot disease in West Africa, Food Agric," *Organ. Plant Protection Bull.,* 19, 1–12.

Kiss, A. and Meerman, F., Eds. 1991, "Integrated Pest Management and African Agriculture," *World Bank, Tech. Paper No. 142,* African Tech. Dept. Ser., Washington, DC.

Klopfenstein, N. B. and Kerl, J. G., 1995, "The potential of biological technological contributions to temperate agroforestry practices," *Agro. Syst.* 30, 29–44.

Klopfenstein, N. B., Chun, Y. W., Kim, M.-S., and Ahuja, M. R. Eds. 1997, *Micropropagation, Genetic Engineering, and Molecular Biology of Populus*, Gen. Tech. Rep. RM-GTR-297. U.S. Department of Agriculture, Forest Service, Rocky Mountain Research Station, Fort Collins, CO, 326.

Lachaud, J. P., Fresnau, D., and Garcia-Perez, 1984, "Etude des strategies d'approvisioenment chez species de fourmis ponerines (Hymenoptera: Formididae)," *Folia Entomol, Mexicana.* 61, 159–177.

Lagerlöf, J., Stark, J., and Svensson, B., 1992, "Margins of agricultural fields as habitat for pollinating insects," *Ag. Ecosyst. Env.* 40, 117–124.

Leius, K., 1967, "Influence of wild flowers on parasitism of tent caterpillars and codling moth," *Can. Ento.*, 99, 444–446.

Le Pelley, R. H., 1968, *Pests of Coffee*, Longman, London, 590.

Litsinger, J. A. and Moody, K., 1976, "Integrated pest management in multiple crop systems," in *Multiple Cropping,* Papendick, R. J., Sanchez, P. A., and Triplett, G. B., Eds., Publication No. 27, American Society of Agronomy, Madison, WI, 293–316.

Liyarage, M. de S., 1993, "The role of MPTS in coconut-based farming systems in Sri Lo," *Agrofor. Today,* July–September, 5, 7–9.

Mahmud, A. W., 1988, "Agroforestry: An option for future timber industry," in *Plantation management for the 21 century.* Incorporated Society of Planters. Pushparajah, E., Ed., 1, 231–255.

Majer, J. D., 1974, "The use of ants in an integrated control scheme for cocoa," in *Proceedings of the 4th West African Cocoa Entomologists*, Legon, Ghana, 181–190.

Mally, C., 1920, "The maize stalk borer, *Busseola fusca* (Fuller)," *Bull. Dept. Ag., South Africa* 3, 1–111.

Marshed-Kharusy, M. N., 1988, *Plant Pathology,* Third edition, Academic Press, San Diego, 803.

Matteson, P. C., Altieri, M. A., and Gagne, W. O., 1984, "Modification of small farmer practices in pest management," *Ann. Rev. Ento.* 29: 383-402.

McClay, A.S., 1990, "Distribution of Leucaena psyllid and its natural enemies in Mexico: Implications for biological control," in *Leucaena psyllid: Problems and Management.* Winrock International/ DRC/NFTA/F-FRED. Napompeth, B. and MacDicken, K. G., Eds., Bangkok, Thailand, 139–143.

Meinke, L. J., 1995, *Adult corn rootworm management*, University of Nebraska Agricultural Research Division Miscellaneous Publication 63-C, 8.

Meinke, L. J., Siegfried, B. D., Wright, R. J., and Chandler, L. D., 1997, "Western corn rootworm resistance to insecticides: Current situation in Nebraska," in *Proc. Illinois Agric. Pesticides Conf. 1997*, Cooperative Extension Service, University of Illinois, Urbana-Champaign, IL, 88–92.

Michon, G. and de Foresta, H., 1995, "Forest resource management and biodiversity conservation: the Indonesian agroforest model, IUCN/Madrid workshop A Biological Diversity conservation outside protected areas: overview of traditional agroecosystems Madrid," in *Alternatives to slash-and-burn in Indonesia, summary report of phase 1. ASB-Indonesia Report Number 4, 1995*, Noordwijk, M. V., Tomich, T. P., Winahyu, R., Murdiyarso, D., Suyanto, Partoharjono, S., and Fagi, A. M., Eds., Bogor, Indonesia, 154.

Minae, S. et al., 1988, *Agroforestry research project proposal for the coffee based system in the bimodal Highlands, Central and Eastern Provinces, Kenya*, ICRAF Report No. 16, Nairobi, Kenya, 1988.

Morse, S. and Buhler, W., 1997, *Integrated pest management: Ideals and realities in developing countries*, Lynne Rienner. Pub., Bouleler.

Moraes, W. B. C., 1984, "Host-parasite Interactions," in The coffee-*Hemileia vastatrix* in *Coffee Rust in the Americas*. Fulton, R. H., Ed., The American Phytopathological Society, St. Paul, MN, 59–74.

Mumford, J. D. and Stonehouse, J. M., 1994, *Pest management policies in less–developed countries*, in proceedings of an international symposium at the Brighton Metropole Hotel, Black, R. and Sweetmore, A., Eds., Monograph No. 61, Nov. 21, 1994, British Crop Protection Council, Farnham, U.K., 11–18.

Muniappan, R., 1993, "Status of *Erythrina* species and the fruit-piercing moth in the Pacific," in *Erythrina in the New and Old Worlds*, Westley, S. B. and Powell, M. H., Eds., Nitrogen Fixing Tree Association (NFTA). Paia, Hawaii, 340–344.

Munyua, H., Bondole, B. M., and Majisu L., 1989, *Agroforestry Literature, A Selected Bibliography on Subsaharan Africa*, International Council for Research in Agroforestry. ICRAF House, Nairobi, Kenya, 110.

Mushala, H. M., 1995, "The moist savannas of East and South Africa: Potential for and constraints to research and development," in *Proceedings of an IITA/FAO Workshop*. Sept. 19-23, 1994, Cotonou, Republic of Benin, 69–86.

Nair, P. K. R., 1993, *An Introduction to Agroforestry*, ICRAF with Kluwer Acedemic Publication. Dordrecht, Netherlands, 499.

Napompeth, B. and MacDicken, Eds., 1990, *Leucaena psyllid: Problems and management*, Winrock International/IDRC/NFTA/FRED, Bangkok, Thailand, 208,

Napompeth, B., 1990, "Leucaena psyllid problems in Asia and the Pacific," in *Leucaena psyllid: Problems and Management*, Winrock International IDRC/NFTA/F-FRED. Napompeth, B. and MacDicken, K. G., Eds., Bangkok, Thailand, 1–11.

Napompeth, B., 1994, "Leucaena psyllid in the Asia-Pacific region: Implications for its management in Africa," *RAPA Publication: 1994/13*. Food and Agriculture Organization of the United Nations (FAO), Bangkok, 27.

Natural Resources Institute (NRI), 1991, *A synopsis of integrated pest management in developing countries in the tropics*. Chatham Maritime, Chatham, U.K., 20.

Natural Resources Institute (NRI), 1992, *Integrated pest management in developing countries, experience and prospects*, National Resources Institute, Chatham, U.K., 77.

Nderitu, J., Burachara. R. A., and Ampofo, J. F. K., 1997, "Relationship between bean stem maggot, bean root rots and soil fertility: literature review with emphasis on research in Eastern and Central Africa," *AHI Technical Report Series No. 4.*

Noordwijk, M. V., Tomich, T. P., Winahyu, R., Murdiyarso, D., Suyanto, Partoharjono, S., and Fagi, A. M., Eds., 1995, *Alternatives to slash-and-burn in Indonesia. Summary report of phase 1*. ASB-Indonesia Report No. 4, 1995, Bogor, Indonesia, 154.

Oduol, P. A. and Aluna, J. R. W., 1990, "The banana (*Musa* Spp,) — *Coffea robusta*: traditional agroforestry systems of Uganda," *Agro. Syst.* 11, 213–226.

Oka, I. N., 1990, *The Indonesian national integrated pest management program — success and challenges*, paper presented at the UNDP Seminar Workshop on Integrated Pest Management. July 1990, New York, 26–27.

Okech, S. H. O. and Gold, C. S., 1996, "Relationship of the banana weevil with its host plant and soil fertility: Literature review with emphasis on studies in Eastern and Central Africa," *AHI Technical Report Series No. 4.*

Okigbo, B. N., 1995, "Major farming systems of the lowland savanna of sub-Saharan Africa and the potential for improvement," in *Proceedings of an IITA/FAO Workshop, Sept. 19–23, 1994,* Cotonou, Republic of Benin, 49–68.

Olofson, H., 1981, "Indigenous agroforestry systems," in proceedings of *Workshop on Environmentally Sustainable Agroforestry and Fuelwood Production with Fast-Growing, Nitrogen-Fixing, Multi-Purpose Legumes*, Nov. 12–20, 1981, East–West Center, Hawaii.

Penot, E., 1995, "Taking the jungle out of the rubber: Improving rubber in Indonesian agroforestry systems," *Agrofor. Today*, July–December, 11–13.

Perfecto, I., 1990, "Indirect and direct effects in a tropical agroecosystem: The maize-pest-ant system in Nicaragua," *Ecology*. 71, 2125–2134.

Perfecto, Ivette, 1977, "Loss of insect diversity in a changing agroecosystem: The case of coffee technification," in *Proceedings of the First Sustainable Coffee Congress*, Rice, R. A., Ashley, M. H., and Jennifer, M., Eds. September 1996, Smithsonian Migratory Bird Center, Washington, DC, 143-155.

Perfecto, I., Rice, R. A., Greenberg, R., and Van der Voort, M. E., 1996, "Shade Coffee: A disappearing refuge for biodiversity," *Biosci.* 46, 598–608.

Pierre, M. B. and Freud, E. H., 1995, "Agricultural innovation in the cotton zone of francophone West and Central Africa," in *Proceedings of an IITA/FAO Workshop, Sept. 19–23 1994,* Cotonou, Republic of Benin, 256–306.

Poswal, M. A. T., Akpa, A. D., and Alabi, O., 1993, "Cultural control of pests and diseases: Prelude to integrated pest management for resource poor farmers in Nigerian agriculture," *J. Sust. Ag.* 3, 5–48.

Raheja, A. K., 1995, "Practice of IPM in South and Southeast Asia," in *Integrated Pest Management in the Tropics: Current States and Future Prospects.* Mengech, A. N., Saxena, K. N., and Gopalan, H. N. B., Eds., John Wiley & Sons, New York, 70–119.

Ramirez, A., Seré, C., and Uguilla, A. J., 1992, "An economic analysis of improved agroforestry practices in the Amazon lowlands of Ecuador," *Agro. Syst.* 17, 65–56.

Rao, B. S., 1973, "Potential threat of South American leaf blight to the plantation rubber industry in Southeast Asia and the Pacific region," *Food Agric, Organ. Plant Protec. Bull.*, 21, 107–113.

Reddy, D. B., 1991, "Integrated pest management: An overview, Asia and the Pacific region," *Tropical Agriculture Research Series. March*, 14, 45–60.

Rice, R. A., Ashley, M. H., and Jennifer, M., Eds., 1977, *Proceedings of the First Sustainable Coffee Congress.* September 1996, Smithsonian Migratory Bird Center, Washington, DC, 428.

Rice, R. A., 1990, *Transforming agriculture: The case of the coffee leaf rust and coffee renovation in southern Nicaragua*, Ph.D. dissertation, University of California, Berkeley, CA.

Roach, S. H. and Thomas, W. M., 1991, "Overwintering and spring emergence of three cocolnellid species in the coastal plain of South Carolina," *Env. Ento.*, 20, 540–544.

Rodenhouse, N. L., Barrett, G. W., Zimmerman, D. M., and Kemp, J. C., 1992, "Effects of uncultivated corridors on arthropod abundance and crop yields in soybean agroecosystems," *Agri. Ecosys. Env.*, 38, 179–191.

Rodrigues, C. J. Jr., 1984, "Coffee rust races and resistance," in *Coffee Rust in the Americas.* Fulton, R. H., Ed., The American Phytopathological Society, St. Paul, MN, 41–58.

Rugalema, G. W., Johnsen, F. H., and Rugambisa, J., 1994, "The homegarden agroforestry system of Bukoba district, North-Western Tanzania. 2. Constraints to farm productivity," *Agro. Syst.*, 26, 205–214.

Ruskin, F. R., Ed., 1992, *Neem A Tree for Solving Global Problems*, report of an Ad Hoc Panel of the Board on Science and Technology for International Development National Research Council, National Academy Press, Washington, DC, 140.

Samedy, J. P., Mitchell, G. C., Engeman, R. M., Bornstein, M. S., and Groninger, N. P., 1986, "Preharvest corn losses to vertebrate pests in Haiti," *An. Mus. Hist. Nat.* Vaparaíso 17. 129–132.

Sanchez, P. A. and Leakey, R. R. B., 1996, *Land-use transformation in Africa: three determinants for balancing food security with natural resource utilization*, paper presented at the 4th Congress of the European Society for Agronomy, July 7–11, 1996, ICRAF, 1–17.

Saxena, K. N., Pala Okeyo, A., Reddy, K. V. S., Omole, E. O., and Ngode, L., 1989, "Insect pest management and socio-economic circumstances of small-scale farmers for food crop production in Western Kenya: A case study," *Insect Science and Its Application.* 10, 443–62.

Schertz, D. L., 1991, "Conservation tillage and environmental issues," in *Agriculture and the Environment: The 1991 Yearbook of Agriculture*, Smith, D.T., Ed., U.S. Government Printing Office, Washington, DC, 15–20.

Schieber, E. and Zentmyer, G. A., 1984, "Distribution and spread of coffee rust," in *Coffee Rust in the Americas*, Fulton, R. H., Ed., The American Phytopathological Society. St. Paul, MN, 1–14.

Schultz, R. C. and Colletti, J., Eds., 1994, *Proc. Third North American Agroforestry Conference.* Ames, IA, Aug. 5–18, 1993, 163–174.

Shi, Z. and Goa, Z., 1986, "On the ecological efficiency of shelter-belt network and its yield-increasing effect in paddy fields," *J. Ecol.*, 5, 10–14.

Showler, A. T., 1995, "Leucaena psyllid, *Heteropsylla cubana* (Homoptera: Psyllidae), in Asia," *Am. Ento.*, 41, 49–54.

Sibuea, T. T. H. and Herdimansyah, D., 1995, "The variety of mammal species in the agroforest areas of Krui (Lanpung), Muara Bungo (Jambi) and Maninjau (West Sumatra), HIMBIO (UNPAD), Bandung, Indonesia," in *Alternatives to slash-and-burn in Indonesia, Summary Report of Phase 1, ASB-Indonesia Report Number* 4, Noordwijk, M. V., Tomich, T. P., Winahyu, R., Murdiyarso, D., Suyanto, Partoharjono, S., and Fagi, A. M., Eds., Bogor, Indonesia.

Siswomihardjo., H., 1991, *Pest control in Indonesia*, Country Report resented at the Seminar on Pest Control. Sept. 24 to Oct. 4, Asian Productivity Organization, 1991, Tokyo, Japan.

Slosser, J. E. and Boring III, E. P., 1980, "Shelterbelts and boll weevils: A control strategy based on management of overwintering habitat," *Env. Ento.*, 9, 1–6, 1980,

Snow, A. A. and Palma, P. M., 1997, "Commercialization of transgenic plants: Potential ecological risks," *Biosci.*, 47, 86–96.

Summers-Smith, J. D., 1988, *The Sparrows. T. and A.D. Poyser.* Calton, U.K., *342.*

Suyin, H., 1959, "The sparrow shall fall," *New Yorker*, 35, 43–50.

Tajuddin, I., 1986, "Integration of animals in rubber plantations," *Agrofor. Sys.*, 4, 55–66.

Thiollay, J. M., 1995, "Are traditional agroforests an alternative for the conservation of rainforest bird diversity? Three case studies in Sumatra," *Cons. Biol.*

Thurold, R. H., 1968, *Pests of Coffee*, Longman, London, 590.

Thurston, H. D., 1984, *Tropical Plant Diseases.* The American Phytopathological Society. St. Paul, MN, 208.

Tian, G., Kang, B. T., Akobundu, I. O., and Manyong, V. M., 1995, "Food production in the moist savanna of West and Central Africa," in *Proceedings of an IITA/FAO Workshop*, Sept. 19–23 1994, Cotonou, Republic of Benin. 107–128.

Toedo, V. M. and Moguel, P., 1977, "Searching For Sustainable Coffee In Mexico: The importance of biological and cultural diversity," in *Proceedings of the First Sustainable Coffee Congress*, Rice, R. A., Ashley, M. H., and Jennifer, M., Eds., September 1996, Smithsonian Migratory Bird Center, Washington, DC, 163–173.

Torres, E. M., 1977, "Outlook on ecological coffee farming: a new production alternative in Mexico," in *Proceedings of the First Sustainable Coffee Congress.* September 1996, Rice, R. A., Ashley, M. H., and Jennifer, M., Eds. Smithsonian Migratory Bird Center, Washington, DC, 211–222.

Torres, J. A., 1984, "Diversity and distribution of ant communities in Puerto Rico," *Biotropica.* 16, 296–303.

Van Den Beldt, R. C., Hatch, C., Sastry, C., Raintree, J., Ghosh, S., Saxena, N. C., and Garrity, D., 1994, "Farm forestry comes of age: The emerging trends in Asia," *Farm Forestry News.* 6, 1, 4–7.

Van Den Beldt, R. C. and Napompeth, B., 1992, "Leucaena psyllid comes to Africa," *Agrofor. Today.* 4 (4), 11–12.

Van Houten, H., Ed., 1995, *1995 Annual Report, ICRAF*, International Center for Research in Agroforestry, Nairobi, Kenya, 93, 97.

Van Houten, H., Ed., 1996, *1996 Annual Report, ICRAF*, International Center for Research in Agroforestry. Nairobi, Kenya. 23–24, 27, 267–268, 272.

Vergare, N. T. and Nair, P. K. R., 1985, "Agroforestry in the South Pacific region — an overview," *Agro. Syst.* 3, 363–379.

Waage, J., 1990, "Exploration for biological control agents of leucaena psyllid in tropical America," in *Leucaena psyllid: Problems and management*, Winrock International/DRC/NFTA/F-FRED, Napompeth, B. and MacDicken K. G., Eds., Bangkok, Thailand, 144–152.

Wang'ati F., 1994, *The African Highlands Initiative: A conceptual framework.* International Center for Research in Agroforestry, Nairobi, Kenya, 24.

Wardhani, M. A., 1991, *Developments in IPM: The Indonesian case*, paper presented at the Conference on IPM in the ASIA–Pacific Region. Sept. 23–27, 1991, Kuala Lumpur, Malaysia.

Waterhouse, D. F. and Norris, D. R., 1987, *Biological Control, Pacific Prospects*, Australian Center for International Agricultural Research (ACIAR), Inkata Press, Melbourne, 454.

Way, M. J. and Khoo, K. C., 1992, "Role of ants in pest management," *Annu. Rev. Entomol.*, 37, 479–503.

Westley, S. B. and Powell, M. H., Eds., 1993, *Erythrina in the New and Old Worlds*, Nitrogen Fixing Tree Association (NFTA), Paia, Hawaii, 358.

Whajon, M. E., Van De Baon, H. E., and Untung, K., 1991, "Resistance Management of brown plant hopper. *Nilaparvata lugens* in Indonesia," in *Pest Management in Rice.* Grayson, B. T., Green, M. B., and Copping, L. G., Eds., Society of Chemical Industry. Elsevier Applied Science, New York.

Williams, D. F., Ed., 1994, *Exotic Ants, Biology, Impact and Control of Introduced Species*, Westview Press, Boulder, CO. 332.

Williams, P. H., 1986, "Environmental change and the distributions of British bumble bees (*Bombus* Latr.)," *Bee World*, 67, 50–61.

Youdeowei, A. and Service, M. W., 1983, *Pest and vector management in the tropics with particular reference to insects, ticks, mites and snails*, Longman, London. 399.

Zenner de Polania, I., 1990, "Biological aspects of the Ahormiga loca, *Patatrechina (Nylanderia) fulva* (Mayr), in Colombia," in *Applied Myrmecology: A World Perspective*. Vander Meer, R. K., Jaffe, K. and Cedeno A., Eds., Westview Press, Boulder, CO. 290–297.

Zethner, O., 1995, "Practice of integrated pest management in tropical and sub-tropical Africa: An overview of two decades (1970–1990)," in *Integrated Pest Management in the Tropics: Current Status and Future Prospects*. Mengech, A. N., Saxena, K. N., and Gopalan, H. N. B., Eds., UNEP. John Wiley & Sons. NY, 1–67.

8 Clonal Propagation of Multipurpose and Fruit Trees Used in Agroforestry

Kenneth W. Mudge and Eric B. Brennan

1-56670-294-1/99/$0.00+$.50

1. INTRODUCTION

One of the keys to the widespread adoption of agroforestry (AF) is the availability of high quality planting stock of appropriate multipurpose and fruit trees (MPFT). Although zygotic seed (i.e., sexual propagation) is often the easiest, cheapest, and most common means of propagating most MPFTs, in this review we have chosen to emphasize the alternative, asexual propagation, as both a traditional and increasingly important contemporary approach to plant propagation. The terms *vegetative* and *clonal propagation* will be used synonymously with *asexual propagation*. The reader is referred to several recent publications on critical aspects of MPFT production from seed, including germplasm conservation and collection, germination, dormancy, storage, distribution, evaluation, etc. (Turnbull, 1990; Sedgley and Griffin, 1989a; Glover and Adams, 1990; Venkatesh, 1988).

Vegetative propagation has traditionally played an essential role in the production of certain MPFT species, and is becoming increasingly important as the scientific discipline of AF focuses on genetic improvement of MPFTs. The purpose of this review is to assess the state of indigenous and more recent knowledge of the art and science of asexual propagation of MPFTs used in AF. Although the review emphasizes woody legumes and other taxa in the broad category of multipurpose trees and shrubs, we also include fruit trees; hence the term MPFT. Asexual propagation of fruit trees will be limited to examples relevant to small-scale AF systems, as opposed to high intensity monocultural orchard production systems. MPFTs are defined broadly to include species like banana (*Musa* sp.) and papaya (*Carica papaya*), which, although lacking secondary growth, function as "trees" in AF systems.

We will discuss the factors involved in determining the most appropriate methods of propagation for a given species, and associated constraints and limitations. Current research on asexual propagation will be reviewed, and outcomes that have or may impact the implementation of modern AF systems will be identified. Last, we will consider areas where future research is most needed for the improvement of AF systems. Table 1 is a glossary of technical terms used throughout the chapter.

2. LIMITATIONS OF SEXUAL PROPAGATION

Across many MPFTs and agroforestry systems, seed is the easiest and cheapest method of propagation and is generally considered the default approach, unless other considerations apply. However, there are a number of circumstances that may limit or even eliminate the option to propagate MPFTs from seed.

2.1 Seed Availability

The most obvious limitation to seed propagation is the unavailability of seed. The most extreme case would be the need to propagate tree species, hybrids, or selections which are biologically

TABLE 1
Glossary of terminology used in the text. Definitions are, in some cases, specific to the context of agroforestry.

Term	Definition
Adventitious	Formation of a root or shoot (bud) from other than naturally occurring sequential branching of a seedling root or shoot system, respectively; i.e., adventitious roots are those which develop on stems or leaves, while adventitious shoots are those which arise on roots (suckering), or on stems, from other than lateral buds. Either adventitious roots and/or buds also may develop *in vitro* from undifferentiated callus tissue in some micropropagation systems.
Allele	The same gene on multiple (two, in the case of diploids) chromosomes within a cell; alleles may be homozygous (identical)or heterozygous (dissimilar)
alleles	One of a pair of genes
aseptic	Plant tissue and associated environment free of microorganisms; characteristic of most micropropagation systems
De novo	Newly developed in response to some external stimulus; not naturally occurring (e.g., root formation on a cutting)
Heterozygosity/ Heterozygous	A preponderance of genes in a plant genome with dissimilar alleles; giving rise to high level of variability in the next generation of seedlings
Homozygosity/homozygous	A preponderance of genes with identical alleles
In vitro	In a tissue culture/micropropagation system; literally "in glass."
Inbreeding	Deliberate self pollination, through several generations, of a naturally outcrossing species, giving rise to increasingly homozygous genotype
Marcottage	Air layering
Meiotic	Cell division giving rise to haploid sex cells (gametes)
Mitotic	Cell division giving rise to genetically identical, diploid, somatic cells, i.e., non gamete cells of the plant body
Ortet	Stock plant from which propagules are obtained for asexual propagation
Precocity	Development of reproductive maturity (ability to flower) sooner than is typical for the species
Propagule	Portion (ramet) of a stock plant (ortet) removed for asexual propagation; i.e., cutting, scion, explant
Ramet	Propagule; portion of an ortet detached for asexual propagation; i.e., cutting, scion, rooted layer, or explant
Recalcitrant	Seed which is intolerant of drying and, in the case of temperate recalcitrant seed, also intolerant low temperature; hence limited storage life.
Scion	Upper portion of a grafted plant which gives rise to the above ground shoot system
Stock (rootstock, understock)	Lower portion of a grafted plant (including the root system) onto which the scion is grafted
Stooling	A type of layering in which rooting occurs at the base of ground level coppice shoots mounded with a moist soil or other medium (e.g., sawdust)
systemic	Occurring throughout the plant body, with the exception of primary shoot and root meristems
vivipary (seed)	Extreme lack of dormancy characterized by germination within the fruit (pod) while still attached to the parent tree.

seedless. Insufficient supply of high quality seed to meet existing demand is a less drastic, though perhaps a more important long term problem. The latter can result from irregular or seasonal (phenological) periodicity of flowering. Monocarpic bamboo (e.g., *Dendrocalamus*) is an extreme example of this since some species require 20 to 50 years of growth to flower and subsequently die. Even annual cycles of seed production combined with poor seed storage characteristics for recalcitrant seeded species like neem (*Azadirachta indica*) may result in seed unavailability.

2.2 Seedlessness

Infertility resulting in seedlessness has arisen in a number of cultivated plant taxa either as a result of domestication in the distant past, or through more recent hybridization. In cultivated banana, seedlessness arose from natural hybridization between species with different ploidy levels (Purseglove, 1972). The resultant increase in palatability contributed to banana's domestication but it necessitated asexual propagation (by division). Similarly, seedless selections of breadfruit (*Artocarpus altilis*), which are more palatable than the seeded varieties of this species, are propagated by root cuttings (Purseglove, 1972). While there are few other naturally seedless MPFTs, there are instances of deliberate selection of seedlessness in tree crop for which either the seed or the fruit is not considered essential to the utility of the tree. For example, seedlessness has been selected in Washington navel orange (*Citrus sinensis*) (Purseglove, 1968), and several ornamental cultivars of honeylocust (*Gleditsia triacanthos*) (Dirr, 1990); both of which are propagated by bud grafting. In such cases selection for seedlessness has been possible because of asexual alternatives to seed propagation.

In the genus *Leucaena*, seedlessness due to triploidy from interspecific hybridization has been exploited in Indonesia. Brewbaker (1988) reported that seedless hybrids are bud grafted for use as shade over coffee (*Coffea* sp.) and tea (*Camellia sinensis*) because they are not prone to weediness, as is the *L. leucocephala* (Hughes and Styles, 1987; Brennan, 1990). Such hybrids are sterile because they are the triploid offspring of crosses between tetraploid (*L. leucocephala*) and diploid species (*L. diversifolia, L. esculenta, and L. pulverulenta*, etc.).

2.3 Insufficient Seed Production to Meet Demand

In some instances, adoption of AF may be hampered by an insufficient supply of MPFT seed (Mugo, 1997). Ironically, the most serious seed shortages may occur with the species that are least prone to weediness and thus more suitable in AF systems (e.g., *Glircidia sepium*, *Inga* spp.). Even in the genus *Leucaena*, which is widely known for prolific seed production and weediness, the demand for seeds of improved varieties exceeds supply. For example, in a 1997 seed catalog, the average price for seed of several available *Leucaena* species and hybrids exceeded $100 per kilo (Agroforester Tropical Seed, Holualoa, Hawaii).

2.4 Seasonal Limitations

The seed of most MPFTs are at least somewhat tolerant of drying and cool temperatures, which are characteristics necessary for moderate to long term seed storage. Consequently, these species can be harvested and stored for later use locally or elsewhere. On the other hand, MPFTs with recalcitrant seed (intolerant of low temperature and/or drying, and hence store poorly; e.g., mango (*Mangifera indica*), neem, citrus, etc.) are only available during a narrow window of time. The woody leguminous genus *Inga* is another example. Several species of *Inga* are used extensively throughout tropical America for fuel wood, coffee shade, and fruit, however seed *recalcitrance* and *vivipary* restrict their broader distribution (transport). Vegetative propagation could provide an alternative, which would permit propagation of *Inga* spp. throughout the year and possibly facilitate distribution of propagules outside of their native range.

2.5 Genetic Variability in Seed Propagated Trees

Sexual outbreeding through cross pollination has evolved as the dominant natural reproductive strategy (breeding system) in trees because it ensures genetic recombination, *heterozygosity*, and the concomitant seedling variation on which natural selection may act to bring about adaptations favorable to survival (Jain, 1976). Domestication of normally outcrossing species, on the other hand, depends on increasing uniformity, not only for its own sake (e.g., synchronous harvest), but also because it makes possible the selection of improved genotypes. Genetic gains associated with deliberate selection of superior genotypes of normally outcrossing species can be captured immediately, via cloning of the selected individual, avoiding the loss of the selected trait in subsequent seedling generations due to segregation of *alleles*.

3. RELATIONSHIPS BETWEEN NATURAL REPRODUCTIVE SYSTEMS AND DEVELOPMENT OF DELIBERATE PROPAGATION STRATEGIES FOR CROP SPECIES

Woody plants have evolved an impressive array of reproductive strategies which have contributed to their survival and proliferation (Hancock, 1993). Several distinct breeding systems have evolved in higher plants including, in order of importance: predominant outcrossing, predominant inbreeding, apomixis, and other vegetative propagation strategies (Jain, 1976). The most common of these in natural populations is outcrossing, which assures the genetic variability on which natural selection acts to bring about adaptive evolutionary change. On the other hand, a shift towards self pollination is regarded as a characteristic of the domestication syndrome of crop species which has resulted in increasing homozygosity, less seedling variability and hence greater crop uniformity (Hancock, 1992). *Inbreeding* has facilitated the fixing of improved genotypes by unconscious or deliberate selection by early agriculturists. This trend has continued in more modern times. Many fruit trees such as apple (*Malus domestica*), grape (*Vitis vinifera*), and citrus, and woody ornamentals such as *Rhododendron* sp. and *Bougainvillea* sp.), and some, elite lines of forest tree species such as Douglas fir (*Pseudotsuga menziesii*) and *Eucalyptus* spp., have been greatly improved genetically as a result of artificial selection programs which have depended, at least in part, on vegetative propagation. Furthermore, the successful future application of genetic engineering to tree crop improvement ultimately will depend on regeneration of intact transformed plants via *in vitro* asexual propagation. Other factors which contribute to the relatively difficult domestication and improvement of MPFT species by classical (sexual) breeding include their long generation times and irregularity in flowering and fruiting (Leakey, Newton and Dick, 1994). In such cases, vegetative propagation can be a powerful tool for capturing genetic gains.

Although sexual reproduction of woody plants has been the dominant theme both evolutionarily and throughout the history of human agriculture, woody plants have evolved a diverse array of vegetative reproductive strategies as well (Sedgley and Griffin, 1989b). Human understanding and modification of these natural vegetative reproductive strategies has been essential to the development of traditional and modern agricultural systems. Throughout the history of crop domestication, farmers and modern plant breeders have inadvertently and/or deliberately selected for ease of propagation. Widespread cultivation of vegetatively propagated crops may actually predate cultivation of seed propagated crops because of the relative ease of propagation of the former (Hancock, 1992).

The relatively "wild," undomesticated state of most MPFTs (Burley, 1993) compared to the world's major seed propagated food crops, such as rice (*Oryza sativa*), maize (*Zea mays*), and beans (*Phaseolus* sp.), has some bearing on the relative difficulty of seed propagation of MPFTs, and hence the relative usefulness of vegetative propagation. With respect to major seed propagated crop species, the domestication syndrome described by Hancock (1992) includes features which facilitate seed collection, like lack of shattering, and rapid germination due to the absence of seed dormancy. On the other hand, the relatively undomesticated MPFT species tend to retain sexual reproductive traits, associated with adaptations to their natural environments (e.g., dormancy, shattering of dehiscent fruit, etc.), that render them more difficult to propagate from seed in sufficient quantity for use in AF systems. For example, whereas seed dormancy has been almost entirely selected against in agronomic legumes like *Phaseolus* beans, seed coat-associated dormancy is a

FIGURE 1. Traditional method of marcottage (air layering) of lychee in the Ranomafana region of Madagascar. A woody branch is wounded by girdling with a knife, and the girdled area is packed with a mixture of cow dung and mud. This is wrapped with the dried leaf of raffia palm and tied with cordage made from the inner bark of haftra (*Dombeya* sp.).

common characteristic of many woody legumes adapted to xeric and/or cold environments (e.g., *Acacia, Leucaena, Robinia*). While such dormancy is advantageous in terms of long-term natural seed bank "storage" or deliberate storage of seed, it may interfere with prompt, uniform germination under cultivation. As a result, a major focus of MPFT seed propagation research has focused on pregermination treatments to overcome dormancy. On the opposite extreme are MPFTs, usually from the humid tropics, which exhibit little if any seed dormancy, and are intolerant of drying and/or low temperature. Such seeds, termed recalcitrant, have very low storage potential (Bonner, 1990; Chin, 1989). Some tropical MPFTs, which produce recalcitrant seeds, include mango, citrus, rubber (*Hevea brasiliensis*), jackfruit, avocado (*Persea americana),* coffee, cocoa (*Theobromo cacao*), *Inga* spp., and neem. Examples of temperate recalcitrant species include nut trees such as hickories (*Carya* sp.*),* pecans (*Carya pecan*), filbert (*Corlyus avellana*), walnut (*Juglans* sp.), and oak (*Quercus* sp.*).*

The observation and utilization by early agriculturists of naturally evolved asexual reproductive strategies must have played an important role in the development of deliberate vegetative propagation techniques. For example, natural rooting of attached vegetative structures (e.g., natural layering of epiphytic *Ficus* species) could easily lead to deliberate *marcottage* (Figure 1) and other layering methods. Cuttage propagation is likely to have arisen from observation of the rooting of wind thrown branches of riparian species, such as willow. Natural division of perennial organs (e.g., tubers, corms, and runners) is the biological basis for vegetative propagation of important crops like cassava (*Manihot esculenta*), potato (*Solanum tuberosa*), banana, etc. Observation of natural grafting at points of branch-to-branch contact may have inspired early attempts approach grafting (Figure 2) and led subsequently to detached scion grafting and budding of fruit and other tree species.

FIGURE 2. Approach grafting. This ancient intact-scion method of grafting allows the graft union to form with minimal water stress because the scion remains attached to its own root system (main tree). After graft union formation, the scion has an alternative source of water and nutrients from the root system of the understock plant in the small pot. It can then be cut away from the main tree (scion donor), with a minimum of water stress.

4. ASEXUAL PROPAGATION STRATEGIES

Asexual propagation may be defined as regeneration of a new individual from a portion (*ramet*) of a stock plant (*ortet*) by processes involving *mitotic* (not *meiotic*) cell division, and subsequent regeneration of complementary cells, tissues, and/or organs or entire plant to replace those "missing" from the ramet. Shoot cuttage, layering and shoot micropropagation involve replacement of roots by the process of *adventitious* root formation, while root or leaf cuttage and some types of micropropagation involve replacement of shoots by adventitious shoot (bud) formation. In the case of grafting, the missing part is replaced by fusing the ramet (*scion*) to a compatible *stock*. "Replacement" may occur via asexual (somatic) embryogenesis in some micropropagation systems. Apomixis is a special case of naturally occurring asexual reproduction in which seed is produced that contains an asexual (not zygotic) embryo which arose from mitotic division of maternal cells in flower-associated tissues, rather than by fertilization of a maternal egg cell by pollination.

4.1 LAYERING

Layering involves induction of adventitious roots along a portion of the shoot of an intact plant so that the newly rooted shoot section can be subsequently detached and transplanted. This may involve root induction on a shoot in the aerial portion of the tree (marcottage), or on a shoot at the base of a tree, either naturally or deliberately brought into contact with soil (tip layering, mound layering, etc.). Since the shoot undergoing rooting has a continuous supply of water from its preexisting root system, layering involves less moisture stress to the shoot being layered than propagation from shoot cuttings. Layering occurs naturally in some species, including purple raspberry (*Rubus occidentalis x idaeus*), many figs (*Ficus*), strawberry (*Fragaria x ananaassa*) (roots associated runner plantlets), some vines (e.g., English ivy, *Hedera helix*), etc. In a broad sense, root suckering could be considered the inverse of shoot layering since it involves adventitious formation of a complementary organ (shoot) on an intact organ (root). In either case, deliberate wounding may or may not be required to induce the process of adventitious root or (bud) formation. In the case of shoot layering (*stooling*, marcottage, tip layering), rooting is facilitated by light deprivation (etiolation) of the shoot by covering a portion of it with a moist medium (e.g., soil, moss, sawdust, etc.

4.2 Cuttage

Unless specified otherwise, cuttage refers to induction and/or elongation of adventitious roots at the basal end of a shoot cutting (ramet), detached from an intact stock plant (ortet). In the case of leaf and root cuttings, both adventitious shoot and adventitious root formation must occur to complement the ramet. With most species, rooting of a shoot cutting involves induction of *de novo* adventitious roots in response to wounding and/or other stimuli. In some species, formation of a new root system may involve elongation of preformed adventitious roots initiated during an earlier stage of normal shoot development (e.g., *Ficus*, bamboo, English ivy). Induction of *de novo* adventitious roots may only require the stimulus of wounding (severance from the ortet) for easy-to-root species, whereas more difficult-to-root species may require the additional stimulus of auxin type plant hormone or growth regulator, and/or additional post severance wounding, moisture, or etiolation. Especially in more difficult-to-root species, avoidance of water stress is critical for rooting cuttings, necessitating precise management of moisture, irradiance, and temperature of the propagation environment.

4.3 Grafting

Grafting involves placing two similar or dissimilar plant organs (stem/stem, stem/root, or root/root) from genetically compatible plants in intimate contact, with sufficient pressure and cambial alignment to induce the formation of an anatomically and physiologically functional graft union between the scion and stock. Just as the presence or absence of an intact root system affects the degree of moisture management required in air layering and cuttage respectively, intact scion grafting such as approach grafting (Figure 2) or inarching requires less intensive moisture management than detached scion grafting or budding. This is because both stock and scion have an intact root system in the case of approach grafting or inarching.

4.4 Micropropagation

In the broad sense, micropropagation refers to *in vitro* plant regeneration involving the use of a relatively small propagule, referred to as an explant, on an artificial nutrient medium within an *aseptic* environment. Development leading to multiplication may involve stimulation of normal (non-adventitious) elongation and/or branching of the original explant (shoot or root), and or *de novo* induction of adventitious organs (shoot and/or root). In the more narrow sense, micropropagation refers specifically to axillary or nodal shoot culture, which is the most common/important type of *in vitro* plant propagation. Shoot culture involves the use of growth regulators (cytokinins and/or auxins) during a multiplication stage to promote repeated cycles of shoot elongation and/or branching, followed by an indefinite number of cycles of subdivision and reculture on fresh medium. Proliferated shoots are then rooted as microcuttings either *in vitro* or *ex vitro*. Rooting may or may not require exogenous auxin-stimulated root induction. Other less common and generally less desirable *in vitro* plant multiplication systems involve proliferation adventitious buds or somatic embryos directly on an explant or from secondary undifferentiated callus or liquid cell suspension cultures.

4.5 Apomixis

Apomixis, or apomictic embryo formation, as described above, is a natural asexual seed formation process, which, unlike more typical zygotic seed arising sexually, involves only maternal (not paternal) genes. Hence, apomictic seedlings are genetically identical (clonal) to the maternal parent. In some cases (e.g., citrus, mango), one or more apomictic embryos may co-occur in the same seed with a zygotic (sexual) embryo. This is known as mixed (zygotic and apomictic) polyembryony. Apomictic seedlings, like sexual seedlings are rejuvenated, capable of flowering only after a somewhat protracted period of maturation, and hence delayed in flowering/fruiting compared to clonal plants arising from cuttage, grafting, or layering. Because of their clonal uniformity, apomictic seedlings are used widely as rootstocks for citrus and mango.

5. PROPAGATION IN TRADITIONAL AND/OR LOW TECHNOLOGY AGROFORESTRY SYSTEMS

Traditional AF systems are defined here as the historical and ongoing integrated use of trees and crops by indigenous peoples using low technology approaches developed more or less *in situ.* To some extent this is the same as defining traditional practices as those which are indigenous, i.e., practiced prior to the relatively recent introduction of "modern" technologies consisting of materials and/or services not available previously. Important propagation-related technologies categorized as modern include those which require the use of electricity, plumbed water and synthetic materials like polyethylene and synthetic rooting hormone (auxin) formulations. As implied above, the MPFTs used as components of traditional AF systems tend to be those which were chosen not only for one or more desirable products or services, but also selected, as least inadvertently, for relative ease of propagation. Many traditional practices involving asexual propagation, on the other hand, have developed for MPFTs, which do not produce seed readily, or at all, such as cassava, banana, or bamboo.

One of the simplest traditional vegetative propagation systems is collection of root suckers, sometimes referred to as wildings. Root suckers originate from adventitious bud formation on roots, either spontaneously, or as a result of wounding roots. Many species naturally root sucker such as *Faidherbia albida* (Ajee and Duhoux, 1994), *Inga feuillei* (Mudge, personal observation), *Robinia pseudoacacia, Ocotea usambarensis* and *Melia volkensii* (Teel, 1984), *Acacia* spp., *Chlorophora* spp, *Cordia alliodora,* and *Melia azedarach* (Longman, 1993). Traditional propagation by division of basal shoots, sometimes referred to as "suckers," but of axillary rather than adventitious origin, is practiced with the tree-like MPFTs, banana and bamboo.

Cuttage is another traditional method of vegetative propagation which requires minimal specialized equipment or skill. The semi-woody perennial root crop cassava is typically propagated from leafless shoot cuttings taken at the time of root harvest (Purseglove, 1968). In many traditional AF systems, large cuttings known as stake or pole cuttings are rooted directly in their final field location for use as living fences, for soil regeneration and production of firewood, medicinals or other products (Gautier, 1995; Jolin and Torquebian, 1992; MacDicken, 1990). For direct field planting, large cuttings are preferred over seedlings and smaller nursery-rooted cuttings, because they are often quicker to produce useful products and services. MPFTs commonly rooted from large stake cuttings include species in the relatively ease-to-root genera *Ficus, Erythrina* (Gautier, 1995; Jolin and Torquebian, 1992), *Gliricidia*, and *Hibiscus* (MacDicken, 1992). Martin (1997) presents an extensive list of species suitable for living fences, many of which are amenable to propagation from stake cuttings. Farmers in Cameroon use seven species of *Ficus*, two of *Erythrina*, and eight other species as hedge rows, for demarcation of boundaries, and containment of livestock (Gautier, 1995). Jolin and Torquebian (1992) observed stake cutting propagation of five species of *Erythrina* and 37 other species in various tropical locations in Central and South America, Africa, and India, as well as *Salix sp., Robinia, Populus, Morus, Cupressus,* and *Castanea* in temperate locations in Europe and North America. Interestingly, several of the genera reported to be successfully rooted from stake cuttings, including *Leucaena leucocephala* (Jolin and Torquebian, 1992; MacDicken, 1990) in the tropics, and the temperate species *Robinia, Tilia, Castanea*, and *Picea* (Jolin and Torquebian, 1992), are generally regarded as moderately difficult to root. It should be noted that in some cases, stake cuttings may erroneously appear to have rooted based on initial shoot growth, however, subsequent decline and excavation may indicate a lack of root formation (Brennan, personal observation).

Propagation from stake cuttings usually begins with pollarding (Gautier, 1995) or coppicing (Jolin and Torquebian, 1992) of established plants, followed by new shoot growth over three or four years, and subsequent harvest and planting of the stakes during the rainy season. The length (0.5–2.5 m) and diameter (5–15 cm) of stake cuttings may vary significantly with species and country (Jolin and Torquebian, 1992; Gautier, 1995). Leaves are removed and stakes may be "conditioned" before planting. This is done by stacking them in a shady location for one to several weeks. Successful rooting of stake cuttings is typically greater than 70% (Jolin and Torquebian, 1992; Gautier, 1995). With some species (*Erythrina burana*) before planting large (2 m long × 10 cm diameter) stake cuttings, the bark is removed from the portion that is to be buried (Teketay, 1990).

Vegetative propagation of MPFTs by methods other than cuttage have been practiced where an advantage can be gained by clonal rather than sexual propagation. Lychee (*Litchi chinensis*), for example, is propagated by marcottage (described below) because it cannot be rooted from cuttings and when planted from seed it takes several years to flower and is not true-to-type.

Grafting is being used in some traditional AF systems. A combination of grafting and cuttage is practiced by Indonesian farmers, to achieve an unusually "intimate" tree/crop association between the Ceara rubber tree (*Manihot glaziovii*) and cassava (*Manihot esculenta*). De Bruijn and Dharmaputra (1974) describe the invention, in 1952, of a unique method of cassava production by a Javanese farmer, Mukibat. Over the last several decades "Mukibat grafting" has been adopted not only by Javanese farmers, but also in Sumatra as detailed by Foresta, Basri and Wiyono (1994). The multipurpose Ceara tree, though displaced for rubber production by *Hevea brasiliensis* (para rubber), is used by farmers as a source of oil for cooking, edible leaves and fodder. Mukibat grafting involves grafting a shoot of Ceara (scion) onto a cassava shoot (understock) which is then rooted. The resulting grafted plant produces the Ceara shoot products while the cassava rootstock produces more than twice the yield of a normal cassava plant (de Foresta et al. 1994). The basis for the increase in cassava yield is the extension of the growing season due to the lack of dormancy of Ceara (E. Fernandes, pers. com.). Cassava, on the other hand, has an annual period of dormancy. This extension of the growing season is inversely analogous to the dwarfing of apple trees which is achieved by the use of selected clonal rootstocks. In that case dwarfing is achieved by a combination of earlier cessation and diminished rate of growth of the fruit variety when grafted onto a dwarfing rootstock (J. Cummins, pers. com.). Tamarind (*Tamarindus indica*) is another example of a traditional grafting practice. In Thailand, farmers have selected numerous sweet tamarind varieties, which are approach grafted onto sour tamarind rootstocks. In addition to approach grafting, some farmers air layer sweet cultivars and then inarch graft several rootstocks onto the transplanted layer for increased stability in windy sites (Brennan, personal observation). Other reports of traditional grafting include the decades-old Indonesian practice of bud grafting seedless Leucaena species and hybrids for use as shade in coffee and tea plantations described by Brewbaker, 1988.

In some cases, minor modifications of a traditional technique may have profound effects on the success of a propagation technique. For example, Tanala farmers in Madagascar have practiced marcottage to clonally propagate lychee, but the traditional practice is only marginally successful (Mudge, personal observation) (see Figure 1). "Modern" air layering, typically involves wrapping a polyethylene moisture barrier around pre moistened moss which surrounds a wounded stem

(Brennan and Mudge, 1997a). The traditional Tanala practice involves packing the wounded stem with a mixture of mud and dung, wrapping it with dried leaves of raffia palm, and securing it with cordage from the inner bark of haftra (*Dombeya sp.*), an indigenous tree. Because the raffia leaf wrapping provides little resistance to evaporation, the dung packing material has a tendency to dry out during periods of low rainfall. Due to repeated wetting and drying of the packing, success is low (<50%), and marcotts which root take from six months to a year before they are sufficiently well rooted to be severed from the parent plant. The recent introduction of polyethylene sheeting, as a moisture barrier, increased the success rate from <50% to >80%, and decreased time to harvest of rooted marcotts from >six months to as little as eight weeks. Access by farmers, world wide, to polyethylene has had a profound effect on their ability to asexually propagate plants. The use of polyethylene for moisture management in cutting propagation is discussed below.

6. CONSIDERATIONS IN THE SELECTION AND USE OF ASEXUAL PROPAGATION TECHNOLOGIES

6.1 Appropriate Technology

The various propagation strategies can be classified along a continuum from passive to active environmental modification, ranging from those that take advantage of natural processes with minimal technological inputs, to those requiring extreme and sophisticated technological intervention. Hence, from low to high technological input, the continuum would be as follows:

harvest of natural suckers or layers < deliberate layering
< cuttage (hardwood < semihard wood < softwood)
< grafting (approach < detached scion)
< micropropagation < *in vitro* embryogenesis.

The point of this ranking is to emphasize the importance of choosing a method of propagation for a given MPFT species that will not only achieve the intended outcome, but also be appropriate for the level of technology available and expertise of the propagator. For example, because of the need for aseptic conditions, it is unlikely that tissue culture techniques will be appropriate for on-farm or nursery production. On the other hand, it is likely that tissue culture will become an increasingly important tool for MPFT breeding/tree improvement programs. Moisture management of leafy cuttings is restricted to the use of polyethylene water vapor barriers when plumbed water and electricity is not available for mist propagation. The specific environmental and physiological constraints, which apply to each method of propagation, will be discussed in Section 6.4.

6.2 Growth Phase Considerations

The ontogenetic development of a woody plant from the point of seed germination to eventual flowering involves a physiological change in receptivity to flower induction by environmental conditions (photoperiod, temperature, etc.). A seedling begins its life cycle in the juvenile phase characterized by its inability to flower regardless of environmental conditions. Subsequently, after a period of as little as a few months to as much as several to many years, the tree undergoes a gradual transition to the adult phase, characterized by the ability to flower under inductive environmental conditions (Hackett, 1985). Obviously, this process of reproductive maturation is essential to the agricultural production of flowers, fruits, or seeds. On the other hand, vegetative propagules (cuttings, scions, explants for micropropagation, etc.) taken from adult phase tissues are significantly more difficult to propagate by cuttage, grafting, layering, or by micropropagation, than juvenile phase propagules.

Recognition and understanding of the temporal and spatial distribution of the juvenile and adult phases of the sexual life cycle has important implications with respect to the outcome of asexual propagation. In this regard, it is important to understand that reproductive maturation (transition from the juvenile to the adult phase) does not occur in differentiated tissues (stems and leaves), but in primary apical and lateral shoot meristems (Zimmerman, 1972). The chronological age of a given meristem at the time of its phase transition is not fixed, but rather it is proportional to its rate of node production (i.e., growth rate) (Hackett, 1985). Hence, a more rapidly growing terminal shoot undergoes meristem maturation (phase change) sooner than a dormant or slower growing lateral or basal shoot meristem. Epicormic buds are buds initiated early in shoot development that lie dormant indefinitely unless renewed bud growth is triggered by coppicing or other stimulus. Epicormic buds laid down early in the development of tree seedlings are juvenile and remain so as long as the buds remain dormant, but the process of reproductive maturation resumes if the buds are stimulated to grow (Kramer and Kozlowski, 1979). In this sense epicormic buds may constitute a reserve of juvenile meristems on an otherwise mature tree. To fully appreciate the significance of growth phase to successful propagation, it is important to understand that shoot tissues (xylem, phloem, vascular cambium, dormant buds) laid down by a juvenile meristem remain juvenile, even after the meristem from which they developed has undergone its phase transition and begins to produce new adult phase shoots. Hence, a given seed derived tree will simultaneously have juvenile tissue (wood, epicormic buds, etc.) near the base of the tree, produced while the meristem was juvenile, while more distal shoots and buds, produced after meristem phase change, will be in the adult phase. The simultaneous existence of both phases on the same tree, known as cyclophysis (Hartman et al., 1997) has important implications for the selection of vegetative propagules from a stock plant. Cyclophysis dictates that propagules taken from the base of a seed grown tree will be easier to propagate asexually, but will flower later than propagules taken from the distal portion of the same tree. Adventitious buds arising from roots (suckering) are apparently also in the juvenile phase, and hence give rise to easily rooted juvenile shoots.

In a minority of woody plant species there are easily recognizable morphological differences between juvenile and adult leaves (heterophylly) and shoots which can facilitate propagule selection. A classic example, which has been the subject of a great deal of growth phase-related research is *Hedera helix*. Juvenile portions of the plant have lobed leaves, a vine growth habit, and performed adventitious shoot born roots, whereas *H. helix* in the adult phase has entire leaves, an upright growth habit and lacks stem born performed adventitious roots (Hackett, 1985). In the case of the Australian phyllodinous acacias (e.g., *A. koa, A. melanoxlon, A. saligna*) juvenile leaves are bipinnately compound whereas adult phase leaves are reduced to a broad flattened petiole (phyllode) without true leaflets (Purseglove, 1968). Juvenile shoots of some citrus species and the temperate woody legume, *Gleditsia triacanthos*, bear spines, whereas adult shoots are spineless. All of these phase-related characteristics can be used by the propagator to select propagules of the appropriate growth phase (juvenile for ease of rooting; adult for rapid onset of flowering) in the relatively few species which exhibit phase-related dimorphisms. Propagule selection for most other species, which do not exhibit phase dimorphism, must be based on the positional considerations described above.

It should be pointed out that a natural process of rejuvenation occurs during embryogenesis, during the formation of zygotic or apomictic seed, or *in vitro* somatic embryos. Apparent rejuvenation associated with micropropagation and serial grafting will be discussed below.

6.3 Cloning as a Tool for Tree Improvement

6.3.1 Exploration, Selection, and Domestication of New MPFTs

In efforts to bring about MPFT improvement through plant exploration and collection of new germplasm, for subsequent evaluation, and eventually domestication, the propagule of choice, collected in the field, is usually seed. This is true for a variety of reasons including the ease of transport and storage, and also it suits the objective of collecting as much genetic variation as possible. Nevertheless, seed availability does not always coincide with the timing of collection trips and hence collection of vegetative propagules may be necessary.

As part of a tropical tree improvement program, Leakey and co-workers at the Institute for Tropical Ecology (ITE) have focused on vegetative propagation technology, mainly via cuttings, for selection of improved genotypes. Many of the species they have worked with are currently or potentially important components of AF systems (Leakey, Newton, and Dick, 1994). This research at ITE, described below, is especially significant due not only to the breadth of species investigated but also to its contribution to our understanding of the physiological factors affecting rooting and the development of techniques for optimizing the propagation environment.

6.3.2 Cloning as a Tool for Tree Seed Orchard Production

With temperate forest species, clonal seed orchards have utilized grafting for several decades (Sweet, 1995) as a means of asexually propagating otherwise difficult to root species such as Douglas fir. One example of asexual propagation being used as a tool for the genetic improvement of MPFTs comes from work with the genus *Leucaena*. At the University of Hawaii, several interspecific *Leucaena* F1 hybrids, produced by hand pollination, have performed well in field trials (Sorensson and Brewbaker, 1994), including psyllid resistance (Wheeler and Brewbaker, 1990). As mentioned earlier, interspecific triploid hybrids between the tetraploid *L. leucocephala* and the diploid species *(L. esculenta, L. diversifolia, L. pulverulenta*, etc.) are seedless and hence free of the problem of self-weediness due to excessive seed production. However, due to the difficulty of clonal propagation there are no large-scale orchards established to produce seed of interspecific F1 hybrid *Leucaena* through open pollination. One approach to achieving this objective would be to exploit the gametophytic self-incompatibility that is characteristic of all diploid species of *Leucaena* and the tetraploid *L. pallida* (Pan, 1985; Sorensson and Brewbaker, 1994). Self-incompatibility (SI) is a mechanism that ensures obligate outcrossing, whereby all individuals comprising a clone of a fertile hermaphroditic genotype are unable to produce seed by self-pollination (Richards, 1986). Since SI operates only within a clone, vegetative propagation is necessary to exploit the SI mechanism which theoretically allows only interspecific F1 hybrid seed production on the SI parent, under open-pollination conditions (Bray, 1984; Brewbaker and Sorensson, 1994; Sorensson and Brewbaker, 1994; Toruan-Mathius, 1992). Wheeler (1991) suggested that grafting could be used to produce seed of interspecific hybrids in open-pollinated orchards. In such an orchard a single clone of a SI species, such as *L. pallida* or diploid *L. diversifolia,* would presumably set seed only from cross-pollination with an interplanted cross compatible species like *L. leucocephala*.

Two studies have reported cloning self incompatible *Leucaena* species by grafting for use in the open pollinated production of interspecific hybrid seed (Bray and Fulloon, 1987; Brennan, 1995). In the former study (Bray and Fulloon, 1987), the cloned species was a diploid (*L. pulverulenta*) which was interplanted with the pollen donor species, (tetraploid *L. leucocephala*) to produce triploid seed for subsequent use as sterile hybrid trees. In the latter study Brennan (1995) cloned tetraploid *L. pallida* by grafting and interplanted these with the pollen donor *L. leucocephala* to produce the fertile F1 hybrid seed (Figure 3). In each case the authors reported only partial success in obtaining hybrid seed production. In both cases vegetative propagation by grafting was necessary to take advantage of intraclonal self-incompatibility. Given to the common occurrence of self-incompatibility that promotes outcrossing among trees of many species, it is likely that similar strategies may be useful in future production of F1 hybrid seed of other MPFT species. One possibility is *Inga*, another genus of tropical woody legumes that exhibits SI (Popenoe et al., 1989). In the first report of asexual propagation of this genus, Brennan and Mudge (1997a) found that *Inga feuillei* rooted well (up to 86%) from leafy cuttings, and 97% by marcottage. Espinal de Rueda (1996) successfully rooted five other species of *Inga* from cuttings.

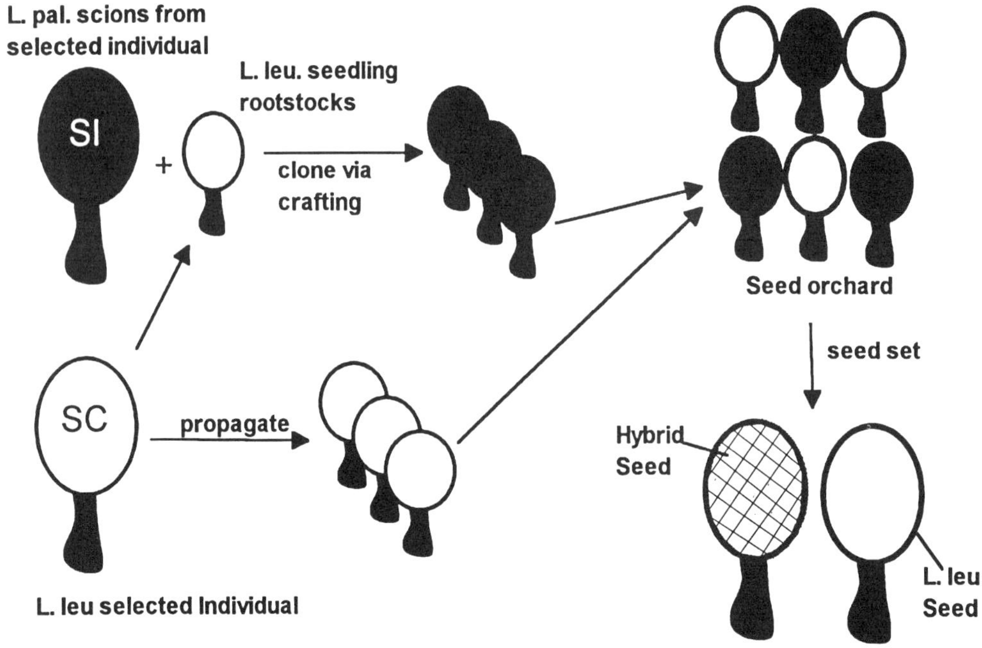

FIGURE 3. Strategy for interspecific hybrid seed production between the intraclonally self incompatible (SI) *Leucaena pallida* and *L. leucocephala.* The *L. pallida* clone, which is propagated by grafting onto seedling understocks of *L. leucocephala,* should produce only F1 hybrid seed.

6.4 Limitations of Asexual Propagation

Although the purpose of the foregoing discussion has been to highlight the advantages and reasons for asexual propagation of MPFTs, there are certain disadvantages associated with cloning. These must, of course, be taken into consideration when formulating a strategy for propagation of MPFTs for any purpose.

6.4.1 Difficult-to-Propagate Asexually

Based on current technology, asexual propagation cannot be accomplished effectively, or at all, for certain taxa. MPFT taxa which are considered difficult to clone include, but are not limited to, *Prosopis, Leucaena, Acacia.* However, due to a lack of research, little is known about the feasibility of clonal propagation of many MPFT species. This was the case with many of the tropical forest species (e.g., *Khaya ivorensis, Triplochiton scleroxylon*) currently under investigation at ITE (Leakey, Newton, and Dick, 1994). Such research has resulted in further gains in the range of species, which can be asexually propagated.

6.4.2 Insufficient Genetic Variability

Another concern which has been raised about asexual propagation is that it will result in a narrowing of the gene pool, which could render a population of a given species more vulnerable to biotic (diseases and pests) or physical stresses. Barnes and Burley (1982) discuss the essential safeguard of using multiple clonal genotypes in any large-scale clonal tree introduction program, and point out that the number of clones necessary may be relatively few. Longman (1993) recommends 10–30 clones in a stand for tropical hardwood species.

6.4.3 Quality Considerations

Questions have been raised from time to time about the quality of clonally propagated planting stock compared to seedlings, especially with regard to root system quality and its impact on subsequent plant performance. Although there has been little research with MPFTs along these lines, Longman (1993) supports the view that cutting root system quality is as high as that of seedlings, if the cuttings have been properly handled in the nursery. Based on anatomical and physiological comparisons, Blake and Filho (1988) found that *Eucalyptus grandis* cuttings were more likely to exhibit drought stress than seedlings, and Sasse and Sands (1996) made a similar finding for *Eucalyptus globulus*. Mohammed and Vidaver (1991) reported that micropropagated Douglas fir *(Pseudotsuga menziesii)* were more prone to water stress than seedlings. Patel et al. (1987) found anatomical irregularities in the root-to-shoot junction of micropropagated black and white spruce (*Picea* sp.) compared to seedlings. On the other hand, Harrision et al. (1989) found that peach (*Prunus persica*) cuttings and seedlings responded similarly to water stress. Regarding temperature stress, Ritchie et al. (1992) reported that Douglas fir cuttings were at least as cold hardy as seedlings. Bender et al. (1987) reported that nutrient uptake and translocation was similar in micropropagated plantlets and seedlings of *Thuja occidentalis*.

6.4.4 Pathogen Transmission

Plant viruses, MLOs (mycoplasma-like organisms), and xylobacteria pose serious problems in many agricultural species. The effects of viral infection on host plants range from outright disease symptoms to non-symptomatic decline in vigor and crop yield. Since viruses, MLOs, and xylobacteria are systemic, they are present in (most) vegetative propagules (ramets), and hence any plant(s) regenerated from an infected ramet is also infected. A virus moves throughout the tree as it grows, including across a graft union. For example, as much as half of the dwarfing effect of M9 clonal apple rootstock on a fruiting variety grafted onto it, is due to viral infection (Ferree and Carlson). Overall, however, the transmission of viruses and MLOs as a result of clonal fruit and nut tree propagation is considered a serious problem world wide on many tree crops including citrus, pome fruits (apple, peach, etc.), walnut (*Juglans nigra*), avocado (*Persia americana*), etc. (Rom and Carlson, 1987). Fortunately part of the solution to this overall problem is associated with the caveat noted above. Viruses occur systemically through *most* of an infected tree, with the important exception of the primary meristems and incompletely differentiated shoot tissue for a short distance behind the meristem per se (generally less than 1 mm). Hence, *in vitro* meristem tip culture can be used to regenerate a plant free of specific viruses. In fact, virus elimination via meristem tip culture is an important component of tree crop improvement programs for many species world wide (e.g., apple, citrus, etc.) (Rom and Carlson, 1987).

7. CURRENT AND FUTURE RESEARCH IN ASEXUAL PLANT PROPAGATION

7.1 Cuttage

As noted above, while a few popular MPFT species (e.g., *Ficus, Erythrina, and Glircidia*) are traditionally propagated from cuttings, most others are seed propagated. However, reports of successful cuttage propagation with species usually seed propagated are worth noting. Some of these include *Acacia* (Badji et al., 1991; Nilum and Verma, 1995; Dick and East, 1992), *Albizia, Calliandra* (Dick et al., 1996) *Leucaena* (Hu and Liu, 1981); Bristow, 1983), *Inga* (Brennan and Mudge, 1997a; Espinal de Rueda, 1996), *Faidherbia* (Danthu, 1991; Nikiema and Tolkamp, 1992), *Prosopis* (Arya et al., 1993; Dick et al., 1994; Felker, 1994; Goel and Behl, 1995; Nilum and Verma, 1995), *Cassia* (Khanna and Arora, 1984), *Azadirachta indica* (Kamaluddin and Ali, 1996; Chander et al., 1996), and Grewia (Shamet and Dhiman, 1991).

Given the relatively undomesticated state of many MPFT species used in tropical AF and forestry systems, research on cuttage propagation of a given species is often ground breaking. When cutting propagation technologies are developed for MPFT production in the rural tropics, one of the primary goals should be to develop propagation systems that are appropriate for small-scale farmers and local nurseries. A comprehensive program should involve optimizing components of the propagation system at each stage including stock plant management, cutting selection, post severance preparation of cuttings prior to sticking, post sticking management of the propagation environment and finally post rooting nursery stock management. As mentioned above, researchers at ITE have investigated factors controlling the rooting of cuttings of several tropical hardwood species, but most intensively with *Triplochiton scleroxylon.* With this particular species their research covers nearly the full range of cutting propagation system components listed above from stock plant related factors to environmental and physiological considerations during rooting. Similar, though less intensive experiments have been performed with other species at ITE including *Acacia tortillis* (Dick and East, 1992), *Prosopis juliflora* (Dick et al., 1994), *Calliandra calothyrsus* (Dick et al., 1996), *Khaya ivorensis* (Tchoundjeu and Leakey, 1996), *Eucalyptus grandis* (Hoad, and Leakey, 1996*), Terminalia spinosa* (Newton et al., 1996), *Milicia excel*sa (Ofori et al., 1996), and *Gnetum africanum* (Shiembo et al., 1996). Much of this research is summarized in a set of general recommendations for cutting propagation of tropical hardwoods (Leakey et al., 1990; Longman, 1993; Leakey et al., 1994).

7.1.1 Stock Plant Management

Several stock plant related factors affecting the subsequent rooting of cuttings include general plant health (vigor, mineral nutrition, pests and diseases), ontogenetic age (juvenile and adult growth phase), growing environment, and position from which the cutting originated. Environmental factors include light (irradiance and spectral quality), temperature, moisture and soil fertility. The influence of these factors on rooting is mediated by their integrated effects on the anatomy and physiology of the cuttings eventually taken from the stock plant, including, but not limited to photosynthesis, carbohydrate metabolism, phenology, water relations and the level and activity of endogenous substances (including phytohormones).

7.1.1.1 Irradiance

In growth chamber experiments, Leakey and Storeton-West (1992) reported that a higher percentage of *T. scleroxylon* cuttings rooted from stock plants grown at lower (250 umol/m^2/s) than at higher irradiance (650 umol/m^2/s). In contrast, 69% of *Prosopis* cuttings rooted when stock plants were grown at 520 umol/m^2/s compared to only 10% at 150 umol/m^2/s (Klass et al., 1985). The positive effect of reduced irradiance on rooting of some MPFTs is consistent with findings from most experiments involving a broad range of herbaceous and woody species (primarily non MPFTs) although there were some exceptions (reviewed by Moe and Andersen, 1988; Maynard and Bassuk, 1988).

7.1.1.2 Etiolation

Another interesting and potentially useful manipulation of the stock plant light environment is the absence of light. Etiolation refers to plant growth in the absence of light, and the same term has been applied to the deliberate practice of growing stock plants in near darkness in order to increase subsequent rooting of cuttings. This practice can have dramatic effects on rooting of cuttings, and is one of the more exciting technological developments to emerge from propagation research in many years, although the practice has been tested with only a few tropical MPFT species. As reviewed by Maynard and Bassuk (1988), and Howard (1994) the practice usually involves enclosing the stock plant in a black cloth or plastic tent for as little as one week, during the period of a new growth flush. In temperate species the treatment is usually applied just as stock plants are emerging from winter dormancy. New shoot growth, which occurs during the dark treatment, is pale yellow due to the absence of chlorophyll development. Etiolated shoots are very tender and cuttings would be easily desiccated if transferred directly to a lighted propagation environment. This is avoided by gradually regreening stock plants in the light, for one to two weeks before taking cuttings. Stock plant etiolation, followed by regreening, dramatically promotes rooting of a number of difficult-to-root temperate species such as apple, filbert (*Corylus avellana*), chestnut (*Castanea s*), oak (*Quercus s*), hornbeam (*Carpinus s*), maple (*Acer s*), and many others. Etiolation-promoted rooted has also been reported for a few tropical species including mango, avocado, and jack fruit (Maynard and Bassuk, 1988). A modification of the etiolation technique, known as banding involves placing a Velcro strip, at the beginning of the regreening period, over the portion of the stock plant stem which will subsequently become the base of the cutting. For some species including avocado, apple, and hibiscus, etiolation plus banding is more effective than etiolation alone. Considering the potential benefits of this low technology approach to rooting of cuttings, additional studies with difficult-to-root tropical MPFTs, especially legumes, are warranted.

7.1.2 Stock Plant Phenology

Another factor influencing the rooting of cuttings is stock plant phenology as influenced by season (time of year). For example, rooting of *Prosopis* cuttings is strongly seasonal, with spring cuttings rooting substantially better that those taken in the fall (Felker and Clark, 1981). Klass et al. (1985) independently investigated two of the components of seasonal variation (temperature and photoperiod), and found that stock plant temperature of 35°C and an optimum photoperiod was best for subsequent rooting of *Prosopis*. The optimum stock plant photoperiod for subsequent rooting of cuttings was only 12 h for rooting in contrast to an optimum of 18 h for the cuttings themselves in the post severance propagation environment. Of course, photoperiod is not the only factor in seasonal differences in rooting of cuttings. Arya et al. (1993) considered seasonal effects from the standpoint of temperature and rainfall. They reported that *Prosopis cineraria* rooted best during the cool/dry period (February–May) which corresponds to the period of most active shoot growth, rather than during the warm/rainy period. Surprisingly, the cuttings taken during the optimal months of February–May had approximately equivalent rooting despite considerable variation in photoperiod and temperature. Low rainfall occurred during these optimal months, suggesting that the limiting stock plant-associated factor controlling seasonal variation in rooting was rainfall rather than photoperiod or temperature. Similarly, Danthu (1992), and Nikiema and Tolkamp (1992) reported that *Faidherbia* cuttings from older (adult phase) trees rooted best when taken at the time of bud break (dry season). However, juvenile cuttings of *Faidherbia albida* rooted best when taken during the period of maximum leaf fall (rainy season) Nikiema and Tolkamp (1992). This suggests an interaction between stock plant phenology and cutting growth phase. *Faidherbia* is well known for its unusual "reverse" phenology of active shoot growth during the dry season and defoliation during the rainy season. *Acacia senegal*, which has more typical phenology, rooted better during the rainy season than the dry season (Badji et al., 1991).

7.1.3 Cutting Selection

The position on the stock plant from which a cutting is taken can influence the rooting of that cutting and growth form in several ways. The position within the canopy from which a cutting is taken will influence the degree of shading under which the cutting develops, which may affect rooting via irradiance and light quality (R/FR). (See Chapter 6.)

Stock plant position from which a cutting is taken will also influence potential rooting via growth phase effects as described earlier, since a seed grown woody plant can simultaneously have both juvenile (basal) and adult (distal) growth phases (cyclophysis) in different locations. The relative ease of rooting of juvenile vs. adult phase cuttings is extensively documented in the woody plant propagation literature (Hackett, 1985), as well as the literature on MPFTs. For example, Nikiema and Tolkamp (1992) compared the rooting of *Faidherbia* cuttings taken from juvenile stump sprouts (coppice shoots) with that of cuttings from adult phase branches and found 53–93% and 27% rooting, respectively. These finding are consistent with those of Danthu (1992) who also worked with *Faidherbia*. For *Prosopis juliflora*, Goel and Behl (1995) reported 61% rooting of cuttings from 1-year-old stock plants, as compared to only 10% from the adult portion of 8-year-old trees.

7.1.4 Post Severance/Pre Sticking Treatment of Cuttings

7.1.4.1 Auxin Application

The discovery of the root promoting activity of the synthetic auxin analogs IBA and NAA (Zimmerman and Wilcoxon, 1935) was undoubtedly one of the most important advances in asexual plant propagation in the twentieth century. Auxins act by triggering early cell division of the root primordia that will eventually develop into adventitious roots (Hartman et al., 1997). Hence, the primary effect of auxin application to cuttings is to increase the number of roots per cutting. Reliance on this criterion for judging the effectiveness of auxin application (on any treatment) may be misleading, however. For example, *Inga feuillei* cuttings were observed to form more than 30 roots per cutting when treated with 20,000 ppm IBA, compared with less than half as many roots per cutting on cuttings treated with 4000 ppm IBA. Roots of the former, however, were short, thickened (brittle) and unbranched, in contrast to longer, narrower, more branched roots on cuttings treated with the lower concentration (Mudge, unpublished observation). In such cases, evaluation of overall root system quality (root number, length, and branching) is more useful than quantification of treatment effect based on roots per cutting alone. Other responses to auxin application which have been noted include reduced time required for satisfactory rooting and increased percentage rooting (Blazich, 1988). In moderate and difficult-to-root species this latter response may ultimately be the most useful benefit. All three responses to applied auxin (roots per cutting, time to rooting, and percentage rooting) were evident in the African mahogany (*Khaya ivorensis*) (Tchoundjeu, and Leakey, 1996).

The most commonly used synthetic auxins, (IBA then NAA) are generally more effective than IAA (Hartman et al., 1997; Nikiema and Tolkamp, 1992), although Shamat and Dihman (1991) found that *Grewia optiva* cuttings rooted better with IAA than IBA or NAA. Combinations of IBA plus NAA are often more effective than either applied separately (Arya et al.1993).

Although the literature on plant propagation contains references to innumerable species where auxin has enhanced rooting, not all species respond, and the response is often somewhat variable depending on interactions with other factors such as phenology and growth phase. The optimal auxin concentration appropriate for a given species varies with the time of year and hardness of wood. Softwood cutting typically require 0.05 to 0.3% (w/w in powder formulation; w/v in quick dips). Semihardwood cuttings respond best in the range of 0.1 to 0.5%, and hardwood cuttings require 0.25 to 1.0%, and occasionally several times higher (Hartman et al., 1997). Seasonal variation in the response of cuttings to auxin is seen in the results of experiments with *Acacia senegal* treated with none, 0.2% NAA, 1.0% IAA, 0.2% or 8.0% IBA at 4 times during the year including the wet and dry seasons. This species is unusual since the most effective auxin treatment was 8.0% IBA, which is extraordinarily high compared to nearly all other reports in the literature. Arya et al. (1993) reported that untreated cuttings of *Prosopis cineraria* did not root, whereas cuttings treated with individual auxins (IAA, IBA, NAA), combinations (IAA+IBA+2,4-D), or auxins combined with the vitamin thiamin (NAA+IBA+thiamine) all resulted in at least 25% rooting. The most effective treatments (>50%) were from the two mixtures. Felker and Clark (1981) also reported that thiamine combined with auxin increased rooting of *Prosopis* cuttings over auxin alone. Auxin-stimulated rooting has been reported with numerous other MPFTs including *Cordia alliodora, Vochysia hondurensis, Albizia guachapele* (Leakey et al. 1990*), Triplochiton scleroxylon, Khaya ivorensis* (Longman, 1993), and *Inga* spp. (Brennan and Mudge, 1997a; Espinal de Rueda, 1996).

As relatively inexpensive as commercial auxin formulations are in developed countries, they are often prohibitively expensive or unavailable to rural farmers in developing countries. Research is needed for developing a low cost source of auxin from materials readily available to farmers who do not have access to commercial sources. Unconfirmed anecdotal reports in the literature describe alleged farmer practices involving stimulating rooting of grape and other hardwood cuttings by imbedding grain seed in a split in the base of the cutting or placing grain seed at the bottom of a planting hole in contact with the cutting base (Hartman et al., 1997). Given that germinating grain seeds are a rich source of both conjugated and free auxin (Cohen and Bandurski, 1982), grain seed-stimulated rooting is a reasonable hypothesis. McCaffery (1998) tested the effect of wheat, oat and maize seed on the rooting of cuttings of *Inga* and carnation (*Dianthus caryophyllus*), under a range of different conditions (cutting phenology, media, environmental management), but grain seed treatments did not stimulate rooting under any conditions. Another low technology approach to developing a readily available source of natural auxin for the rooting of cuttings might be to test the effects of alcohol (e.g., rum) extracts of herbaceous or woody plant shoot tips on rooting. Given that shoot tips are the principal source of auxin biosynthesis in most plants (Salisbury and Ross, 1978), and that auxins are readily soluble in alcohol, this approach might be successful, but has not been reported in the literature.

7.1.4.2 Leaf Area Reduction

The rooting of a cutting depends on maintaining a positive water balance and a sufficient carbohydrate reserve to support root development. Both of these are affected by transpiration, but in opposing ways; increasing transpiration results in water loss potentially leading to desiccation, but transpiration also results in CO^2 uptake which may be a limiting factor in photosynthesis. On the one hand, transpiration must be minimized to avoid excessive water loss, since the cutting initially lacks a root system. On the other hand, transpirational CO^2 uptake may favor rooting since photosynthesis has been shown to occur during rooting of cuttings (Leakey and Storeton-West, 1992; Hoad and Leakey, 1992; Machida et al., 1977; Okoro and Grace, 1976; Davis and Potter, 1981) and indirect inference (weight gain during rooting, stomatal conductance: Loach and Gay, 1977; Newton and Jones, 1993b; Leakey and Coutts, 1989; Davis and Potter, 1981). In some cases photosynthesis may be a limiting factor in rooting (Davis and Potter, 1981; Leakey and Coutts, 1989). Hence, a seemingly reasonable strategy would be to manage leaf surface area to achieve a balance between sufficient transpiration to satisfy photosynthetic demand for CO^2 uptake, but avoid excessive moisture loss. Leaf area reduction, either by removal of entire leaves or partial trimming of each leaf is one way to manage this balance to optimize rooting. Kamaluddin and Ali (1996) found that leaf area was positively correlated with the number of roots per cutting, root dry weight and shoot growth of neem cuttings, although it did not affect the percentage of cuttings rooted. African mahogany (*Khaya ivorensis*) cuttings were seemingly moisture-limited since rooting was inversely proportional to leaf area (Tchoundjeu and Leakey, 1996). Cuttings of *Triplochiton scleroxylon*, on the other hand, were shown to be dependent on post severance photosynthesis and rooted best with intermediate leaf area (Leakey and Coutts, 1989). In the latter case, dry weight gain during propagation was directly proportional to leaf area, but leaf water potential (an indicator of water deficit) was inversely proportional to leaf area (i.e., greater leaf area resulted in more water deficit). Hence, it appears that in this species, unlike *Kahaya ivorensis*, optimal rooting was a compromise between maximizing photosynthesis and minimizing water loss. Brennan and Mudge (1997a) found that *Inga feuillei* rooted well (up to 86%) from leafy cuttings but rooted poorly or not at all from defoliated cuttings.

7.1.5 Management of the Propagation Environment: Moisture, Temperature, and Light

Once cuttings have been properly selected and treated prior to sticking, management of the propagation environment is critical for successful rooting. Environmental management of leafy cuttings is to a very great extent a matter of avoiding excessive moisture loss (Loach, 1988), although in some species, CO^2 uptake is a consideration as well, as discussed above.

Maintenance of favorable cutting water relations depends on management of three critical environmental parameters — irradiance, humidity, and temperature. These must be carefully managed regardless of whether cuttings are propagated in a high tech greenhouse with sophisticated modern environmental controls, or in simpler systems that use polyethylene moisture barriers and/or shading. A practical understanding of the physiological response of leafy cuttings to these parameters may make the difference between success and failure.

Water loss from leafy cuttings via transpiration is controlled primarily by the interactive effects of atmospheric moisture (approximated by relative humidity), temperature, and light. If water loss by the cutting exceeds water uptake, water stress will occur, and, if severe enough, the cutting will die and/or not root (Loach, 1989). The potential rate of water loss increases as the difference in atmospheric moisture between the interior of the leaf and the surrounding air increases. Similarly, water loss increases as the difference in air temperature between the leaf and the surrounding air increases. Consequently, the principal management strategies for minimizing moisture loss from cuttings are increasing the relative humidity of the propagation and decreasing the temperature, and particularly decreasing the leaf to air temperature difference. The effect of light on moisture loss from cuttings is mediated by its effect on leaf and air temperature, and on stomatal aperture. In practical terms water loss is reduced by decreasing light levels because of the reduced leaf and air temperatures associated with shading. At higher light levels, leaf temperature is often somewhat warmer than surrounding air, in some cases by as much as several degrees (Mudge et al., 1995; Loach, 1989), creating a considerable driving force for water loss even at high atmospheric humidity.

Humidity is managed in most propagation systems either by the use of a polyethylene barrier or intermittent mist. Although polyethylene is often more appropriate when electricity and plumbed water are lacking, one must be careful to manage the system to avoid a temperature increase. In polyethylene systems, shade must be used to control temperature. Grange and Loach (1983) recommended sufficient shading to give a maximum irradiance of 100 W/m^2 which is ≥70% shading under "summer" (warm, high light) conditions. Under mist, on the other hand, the mist itself achieves cooling, so that higher levels of irradiance are tolerable.

Due to the adverse effect of light on leaf temperature, polyethylene systems are generally more effective than mist systems under relatively cool, low light (Loach, 1977; Mudge et al. 1995) than under warmer, higher irradiance conditions where mist systems are favored (Loach, 1988).

Another factor, which may influence the effectiveness of polyethylene and mist systems in cool climates, is their effect on night temperatures. Although minimizing leaf and air temperature during the daytime is a key consideration in successful cutting propagation, nighttime temperatures can have an important but opposite effect. In the relatively cool Kenya highlands, near the equator where daytime irradiance levels were high, hibiscus cuttings rooted better under polyethylene than under mist, even though daytime leaf and air temperatures were several degrees higher under polyethylene than under mist (Mudge et al., 1995). At night, however, leaf and air temperatures decreased to as low as 6°C for cuttings which had been misted during the day (mist off at night). In contrast leaf and air temperatures were higher by approximately 4°C under polyethylene. The lower rooting under the mist treatment, compared to those under polyethylene, was probably due, at least in part, to suboptimal night temperatures for the former.

Leakey et al. (1990) described an easily constructed polyethylene propagation chamber approximately 2 to 4 m long × 1 m wide × 0.5–1.0 m high consisting of a wooden frame with a hinged cover for access to the cuttings. One drawback of this relatively large raised polyethylene system is that the relative humidity inside the chamber dropped considerably upon opening, and reequilibrated slowly due to the large chamber volume. Brennan and Mudge (1997a) described a smaller (modular) raised polyethylene system made of a 10 × 36 × 51 cm plastic flat, with an attached wire frame which are enclosed in a clear polyethylene bag. This system requires opening less frequently since it is usually filled to capacity with cuttings initially and so is not reopened subsequently for addition of new cuttings as a larger system might be. Consequently, humidity remains more uniform, and when the bag is opened, humidity equilibrates faster due to its lower volume.

An even simpler polyethylene system, refereed to as contact polyethylene, does not require any frame. It involves merely placing a sheet of polyethylene on top of the cuttings and sealing the edges with a rubber strip or lath strip. Contact polyethylene has been shown to improve the rooting of cuttings compared to raised polyethylene systems (Loach, 1977; Mudge et al., 1995; Grange and Loach, 1983,). Grange and Loach (1983) speculated that the improved rooting under contact polyethylene compared to raised polyethylene was due to the lower leaf to air temperature difference, which they observed in the former. Mudge et al. (1995) also reported higher rooting with contact compared to raised polyethylene system despite the fact that there was no difference in the leaf to air temperature gradient between the two systems.

7.2 Grafting

Research on propagation of fruit trees by grafting has proceeded steadily, ever since its adoption by early agriculturists several thousand years ago. Mango, which is traditionally propagated in India by approach grafting (Kanwar and Bajwa, 1974), has received considerable attention as a subject of grafting research. Unlike the majority of grafting techniques which involve detaching the scion (ramet) from the stock plant (ortet), approach grafting has the advantage of the scion remaining attached to its own root system until the graft union is formed (see Figure 2). Approach grafting allows a relatively low rate of multiplication due to the labor and spatial constraints inherent in tying seedling rootstocks into the canopy of established scion donor trees. Research on detached scion grafting for mango has focused on side grafting (Kanwar and Bajwa, 1974), bench grafting (Majumdar and Rathore, 1970), epicotyl grafting (Chakrabarti and Sadhu, 1983), and saddle grafting (Thomas, 1981), and the effects of defoliation and bud stick storage on veneer grafting (Singh and Srivastava, 1979). Grafting is used extensively for clonal propagation of a wide range of temperate fruit and nut species including apple, pear (*Pyrus communis*), grape (*Vitis vinifera*), and stone fruits including peach, cherry (*Prunus avium*), almond (*P. dulcis*), and for several important tropical fruit species including mango, citrus, avocado, etc. The primary focus of research related to grafting of these important fruit crops is on selection of clonal or seedling rootstock for adaptation to particular soil characteristics (moisture content, salinity, pH, etc.), as well as disease and pest resistance (Rom and Carlson, 1987).

Another interesting use of grafting is in papaya breeding. Although papaya is commercially propagated from seed, bud grafting has been used in papaya breeding (Sookmark and Tai, 1975) because it allows selection of clones not only of known genotype but also gender, since the seedling gender of this dioecious species is unknown prior to flowering (Singh et al., 1985).

Compared to tree species grown primarily for their fruit, grafting is seldom used to propagate multipurpose trees grown for other purposes. However, there are a few cases where grafting has been used to clone desirable selections of MPFTs such as seedless *Leucaena* (Brewbaker, 1988) and elite Haitian selections of *Prosopis juliflora* (Wojtusik et al., 1993). Wojtusik and Felker (1993) compared several different grafting techniques and reported invigorating effects of specific rootstocks on scion biomass production and a wide range of interspecific graft compatibility within the genus *Prosopis*.

As described above in Section 6.3.2, grafting has shown promise as a method of propagating selected genotypes to take advantage of self-incompatibility in leucaena seed orchards. *Leucaena* species have been successfully grafted using several standard methods including whip and tongue, and cleft grafting (Versace, 1982), approach grafting (Brennan, 1992) and more recently by a somewhat novel technique called single bud splice grafting (Brennan and Mudge, 1997b). This technique relies on a plexiglass grafting tool that allows grafting of small (3–5 mm diameter) single node scions onto two- to three-month-old seedling rootstocks. Using this method, the authors reported a high degree of interspecific graft compatibility between 18 Leucaena scion species and 2 rootstock species.

There are few reports on grafting other woody legume species with the exception of the temperate species *Gleditsia triacanthos*. A thornless variety *G. triacanthos* and selected seedless cultivars (e.g., Moraine) are routinely bud grafted for use as ornamentals (Dirr, 1990). Thornlessness and seedlessness are characteristics that could be useful in temperate AF systems involving this genus.

Acacia spp. and the closely related *Faidherbia albida*, are examples of tropical MPFTs, which have proven extremely difficult to propagate asexually, particularly from adult trees. In such cases approach grafting may be possible, as has been the case with *Acacia koa* (Brennan, 1995, unpublished data). In addition, there are reports on successful micrografting performed *in vitro* (tissue culture) (Palma, 1996; Detrez, et. al, 1992; Detrez, 1995; Monteuuis, 1996) or *ex vitro* (post micropropagation) using micropropagated scions (Palma et al., 1996). The success of these *in vitro* methods is related to the apparent rejuvenation of reproductively mature tissue resulting from *in vitro* culture. *In vitro* micrografting will be further discussed below in the section on micropropagation.

Aside from the selection and use of clonal rootstocks for few species of fruit trees including apple, citrus, and some stone fruits, most tree grafting, including most MPFTs, involves the use of seedling rootstocks. Because of the high degree of heterozygosity associated with naturally outcrossing tree species, seedling rootstock effects tend to be more variable than clonal rootstocks. Nevertheless, Wojtusik and Felker (1993) reported that scions of several *Prosopis* spp. were invigorated (produced more biomass) when grafted onto seedling rootstocks of species other than themselves. Brennan (1995) evaluated the effects of two lines *of Leucaena leucocephala* (weedy and the giant type, K636) and *L. diversifolia* as rootstocks for several scion species (*L. leucocephala, L. diversifolia, L. esculenta, L. pallida* and the hybrid KX3). Although the effects of rootstock genotype on scion growth were complicated by significant genotype x environment interactions, there were significant effects of rootstock genotype on growth of several scions. Although these early investigations of rootstock genotype effects on scion growth and development of *Prosopis* and *Leucaena* are encouraging, they are far from the sophisticated use of rootstock genotype to influence scion vigor, root system adaptation to specific soil types, and other characteristics in apple and citrus. It is obvious that research on MPFT rootstocks is in its infancy. Clonal rootstock selection for MPFTs is an area for future research, which could contribute significantly to MPFT improvement. For example, it may be possible to select for MPFT rootstock architecture that minimizes below ground competition with interplanted crops. Another important area well worth investigating is selection of MPFT rootstocks for adaptation to specific soil types and soil born pests and diseases.

7.3 Micropropagation

Micropropagation has been useful for the large-scale commercial production and genetic improvement of a few traditional forestry species such as eucalyptus, pine (*Pinus*), and Douglas fir, as well as several temperate woody ornamentals such as *Rhododendron* sp., lilac (*Syringa vulgaris*), and *Kalmia sp.* Despite the intriguing research results with microprogation of MPFTs during the past two decades, it seems unlikely that micropropagation will be used for the routine production of MPFTs for direct use in AF systems, given the emphasis on the use of resource appropriate technologies. Micropropagation may, however, be useful in MPFT breeding/genetic improvement programs where lower technology propagation approaches are less suitable.

Most research on the micropropagation of MPFTs has focused on explant selection and manipulation of the auxin and cytokinin phytohormones to optimize *in vitro* shoot and subsequent root production, respectively. Trigiano et al., (1992) recently reviewed the status of micropropagation of both tropical and temperate woody legumes. The majority of research on micropropagation of tropical woody legumes has involved the genera *Acacia* and *Leucaena*. More than 10 different species of *Acacia* have been investigated, while most research on *Leucaena* has focused on *L. leucocephala* (Table 2).

TABLE 2
Research reports of micropropagation of multipurpose and fruit trees used or potentially useful in agroforestry.

Species	Approach[1]	Growth Phase	Regeneration	Reference
A. saligna	ASC	adult	Yes	Barakat et al., 1992
A. saligna	ASC	7 mo.	Yes	Jones et al., 1990
A. nilotica	ASC	juvenile	Yes	Dewan et al., 1992
A. senegal	C	ND	No	Hustache et al., 1986
A. senegal	MG	4 yr.	Yes	Palma et al., 1996
A. senegal	ASC	juvenile	Yes	Badji et al., 1993
A. koa	C	juvenile	Yes	Skolmen and Mapes, 1976
A. nilotica	SE	juvenile (3N)	Yes	Garg et al., 1996
A. mangium	ASC	juvenile	Yes	Galiana et al., 1991
A. mangium	MG	adult	Yes	Monteuuis, 1996
A. mearnsii	ASC	juvenile	Yes	Huang, 1994
A. bivenosa	ASC	7 mo.	Yes	Jones et al., 1990
A. holosericea	ASC	7 mo.	Yes	Jones et al., 1990
A. salicina	ASC	7 mo.	Yes	Jones et al., 1990
A. sclerosperma	ASC	7 mo.	No	Jones et al., 1990
A. catechu	SE	juvenile	Yes	Rout et al., 1995
A. melanoxylon	C	juvenile	Yes	Meyer and van Staden, 1987
A. auriculiformis	ASC	ND	Yes	Reddy et al., 1995
A. auriculiformis	C	juvenile	Yes	Ranga Rao and Prasad, 1991
A. auriculiformis	ASC	juvenile	Yes	Mittal et al., 1989
Faidherbia albida	ASC	juvenile	Yes	Duhoux and Davies, 1985
Leucaena leucocephala	DO, C	juvenile	No	Nagmani and Venketeswaran, 1983
L. diversifolia	DO, C	juvenile	No	Nagmani and Venketeswaran, 1983
L. retusa	DO, C	juvenile	No	Nagmani and Venketeswaran, 1983
L. leucocephala	C	juvenile	No	Mazari and Rubluo, 1984
L. leucocephala	ASC, DO	juvenile	Yes	Hossain et al., 1993
L. leucocephala	ASC	14 mo.	Yes	Goyal and Felker, 1985
L. leucocephala	ASC	juvenile	No	Glovak and Greatbatch, 1981
L. leucocephala	ASC	adult	Yes	Dhawan and Bhojwani, 1987
L. leucocephala	ASC	juvenile	Yes	Ravishankar et al., 1983
L. leucocephala	ASC	juvenile/adult	No	Toruan-Mathius, 1992
L. pallida	ASC	juvenile/adult	No	Toruan-Mathius, 1992
L. Kx3	ASC, SE	juvenile/adult	No	Toruan-Mathius, 1992
Prosopis cineraria	ASC	ND	ND	Goyal and Arya, 1984
Albizia procera	ASC, DO	juvenile	Yes	Hossain et al., 1993
Albizia lebbeck	C	adult	Yes	Gharyal and Maheshwari, 1990
Cassia fistula	C	adult	Yes	Gharyal and Maheshwari, 1990
C. siamea	C	adult	Yes	Gharyal and Maheshwari, 1990
Robinia pseudoacacia	C	adult	Yes	Han et al., 1990
Gymnocladus	?	juvenile/adult.	ND	Smith and Obeidy, 1991

[1] Abbreviations: ASC = axillary shoot culture, C = callus, DO = direct organogenesis, SE = somatic embryogenesis, MG = micrografting, ND = no data

In an extensive survey of woody plant taxa including 126 species across 67 genera, 33 families, 16 orders and six subclasses, Einset (1991) found that species in the order Fabales (including Leguminosae) tended to exhibit a positive *in vitro* growth response to cytokinins. Because of considerable interspecific differences and even lab-to-lab differences, few generalizations can be made about experimental attempts to optimize cytokinin-induced axillary shoot proliferation, except that benzyl adenine (BA) and kinetin (Kin) in the range of 1 to 10 μM are most commonly employed for this purpose. Two other cytokinins, zeatin and thiadiazuron (TDZ), are regarded as more potent cytokinins than either BA or kinetin in a wide range of species (Heutteman and Preece, 1993) including legumes (Chalupa, 1987; Yusnita et al., 1990; Smith and Obeidy, 1991). For the micropropagation of ericaceous species, a third alternative cytokinin, isopentynl adenine (2iP), is preferred to BA or Kin. These three alternatives to BA and Kin have been tested in only a few woody legumes. Toruan-Mathiius (1992) found BA to be more effective for axillary shoot production than either Kin or 2iP in *L. leucocephala, L. pallida*, or the *Leucaena* hybrid, KX3. Similarly, shoot production in *Acacia nilotica* was higher in response to BA than either Kin or zeatin. Considering their usefulness in other plant taxa, evaluation of TDZ and zeatin as possible alternatives to the more traditionally used BA or Kin, is advisable across a wider range of other MPFTs.

The ease of micropropagation, like any asexual propagation approach, declines with increasing ontogenetic age of stock plant; i.e., adult phase explants are more difficult to micropropagate than juvenile explants. From Table 2, it is apparent that most woody legume micropropagation systems have involved culture initiation from juvenile explants, in most cases from seedlings or ungerminated zygotic embryos. When starting from older, reproductively mature stock plants of seedling origin, successful micropropagation can be facilitated by explant selection from the juvenile portion of a tree. As explained earlier, such juvenile tissue may be present in the form of dormant epicormic buds at the base of an otherwise mature stock plant. Epicormic shoots arising from coppicing (or natural environmental stress) give rise to juvenile shoots. Meyer and van Staden (1987) reported that shoot cultures established from coppice shoot explants of *Acacia melanoxylon* proliferated more rapidly and gave rise to organogenic callus to a greater extent than cultures initiated from adult shoot explants. Similarly, Smith and Obeidy (1991) reported that *Gymnocladus* shoots initiated from juvenile basal trunk explants proliferated comparably to seedling explant-derived cultures. Similarly, Sanchez and Vietez (1991) reported that explants from juvenile basal sprouts of chestnut exhibited higher establishment, growth and proliferation than explants from the adult phase canopy. Since adventitious buds formed on roots of mature trees are presumed to be in the juvenile growth phase (Hackett, 1985), explants from root suckers might be expected to perform better than adult explants from other locations on the same tree. Barghchi (1987) successfully initiated cultures from juvenile adventitious shoots arising on root cuttings of *Robinia pseudoacacia*. These root sucker explants proliferated and rooted better than explants from shoots of adult phase trees. Other MPFTs which have a natural propensity to sucker from intact roots include *Faidherbia albida* (Ajee and Duhoux, 1994), *Inga feuillei* (Mudge, personal observation), breadfruit, etc. In such species, root suckers could potentially be used as a source of juvenile shoot explants for a shoot culture-based micropropagation systems. On the other hand, Gassama (1989, cited in Detrez et al., 1992) reported that microcuttings of *Faidherbia albida* from root suckers of adult phase trees were difficult to root. Han et al. (1997) reported that there was no difference in *in vitro* performance of basal (juvenile) and more distal (adult) explants of *R. pseudoacacia*. These later two studies suggest that the generalization that juvenile explants perform better than adult explants does not always hold true.

Root organ culture (ROC) is another related approach to micropropagation of species, which have a natural tendency to sucker. ROC involves culturing root explants, usually in the presence of auxin (without cytokinin) to encourage more or less normal root growth and development (including branching of lateral roots). ROC, has the potential advantage that subculture, by cutting one root into several pieces, requires less precision and time than subdivision of multiple shoot cultures into individual microshoots. In contrast to shoot organ culture, which is dependent on plantlet formation via adventitious root formation, ROC requires adventitious shoot formation. This is much more easily accomplished with the relatively few species that have a natural tendency to form root suckers *in situ,* like raspberry, and poplar (Borgman and Mudge, 1984 and references cited there in), and *Faidherbia albida* (Ahee and Duhoux, 1993). Given the relatively few woody species with the habit of root suckering, the *in vitro* ROC approach would seem to be less broadly applicable across MPFT species than the more common shoot culture-based micropropagation systems. ROC has been attempted with *Faidherbia* albida, which suckers naturally, and for which *in vitro* shoot culture has had only limited success (Duhox and Davies, 1985; Gassama and Duhox, 1987; Ahee and Duhoux, 1993). Although seedling radicle-derived ROC of *Faidherbia* was successfully initiated, it could only be subcultured for a limited period. Nonetheless, during that period, adventitious shoot formation did occur, and rooted plantlets were established *ex vitro* (Ahee and Duhoux, 1993).

Another *in vitro* technique that may be useful for overcoming the difficulty of clonal propagation of adult phase explants is micrografting. *In vitro* micrografting is reported to bring about the rejuvenation of adult phase explants (Jonard, 1986). The technique involves grafting small adult phase scions onto an *in vitro* germinated seedling rootstock (Trigiano et al., 1992). This approach has potential advantages over either direct micropropagation (without grafting) or conventional *in situ* grafting (without micropropagation). *In vitro* micrografting was originally used to eliminate viruses by culturing citrus shoot tip explants that would otherwise not survive *in vitro* (Jonard, 1986). Subsequently, *in vitro* micrografting of adult phase explants onto seedling rootstocks has been used as a means of cloning Douglas Fir (Monteus, 1995) and *Acacia mangium* (Monteus, 1996). In both cases, a single explant yielded only one grafted individual upon *ex vitro* reestablishment, i.e., there was no multiplication per se. Nonetheless this is analogous to the 1:1 ratio of scion to new plant achieved by conventional grafting.

A particularly interesting use of micrografting has been used to take advantage of the apparent rejuvenating effects of *in vitro* culture per se (Trigiano, 1992), and/or the rejuvenating influence of juvenile seedling understock on adult scions (Hartman et al., 1997). Detrez et al., (1992) achieved rapid multiplication of *Faidherbia albida*, by micrografting adult shoot tip explants onto seedlings. Using what they called chain micrografting, explants were grafted onto seedlings. The shoot explants (microscions) elongated into multinode shoots, which could then be cut into individual nodes for regrafting onto a new *in vitro* seedling understock. Repeating the process over three cycles, they found that the multiplication coefficient (the number of single node scions harvested from a multinode shoot grown from one original single node scion) increased from 5 with the first cycle to 8.5 and 10.2 with the second and third cycles, respectively. This increase in multiplication was attributed to rejuvenation. Similarly, Perrin et al. (1994) showed that adult scions of *Hevea brasiliensis* micrografted onto seedling rootstocks could be multiplied *in vitro* by successive subculturing of new scion growth, but only if a small piece of the original seedling rootstock was included with each subculture. Subsequent rooting of these subcultured shoots (originating from an adult phase microscion explant) was as high as microshoots grown from a seedling (juvenile) explant, suggesting that rejuvenation had occurred. In an experiment with *A. senegal,* Palma et al. (1996) reported that conventional shoot micropropagation of adult explants resulted in *in vitro* proliferation of microshoots, which could then be successfully grafted *in situ* (nursery). These encouraging results from *in vitro* micrografting of several MPFT species suggest that it is an approach worthy of further exploration as means of achieving rejuvenation and multiplication of MPFTs which are otherwise difficult to propagate asexually (e.g., *Prosopis, Leucaena,* etc.).

Micropropagation is becoming an important tool for conservation of plant species including MPFTs. *In vitro* germplasm conservation may be especially important for MPFTs with recalcitrant seeds, and may save space compared to conventional *in situ* collections. Cryopreservation techniques which hold tissue cultures, and even intact buds at reduced (>0°C), and even ultralow (–196°C) temperatures, also may be useful in conservation of MPFTs (Sakai, 1985). In addition, transport of germplasm across international boundaries can be facilitated by *in vitro* cultures since they are soil-free (Engmann, 1995) and often free of pests and diseases.

Since the beginning of the use of *in vitro* culture for clonal propagation, there have been concerns about the loss of genetic fidelity via mutation and/or induction of aneuploidy, and/or "physiological decline" resulting from *in vitro* culture (Evans and Bravo, 1986). These deleterious changes are more likely when explants have gone through an undifferentiated stage, i.e., callus or suspension culture, prior to adventitious organogenesis (Muashige, 1974). There have been claims that even callus-free shoot culture of banana results in off types (Smith and Drew, 1990). Krikorian et al. (1993) reinvestigated this situation and found that off types were associated with the pre-severance condition of the shoot from which the explants were collected, rather than the micro-propagation process, per se. Clearly, future exploitation of micropropagation for MPFT improvement will require careful attention to problems of unintended variation *in vitro*.

An application of *in vitro* clonal propagation that is becoming extremely important is its use as a component of plant genetic engineering. This refers to recombinant DNA technology involving the transfer of a (useful) gene or genes, promoters, etc., from one organism (plant, bacterium, fungal or even animal) to another plant, in order (eventually) to confer some desirable genetic trait such as disease or pest resistance, altered ripening, drought tolerance, etc. Essentially all current transformation strategies including plasmid, electroporation [biolistic (gene gun) mediated] involve *in vitro* culture and subsequent regeneration of transformed plants. In the last decade many woody species and herbaceous perennial species have been successfully transformed including (but not limited to) apple (Korban and Chen, 1992), poplar (Confalonieri et al., 1995), grape (Scorza et al., 1995a), Juneberry (Hajela et al., 1993), pine (Aronen et al., 1995), papaya (Tennant et al., 1994), apricot (Camara et al., 1992), plum (Scorza et al., 1995b), almond (Archilletti and Damiano, 1995), *Allocasuarina* (Phelep et al., 1991), spruce (Hood et al., 1990), and eucalyptus (Macrae and Van Staden, 1993).

Although, in most cases, woody plant transformation research has not yet yielded agriculturally useful improvements, this technology will probably make important future contributions to the genetic improvement of tree crops, including some species useful in agroforestry. As practical woody plant transformation systems begin to come on line, there are concerns about the potentially undesirable consequences of widespread unintended dispersal ("escape") of foreign DNA from the transformed crop species to related wild species through pollen transfer. One solution to this problem may be deliberate genetic engineering of (male) sterility into otherwise transformed crop species, rendering them incapable of pollen-mediated gene transfer (Strauss et al., 1995).

8. CONCLUSION

Clonal propagation has played an important role in traditional agroforestry practices, particularly for staple food crops like cassava, banana, taro, potato, sweet potato and others. In addition to these herbaceous components, asexual methods have been used to propagate some of the woody components of agroforestry systems as well. Farmers have employed easily rooted tree species as boundary plantings, and grafting and layering have been used for fruit tree propagation and selection for centuries. Important advances in cutting propagation technology over the last several decades include the introduction of polyethylene and automatic mist systems for moisture management, and the use of synthetic auxins for promotion of adventitious rooting. Although these have been widely utilized to expand the palate of woody species and improved cultivars used for ornamental purposes (especially in temperate areas), they have only recently begun to be applied to the genetic improvement and nursery production of MPFTs and for the domestication of new species. Recent systematic approaches to optimization of pre and post severance environmental conditions for the rooting of cuttings has demonstrated that cutting propagation is feasible for a much broader range of MPFT species than previously considered. Likewise, recent research in micropropagation is beginning to have an impact on genetic improvement of MPFTs. Regardless of asexual propagation technique, the difficulty of propagating adult phase plant material will have to be overcome so that selection of improved genotypes can reach its full potential. New approaches like etiolation of cuttings and *in vitro* micrografting have shown great promise for improved asexual propagation of a wide range of difficult-to-root adult phase woody crops, and the experimental evaluation of these methods with many as yet untested MPFT species is likely to bring further success. *In vitro* micropropagation will have an increasingly important role as genetic transformation systems for more MPFT species are developed. Continued improvements at the low technology end of the spectrum, such as improved moisture management systems for cutting propagation, and natural alternatives to commercial rooting hormones should not be neglected, so that farmers will have access to improved clonal selections locally. As advances are made at both the low and the high end of the technological spectrum, asexual propagation will increase in importance as a powerful tool for the development of sustainable AF systems.

REFERENCES

Ahee, J. and Duhoux, E., 1994, "Root culturing of Faidherbia-*Acacia albida* as a source of explants for shoot regeneration," *Plant Cell, Tiss. Org. Cult.*, 36, 219.

Altman, A., Ya'ari, A., Pelah, D., Gal, A., Tzfira, T., Wang, W. X., Shoseyov, O., Vainstein, A., and Riov, J., 1995, "*In vitro* organogenesis, transformation and expression of drought-related proteins in forest tree cultures," *Current issues in plant molecular and cellular biology: proceedings of the VIIIth International Congress on Plant Tissue and Cell Culture*, Florence, Italy, June 12–17, 1994, 87–94, Kluwer Academic Publishers, Dordrecht, Netherlands.

Archilletti, T., Lauri, P., and Damiano, C., 1995, "*Agrobacterium*-mediated transformation of almond leaf pieces," *Plant Cell Reports*, Springer International, Berlin, Germany, 1995, 14 (5), 267–272.

Aronen, T., Hohtola, A., Laukkanen, H., and Haggman, H., 1995, "Seasonal changes in the transient expression of a 35S CaMV-Gus gene construct introduced into Scots pine buds," *Tree Physiol.*, Heron Publishing, Victoria, BC, Canada, January 1995. 15 (1), 65–70.

Arya, S., Toky, O. P., Tomar, R., Singh, L., and Harris, P. J. C., 1993, "Seasonal variation in auxin-induced rooting of *Prosopis cineraria* stem cuttings," *The Int. Tree Crops J.*, 7,249.

Badji, S., Ndiaye, I., Danthu, P., and Colonna, J. P., 1991, "Vegetative propagation studies of gum arabic trees. 1. Propagation of *Acacia senegal* (L.) Willd. using lignified cutting of small diameter with eight nodes," *Agro. Syst.*, 14, 183.

Badji, S., Mairone, Y., Ndiaye, I., Merlin, G., Danthu, P., and Colonna, J. P., 1993, "*In vitro* propagation of the gum arabic tree (*Acacia senegal* (L.) Willd.) 1. Developing a rapid method for producing plants," *Plant Cell Reports,* 12, 629.

Barakat, M. N. and El-Lankany, M. H., 1992, "Kernel propagation of *Acacia saligna* by shoot tip culture," *Euphytica,* 59, 103.

Barghchi, M., 1987, "Mass clonal propagation *in vitro* of *Robinia pseudoacacia* (Black locust) cs. 'Jaszkiseri'," *Plant Sci.*, 53, 183.

Barnes, R. D. and Burley, J., 1982, "Vegetative propagation in the improvement of tropical forest trees," in *Improving vegetatively propagated crops,* Abbott, S.J. and Atkin, R.K., Academic Press, London, 212.

Bender, L., Harry, I. S., Yeung, E. C. and Thorpe, T. A., 1987, "Root histology, and nutrient uptake and translocation in tissue culture plantlets and seedlings of *Thuja occidentalis* L.," *Trees,* 1,232.

Bhogwani, S. S. and Razdan, M. K. 1983, *Plant tissue culture: Theory and practice,* Elsevier Scientific Pub., Amsterdam, Netherlands.

Blake, T. J. and Filho, W. S., 1988, "Drought tolerance, growth partitioning and vigor in eucalypt seedlings and rooted cuttings," *Tree Physiol.,* 4, 325.

Blazich, F., 1988, "Chemicals and formulations used to promote adventitious rooting," in *Adventitious Root Formation in Cuttings*, Davis, T., Hassig, B.E., and Shankla, N., Eds., Disocorides Press, Portland, OR.

Bonner, F. T., 1990, "Storage of seeds: potential and limitations for germplasm conservation," *For. Ecol. Mgt.*, 35: 35–43.

Borgman, C. A. and Mudge, K. M., 1986, "Factors affecting the establishment and maintenance of 'Titan' red raspberry root organ cultures," *Plant Cell, Tiss. Org. Cult.*, 6, 127.

Bray, R. A. and Fulloon, M. G., 1987, "Producing F1 seed of *Leucaena pulverulenta* x *Leucaena leucocephala* hybrids," *Leucaena Res. Rep.*, 8,19.

Bray, R. A., 1984, "Evaluating hybrids between Leucaena leucocephala and L. pulverulenta as potential forage plants," *Aus. J. Exp. Agri. Anim. Husb.*, 24, 379.

Brennan, E. B., 1995, *Vegetative and seed propagation of Inga and Leucaena,* M.S. Thesis, Cornell University, Ithaca, NY.

Brennan, E. B. and Mudge, K.W., 1997a, "Vegetative propagation of *Inga fueillei* D.C. from cuttings and air layers," *New Forests*,15: 37–51.
Brennan, E. B. and Mudge, K. W., 1997b, "Single but splice graft: A novel grafting technique and tool for clonal propagation of Leucaena," *New Forests.*
Brennan, E. B., 1992, "A new grafting technique for *Erythrina, Leucaena*, and possibly other nitrogen fixing tree species," *Nitrogen Fixing Tree Research Reports*, 10, 85.
Brewbaker, J. L. and Soensson, C. T., 1988, "New tree crops from interspecific Leucaena hybrids," in *Advances in New Crops*, Janick, J. and Simon, J. E., Eds., Timber Press, Portland, OR, 283.
Brewbaker, J. L. and Sorensson, C. T., 1992, "Domestication of lesser-known species of the genus *Leucaena*," *Tropical trees, the potential for domestication and rebuilding of forest resources*, Leakey, R. B. B. and Newton, ITE Symposium No. 29 Edinburgh Centre for Tropical Forest Symposium No. 1. U.K., HMSO Publishing, U.K., 195.
Brewbaker, J. L., 1990, "*Leucaena*: a multipurpose tree genus for tropical agroforestry," *Agroforestry, a Decade of Development*, Steppler, H. A. and Nair, P. K. A., ICRAF, Nairobi.
Brewbaker, J. L., 1988, "Cloning of seedless *Leucaenas* for plantation use," *Leucaena Res. Rep.*, 9,111.
Bristow, S., 1983, "Propagation of *Leucaena* by cuttings," *Nitrogen Fixing Tree Research Reports*, 1,28.
Burley, J.," 1993, Biological potential for genetic improvements of trees used in agroforestry systems," in *Agroforestry in South Asia*, Beathey, W. R., Koshla, P. K., and Seckler, K., Eds., Winrock International, Arlington, VA.
Camara-Machado, M. L., Camara-Maahado, A., Hanzer, V., Weiss, H., Regner F., Steinkellner, H., Mattanovich, D., Plail, R., Knapp, E., Kalthoff, B., and Katinger, H., 1992, "Regeneration of transgenic plants of *Prunus armeniaca* containing the coat protein gene of Plum Pox Virus," *Plant Cell Rep.*, Springer International, Berlin, Germany, 1992, 11, 1, 25–29.
Castle, W. S., 1987, "Citrus Rootstocks," in *Rootstocks for fruit crops*, John Wiley & Sons, New York, Ch. 11, 361–400.
Chakrabarti, U. and Sadhu, M. K., 1983, "Effect of grafting time, variety and nature of scions on the success of epicotyl grafting in mango," *Indian J. Agric. Sci.*, 53, 637.
Chalupa, V., 1987, "Effect of benzylaminopurine and thidiazuron on *in vitro* shoot proliferation of *Tilia cordata* Mill., *Sorbus aucuparia* L. and *Robinia pseudoacacia* L.," *Biol. Plant*, 29, 425.
Chander, H., Singh, R. R. and Mandal, B. S. 1996, "Effect of auxin on number of leaves per cutting and length of main branch in neem (*Azadirachta indica*)," *Ann. Biol.*, 12, 57.
Chin, H. F., 1989, "Storage of recalcitrant seeds: Past, present and future." In *Tropical Tree Seed Research*, J. W. Turnbull, Ed., 89–92.
Australian Centre for International Agricultural Research, Forestry Training Centre, Gympie, Australia.
Cohen, J. D. and Bandurski, R. S., 1982, "Chemistry and physiology of the bound auxins," *Ann. Rev. Plant Physiol.*, 33, 403.
Confalnieri, M., Balestrazzi, A., Bisoffi, S., and Cella, R., 1995, "Factors affecting *Agrobacterium tumefaciens*-mediated transformation in several black poplar clones," *Plant Cell, Tiss. Org. Cult.*, Kluwer Academic Publishers, Dordrecht, Netherlands, December 1995, 43 (3), 215–222.
Danthu, P., 1992, "Vegetative Propagation of adult Faidherbia albida by branch and root cuttings," in *Faidherbia albida in the West African Semi-Arid Tropics: Proceedings of a Workshop*, Vandenbeldt, R. J., Ed., ICRAF, Nairobi, Kenya.
Davis, T. D. and Potter, J. R., 1981, "Current photosynthate as a limiting factor in adventitious root formation on leafy pea cuttings," *J. Am. Soc. Hortic. Sci.*, 106, 278.
Davis, T. D. 1988, "Photosynthesis During Adventitious Rooting," in *Adventitious Root Formation in Cuttings,* 2, Davis, T. D., Haissig, B.E., and Sankhla, N., Eds., Dioscorides Press, Portland, OR.
de Bruijn, G.H. and Dharmaputra, T.S., 1974, The Mukibat system, a high yielding method of cassava production in Indonesia. *Netherlands J. Ag. Sci.*, 22, 89.
de Foresta, H. A. Basri and Wiyono, 1994, A very intimate agroforestry association, cassava and improved homegardens: the Mukibat technique., *Agrofor. Today*, 6.
Detrez, C. 1995, Shoot production through cutting culture and micrografting from mature tree explants in *Acacia tortilis* (Forsk.) Hayne subsp. *raddiana* (Savi) Brenan, *Agro. Syst.*, 25, 171.
Detrez, C., Ndiaye, S., Kerbellec, F., Dupuy, N., Danthu, P., and Dreyfus, B. 1992, 91. Meristem micrografting of adult *Faidherbia albida*, in *Faidherbia albida in the West African Semi-Arid Tropics: Proceedings of a Workshop,* Vandenbeldt, R.J., ICRAF, Nairobi, Kenya.

Dewan, A., Nanda, K. and Gupta, S.C. 1992, *In vitro* micropropagation of *Acacia nilotica* subsp. indica Brenan via cotyledonary nodes, *Plant Cell Reports,* 2, 18.

Dhawan, V. and Bhojwani, S.S. 1987, *In vitro* nodulation of micropropagated plants of *Leucaena leucocephala* by Rihizobium, *Plant and Soil,* 103.

Dick, J. McP. and East, K. 1992, Influence of light quality on growth and rooting of *Acacia tortilis*, a tree species of semiarid east Africa, *Nitrogen Fixing Tree Res. Reports,* 10, 97.

Dick, J. McP., McBeath, C., Bissett, H. and Pottinger, A. 1996, Rooting ability of *Calliandra calothyrsus* leafy stem cuttings in a non-mist propagator, *Agro. Syst.,* 33,187.

Dick, J. McP., Blackburn D.G. and McBeath, D. 1994, Stem respiration in leafy cuttings of *Prosopis juliflora* during the rooting process, *New Forests,* 8, 179.

Dirr, M.A. 1990, *Manual of Woody Landscape Plants*, Stipes Publishing Company, Champaign, Illinois.

Einset, J.W.1991, Woody Plant Micropropagation with Cytokinins, *Biotechnology in Agriculture and Forestry,* 17, 190.

Espinal de Rueda, D. 1996, *Vegetative propagation of the tropical tree Inga sp. by cuttings and tissue culture.* MS Thesis, College of Agriculture and Life Sciences at Cornell University, Ithaca, NY.

Evans, DA and Je Bravo, 1986, Phenotypic and genotypic stability of tissue cultured plants, *Tissue culture as a plant production system for horticultural crops*, Zimmerman, R.H., Griesbach, R.J., Hammerschlag, F.A. and Lawson, R.H., Martinus Nijhoff, Amsterdam.

Felker and Clark, 1981, Rooting of mesquite cuttings, *J. Range Mgmt.* 34,466.

Felker, P. 1994, 72. Capturing and managing the genetic variation in *Prosopis* sp. for economically useful characters. *Tropical trees, the potential for domestication and rebuilding of forest resources,* Leakey, R.R.B. and Newton, A.C., Eds., ITE Symposium No. 29, Edinburgh Centre for Tropical Forest Symposium No. 1, August 23–28, 1992, HMSO Publishing, London.

Engelmann, F., 1995, "Brief overview of IPGRI's research activities on *in vitro* conservation of plant species," *Botanic Gardens Micropropagation News*, Kew Gardens, World Wide Web URL http://www.rbgkew.org.uk/science/micropropagation/bgmn2-1-3.html.

Galiana, A., Tibok, A., and Duhoux, E., 1991, "*In vitro* propagation of the nitrogen-fixing tree-legume *Acacia mangium* Willd.," *Plant and Soil,* 135, 151.

Garg, L., Bhandari, N. N., Rani, V., and Bhojwani, S. S., 1996, "Somatic embryogenesis and regeneration of triploid plants in endosperm cultures of *Acacia nilotica*," *Plant Cell Reports,* 15, 855.

Gassama, Y.K. 1989, "Culture *in vitro* et amelioration symbiotique chez *Acacia albida* (Leguminosae) adulte," *Trees for development in sub-Sahara Africa: proceedings of a regional seminar held by the IFS, Kenya,* Feb. 20–25, International Foundation for Science, Kenya.

Gautier, D., 1995, "The pole-cutting practice in the Bamileke country (Western Cameroon)," *Agro. Syst.* 31, 21.

Gharyal, P. K. and Maheshwari, S. C., 1990, "Differentiation in explants from mature leguminous trees," *Plant Cell Reports,* 8, 550.

Glovak, L. and Greatbatch, W., 1981, "Successful tissue culture of *leucaena*," *Leucaena Research Reports,* 3, 81.

Goel, V. L. and Behl, H. M., 1995, "Propagation of Prosopis juliflora from rooted stem cuttings," *Int. Tree Crops J.,* 8, 193.

Goyal, Y., Bingham, R. L., and Felker, P., 1985, "Propagation of the tropical trees, *Leucaena leucocephala* K67, by *in vitro* bud culture," *Plant Cell Tissue Organ Culture,* 4, 3.

Grange, R. I. and Loach, K., 1983, "Environmental factors affecting water loss from leafy cuttings in different propagation systems," *J. Hort. Sci.* 58, 1.

Shamet, G. S. and Dhiman, R. C., 1991, "Effect on auxins (IAA, IBA and NAA) on rooting behavior *Grewia optiva* burrett stem cuttings under intermittent MI," *Indian Forester,* January, 44.

Grossnickle, S. C. and Russell, J. H., 1990, "Water movement in yellow cedar seedlings and rooted cuttings: comparison of whole plant and root system pressurization methods," *Tree Physiol.,* 7, 57.

Hackett, W. P., 1985, "Juvenility maturation and rejuvenation in woody plants," *Horticultural Reviews*, Janick, J., AVI Publ., Westport, CT, 7, 109.

Hajela, R. K., Hajela, N., Bolyard, M. G., Barnes W. M., and Sticklen M. B., 1993, "A simple transformation system using adventious shoot multiplication of Juneberry," *Hort. Sci.*, April 1993, 28 (4), 330–332.

Han, K. H., Davis, J. M., and Keathley, D. E., 1990, "Differential responses persist in shoot explants regenerated from callus of two mature black locust trees," *Tree Physiol.,* 6, 235.

Han, K. H., Shin, D. I., and Keathley, D. E., 1997, "Tissue culture responses of explants from branch sources with different degrees of juvenility in mature black *locust (Robinia pseudoacacia)*," *Tree Physiol.*, 17, 671.

Hancock, J. F., 1992, "Plant evolution and the origin of crop species," Prentice Hall, Englewood Cliffs, NJ.

Harrision, R. D., Daniell, J. W., and Cheshire, J. M., 1989, "New photosynthesis and stomatal conductance of peach seedlings and cutting in response to changes in soil waste potential," *J. Am. Soc. Hortic. Sci.*, 114, 986.

Hartman, H. T., Kester, D. E., Davies, F. T., and Geneve, R. L., 1997, in *Plant Propagation: Principles and Practices,* 6th Edition, Prentice-Hall, Inc., Upper Saddle River, NJ.

Hossain, M., Hanus, D., Kevers, C., and Gaspar, T., 1993, "*In vitro* multiplication of the woody legumes *Leucaena leucocephala* and *Albizzia procera*," *Saussurea,* 0, 1.

Howard, B. H., 1994, "Manipulating rooting potential in stock plants before collecting cuttings," in *Biology of Adventitious Root Formation*, Davis, T. D. and Hassig, B.E., Eds., Plenum Press, New York.

Huetteman, C. A. and Preece, J. E., 1993, "Thidiazuron: a potent cytokinin for woody plant tissue culture," *Plant Cell, Tiss. Org. Cult.*, 33, 105.

Hughes, C. E. and Styles, B. T., 1987, "The benefits and potential risks of woody legume introductions," *Int. Tree Crops J.*, 4, 209.

Hoad, S. P. and Leakey, R. B., 1996, "Effects of pre-severance light quality on the vegetative propagation of *Eucalyptus grandis* W. Hill ex Malden," *Trees,* 10, 317.

Hu, T. W. and Liu, C. C., 1981, "Vegetative propagation of leucaena by leafy cuttings under mist spray," *Leucaena Res. Rep.*, 2, 50.

Huang, F. H., Al-Khayri, J. M., and Gbur, E. E., 1994, "Micropropagation of *Acacia mernsii*," *In vitro Cell Dev. Biol.*, 30, 70.

Hustache, G., Barnoud, F., and Joseleau, J. P., 1986, "Callus formation and induction of a cell suspension culture from *Acacia senegal*," *Plant Cell Rep.*, 5, 365.

Jain, S. M., 1976, "The evolution of inbreeding in plants," *Ann. Rev. Ecol. Systemat.*, 7, 469.

Jolin, D. and Torquebian, E., 1992, "Large cuttings," *Agrofor. Today*, Oct.-Dec., 15.

Jonard, R., 1986, "Micrografting and its application to tree improvement," in *Biotechnology in Agriculture and Forestry, Trees I*, Bajaj, Y. P. S., Ed., Springer-Verlag, New York.

Jones, T. C., Batchelor, C. A., and Harris, P. J. C., 1990, "*In vitro* culture and propagation of Acacia species (*A. Bivenosa, A. Holosericea, A. Salicina, A. Saligna and A. Sclerosperma*)," *The Int. Tree Crops J.*, 6, 183.

Kamaluddin, M. and Ali, M., 1996, "Effects of leaf area and auzin on rooting and growth of rooted stem cuttings of neem," *New Forests,* 12, 11.

Kanwar, J. S. and Bajwa, M. S., 1974, "Propagation of mango by side grafting," *Indian J. Of Ag. Sci.*, 44, 270.

Khanna, K. and Arora, J. S., 1984, "Note on propagation of *Cassia fistula* (Amaltas) by cutting and layering," *Indian J. Hort.*, 41, 137.

Klass, S., Bingham, R. S., Finker-Templeman, L., and Felker, P., 1985, "Optimizing the environment for rooting cuttings of the highly productive clones of *Prosopis alba*," *J. Hort. Sci.*, 60, 275.

Klopfenstein, N. B. and Chun, Y. W. Eds., 1997, *Micropropagation, genetic engineering, and molecular biology of Populus.*

Korban, S. S. and Chen, H., 1992, "Apple," *Biotechnology in Agriculture*, CAB International, Wallingford, U.K., (8), 203–227.

Kramer, P. J. and Kozlowski, T. T., 1979. *Physiology of woody plants*, Academic Press,San Diego, CA.

Krikorian, A. D., Irizarry, H., Cronauer-Mitra, S. S., and Rivera, E., 1993, "Clonal fidelity and variation in plantain (Musa AAB) regenerated from vegetative stem and floral axis tips *in vitro*," *Ann. Bot.*, 71, 519.

Lambert, C. and Faculte des Sciences, Angers, France, 1992, "Use of *Agrobacterium rhizogenes* to create transgenic apple trees having an altered organogenic response to hormones," *Theor Appl Genet*, Springer International, Berlin, Germany. October 1992, 85, 1, 105–109.

Leakey, R. B., Newton, A. C., and Dick, J. McP., 1994, "Capture of genetic variation by vegetative propagation: processes determining success," in *Tropical trees, the potential for domestication and rebuilding of forest resources,* Leakey, R. R. B. and Newton, A. C., Eds., ITE Symposium No. 29 Edinburgh Centre for Tropical Forest Symposium No. 1, Aug. 23–28, 1992, HMSO Publishing, U.K., 72.

Leakey, R. R., and Coutts, M. P., 1989, "The dynamics of rooting in *Triplochiton scleroxylon* cuttings: their relation to leaf area, node position, dry weight accumulation, leaf water potential and carbohydrate composition," *Tree Physiol.*, 5, 136.

Leakey, R. R. B. and Storeton-West, R., 1992, "The rooting ability of *Triplochiton scleroxylon* cuttings: the interactions between stock plant irradiance, light quality and nutrients," *For. Ecol. Mgt.*, 49, 133.

Leakey, R. R. B., Mesen, J. F., Tchoundjeu, Z., Longman, K. A., Dick, J., Newton, A., Matin, A., Grace, J., Munro, R. C., and Muthoka, P.N., 1990, "Low-Technology Techniques for the Vegetative Propagation of Tropical Trees, *Commonwealth For. Rev.*, 69, 247.

Loach, K. and Gay, A. P., 1977, "The light requirement for propagating hardy ornamental species from leafy cuttings," *Scientia Hortic.*, 10, 217.

Loach, K., 1988, "Water relations and adventitious rooting," *Adventitious root formation in cuttings*, Davis, T. D., Hassig, B.E., and Sankhla, N., Eds., Dioscorides Press, Portland, OR.

Loach, K., 1977, "Leaf water potential and the rooting of cuttings under mist and polyethylene," *Physiologia Plantarum*, 40, 191.

Longman, K. A., 1993, "Tropical trees: propagation and planting manuals," *Rooting cuttings of tropical trees*. Commonwealth Science Council, London.

MacDicken, K. G. 1990, "Agroforestry management in the humid tropics," in *Agroforestry, classification and management*, MacDicken, K. G. and Vergara, N. T., Eds., John Wiley & Sons, New York, 98.

Machida, A., Ooshi, A., Hosoi, T., Komatsu, H., and Kamota, F., 1977, "Studies on photosynthesis in cuttings during propagation. I. Changes in the rate of apparent photosynthesis in the cuttings of several plants after planting," *J. Jap. Soc. Hotic. Sci.*, 46, 274.

Macrae, S. and Van Staden, J., 1993, "*Agrobacterium rhizogenes*-mediated transformation to improve rooting ability of eucalypts," in *Tree physiology*, Heron Publishing, Victoria BC, Canada, June 1993, 12 (4), 411-418.

Majumdar, P. K. and Rathore, D. S. 1970, "Bench grafting in mango," *Indian Hort.*, 14, 11.

Manders, G., Davey, M. R., and Power, J. B., 1992, "New genes for old trees," *J. Exp. Bot.*, Oxford University Press, September 1992, 43, 254, 1181–1190.

Martin, F. W., 1997, *The living fence, its role on the small farm*. Educational Concerns for Hunger Organization (ECHO), World Wide Web, http://www.xc.org/echo/tnfentbl.htm.

Maynard, B. K. and Bassuk, N. L., 1988, "Etiolation and banding effects on adventitious root formation," *Adventitious Root Formation in Cuttings*, 2, Davis, T. D., Haissig, B. E., and Sankhla, N., Eds., Dioscorides Press, Portland, OR.

Mazari, A. and Rubluo, A., 1984, "*In vitro* responses of *Leucaena leucocephala* c.v. Peru," *Bull. IGSM*, 12, 35.

McCaffery, K. A., 1997, *Evaluation of grain seed as a rooting stimulus in cuttings of Inga Mill. and Dianthus caryophyllus L.: Putting a centuries old practice to the test*, MS Thesis, Cornell University, Ithaca, NY.

Meyer, H. J. and van Staden, J., 1987, "Regeneration of Acacia melanoxylon plantlets *in vitro*," *S. Afr. Tydskr. Plantk.*, 53, 206.

Mittal, A., Agarwal, R., and Gupta, S. C., 1989, "*In vitro* development of plantlets from axillary buds of *Acacia auriculiformis* — a leguminous tree," *Plant Cell, Tiss. Org. Cult.*, 19, 65.

Moe, R. and Andersen, A. S., 1988, "Stock plant environment and subsequent adventitious rooting," in *Adventitious Root Formation In Cuttings*, 2, Davis, T. D., Haissig, B. E. and Sankhla, N., Eds., Dioscorides Press, Portland, OR.

Mohammed, G. H. and Vidaver, W. E., 1991, "Early development of Douglas-fir plantlets following transfer to the greenhouse," *Plant Sci.*, 76, 259.

Monteuuis, O., 1995, "Influence of the grafting technique on meristem micrografting of Douglas-fir," *New Forests*, 10, 267.

Monteuuis, O., 1996, "*In vitro* shoot apex micrografting of mature *Acacia mangium*," *Agro. Syst.*, 34, 213.

Mudge, K. W., Mwaja, V. N., Itulya, F. M., and Ochieng, J., 1995, "Comparison of four moisture management systems for cutting propagation of *Hibiscus rosa-sinensis* L., *Bougainvillea x Buttiana* Holtt. & Standl., and *Dovyalis caffra* (Hook. f. et Harv.) Warb. *J. Am. Soc. Hort. Sci.*, 120, 366.

Muashige, T., 1974, "Plant propagation through tissue cultures," *Ann. Rev. Plant Physiol.*, 25, 135.

Mugo, F., 1997, *Factors contributing to woodfuel scarcity and the consequent use of crop residues for domestic energy in rural Kenya*, MS Thesis, Cornell University, Ithaca, NY.

Nagmani, R. and Venketeswaran, S., 1983, "*In vitro* culture of hypocotyl and cotyledon segments of *Leucaena*," *Leucaena Res. Rep.*, 4, 89.

Newton, A. C, Dick, J. McP., McBeath, C. and Leakey, R. R. B., 1996, "The influence of R:FR ratio on the growth, photosynthesis and rooting ability of *Terminalia spinosa* Engl. and *Triplochiton scleroxylon* K. Schum," *Ann. App. Biol.*, 128, 541.

Newton, A. C. and Jones, A. C., 1993, "Characterization of microclimate in mist and non-mist propagation systems," *J. Hort. Sci.*, 68, 421.

Newton, A. C. and Jones, A. C., 1993, "The water status of leafy cuttings of four tropical tree species in mist and non-mist propagation systems," *J. Hort. Sci.*, 68, 653.

Newton, R. J., Marek-Swize, K. A., Magallanes-Cedeno, M. E., Dong, N., Sen, S., and Jain, S. M., 1995, "Somatic embryogenesis in slash pine *(Pinus elliottii Engelm.)," Forestry Sciences*, Kluwer Academic Publishers, 1995, 44, 183–195.

Nikiema, A. and Tolkamp, G. W., 1992, "Propagation of Faidherbia albida by cuttings: experience of the National Forests Seed Center," in *Faidherbia albida in the West African Semi-Arid Tropics: Proceedings of a Workshop,* Vandenbeldt, R. J., Ed., ICRAF, Nairobi, Kenya.

Nillson, O., Mortiz, T., Sundberg, B., Sandberg, G., and Olsson, O., 1996, "Expression of the *Agrobacterium rhizogenes* rolC gene in a deciduous forest tree alters growth and development and leads to stem fasciation," *Plant Physiology*, American Society of Plant Physiologists, Rockville, MD, October 1996, 112 (2), 493–502.

Nilum, S. P. and Verma, R. C., 1995, "Vegetative propagation of *Acacia catechu*, *Dalbergia sissoo* and *Prosopis cineraria* by cuttings," *Int. Tree Crops J.,* 8, 139.

Norelli, J., Mills, J. A. and Aldwinckle, H., 1996, "Leaf wounding increases efficiency of *Agrobacterium*-mediated transformation of apple," *Hort. Science*, October 1996, 31 (6), 1026–1027.

Ofori, D. A, Newton, A. C., Leakey, R. R. B., and Grace, J., 1996, "Vegetative propagation of *Milicia excelsa* by leafy stem cuttings: effects of auxin concentration, leaf area and rooting medium," *For. Ecol. Mgt.*, 84, 39.

Okoro, O. O. and Grace, J., 1976, "The physiology of rooting *Populus* cuttings. I. Carbohydrates and photosynthesis," *Physiol. Plant,* 36, 133.

Palma, B., Vogt, G. F., and Neville, P., 1996, "A combined *in vitro/in vivo* method for improved grafting of *Acacia senegal* (L.) Willd," *J. Hort. Sci.,* 71, 379.

Pan, F., 1985, "Systematics and genetics of the *Leucaena diversifolia* complex," Ph.D. Dissertation, University of Hawaii, Honolulu, HI.

Patel, K. R., Rumary, C., and Thorpe, T. A., 1987, "Plantlet formation in black and white spruce. III. Histological analysis of *in vitro* root formation and the root-shoot union," *NZ J. for Sci.*

Perrin, Y., Lardet, L., Enjalric, F., and Carron, M. P., 1994, "Rajeunissement de clones matures *d'Hevea brasiliensis* (Mull. Arg.) par microgreffage *in vitro*," *Can. J. Plant Sci., 623.*

Phelep, M., Petit, A., Martin, L., Duhoux, E., and Tempe, J., 1991, "Transformation and regeneration of a nitrogen-fixing tree, *Allocasuarina verticillata Lam*," *Bio/technology*, Nature Publishing, New York, May 1991, 9 (5), 461–466.

Purseglove, J. W., 1968, *Tropical Crops, Dicotyledons*, John Wiley & Sons, New York.

Purseglove, J. W., 1972, *Tropical Crops, Monocotyledons*, John Wiley & Sons, New York.

Ranga Rao, G. V. and Prasad, M. N. V., 1991, "Plantlet regeneration from the hypocotyl callus of *Acacia auriculiformis* — multipurpose tree legume," *J. Plant Physiol,* 137, 625.

Ravelonandro, M., Bachelier, J., Dunez, J., Scorza, R., Callahan, A., Cordts, J., and Gonsalves, D., 1995, "Genetic engineering plum pox virus coat protein gene in plants," *Acta Horticulturae*, International Society for Horticultural Science, Wageningen, Netherlands, June 1995, (386), 327–330.

Ravishankar, G. A., Wali, A., and Grewal, S., 1983, "Plantlet formation through tissue cultures of *Leucaena leucocephala*," *Leucaena Research Rep.,* 4, 37.

Reddy, P. C., Patil, V., Prasad, T. G., Padma, K., and Udayakumar, M., 1995, "*In vitro* axillary bud break and multiple shoot production in *Acacia auriculiformis* by tissue culture technique," *Current Sci.,* 69, 495.

Richards, A. J., 1986, *Plant Breeding Systems*, Chapman & Hall, London.

Ritchie, G. A., Tanaka, Y., and Duke, S. D., 1992, "Physiology and morphology of Douglas fir rooted cuttings compared to seedlings and transplants," *Tree Physiol.,* 10, 179.

Rom, R. C. and Carlson, R. F., 1987, *Rootstocks for fruit crops*, John Wiley & Sons, New York.

Rout, G. R., Samantaray, S., and Das, P., 1995, "Effect of growth regulators and culture environment on somatic embryogensis of *Acacia catechu* willd.," *Israel J. of Plant Sciences,* 43, 263.

Sakai, A., 1986, "Cryopreservation of germplasm of woody plants," in *Biotechnology in Agriculture and Forestry 1, Trees I*, Bajaj, Y., Ed., Springer-Verlag, Berlin, 113–129.

Salisbury, F. B. and Ross, C. W., 1978, *Plant Physiology*, 2nd ed., Wadsworth Publishing Co. Belmont, CA.

Sanchez, M. C. and Vietez, A. M., 1991, "*In vitro* morphogenetic competence of basal sprouts and crown branches of mature chestnut," *Tree Physiol.* 8:59-70.

Sasse, J. and Sands, R., 1996, "Comparative responses of cuttings and seedlings of *Eucalyptus globulus* to water stress," *Tree Physiol.*, 16, 287.

Scorza, R., Cordts, J. M., Ramming, D. W., and Emershad, R. L., 1995a. "Transformation of grape (*Vitis vinifera L.)* zygotic-derived somatic embryos and regeneration of transgenic plants," *Plant Cell Reports*, Springer International, Berlin, Germany, 1995, 14 (9), 589–592.

Scorza, R., Levy, L., Damsteegt, V., Yepes, L. M., Cordts, J., Hadidi, A., Slightom, J., and Gonsalves, D., 1995b, "Transformation of plum with the papaya ringspot virus coat protein genc and reaction of transgenic plants to plum pox virus," *J. Am. Soc. Hort. Sci.*, Alexandria, VA, November 1995, 120 (6), 943–952.

Sedgley, M. and Griffin, A. R., 1989a, "Tropical Tree Seed Research," in *Tropical Tree Seed Research*, Turnbull, J. W., Ed.), 89–92, Australian Centre for International Agricultural Research, Forestry Training Centre, Gympie, Australia.

Sedgley, M. and Griffin, A. R., 1989b, *Sexual Reproduction of Tree Crops*, Ch. 3, "Floral physiology and breeding systems;" Ch. 8, "Genetic consequences of variation in reproductive biology," Academic Press, New York.

Shiembo, P. N., Newton, A. C., and Leakey, R. R. B., 1996, "Vegetative propagation of *Gnetum africanum*, a leafy vegetable from West Africa," *J. Hort. Sci.*, 71,149.

Singh, N. P. and Srivastava, R. P., 1979, "Effect of storage of bud-sticks on succession veneer grafting in mango," *Ind. J. Hort.*, 23, 6.

Singh, R. N., Singh, G., and Rao, O. P., 1985, "Vegetative propagation of papaya through budding," *Indiana J. of Hort.,* 1.

Skolmen, R. G. and Mapes, M. O., 1976, "*Acacia koa* gray plantlets from somatic callus tissue," *J. Hered.,* 67, 114.

Smith, M. K. and Drew, R. A., 1990, "Growth and yield characteristics of dwarf off-types recovered from tissue-cultured bananas," *Aus. J. Exp. Ag.*, 30, 575.

Smith, M. A. L. and Obeidy, A. A., 1991, "Micropropagation of a mature, male Kentucky coffee tree," *Hort. Science,* 26, 1426.

Sookmark, S. and Tai, E. A., 1975, "Vegetative propagation of papaya by budding," *Acta Hort.,* 49, 85.

Sorensson, C. T., Shelton, H. M., and Austin, M. T., 1994, "Seedling vigor of some Leucaenas and their hybrids," *Tropical Grasslands.*

Sorensson, C. T. and Brewbaker, J. L., 1994, "Interspecific compatibility among 15 *Leucaena* species (*Leguminosae: Mimosoideae*) via artificial hybridization," *Am. J. Botany*, 8, 240.

Strauss, S. H., Rottmann, W. H., Brunner, A. M., and Sheppard, L. A., 1995, "Genetic engineering of reproductive sterility in forest trees," in *Molecular breeding: New strategies in plant improvement*, Kluwer Academic Publishers, 1995, 1 (1), 5–26.

Sweet, G. B., 1995, "Seed orchards in development," *Tree Physiol.*, 15, 527.

Tchoundjeu, Z. and Leakey, R. R. B., 1996, "Vegetative propagation of African Mahogany: Effects of Auxin, node Position, leaf area and cutting length," *New Forests*, 11, 125.

Teel, W., 1984, *A pocket directory of trees and seeds in Kenya*, KENGO, Nairobi.

Teketay, D., 1990, "*Erythrina burana*, Promising multipurpose tree from Ethiopia," *Agrofor. Today*, 2, 13.

Tennant, P. F., Gonsalves, C., Ling, K. S., Fitch, M, Manshardt, R., Slightom, J. L., and Gonsalves, D., 1994, "Differential protection against papaya ringspot virus isolates in coat protein gene transgenic papaya and classically cross-protected papaya," *Am. Phytopath. Soc.*, St. Paul, MN. November 1994, 84 (11), 1359–1366.

Thomas, C. A., 1981, "Research note: Mango propagation by saddle grafting," *J. Hort. Sci.,* 56, 173.

Toruan-Mathiius, N., 1992, "Micropropagation of *Leucaena* species by tissue culture techniques," *Leucaena Res. Rep.,* 13, 49.

Trigiano, R. N., Geneve, R. L., Merkle, S. A., and Preece, J. E., 1992, "Tissue and Cell Cultures of Woody Legumes," *Hort. Rev.,* 14, 265.

Versace, G., 1982, "Propagating leucaena by grafting," *Leucaena Res. Rep.,* 3, 3.

Wheeler, R. A. 1991, *Guide to management of Leucaena seed orchards*, Winrock International and F/FRED, Bangkok, Thailand.

Wheeler, R. A. and Brewbaker, J. L., 1990, "An evaluation of results from the *leucaena psyllid* trial network," *Leucaena Res. Rep.*

Wojtusik, T. and Felker, P., 1993, "Inter-specific graft incompatibility in *Prosopis*," *For. Ecol. Mgt.*, 59, 329.

Wojtusik, T., Felker, P., Russell, E., and Benge, M. D., 1993, "Cloning of erect thornless, non-browsed nitrogen fixing trees of Haiti: principal fuelwood species (*Prosopis juliflora*)," *Agro. Syst.*, 21, 293.

Yao, J. L., Cohen, D., Atkinson, R., Richardson, K., and Morris, B., 1995, "Regeneration of transgenic plants from the commercial apple cultivar Royal Gala," *Plant Cell Rep.*, Springer International, Berlin, Germany. 1995, 14 (7), 407–412.

Yusnita, S., Geneve, R. L., and Kester, S. T., 1990, "Micropropagation of white flowering eastern redbud (*Cercis canadensis* var. *alba* L.)," *J. Environ. Hort.,* 8, 177.

Zimmerman, R., 1972, "Juvenility and flowering in woody plants: A review," *Hort. Science*, 7, 447.

Zimmerman, P. W. and Wilcoxon, F., 1935, "Several chemical growth substances which cause initiation of roots and other responses in plants," *Contrib. Boyce Thompson Instit.* 7, 209.

9 Confronting Complexity, Dealing with Difference: Social Context, Content, and Practice in Agroforestry

Dianne Rocheleau

1-56670-294-1/99/$0.00+$.50

1. INTRODUCTION

Agroforestry burst onto the international development scene in the 1970s amid a rush of bold promises to eradicate poverty and hunger and to stem the tide of land degradation in "the tropics" or "developing countries." However, Agroforestry is no invention of the international scientific community or the international development agencies, no more so than agriculture or livestock management. It is a relatively new word for a variety of land use practices familiar to millions of farmers, herders, forest dwellers and gardeners the world over. Formally, agroforestry has been defined as a holistic approach to land use, based on the combination of trees and shrubs with crops, pasture or animals on the same land unit, whether simultaneously or in sequence (Lundgren, 1987). Informally, scientific agroforestry has come to mean the widespread promotion of a few species of "miracle trees" as hedgerow intercrops with maize to provide nitrogen, organic matter, soil cover and fuelwood, or the intensification of commercially oriented tree crops. In practice, most farmers, herders, forest dwellers and gardeners see agroforestry as the integration of woody plants and woodlands within pastoral, agrarian and even urbanizing landscapes or the accommodation of crops and animals in savannas, woodlands and forests (Rocheleau et el. 1988, 1989).

The distinction between these images is indicative of a significant underlying difference in orientation between "global" and "local" sciences of agroforestry. There is a vast discrepancy between the richness of established and evolving practice, the ambitious promise of new agroforestry research and development programs, and the poverty of social process and results in many of the experimental initiatives. Much of the discrepancy is bound up in the conventional scientific vision of agroforestry as a technological abstraction outside of time and place and the consequent attempts to develop new technologies in a social and ecological vacuum. In contrast, most established agroforestry practices have emerged out of ecological and cultural particularism, as is the case for most complex farming systems in diverse environments (Richards, 1985).

Global science, however, is not alone in failing to serve the interests of all interested people and the myriad of other species who live and die in their midst or in their wake. Local sciences also reflect the existing relations of power and the distinct, socially constructed visions of home, habitat and livelihood ecologies of particular groups in a given context. These visions, whether shared or disparate, are shaped by the coevolution of social and ecological dimensions of land use (Rocheleau et al., 1997b). Agroforestry systems, whether imported or locally developed, are nested within complex social relations of power, including both conflict and solidarity. This chapter proposes the pursuit of hybrid socio-ecological sciences of AF, based on the premise that both the dominant scientific paradigm we have come to accept as "science" and the myriad of local sciences sometimes enshrined as culturally perfect, site-specific practices, are all finite, local and limited. Each is embedded in and actively reshaping particular social, ecological and historical contexts.

This chapter seeks to address the gap between these distinct models for building agroforestry knowledge and improving agroforestry practice. The hybrid science advocated here goes beyond a model based on expanding one agroforestry science to incorporate many local knowledges, to a coalition model that seeks to reconcile a multiplicity of agroforestry sciences, as needed for particular cases, and to orchestrate useful and respectful interaction between them. Because this is such a broad task, the chapter will deal specifically with the social science mandate (Section 2.0) for a broader emerging agroforestry research paradigm that integrates social, ecological and economic elements across disciplines.

The social science mandate encompasses the social context, content, participation and practice in agroforestry research. This reflects a strong commitment to more depth as well as relevance of the social analysis conducted with, about and for agroforestry technology initiatives. All four social dimensions (context, content, participation and scientific practice) provide distinctive insights and tools to describe and address the complexity and differences in agroforestry interests within and between places, and within and between groups of people. A discussion of the promises and pitfalls of agroforestry and social science research (Section 3.0) outlines some of the potentials as well as the institutional and conceptual challenges and complications encountered in the last three decades. The elements of an alternative research paradigm emerge in a review of participation and institutional models, with a focus on specific participatory practices and related social methods in agroforestry research (Section 4). This section provides a historical overview, a comparative evaluation and diverse examples of methods, research design, and institutional arrangements. The methods are presented and synthesized so as to extend this interdisciplinary and "international" agroforestry science to engage more actively with the social, ecological and economic sciences grounded in the distinct experiences of particular places and groups of people. The challenges of diverse and complex social and ecological systems demand more robust and flexible agroforestry sciences. The objective of the entire review and discussion is to stretch our social and ecological imaginations in the interest of more equitable and effective agroforestry research.

2. A SOCIAL SCIENCE MANDATE FOR COLLABORATIVE AGROFORESTRY RESEARCH

Social science can help agroforestry research and development to confront complexity and to deal with differences within and between contexts ranging from the laboratory benches of biotechnologists to the forests, fields and rangelands of rural people. There are two compelling reasons to do this. First, agroforestry research and planning needs to address concerns about equity and justice which often constitute the stated focus and funding rationale of agroforestry development programs. Second, agroforestry science needs to confront complexity and deal with difference in order to ensure the practical fitness and effectiveness of technologies. Both social and technical imperatives demand a better understanding of the multiple and complex actors at work in the landscape. Agroforestry systems affect and are affected by lifeways and livelihoods. They are shaped by uneven relations of power, including both conflict and affinity. Agroforestry technologies are also an integral part of the complex web of social relations that tie people to each other and to particular ecologies, economies and cultures from local to global scale.

This socially embedded perspective on agroforestry implies a fourfold mandate for social science. First social science can describe and explain the social context of agroforestry technologies and land use systems. The nested social framework for agroforestry ranges from individual lives, households and local communities to regional, national and international social structures, processes and power relations. Second, there is a need for continuing analysis of the social content of agroforestry technologies and land use systems, and their impact on various groups and places within the contexts defined above. Third, the separation of global and local agroforestry sciences challenges an emerging socio-ecological science to bridge the gaps and to create a robust coalition across different traditions of intellectual inquiry and empirical exploration. Finally, there is a major role for social science and related practical skills to facilitate and analyze participatory processes and practices in agroforestry research and planning.

2.1 Social Context

Social context is more than the rich description of cultural differences. It encompasses such disparate elements as land, labor and commodity markets, and political and economic restructuring at national and international scales. Social context also includes the articulation of local and larger systems that maintain uneven power relations on the basis of gender, class, age, ethnicity, religion, occupation or other axes of difference. Cultural ecology approaches describe the relation between cultural practices and landscape as an adaptation to pre-existing environmental conditions, or

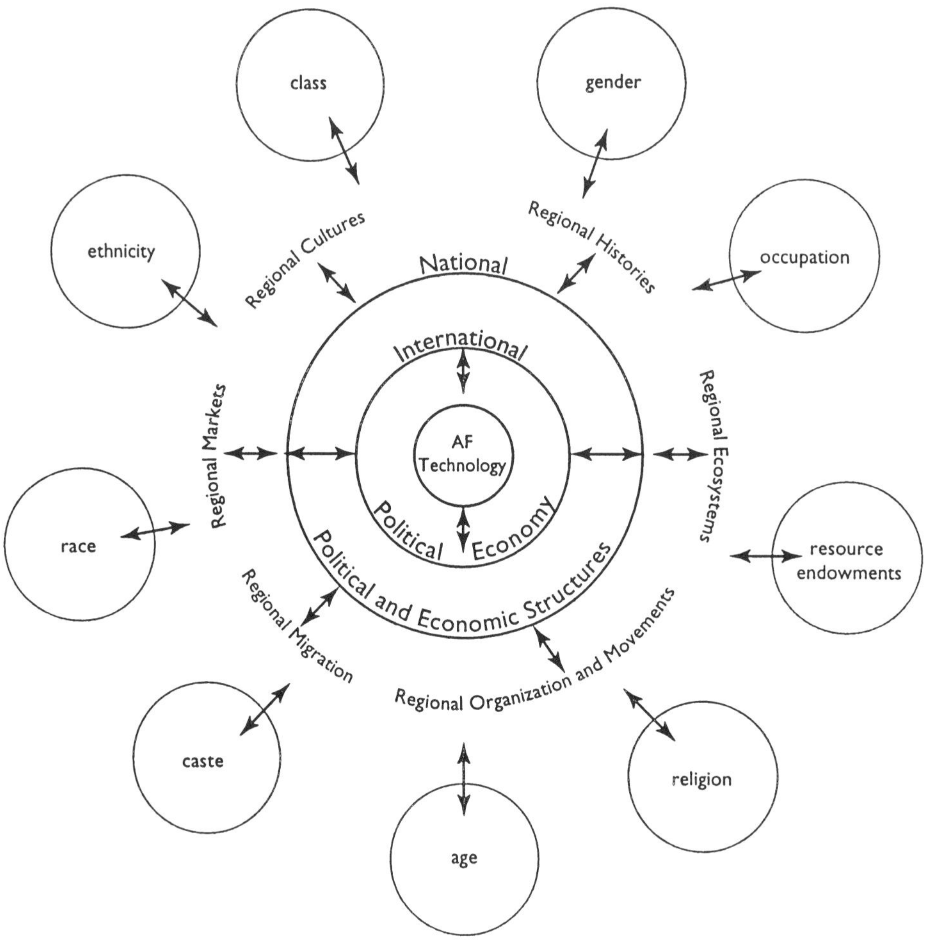

FIGURE 1. Social context of agroforestry.

conversely, they may emphasize the shaping of landscape by "traditional" cultural practices. Political ecologists focus on the interplay of livelihoods, landscapes and lifeways (economies, ecologies and cultures) under uneven relations of power. Others "map" the social location of various actors within "power grids" that integrate cultural, political, economic and other elements that affect privilege and disadvantage in particular places. Applied social science in agroforestry development often describes the social "landscape" in terms of the receptivity or demands of various groups and localities with respect to particular technologies or land use changes. All of these approaches can contribute to agroforestry research, although the latter predominates in technical research and policy circles, for obvious reasons. There is good reason to join the insights of careful, constructive critique with the more facilitative and service roles commonly assigned to social science.

The contemporary social context for agroforestry research reflects the global restructuring of economies, ecologies, cultures and political systems (Figure 1). Each of these elements constitutes a nexus of social relations between groups of people and between people and their physical and biological surroundings. We are witnessing the reintegration of each at different scales and under new relations of power. Local and regional changes in landscapes, livelihoods and land use are embedded in global processes of economic, environmental and social change. These changes in turn are both anchored and expressed in everyday life within communities and households. Diverse actors are at work, often simultaneously, as producers, processors, resource users, resource managers and consumers. All interact within complex social relations of power based on both conflict and affinity (Rocheleau et al., 1997b).

The repositioning of households within the global economy is affecting the position of the household in the community, of individuals within the household, and of both within local ecosystems. In the past, development planners have treated the economy as a global and national entity, environment as global and regional, and class and ethnic differences as national categories. They have often relegated gender to the "micro-level" as a detail of social life expressed solely within households and outside of the economy. In fact the process and pattern of the current restructuring of livelihoods and landscapes is necessarily both global and local. It hinges in no small part on a renegotiation of the gender and class division of land, labor, cash and commodities, capital, markets and social organization within local communities and households (Rocheleau, 1997).

This restructuring also has very concrete material manifestations relevant to agroforestry research and development. Only a few decades ago, many rural people integrated all or most of their production and consumption activities at the level of local landscapes, through household and community social units. With the commercialization of agriculture and the expansion of land markets, rural people have increasingly concentrated their production activities within private landholdings on small farms. They have accessed some commodities through local exchange and have traded some cash crops for consumer goods through national and international markets. The result has been a simultaneous concentration and micro-localization of production activities in household plots and an expansion of trade and market linkages to national and international systems (Rocheleau, 1997).

The latest wave of economic restructuring has pulled many more smallholders into more specialized monocrops or simplified agroforestry cash crop production, often displacing complex intercropping systems. This trend has, in turn, linked many farm households and entire communities more tightly into global markets as consumers of imported staple foods and manufactured goods. This is true for rural people in agrarian, rangeland and forest landscapes. As a consequence farm households have decreased the flow of energy and cycling of materials among their farms and between their farms and the surrounding landscape. They have dramatically increased the exchanges of energy and materials between their plots and larger national and international systems, shaped by economic processes integrated at a planetary scale.

Households vary considerably in their ability to enter into these restructured economies and ecologies on favorable or at least viable terms of control and exchange. Even those who cannot or choose not to participate in new cash crop initiatives, may find themselves enmeshed in local landscapes and economies transformed by the spread of new commodities and land use systems to the farms of their neighbors. The processing and marketing channels for their established cash crops may disappear or may offer less favorable terms. The use and management of land, water and forest resources also may change dramatically with the widespread adoption of new monocrops by surrounding farm households and large scale commercial producers. This in turn may pressure non-participating smallholders to sell out and migrate or to enter into a land use system that puts them at a disadvantage relative to their neighbors and relative to their own prior land use and livelihood system (Rocheleau, 1997). Whatever the existing role of agroforestry in changing land use systems, the prior social context and the trends in the social terrain, across scales, map the moral and technical coordinates for agroforestry land use systems and technology innovation. (Box 1 illustrates this concept.)

BOX 1. INFLUENCE OF SOCIAL RELATIONS ON ECOLOGICAL OUTCOMES

A graphic example of this phenomenon has been documented in two regions in the Dominican Republic. In the Sierra region a sustainable development program based on the replacement of shifting cultivation with improved varieties of tree crops (coffee) led to exactly this response. Largeholders expanded their already substantial coffee acreage with the new caturra (Brazilian Dwarf variety, non-shaded) coffee on the best lands. They also expanded and displaced food crop plots to forest and pasture lands (often on steep slopes) to produce food for their hired labor (both local smallholders and seasonal migrant labor from the lowlands). Smallholder participants also planted the non-shaded coffee on their best lands. They, too, displaced food crops to forest, pasture and steeper slopes, or produced less food and bought more from the small store of the largeholder family.

Ecological models of land use change in this case projected that an increase in coffee acreage under these social circumstances would result in a net increase in erosion at the watershed level. It would also create an increased dependency of smallholders on largeholders for coffee processing and market connections as well as wage labor and food supply. Both results ran counter to the program's intent to reduce erosion and increase smallholder income. The outcome was influenced by the structure of inter-household relations between smallholder and largeholder farmers in the community and the within-farm structure of food production for the work force in both types of households (Rocheleau, 1984). Social variables in this case affected both the type and the degree of environmental impact, and reversed the intended outcome. Eventually, the falling price of coffee reversed the trend by discouraging the expansion of coffee in the area.

2.2 THE SOCIAL CONTENT OF AGROFORESTRY TECHNOLOGIES

Whatever their "target" or collaborator groups, agroforestry initiatives inevitably affect individuals, households, and communities and the terms of peoples' interactions with their environment and with each other. New technologies and land uses will, however, have very different effects on distinct groups of households and of individuals within and across households and communities. The differences may go beyond degree, to the point that what is positive for one group may spell disaster for another, in social, environmental or economic terms.

2.2.1 Implications of Agroforestry Environmental Impacts

To frame the discussion of the social implications of agroforestry technology, we first must define the nature of sustainability, in terms of environmental criteria, starting with the physical, chemical and biological elements most subject to change by rural land use and resource management practice. Table 1 lists elements of sustainability in rural landscapes. The interests of any one group may be tied more to one or another of these aspects of sustainability within their local ecosystems. As agroforestry research and development programs evaluate the effects of a particular agroforestry system, it is possible to note not only which of these elements will be affected, but the relative importance of each of these to various groups of resource users and otherwise interested parties.

The use values of ecosystem products and services provide another reliable guide to the link between specific elements of the agroforestry systems, the local environment and the interests of particular groups of people (Raintree, 1987a). A list is provided in Table 2 which includes both goods and services that support subsistence as well as commercial production. The relative value attached to a given good or service on the list will vary by land user group. For example, at the community level, the members of a cooperative bakery might value fuelwood supply more than food supply if they currently purchase food with their bakery earnings. Likewise, in decisions between particular agroforestry species and technology options, households with substantial cash income that purchase

TABLE 1
Elements of Sustainability in Rural Landscapes

Physical integrity
- Water supply quantity, quality and timing
- Runoff quantity and quality
- Erosion and sedimentation
- Slope stability

Chemical integrity and nutrient cycles
- Agrochemical pollution (fertilizers, pesticides, herbicides)
  - Pollution of air, water and soil
  - Exposure of rural people to spraying, tainted water, stored agrochemicals, tainted foods
  - Movement of nutrients in biogeochemical cycles

Biological health of ecosystems at global, local and plot scales
- Biodiversity for its own sake
- Biodiversity for agricultural production
  - Crop and livestock genetic diversity
  - Wild plants and animals as gene pools for domesticates
  - Wild plants and animals as potential future domesticates
  - Companion plants and animals (pollinators, seed treaters and seed dispersers of plants; pesticidal plants and pest predators;
  - Fodder for livestock (from cattle to bees)
  - Soil micro fauna and flora that foster crop/fodder production

Biodiversity for local non-agricultural production

Commercial extractive products (honey and nuts, medicinal herbs, regular fruits and vegetable supplements — wild leafy vegetables, roots, fruits); famine foods; specialty foods for sale (mushrooms, caterpillars, insects, rare fruits, nuts, flowers); game animals for food or for sale of meat, hides; game animals for sale of other products (horn, shell, bone etc.)

most of their food supply might value stable cash income from cash crops such as coffee or timber more than a favorable seasonal distribution of food supply. A poor family with no off-farm income that grows their own food might prioritize seasonal food availability in their choice of species in agroforestry systems. Herbalists or religious leaders might choose to designate a particular patch of forest or range land as a local reserve for medicinal plants and/or religious observances, rather than allow new settlers to convert the area to food crop plots. They might care little whether the proposed new plots were to be converted to intensive agroforestry or monocropped annuals (Rocheleau, 1997).

2.2.2 Implications of Production Strategies

Throughout the world people in agrarian and forest landscapes face similar choices with respect to production and resource management problems and a limited array of commonly employed options. These include several strategies to maintain or improve production and/or resource management. Some of these options are compatible with agroforestry and can be combined while others are incompatible with agroforestry and mutually exclusive. Each of these choices implies a distinct set of costs and benefits for different groups of resource users and agroforestry managers/adopters:

1. Intensify labor. Produce more on same land with more labor.
2. Intensify and improve management. Change specific soil, water, crop and livestock management practices to get better economic returns and environmental quality, with same land and labor.

TABLE 2
Elements of Use Value Related to Sustainability

- Cash Income (through environmental production and service)
  - Employment
  - Sale of Products
  - Reduced spending on otherwise purchased items
  - Direct exchange of products for other goods
- Food Supply (for people and for livestock)
  - Amount
    - total annual supply
    - seasonal distribution
  - Quality
    - nutrition
    - taste/texture/dietary preference
    - ease of processing/preparation
- Medicinal Products (human and veterinary remedies)
  - Amount — quantities available
  - Quality — species composition of wild plants
  - Accessibility — distance, location, secure access
- Energy Supply
  - Uses (cooking, home heat, light, amenity)
  - Amount
  - Quality
  - Ease of Access
  - Cost Savings (amount cash saved)
- Shelter/Infrastructure
  - Building materials
  - Shade
  - Protection (from weather, animal/insect pests, intruders)
  - Definition of boundaries (fences, markers)
  - Cultural and aesthetic expressions
- Savings and Investments
  - Livestock
  - Land
  - Buildings
  - Business
- Industry/Business
  - Raw materials for crafts, local industries
    - Amount
    - Quality
    - Accessibility
- Social production
  - Funds, goods, services to meet social obligations
  - Species, places or products for religious or cultural observances

(Raintree, 1983, 1987b)

3. Shift labor to non-farm pursuits, wage labor; reallocate household labor from domestic or self-employment to cropping; convert croplands to forest, pasture, gardens; rent out croplands.
4. Commercialize. Shift toward major cash crops; adopt non-traditional agricultural export crops to increase farm earnings; simplify labor allocation, production and marketing.

5. Shift from annuals to perennials for more soil and canopy cover, less tillage and different labor demands; adopt agroforestry, social forestry, or tree cash cropping systems.
6. Diversify or maintain multiple crops within fields and gardens to improve yields, pest control, food self-sufficiency, and maintain plants of medicinal, cultural and religious significance.
7. Rearrange land use in same holdings according to ecological opportunism: selectively intensify in least fragile, most fertile and well-watered areas; reforest degraded slopes, ridges, stream banks; plant windbreaks; combine species for positive mutual effects in pest control, soil fertility, weed suppression, shading.
8. Rearrange plants in same holdings into more simplified, specialized (and less diverse) blocks: salad vegetables in small gardens, staples in croplands, cash crops in separate monocrop plots, commercial timber trees in monocrop woodlots (Rocheleau, 1997).

Distinct environmental consequences follow from each of the production and related resource management options exercised by rural people as individuals and members of households, communities and cross cutting groups. Environmental outcomes are based on land and labor allocation decisions, the type and degree of connection to local, national and international labor and commodity markets, and the strength and structure of people's affiliations with social organizations of various types. The concrete expression of these social and economic implications of agroforestry or other technology choices include positive or negative changes in: soil and water pollution from agricultural chemicals and organic wastes; quality, quantity and timing of water supply; erosion rates in fields and streambanks; sedimentation rates of rivers, lakes and dams; stability of hillslopes; and biodiversity, including species and genetic diversity of crops, livestock and the biota of the surrounding ecosystems. The differential constraints, opportunities, decisions and performance of the many distinct groups in any one place contribute to the net effects of global markets on local, regional and global ecosystems. Agroforestry technologies have social content that reflects back on social context and also conditions ecological changes across scales (Figure 2).

The social content and impact of specific agricultural technologies has been well documented historically. Green Revolution technologies and highly mechanized approaches were both differentially adopted and had distinct effects on people of different wealth classes and on women and men. National and international planners measured the success of the Green Revolution agricultural technologies in terms of national per capita food production. However, many critics traced its failures to the selective exclusion of poor smallholders and women-headed households and the displacement of local foods, resources and employment opportunities of particular importance to the poor and women in general. Success at the national level was experienced in very divergent ways within rural communities. The net results for rural people ranged from profit and growth to hunger and privation, based on differences between and within rural households (Oasa and Jennings, 1982).

In spite of this prior experience, many researchers and policy analysts seem to have assumed that agroforestry, sustainable agriculture and locally based resource management were exempt from this effect. They did not expect these technologies and resource management strategies to differentiate between and within households, either in their adaptability or in their effects. However, differences between communities, between households and within households have affected their ability not only to adopt the high technology packages of the Green Revolution, but also their capability to participate in organic agricultural production, forestry, agroforestry, soil conservation and a host of similar "sustainable" technology and land use changes (Schroeder and Suryanata, 1997; Sachs, 1993; Poats et al. 1988; Fortmann and Rocheleau, 1984).

Household abilities to enter into new sustainable land use systems and their experience of broader changes adopted by their neighbors, depend on endowments of land and related resources, available labor, access to cash, commodities and markets, and organizational affiliation. For example, to enter into new tree cash crops, households with little land might need to replace their food

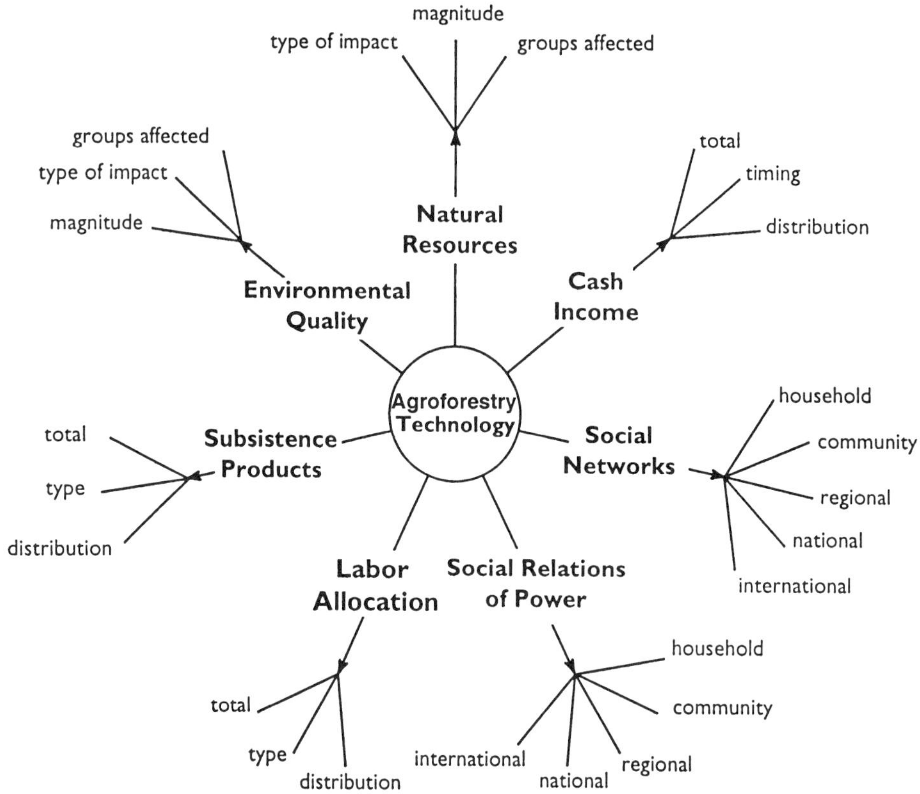

FIGURE 2. Social content and impacts of agroforesty.

crop lands, patio gardens, forest and pasture plots with new tree cash crops (Schroeder, 1993; Rocheleau et al. 1996 a and b; Fairhead and Leach, 1996). Or, they might replace food crops with the new cash crop on their best plots, then displace their food crops to more fragile lands or to rented lands formerly in forest and pasture, with a net negative effect on the surrounding watershed. Largeholders might have to make less difficult decisions in terms of land conversion, but might also expand both the cash crop and additional food for their hired labor onto steeper lands currently in pasture or forest (Rocheleau, 1984).

Each of the eight land use and technology strategies noted above affects control and access to the following six domains of resource use and management within communities as identified in Table 3: land and related; labor; cash and commodities; other forms of capital; markets; and organizations. The effects can be tabulated based on differences between households, as well as between individuals within households or in groups that cut across households within communities. For example, the adoption of export crops in smallholder households with men engaged in wage labor would probably depend on both reallocation and absolute increases in women's labor. In land-limited households this would also lead to the partial conversion of food crop lands to export monocrops and a decrease in the amount and quality of food produced. Depending on the crop — and the required or recommended management practices — the change could involve either a reduction or increase in water pollution and soil erosion, with impacts distributed unevenly between and within households, based on the location and quality of their landholdings and their water sources. A change of cash crop might also affect the gender structure of local markets or of local and regional organizations (Box 2).

TABLE 3
Six Domains of Inter- and Intra-Household Differences in Resource Use and Management

1. Natural Resources
   Land, Water, Crops, Livestock (meat, milk, skins) Wild plants, Wildlife
2. Labor
   Own, Family/Household, Exchange, Shared, Hired
3. Cash and Commodities
   Cash
   Products of labor and natural resources
      Food, Fuel, Fodder, Buildings Materials, Craft Materials, Medicine
      Commodities for sale
4. Other Capital
   Infrastructure
      Houses, Barns, Corrals, Fences, Other Buildings, Water tanks, Irrigation Works
   Equipment
      Carts, Vehicles, Plows, Tools, Looms, Sewing Machines, Ovens, Stoves
5. Markets
   Scale
      Local, Regional, National, International
   Types of Commodities
   Terms of Exchange
6. Organizations that Mediate/Facilitate
   Access to/control of public services
   Access to/control of resources
   Access to/control of labor
   Access to/control of markets
   Access to/control of savings and investments

(Adapted from Rocheleau, 1997)

2.3 Participatory Processes and Practices

The social science mandate to address participation and research methods raises two questions. First, how has social science contributed to and learned from participatory processes and practices in agroforestry research? Second, how can social science help to evaluate and shape the future nature of agroforestry participation and research practice to better address complex social and ecological contexts?

2.3.1 Historical Role of Participation in Agroforestry

Agroforestry research surfaced in scientific circles at a time when many international and national agricultural research programs were experimenting with social science components and exploring more participatory approaches in research planning, technology design, dissemination and evaluation (Rhoades and Booth, 1982: Rhoades, 1989; Chambers, 1981). Initially, experimental agroforestry programs, like most agricultural FSR, sought to harness social science skills in the service of pre-defined technological imperatives (Raintree, 1987b). In spite of difficult beginnings with social science in a strictly service role, many agroforestry institutions have expanded their perspectives to better address the social, economic and environmental concerns of land users (Raintree, 1987a and b; Rocheleau, 1987; Rocheleau et al. 1988,1989; Sumberg and Okali, 1989; Davis-Case, 1989).

Some organizations embedded agroforestry within a more participatory farming systems research and extension framework at the household level. Others promoted it within community

Box 2. Gender, Class, and Community at Work in the Forest Enterprise Project

Social stratification may have important social and ecological implications for sustainable development initiatives. In a hilly region in the Cibao Valley of the Dominican Republic, an international NGO, ENDA-Caribe, conducted a farm forestry program with the Rural Federation of Zambrana-Chacuey. Between 1990 and 1993 hundreds of farmers planted small blocks and rows of timber trees on their farms and constructed a community sawmill to process and market their timber. The Federation consisted of roughly 800 members in 500 households who shared a thirty year history of successful land struggles. They (or their parents) had engaged in popular campaigns for land-to-the-landless, including "land invasions" and other forms of civil disobedience (Rocheleau and Ross, 1995). While many well-organized communities focus on identity, the Federation built successfully on affinities between very distinct groups across race, class and gender lines. As of 1993 the members were almost all smallholders by national standards, yet they were substantially different in ways that matter for both adoption and outcomes of new forestry technologies and enterprises.

The research program of the project discovered a gender and class based difference in receptivity to planting *Acacia mangium* timber plots on small farms. Households connected to the Federation solely through women's groups did not have full access to technical assistance and information. Many women in male-headed households were unable to secure land to plant *Acacia* timber lots while others were unable to protect their garden or intercropped food plots from conversion to men's timber lots. Families with very small holdings had to choose between garden and food crops and timber cash crops, while people with larger holdings could have all three. Tobacco growers could substitute monocrop timber for monocrop tobacco and still keep gardens and food crops. Women as a group and near-landless men needed either timber tree species amenable to intercropping, intercropped non-timber tree products, or access to off-farm group or leased plots for *Acacia* timberlots. Both groups also required more direct representation in the decisions of the project and the newly formed Federation offshoot, the Wood Producers' Association. In spite of strong community and institutional organization, gender and class have influenced the selection or self-selection of participants, the choice of products included in the program, the structure of public participation and decision-making, and the distribution of costs and benefits of these forest based enterprises. Households presented distinct circumstances for the new timber cash crop based on landholding size, quality, and location; livelihood strategies; and type and degree of Federation affiliation. Individuals differed in their ability and interest to participate on the basis of gender, personal income, occupation and organizational affiliation (Rocheleau and Ross, 1995).

forest programs focused on common, public or other shared sites managed by larger groups of local people (Buck, 1990, 1993; Bradley, 1991; Feldstein et al., 1990). Yet in both cases, in spite of the presence of social scientists or social objectives, the "target groups" often failed to adopt the recommended agroforestry technologies or to adhere to rigid research protocols and to participate enthusiastically in formal tests of agroforestry technologies still under scientific scrutiny. Clearly it was not enough to just "add social science, participate, and stir." However, several notable success stories and two decades of cumulative experience suggest potential for fruitful collaboration across the boundaries between scientific research organizations and local communities (Buck, 1993; Hoskins, 1982, 1996; Fortmann, 1996; Scherr, 1990; Rocheleau et al. 1988, 1989; Rocheleau, 1991 a and b; Colfer, 1989; Sumberg and Okali, 1989; Campbell, 1996). The most promising of these avenues is the trend toward greater emphasis on participation, partnership and negotiation, both as elements of research process and subjects of study.

2.3.2 Social Science and Methodological Innovations

There are several ways in which social science can contribute to methodological innovations for participatory agroforestry research. One model is a variant of the traditional technology transfer model, in which tried and true social tools of participatory research are transferred to agroforestry researchers. Social science can contribute technical skills and experience with established methods for data collection, including qualitative social data collection and analysis and collaborative methods for collecting and analyzing both quantitative and qualitative data, whether ecological, social or economic in content. The paradigms and processes may range from those based in community organization and empowerment (Paolo Friere, 1972; Uphoff, 1992) to polling methods and approaches used to explore markets for new products and to predict and explain election results (for agricultural adaptations see Sperling, 1994; Ashby and Sperling, 1995). However, this represents just the surface of a very deep well of potential contributions. The prior analysis of social (and ecological) context can inform the choice of methods and the design of a broader collaborative process appropriate to the particular places and people as well as to the institutions involved.

A consultant model constitutes one alternative to the technology transfer model for social science input into participatory process and methods for agroforestry. An advisory approach could tap the expertise of social scientists to better gage the social context, to evaluate the social content of proposed technologies and to predict the likely social implications of specific technology options. Based on an expert social science judgment of the situation, the social research design, from basic questions and hypotheses, to methods, research protocol and structure of collaboration would be developed and presented as a consultant's recommendation to an agroforestry research planner or manager (Raintree, 1987a and b).

Beyond training or advising technical agroforestry research staff to use prescribed and "canned" social methods, a collaborative model may be possible and more desirable. Social scientists can also work directly with other agroforestry scientists and communities of land users and other interested parties to choose useful and effective methods. Together they can devise suitable plans of collaborative activities that address the needs of both groups and give the greatest scope to their collective imagination and skills. This is the route most likely to lead not only to methodological innovation but also to a fusion of theoretical and applied advances in understanding of agroforestry systems, social processes and research practice (Campbell, 1996; Carter, 1996; Slocum et al. 1995; Scoones and Thompson, 1994; Buck, 1990, 1993; Rocheleau, 1991 a and b; Davis-Case, 1989; Barrow, 1992; Rocheleau et al. 1988).

2.4 Bridging Multiple Sciences

The last of the four mandates is the least researched and perhaps the most needed in agroforestry research at present and for the foreseeable future. A social science attuned to the cross currents of diversity among many publics and participants as well as among distinct scientific traditions can help to facilitate a fruitful encounter across several lines of difference. First there is the long-standing and much discussed need for greater integration of social, economic and ecological sciences in formal research settings (Raintree, 1987b; Lundgren, 1987; Nair, 1989). Second, researchers need to address the distinct ways of knowing and learning across culture, class, gender and other bases of identity and affinity, both among a diverse array of constituencies within the general public as well as among researchers in international and national scientific organizations. Finally, the structures and processes for bridging those differences need to be negotiated in each and every case to some extent. Social scientists can help to elucidate the general principles of those processes (Haraway, 1991; Harding, 1986; Richards, 1985, 1989). They can suggest broad practical guidelines for procedures, and describe options. Ideally, they should facilitate the actual negotiations to define particular terms, boundaries and objectives of collaboration across scientific traditions and practices (Desloges, and Gauthier, 1997; Cline-Cole, 1996; Fortmann, 1996; Kandeh and Richards, 1996; Posey, 1985; Balee, 1989; Brookfield and Padoch, 1995; Anderson, 1990; Anderson et al. 1991; Carter, 1996).

3. PROMISES AND PITFALLS OF AGROFORESTRY AND SOCIAL SCIENCE

The early emergence of agroforestry as a science was marked by widespread optimism. The promise of agroforestry captured the interests of ecologists and social scientists as well as many development practitioners and planners, economists, agronomists and foresters, with good reason. Agroforestry has the potential to increase effective production space through vertical expansion of three-dimensional arrangements and through functional expansion of niche space in rural land use systems. It can achieve synergy through creative placement and management of trees, crops, pasture and animals in more tightly integrated systems, through more efficient capture of and recycling of nutrients by plants, creation of hospitable micro-climates for understory plants, staggered timing of management tasks and multiple product harvests, and greater overall value and diversity of products. It seemed especially well-suited to address both production and conservation goals in smallholder farming and forest farming communities.

3.1 Agroforestry Research

Extrapolated from the very potential of agroforestry came the widely developed belief that agroforestry is inherently a "good thing." In part, advocates of agroforestry often assumed that it would constitute a net improvement based on its application in a hypothetical degraded, treeless landscape of universally impoverished people. Agroforestry seemed poised to automatically reconcile productivity, equity, and environmental sustainability objectives in rural areas faced with both poverty and degradation of natural resources. Millions of farm workers and displaced forest dwellers and farmers in agroforestry plantation systems throughout the world might differ with this facile perception. Many critics have asked: sustainability of what? And for whom?

A second assumption widely shared in technology transfer research circles was that scientists working under controlled conditions in laboratories and research stations could isolate, replicate and package the most desirable elements of agroforestry systems for vast tracts of rural farm, range and forest land. However, some agroforestry social and biophysical scientists have asked not only if this was possible, which was (and still is) dubious, but also whether it was desirable (Raintree, 1987 a and b; Rocheleau et al., 1989; Buck, 1990; Atta-Krah and Francis, 1987; Sumberg and Okali, 1989). The prevailing social, economic and ecological conditions are both variable and dynamic, with highly differentiated publics for any new agroforestry technology. Good science will also need to take account of the existing land use systems and processes of innovation as well as the opportunity cost of committing substantial scientific, development and community resources to any single, narrowly conceived and tightly focused path to land use change and technology innovation (Poffenberger, 1990; Peluso, 1992; Katz, 1993; Colfer 1989; Fernandes et al. 1984; Nair, 1989; Denovan et al. 1985; Michon et al. 1989).

3.2 Social Science in Agroforestry

agroforestry research and development institutions sought out social science expertise in response to the social critiques and their own practical problems with field based research in complex social and ecological terrains. Initially, many social and biophysical scientists shared a prevailing assumption that the inclusion of social variables and social scientists in agroforestry research and development programs would automatically guarantee better social outcomes, both in the research process itself and the resulting technologies. This has proven to be an erroneous and dangerous assumption. The social sciences are no less fallible than biology and physics and can also, like nuclear physics, serve various ends.

Social science knowledge and skills can also be misused in the interest of researchers or the broader technology transfer process (Escobar, 1995). If they are not attentive to the broader context of the research process, social scientists may identify, select and groom participants for experimental success rather than for active representation of community interests. This may occur as an intended by-product of uncritical social science in service of otherwise legitimate agroforestry technology objectives. In a similar vein, research organizations also call upon social science to describe and poll the potential consumers of new technologies, and to inform product development. Quite often this very legitimate, if limited, mandate strays into consumer profiles to inform marketing strategies, regardless of the intrinsic merit or appeal of a predefined product.

Once we know something about the social context and content of agroforestry technologies and about the social techniques to facilitate the research process, there remains a value and ethics based decision about how to apply that insight and skill. Social science does not resolve the question of what to do in response to a better understanding of social context, content and process and the logics of local sciences. Interventions may address the people, the research and planning process, the technology, or the context, and the ends of those interventions may range from social justice to environmental conservation and from greater bureaucratic control of production to personal or corporate profits. The challenge is for social scientists to clarify and clearly state not only the social implications of biophysical research but also to examine the social implications of their own actions within agroforestry research (Rocheleau and Slocum, 1995; Scoones and Thompson, 1994).

3.3 Social Science and Agroforestry

The development of a new hybrid agroforestry science may benefit from a review of the historical relationship of social science and agroforestry and the contemporary varieties of collaboration. The types of affiliation range from social science and skills in the service of agroforestry science, to partnerships weighted in favor of one or the other, to agroforestry in the service of socially defined objectives and processes, to social critique and reform of agroforestry science. Figure 3 depicts these distinctions. The next decade may well see the emergence of a socially and ecologically informed integrated science of agroforestry.

The history of research themes in agroforestry social science has spanned a range of topics and approaches. The most successful achievement of the decade from 1977–1987 was the focus on smallholders and "local people," land use, land tenure and, eventually land users. The period from 1987–1997 brought a further focus on land users with the emphasis on multiple land user groups, defined by gender, class, age, and ethnicity as well as occupation and livelihood (farmers, herders, forest gatherers). Some agroforestry programs specifically "targeted" women, though the decade was marked by a transition from agroforestry projects that used women to implement fixed technology packages to initiatives that consulted women and attempted to serve them with "suitable" technologies. In some cases agroforestry programs and related programs sought to address the larger issues of gendered labor, landscapes, tenure, trade and organizations (Fortmann and Rocheleau, 1984). Box 4 illustrates the concept of gendered landscapes.

Simultaneously, many agroforestry researchers focused on local knowledge as well as local maps of resources, tree and other plant species, and land use potentials. The concept of multi-dimensional tree tenure and tenure niches provided a more complex model of resource access and control applicable to various types of claims and control over land, animals, plants and their products (Fortmann, 1995; Fortman and Nabane, 1992; Fortmann et al. this volume; Bruce et al. 1993; Rocheleau, 1987). This, in turn, allowed social scientists and field practitioners to recognize and address the multiple, nested and overlapping rights of various groups involved in agroforestry systems, whether integrated at plot, landholding or landscape scale (Rocheleau and Edmunds, 1997a). Box 5 provides an example of trees and tree products.

4. THE NEXT TEN YEARS — JOINING SOCIAL SCIENCE AND AGROFORESTRY RESEARCH PRACTICE

The future of this sometimes fitful and sometimes fruitful association between agroforestry and social science carries a challenge to better incorporate complexity, contingency and flexibility. Not only are there multiple user groups, but any one person in a given group is multiply located. That

Box 4. Complex Tenure In Gendered Landscapes: An Example From Machakos District, Kenya.

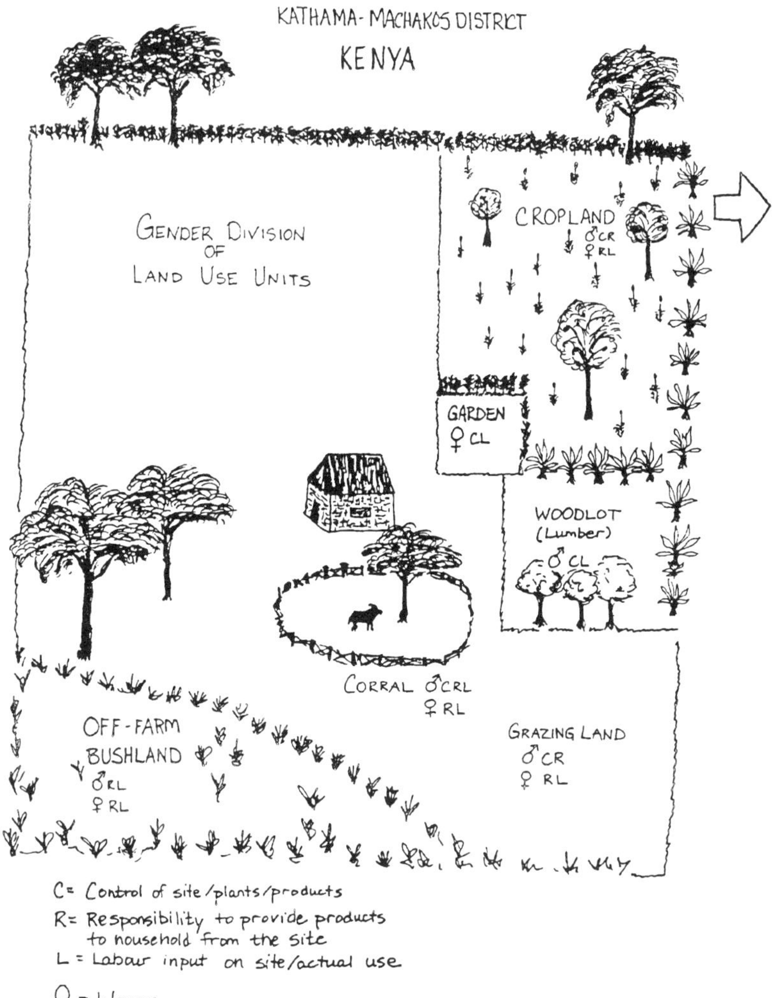

is, a woman farmer in a particular place might also be wealthy or poor, young or old, Christian or Muslim, and a keeper of cows or goats. Each of these elements might be expressed in some affiliation and might carry with it a particular set of interests or perspectives that impinge on the acceptance, rejection or modification of a given agroforestry technology, let alone a whole new agroforestry system. Just to further complicate the matter, the same woman might stop raising goats and join a gardening group, depending on shifts in land, labor and commodity markets, or in response to the migration of other household members. So peoples' identities are not only complex, but also very fluid. This suggests that a single "recommendation domain" (a whole region and class of farms/farmers as uniform technology "recipients") never was and never will be a realistic goal.

Box 5. Gendered Sharing and Division of Trees and Tree Products: A Detail From a Machakos Landscape. Excerpted from Rocheleau, Thomas-Slayter and Edmunds, 1995, in Cultural Survival Quarterly.

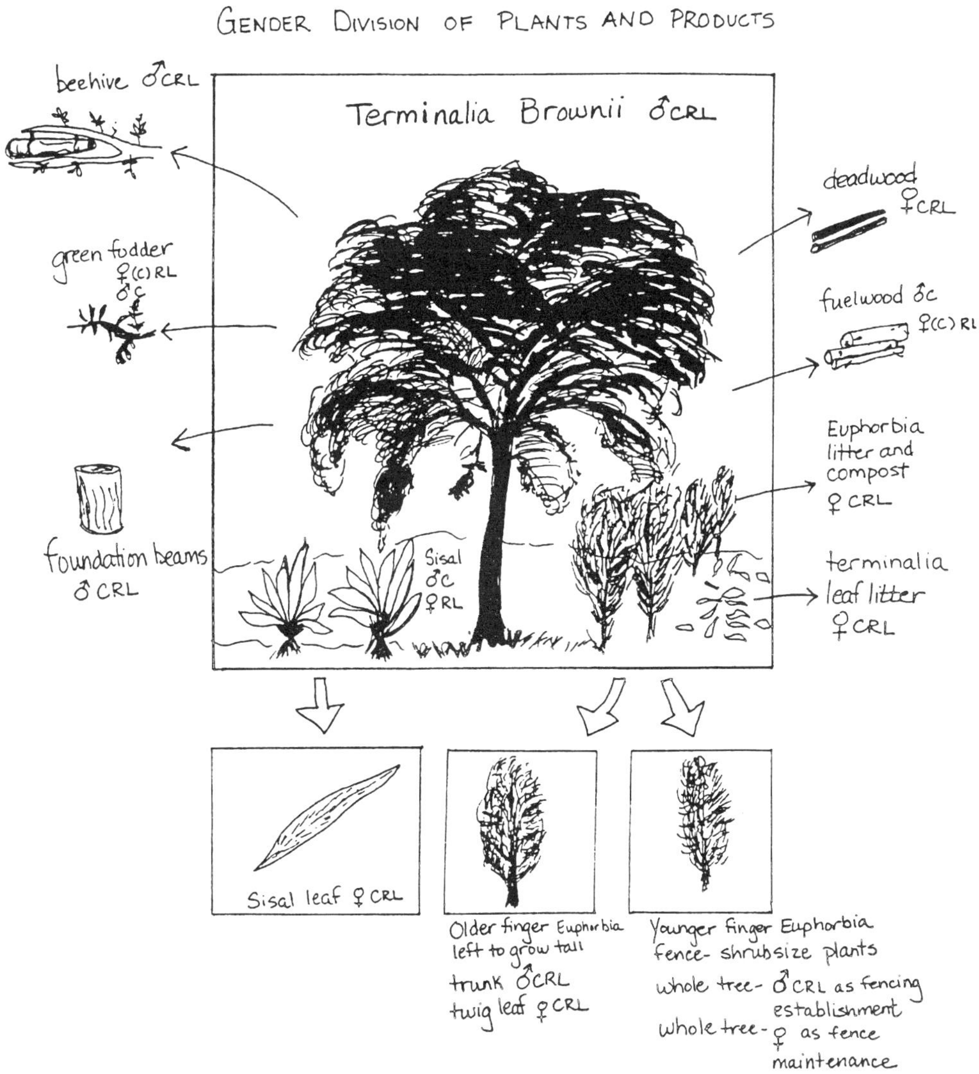

4.1 A New Hybrid Science — Joining Social Science and Agroforestry Research Practice

Questions and critique of participatory approaches to agroforestry seem more than justified, since the consequences of "maldevelopment" are both real and serious (Shiva, 1988; Sachs, 1993; Escobar, 1995). A careful critique can reveal possibilities for an alternative practice that seeks to promote broader social and ecological options, combining livelihoods and life support in local landscapes that are at once home, habitat and workplace to those who live there. There is hope to

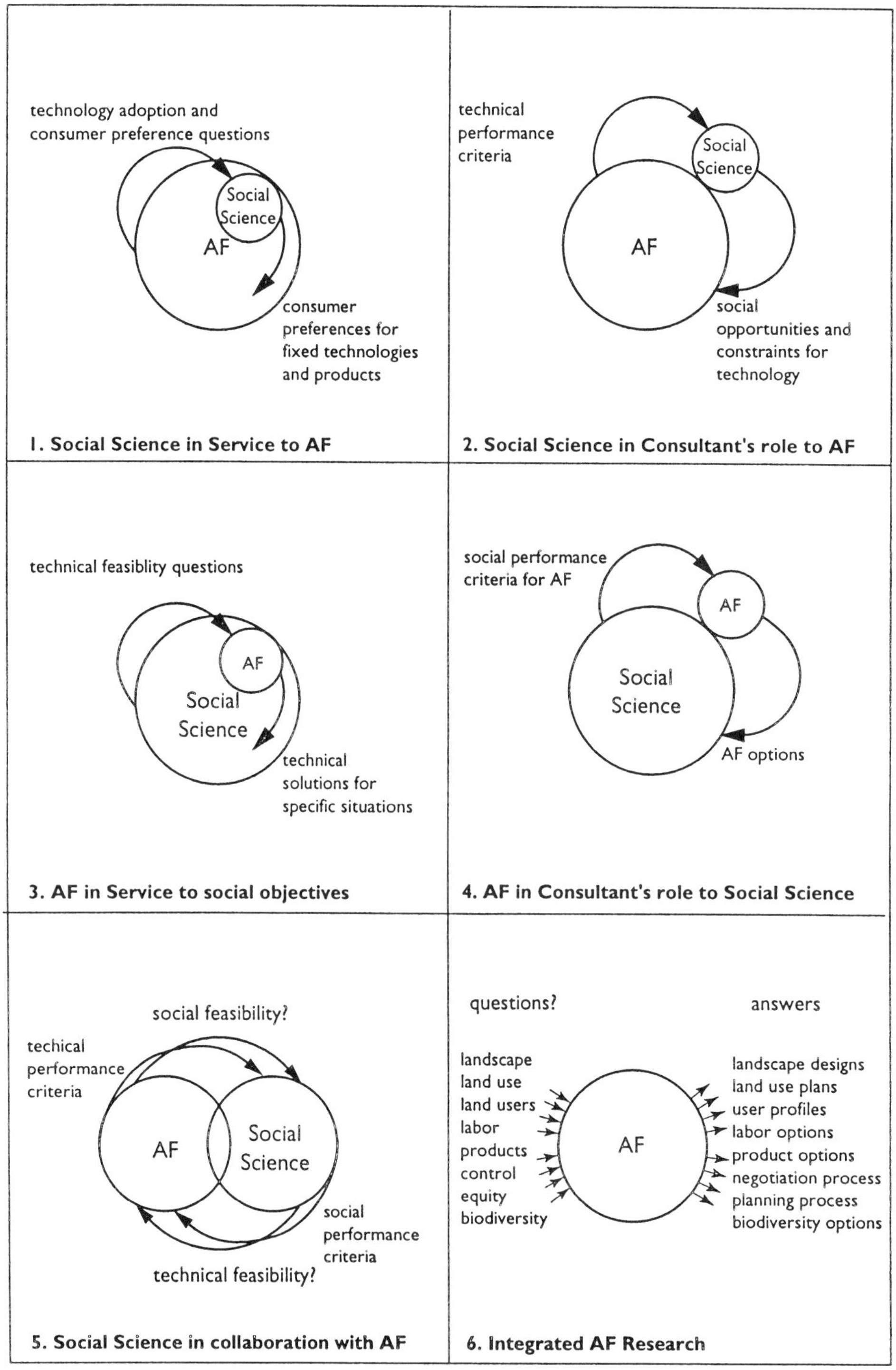

FIGURE 3. An illustrated typology of agroforestry and social science collaboration.

share a broad range of experience and expertise from the past, a wide array of evolving agroforestry sciences and a multiplicity of visions of the future, across the permeable boundaries defined by gender, class, race, culture, and nationality. Neither participation nor environmental criteria automatically guarantee just, equitable and ecologically viable futures, but both constitute essential ingredients of a common future worth sharing. The discussion which follows is presented in the interest of furthering such a science and practice. It proceeds from a belief that a democratization of the art and science of agroforestry can engage the social and ecological imaginations and practical skills of a multiplicity of actors in the interest of more just and humane ecologies.

A multiple scale, "telephoto" analysis, coupled with a focus on the diversity of interests and experiences at each stop along the way can make invisible processes and people more visible. More importantly, once we "map" the diverse social terrain for sustainable resource management we can involve the full range of user groups and interest groups in any one place in the planning and evaluation of land use changes. The major premise of this approach diverges from past resource management traditions — there is no single optimal land use or technology mix in a given place. Rather, there are many distinct visions of optimal environmental and economic futures and many possible combined outcomes. Both the visions and the composition of the various interest groups change over time, based on their own internal dynamics and in response to changes in the local, national and international contexts. The objective then shifts from a single optimal solution to a robust (fair and effective) process for the constant negotiation of acceptable solutions among competing and converging interests in any one place and between local and larger systems (Rocheleau, 1997).

The history of households and user groups in resource management and sustainable development has set the stage for development of new perspectives on agroforestry technologies and land use systems focused on a diverse constellation of changing land user groups. The recent focus on complexity and difference (Guijt and Shah, 1997: Rocheleau and Edmunds, 1997a; Rocheleau et al. 1997b; Peluso, 1992,1995; Schroeder and Suryanata, 1997;) as well as conflict management (Hoskins, 1996; Sarin, 1996; Desloges and Gauthier, 1997) in agroforestry and community forestry promises to usher in an approach that goes beyond participation to negotiation, based on coalition building as a metaphor for good scientific practice (Harding, 1986; Rocheleau, 1997). Some emergent paradigms of policy formation (Leach and Mearns, 1996) and community based research practice (Scoones and Thomson, 1994; Rocheleau and Slocum, 1995) propose that distinct actors (both "insiders and outsiders") agree to disagree on some things, get very clear on what they can and do agree upon, under what conditions and to what ends, then proceed to work at agroforestry and related research and development goals in an agreed upon, mutually constructed process. These approaches are predicated upon the same premise of this chapter, that distinct sciences and practices are not necessarily mutually exclusive. The most effective agroforestry planning and research tools will not ignore or even avoid conflict, nor will they provoke or fan it, but rather acknowledge and work with it. A hybrid science of agroforestry need not search for a single optimal solution. Rather, it seeks to define a process for reaching contingently acceptable options, always subject to revision and further negotiation (under agreed processes and terms).

The recognition of conflict, power and negotiation as practical concerns of agroforestry research and development also opens the way to incorporation of more academic analyses such as those produced within academic schools of thought such as political ecology (see Peet and Watts, 1997; Rocheleau et al. 1996 a and b; Bryant and Bailey, 1997; Schroeder, 1993; Peluso, 1992). These studies, often critical of agroforestry research and planning practice, have focused on social movements, social stratification and struggles over resources, profits and decision making power. An applied or practical variant of political ecology could, however, help to inform, reform or transform agroforestry technology design and testing within complex social terrains formed by the workings of power, expressed as both solidarity and conflict. Box 6 suggests an application of this approach.

Complex publics living in equally diverse and dynamic ecologies require multi-dimensional maps that reflect the complexity and fluidity of tenure, use and management as well as land cover patterns, habitats and species. It is not only the social landscape that now presents a more complicated image and demands more robust models. Our vision of landscapes, ecologies, farms and forests has also advanced from the gaze of the telescope to the microscope to the wide angle and telephoto lenses, to the kaleidoscope.

Box 6. Political Ecology: Agroforestry in Social Context in Machakos District, Kenya

The households and individuals who are casualties of the privatization of land and related resources, of land shortages, and even of the "sustainable" intensification of agroforestry land use, have in many cases remained invisible. This process has recently been documented and debated by researchers in Machakos District, Kenya, where privatization and land use intensification in a semi arid wooded savanna displaced some smallholder farmers from use of the common woodlands and pastures for grazing and gathering (Wisner, 1990). Without this crucial supplement to their subsistence base they were unable to make ends meet and sold or abandoned their holdings (Mbogoh, 1991). Their former lands may even have passed into the hands of wealthier farmers to be converted to orchards or other environmentally friendly and economically productive agroforestry land uses (Rocheleau et al. 1995).

While some have heralded this phenomenon in Machakos as economic and environmental recovery (Tiffen et al. 1994), many of the displaced households went on to find and clear dry forest lands in "open" frontier areas, to seek work in plantations or to find wage labor or refuge on the fringes of cities. Many have been rendered homeless in the terms of their own culture. Neither their poverty nor their environmental impact were erased by their displacement. Both problems were simply displaced to another jurisdiction, to be counted by another census or another environmental study. Meanwhile, among those who stayed, the hunger of some and the poverty of many remain, albeit shaded by the tree crops and woodlots of their wealthier neighbors and employers (Rocheleau et al. 1995, forthcoming). This same socially divergent, and sometimes polarized story has played out in many variations as documented by recent studies in Central America, South America and the Caribbean (Stonich, 1993; Anderson et al. 1991; Arizpe et al. 1993; Schmink and Wood, 1992; Townsend, 1995; Rocheleau and Ross, 1995) Africa (Schroeder, 1993) and Southeast Asia (Schroeder and Suryanata, 1997; Peluso, 1992).

4.2 Learning from Experience

As researchers consider new initiatives in social science and agroforestry they can learn much from both the successes and failures of local participation in the past 30 years of agroforestry technology and land use research as well as from socially focused agroforestry development programs. The recent history of agroforestry research and development illustrates the distinct approaches and converging experience of technical and social development programs in pursuit of broader participation and environmental objectives. The 1980s brought a profusion of research and extension programs in Farming Systems, Sustainable Agriculture, Agroforestry and Social Forestry. Over the last decade many of these programs have reached beyond the confines of professional scientific traditions. They have experimented with more direct collaboration with rural people, and/or with rural development, social service and relief agencies already well-established in rural communities. In turn, community development agencies of many types have also expanded their efforts to test and develop agricultural, forestry and conservation technologies or have sought to collaborate with research organizations in the field.

The converging experience of participatory agroforestry research initiatives in research and development institutions can provide a more advanced point of departure for the sustainable development initiatives of the next century. Two decades of intensive documentation, research and development in agriculture, agroforestry, social forestry and conservation has taught us to look beyond our traditional research models. Experience suggests that the scientific establishment is too small and too specialized to generate fixed "packages" of production and resource management technologies for the multiplicity of diverse environments in the world. Fortunately, there is no need to do so, since farmers, pastoralists and forest dwellers already have substantial knowledge as well as the ability to conduct both collaborative and independent research. Rural people often possess an inherent advantage over research institutions when dealing with trials of complex land use systems, as systems, *in situ* (Chambers, 1989). Experience has shown that there is no single best, fixed land use "package" for any given region or group of people, but rather a vast array of principles and components that can be constantly recombined, tested and modified to suit changing social, economic and ecological conditions for individuals, households, communities and nation-states (Rocheleau, 1991a).

Participatory research represents one way to expand sustainable development research capabilities in the complex conditions faced by rural people (Uphoff, 1992). For some professional scientists "participatory research" implies that "we" allow "them" (rural people) to participate in "our" research. For community organizers or rural communities it may mean that "they" allow outsiders (us) to take part in local land use experiments and their interpretation. What we all imply, but seldom discuss, is that we propose to join together people and institutions with very distinct traditions of acquiring and testing knowledge, in order to develop sustainable land use practices of interest to both. The objective of this collaboration is to develop a socially informed ecology and production science and an ecologically and livelihood oriented social science, both cognizant of difference and diversity.

Agroforesters cannot expect to achieve this through simple addition of conventional research methods and a new interest in local participation. Many participatory research "recipes" suggest taking one agroforestry technology, one standard research trial, one part good will, one plot of land, add local participants and stir. The results rarely meet the expectations of either outside researchers or participating communities, despite considerable effort on both sides. A really honest joint effort will require everyone involved to stretch their imaginations, their skills and their definitions of science. Although the question has many facets this section will focus on: (1) broadening the definition of agroforestry research and science; (2) widening the scope for who participates, where, on what terms; and (3) an expanded repertoire of practical research methods and flexible institutional arrangements that deal with social and ecological complexity.

4.2.1 Formal Research Models

Most agroforestry research in the 1980s followed a linear model that tested species, interactions and prototypes for new technologies first on-station, then later evaluated and refined the "winners" on-farm or in-the-forest (Figure 4). Researchers often asked the "basic" questions on-station and the "practical" questions on-site. They tended to assign "participation" to discrete steps within the research process. Farming Systems approaches often relegated participation entirely to the first and final phases in the technology testing process, as "problem diagnosis" and "adaptive research," respectively. For example, researchers consulted selected farmers about their production problems in a particular crop or livestock system. The research team designed a technology to address the problem(s) and on-station trials to test their ideas, as illustrated in the first part of Figure 4. Later, in "adaptive" on-farm research, participating farmers (usually men heads of household) contributed plots, their own and family labor, and perhaps their opinions as to the performance of a particular tree species, or an entire agroforestry package. The research was deemed to be participatory by virtue of its response to farmer problems, its location in a farmer's field (off-station), the farmer's presence, or the farmer's judgment of the technology. This view of "participatory research" limited formally recognized scientific research and land users' knowledge to fixed times and places and diminished their joint capacity as innovators and experimenters.

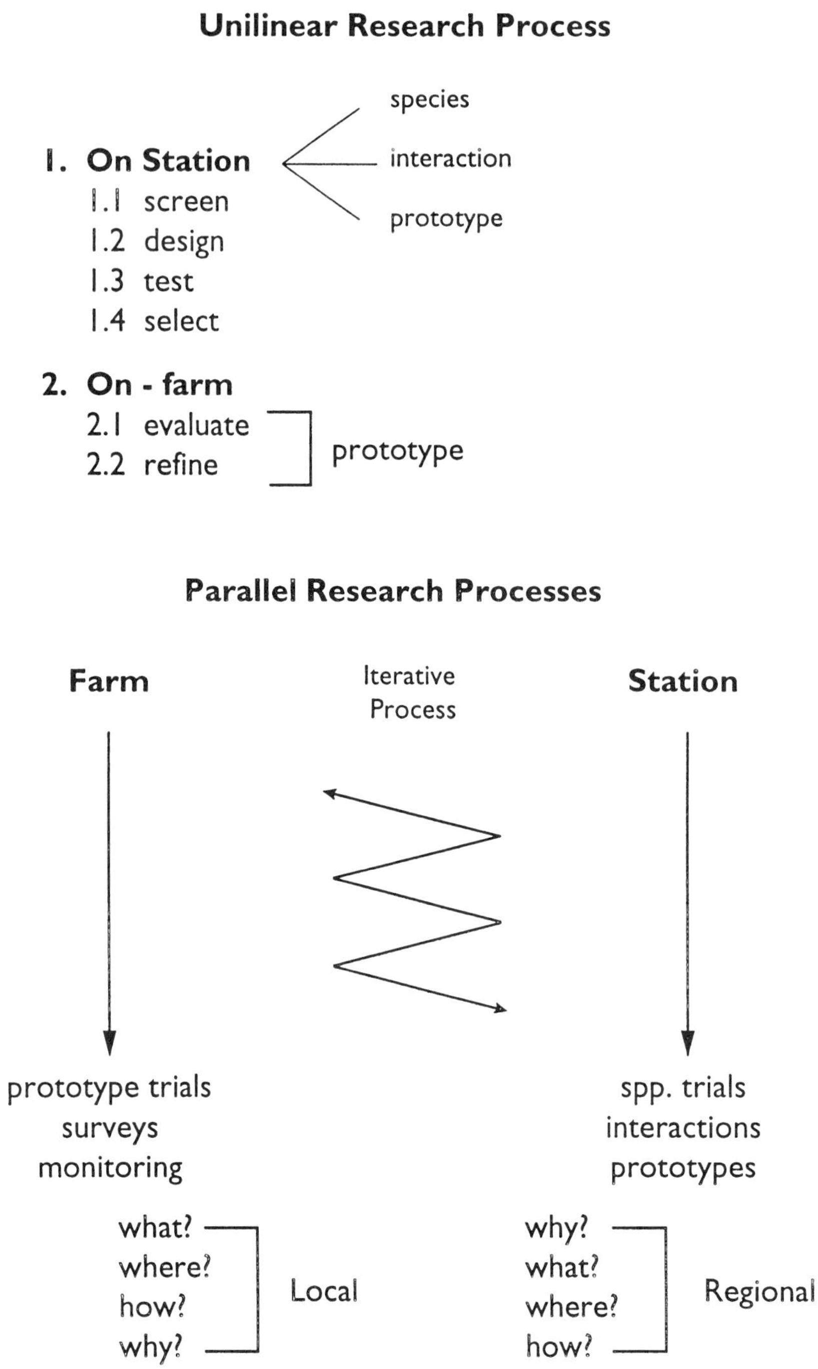

FIGURE 4. Alternative models for linking on-farm and on-station research.

However, some Farming Systems researchers ventured beyond the traditional linear research model (Hildebrand and Poey, 1985) with parallel lines of research on-station and on-farm (Figure 4). They proposed that researchers pursue parallel, independent, non-participatory projects on-station and on-farm, and further suggested tying the activities at both sites into an interactive process. Farmers and researchers may exchange information, planting material and evaluation of experiments at both sites. A farmer's exploratory trial on-farm could inspire an experiment on-station to monitor tree-pest-crop interactions, while a researchers' idea from a species screening trial on-station might lead to the introduction of an exotic tree into traditional home gardens on several farms. In such cases participation was to permeate all research activities, joining research station and on-farm endeavors (Biggs, 1988).

Later this research model and agroforestry technologies were extended and expanded to address conservation issues in forest and park boundary areas as well as degraded farm and rangelands. In most such cases rural people did not participate in problem definition, as they themselves were often already identified by outsiders as part of the problem. Local residents usually entered the picture to make a choice between pre-defined options or perhaps to help implement pre-arranged solutions to hunting, habitat destruction, land degradation and other threats to wildlife, woodlands and watersheds. The research in such cases sometimes served primarily to inform national and international environmental organizations about the "perceptions" of local people, their "receptivity' to conservation initiatives, and their potential as allies and collaborators in a pre-set agenda.

4.2.2 The Broader Potentials of Research

"Scientific" agroforestry research need not be synonymous with a randomized block design field trial or a multiple transect survey of plants and wildlife; nor must it imply a statistically analyzed survey questionnaire administered to a "random sample" of a population. While all of these research types are valid and obviously useful, none of them possesses an inherent advantage for all research questions and circumstances. Agroforestry incorporates several valid and complementary categories of research topics and activities, as well as place, scale, timing and methods of research (Rocheleau, 1987, 1991a; Chambers et al. 1989; Muller and Scherr, 1990).

In fact, the range of choices is far wider than formal research publications suggest. Land use research on production and conservation technologies can include observation, measurement, description (qualitative or quantitative), data and sample collection, design, testing, analysis and evaluation. Our mandate may be prediction, explanation or technology development. Our analyses may be static or dynamic with respect to time, and they may focus on problems at varying scales, from nutrient uptake by plants to production processes in whole landscapes to division of labor, land and authority at household and community level. We may conduct controlled, semi-controlled or un-controlled experiments, or even transform a structured observation of existing processes into an "insinuated experiment."

Moreover, sustainable land use research should transcend our convenient dichotomy between on-station and on-site research; it also may take place in the laboratory, on a "model" farm, in a park, on open rangeland, in-the-forest, or at a combination of these sites. We can also refrain entirely from experimentation and conduct survey research, as in ecological sampling of tree species in a forest, a sociological questionnaire to determine community structure, or an ethnographic survey to explore and document local botanical science (both popular and specialized). In short, research not only extends beyond the research station, it encompasses substantially more than can be held within the confines of controlled experimental plots on any property.

4.2.3 The Varieties of Participation

Participation is likewise subject to a broad range of interpretation. It has been variously construed to catalyze, facilitate, assimilate and suppress the initiative of rural people, depending upon the context and the players (Langley, 1986; Oakley, 1987). For the sake of simplicity Oakley (1987) reduces the varieties of participation to two basic forms: mobilization and empowerment. Where research is concerned we can also distinguish between extractive and interactive approaches.

Agricultural and environmental research programs often equate participation with mobilization and extract contributions of work, knowledge and other resources from participants. Many scientists rely on rural people mainly to provide land and "authentic" labor for experiments and to indicate "consumer preferences." In fact, rural people are often well-placed to identify problems, formulate solutions and devise tests of complex innovations *in situ*. They may participate in land use research in roles ranging from free labor on-farm to board members of research stations. Land users may play several distinct roles in the research process:

1. Labor (free, paid);
2. Hosts (to guests, to parasites);
3. Informants (representative, specialized);
4. Evaluators (of technology, of research process);
5. Collaborators (occasional, regular);
6. Partners (senior, equal, junior);
7. Advisors (informants with authority);
8. Board members (participants with power).

There is ample precedent for these roles and for interactive approaches in health, literacy and agriculture (Bunch, 1985; Jiggins,1988; Feuerstein, 1986).

A land user focus can accommodate all of the groupings which define privilege, power and poverty in a given time and place. We may distinguish between land user groups based on activity, tenure (terms of access, use and ownership), and social unit of organization. For example, we can determine which groups use specific land areas, plants, products or services in a given place, and the importance of the resources to them, as an indicator of their stake in land use change. This, in turn, can help us to identify convergent, complementary and conflicting interests of affected groups in the process and the eventual results of sustainable land use research and development.

4.2.4 Participatory Research Expertise: Non-Existent or Just Invisible?

Participatory field research in agroforestry initially focused on rapid appraisals for research or development planning, or on surveys (quantitative, qualitative and combined). Of the technology trials conducted on-farm most of the documented cases involved farmers in controlled experiments designed by outside researchers. The more collaborative trials were often *ad hoc* and usually unpublished prior to the late 1980s with a few exceptions (Chambers et al. 1989; Okali and Sumberg, 1989; Atta-Krah and Francis, 1987; Rocheleau et al. 1989; Scherr, 1990).

Eventually, researchers and field workers began to experiment more widely and to pool their collective experience with the design and management of on-site experiments or sampling and monitoring programs in partnership with rural people. Detailed documentation and analysis of locally initiated trials and experiments also began to emerge (Richards, 1985; Scoones, 1988; Wilson, 1988, 1989; Gumbo et al. 1988; Juma, 1989). The combination of any sort of trial or experiment with historical documentation and analysis was slower to surface and is still less common. Most reporting of rural people's production and conservation science was limited to descriptions of existing and/or traditional practice as an accomplished fact. Within forest and wildlife management, rural people's agroforestry knowledge was increasingly recognized by outsiders, only to be cast as "timeless and unconscious ecological wisdom" or as remnants of "traditional" practice. However, a growing body of literature appeared that treated local science and practice as the latest expression of a continuing process of learning and discovery (Anderson, 1990; Anderson et al. 1991; Arnold and Dewees, 1997; Chambers and Jiggins, 1986; Colfer, 1989; Dewees, 1989; Gupta, 1989; Oldfield and Alcorn, 1991; Denevan et al. 1984; Posey, 1985; Rocheleau, 1991b; Richards, 1985; Scherr, 1990; Scoones and McCracken, 1989; Scoones and Thompson, 1994; Warren et al. 1994).

Much of the best agroforestry participatory research and experience has not been recorded. Numerous field workers continue to conduct isolated, undocumented research within agroforestry programs. Likewise, researchers often participate in community organization and institutional innovation to improve their research and attune it to local conditions. However, they are unlikely to report even the fact itself, let alone the process. This is particularly crucial in the rapidly growing number of forest, biodiversity and wildlife management projects which address complex socio-ecological relationships through separate programs of biological research and public relations. Social research and management programs in this context are often couched in terms of social engineering to achieve separate agroforestry production and conservation objectives. The ecology embedded in local society and the cultural threads that run through the surrounding ecosystems are seldom addressed formally nor documented.

Beyond the research and participation dichotomy, we face a braided institutional divide along social/biological, production/conservation and government/NGO lines which constitutes a substantial barrier to shared knowledge. The existing institutional structure encourages silence on work at the boundaries between research, development and participation by those who actually know the territory best. As long as the more integrative work is submerged it is also inaccessible to review, constructive criticism and progressive improvement through collective learning and innovation.

Alternatively, we can make the most of opportunities to link these non-reinforcing cycles of research and development, social process and technology innovation, to stop spinning our wheels and get somewhere. Some of our best data and insights are transmitted through stories, a professional oral tradition, and through the skills of our trades. The challenge will be to distinguish significant stories from mere anecdotes (Rocheleau, 1991b) and to combine them with classification and description of possible field methods. From these we need to build a coherent, larger body of shared knowledge and practice accessible to our various domains of science, practice and critique, including those of rural people.

4.3 Expanding Our Repertoire

We can improve our capabilities for participatory research if we abandon fixed packages of research methodology and broaden our horizons to include a wide variety of principles, methods and other people's field experience. The broad principles presented above, the "stories" of colleagues from the field and the partial list of specific methods summarized below, represent tools and raw materials. From these, individuals and institutions can develop appropriate participatory research programs for sustainable production and conservation within various local and national conditions.

Most of the methods or techniques listed below can be used in an interactive or extractive way and most of them could in fact help to describe, plan, test, monitor or evaluate a technology, to document an existing system of resource use and management, or to facilitate the development of a new one. Some of the methods listed are actually labeled packages of methods, but need not be kept intact. The list is meant to convey a sense of the wide range of possibilities, a history of development and application for each and an invitation to modify and combine these tools to suit the problem, the participants, the diversity of constituencies, and the institutional opportunities in a given case. Each entry carries a brief descriptive note and a selected list of references to facilitate access to the relevant literature.

4.3.1 Appraisal Methods

Rapid Rural Appraisal (RRA) consists of short, intensive, informal field surveys that focus on consultation between teams of outside "experts" and rural people to define research and development problems and solutions. Researchers, planners, administrators or technical advisors travel to rural communities for a few days to a few weeks to meet individuals, households and community groups. They discuss local views of social, economic and technological problems and determine priorities for research, development or policy intervention. The early versions of RRA were developed and widely used to identify household level problems and research priorities in farming systems (Collinson, 1981; Hildebrand, 1981) and agricultural development research (Chambers, 1981, 1983; Chambers and Ghildyal, 1985; Chambers and Jiggins, 1986). Several recent publications summarize the methods now commonly used in Farming Systems Research and agroforestry research and extension (Muller and Scherr, 1990; Khon Kaen University, 1987; Chambers et al., 1989; Rocheleau et al., 1988; Jiggins and Feldstein,1994; Buck 1990).

Through use in participatory research agendas, the tools and the practice of RRA have expanded to better meet the needs of researchers, development workers and rural people (RRA Notes Newsletter; I.L.E.I.A. 1988, a, b.; Chambers et al. 1989). Spin-offs from this approach include research on farmers' prior knowledge and experience, as well as farmers' innovations and their modification of researcher-designed "packages" in agriculture (Rhoades and Booth, 1982; Rhoades, 1989; Jiggins, 1986 a, b; Fernandez and Salvatierra, 1989), agroforestry (Anderson 1990; Buck, 1990, 1993; Rocheleau, 1987, 1991 a and b; Rocheleau et al., 1988; Scherr, 1990 a and b; Scoones, 1989; Wilson, 1987, 1989), soil and water conservation (Kiriro and Juma, 1989) and national park management. Field practice has extended the process within and beyond "the household," including specific techniques for gender analysis (Ashby, 1987; Poats et al., 1988, 1989; Thomas-Slayter et al. 1993; Jiggins and Feldstein, 1994; Slocum et al. 1995; Rocheleau, 1988 a and b, 1995; Rocheleau et al. 1997c), and for group and community level interviews and workshops (Barrow, 1992; Norman et al., 1988; Sutherland, 1987; Kean, 1987; Jiggins, 1986 a, b, 1988; Rocheleau et al., 1988; Fernandez and Salvatierra, 1989; Gupta et al., 1989; Lightfoot et al., 1988, 1989).

Field research and development workers have further stretched and reshaped this robust set of techniques to address issues of sustainability and the larger landscape beyond the farm boundaries. Agroecosystems Analysis (Conway, 1985, 1987) focuses on villages, communities, or watershed units and deals explicitly with long term ecological concerns and environmental management in rural farming systems. Researchers and local representatives walk along transects through the landscape, and conduct interviews with individuals, households and groups. The team maps whole communities, ecosystems and specific plots with residents and key informants. This approach has been developed and applied primarily in studies of watershed management and water use within agricultural systems, from hillslope farms to lowland rice paddies. Another example is Total Catchment Management, and the Land Care movement as documented by the University of Western Sydney at Hawkesbury. This approach begins with RRA, then emphasizes action research and land user participation in resource management under complex, changing and highly uncertain conditions.

Diagnosis and Design adapts farming systems approaches and RRA for agroforestry design and testing (Raintree, 1987 a, b; Muller and Scherr, 1990). Reconnaissance surveys, informal household level interviews and alternating cycles of survey and technology testing allow for design of agroforestry technologies that address local problems and fit within the larger farming system (Raintree, 1987 a, b). This approach can be expanded to include land user groups at community level (Rocheleau, 1985, 1987, 1991a and b) or research and development interests at national level (Scherr, 1987). Some researchers have combined diagnosis and design, agroecosystems analysis and related approaches to fit the needs of community-based agroforestry research and extension programs (Abel et al., 1989; Davis-Case, 1989; Feldstein et al. 1989; Rocheleau, 1995; Rocheleau et al., 1996a, 1989, 1988; Scoones, 1988; Hoskins, 1996; Buck, 1989,1990; Scherr, 1987,1990).

Community-based ecological research focuses initial appraisal on local knowledge systems, particularly knowledge that links livelihood to ecology. These approaches have special relevance for agroforestry programs. The science of everyday life in rural landscapes often involves the integration of wildlife, water sources, crops, livestock and woody plants, within forests, rangeland, croplands, and gardens. It also encompasses the invisible food, fodder and wood production systems in the "spaces between": roadsides, fences, fallows, gullies, and stream banks (Rocheleau et al., 1988). Field researchers affiliated with ENDA Zimbabwe and the Zimbabwe Forestry Commission have developed participatory approaches to forestry and agricultural research which build on local science and practice (Wilson, 1988; Scoones, 1989; Gumbo et al., 1989; Matose, 1993; Clarke, 1990; Seitz, 1993). The process begins with a fairly lengthy exploration by a resident action research team. From the outset, researchers and local participants compile written records of local ecological history and science, as well as a set of action research proposals.

Similar methods have been elaborated in veterinary research with pastoralists in Kenya, and by Fernandez and Salvatierra (1989) in livestock management and veterinary research with women's groups in Peru. Several researchers (Gupta et al., 1989; Rocheleau, 1995; Rocheleau et al., 1996 a and b; Rocheleau et al., 1988; Davis-Case, 1989) have combined group discussions and mapping exercises based on local knowledge of ecosystems and livelihoods. Carney (1988) has combined participant observation and ethnographic survey with appraisal of class and gender division of knowledge and resource use in an irrigation project in the Gambia. Peluso (1995) has used "counter-maps" to make visible the presence of people or their resources in Indonesian forests.

Research on local ecological science is richest if it combines the study of both popular and specialized knowledge. Many researchers start from interviews with a broad base of representative community groups, to determine what is "common knowledge." They can then ask the groups to identify knowledgeable group members and other specialists in the community (Rocheleau et al., 1988, 1989). Eventually researchers learn enough about the topics at hand and about the identity of specialists to allow them to record the knowledge of the eldest or the most skilled members of the community. They can also identify and record the emerging knowledge and practice of the rising generation or the distinct science and practice of particular groups, whether by gender, ethnicity, class, occupation, or locality. Box 7 illustrates this approach.

Box 7. Bungoma District Indigenous Plants Project

Kenyan researchers at the Africa Centre for Technology Studies (ACTS) documented local use and breeding of indigenous plants in Bungoma District. A mix of participatory and standard research methods from social and biological sciences provided information for agroforestry and land use researchers as well as for farmers, herbalists and a group of young people interested in reintroducing and disseminating some of the food and medicinal plants in local cropping systems and gardens.

In a creative synthesis of rapid appraisal, ethnobotany and sociological survey techniques Calestus Juma (1989) conducted community-based surveys using a combination of community meetings, key informant interviews, and survey questionnaires. Locally nominated participants developed and refined the formal questionnaires in a workshop setting. The workshop served as an extended two-way key informant "interview" that shaped both the content and format of the subsequent survey. Representatives of the research project and local residents learned about each other as well as about indigenous plants, their uses and their habitats. The process made the survey effective as a learning tool for both researchers and the participating communities. Some of the local project staff and young trainees continued work on their own with the more interesting species identified during the surveys.

If the work is applied within an action research approach, then this same understanding and information can be mobilized within the community's own research and development efforts (Thrupp, 1989; Bebbington, 1990). The involvement of local residents as researchers, recording their own community's knowledge, can also serve as a catalyst to organization and educational initiatives. When women, children and the poor formally record their own experience the research effort can also strengthen their position within the larger community (Fortmann, 1996).

Participatory Rural Appraisal (PRA) has grown out of a synthesis of RRA, Agroecosystems Analysis, Diagnosis and Design and other appraisal methods with action research and community organization techniques. It has coalesced from a number of centers of innovation, including:

International Institute for Environment and Development (Scoones and Thompson, 1994; McCracken et al. 1988). The Institute for Development Studies at the University of Sussex (Chambers et al. 1989; Chambers, 1992); Clark University, World Resources Institute, the National Environment Secretariat of Kenya and Egerton University (NES et al., 1991; Kabutha et al. 1990; Ford et al., 1993; Thomas et al., 1993; Slocum et al., 1995; Rocheleau, 1995; Thrupp, 1989; Zazueta et al. 1994); several NGO networks (ILEIA, 1988a and 1988b; Bunch, 1985), international agricultural centers (Rocheleau et al. 1988; Lightfoot et al. 1991; Jiggins and Feldstein, 1994; Ashby and Sperling, 1995) and United Nations organizations, from FAO (Hoskins, 1996; Bruce, 1989; Davis-Case, 1989) to UNICEF (Kabutha et al. 1991).

Like other forms of rural appraisal, PRA normally consists of a one to three week exercise based on collaboration between rural people from someplace and outside "experts" from somewhere else. The difference is in the structure and tone of interaction between them, and in the shift from researcher-subject to joint exploration as equals. In some cases, outside researchers function mainly as facilitators of a community-run process, and increasingly, as trainers of facilitators from rural communities. Overall PRA relies more heavily on the judgment and analytical capabilities of rural people, rather than simply "tapping" their knowledge in bits and pieces to fill-in-the-blanks in the analytical frameworks of outsiders. The methods of inquiry are more explicitly interactive and tend to be more visual in orientation than earlier RRA approaches. In addition PRA includes leadership and technical expertise from local residents and sometimes public servants (Chambers et al. 1989; Kabutha et al. 1991).

The "steps" have been recorded and described by some practitioners as a discreet package of activities, with a given order (NES et al., 1991). Others (Scoones and Thompson, 1994; Cornwall et al. 1994) eschew any attempt to regularize the process and prefer to discuss principles, document specific techniques and report on particular case studies, to inform further work in a variety of contexts. One element which characterizes many versions of PRA is greater attention to history and to the possible futures of rural people (Rocheleau, 1987; Rocheleau et al., 1988, 1989; Chambers, 1992). PRA includes matrix ranking exercises, community histories, diagrams of organizations and institutions, maps of farms, landscapes or watersheds, and diagrams of production systems, ecosystems and social processes. Often people will rank themselves and their neighbors with respect to wealth and well-being, to clarify social structure and process and to situate their own knowledge and perspective for themselves and outsiders. Ranking exercises can also clarify the use and relative importance of various water sources, soil types, land types, crops, livestock, trees, fodder, energy sources, insects, wild plants and wild animals. They may discuss and rank these elements of livelihood and landscape as problems, as resources or as desired characteristics of a possible future.

While PRA has often been used to mobilize community discussion, planning and immediate action on AF, it can also facilitate development of longer term action research. There is ample scope for use of PRA in evaluation of on-going projects and programs, whether large scale or local in nature. The emerging techniques of PRA could facilitate widespread adoption by agroforestry research and development programs of participatory methods previously limited primarily to literacy, health and community organization programs (See works by Freire, 1972; Feuerstein, 1986; H. Richards, 1985; Slocum et al. 1995).

All of these appraisal methods can help to describe a particular place and situation, and to direct research and development plans that fit rural people's realities and aspirations. Some work best within a recurring cycle of surveys and trials of various types, including the experimental approaches listed below.

4.3.2 Field Experiments and Trials

Much of the literature on on-farm agricultural research discusses "how to" reconcile statistically valid experimental designs with field conditions (Hildebrand, 1986; Shaner et al., 1982; Collinson, 1981; Tripp, 1984; Ashby and Sperling, 1995). The most frequently used methods are those which allow for control plots, some variation in treatment, and statistical analysis of variable performance by different treatments within or between farms.

Farmer and researcher preferences for different research designs may differ substantially. While outside researchers can gain substantial information by varying treatment between farms, farmers may gain more insights by having controls and/or a range of treatments on their own farms to compare close-up (Robert Hart, personal communication, 1984). However, farmers may not appreciate the placement of these various treatments within a randomized block design. This is especially impractical for forestry, water management, wildlife management and conservation practices, which are not divisible into small fractions of a single plot. Group-focused trials at multiple sites provide an alternative, by combining different real-scale treatments on various members' lands with regular group meetings to observe and compare all treatments in the multi-site experiment.

Researchers can also use informal trials to explore technology design prior to more formal, elaborate trials (Sumberg and Okali, 1989; Atta-Krah and Francis, 1987; Rocheleau et al. 1989) or simply as a way to learn more about the detail of farming practice (Edwards, 1987 a, b, c) before committing local residents and research institutions to a substantial research effort. It is also possible (though professionally risky) to pursue formal but not controlled experiments on complex land use systems with farmers. This applies particularly to technology innovations for home gardens and similarly complex systems both on-farm and in the larger landscape. See Box 8.

Box 8. Collaborative Livestock and Cropping Research in Aramachay, Peru

Herders and farmers of the Aramachay Women's Production Committee in Peru requested a multi-site, group based research design in a collaborative livestock and cropping research project. In veterinary experiments based on local and outsiders' science they separated blocks and treatments by family herds, rather than mix treatments across herds. Farmers took an active role in the design and evaluation of the cropping systems research process, as well as participating in technology evaluation. As a result, the project team developed a robust, statistically valid research design that was convenient for farmers and herders (Fernandez and Salvatierra, 1989). The control groups for veterinary treatments shifted from a within-herd to between-herd scale in response to farmer demands. They also expanded the experimental design to compare local herbal remedies, pharmaceuticals and control groups.

A few sources also treat the issue of farmer participation and the quality of interaction between outside researchers and farmers, as well as research designs that fit the needs of both (Hildebrand and Poey, 1985; Scoones, 1989). The possible terms of collaboration on research trials range from researcher-designed trials on research stations and in parks to rural peoples' own experiments that are "discovered" and documented by research institutions (Rocheleau et al., 1989). The following typology (Rocheleau and Malaret, 1987; Poats et al., 1989) presents a spectrum of collaborative arrangements between local science and practice and "the scientific establishment":

(1) Researcher designed and managed trials, usually on station or special plots. Land users are consulted and their problems are addressed, but their resources, management practices and evaluation are not part of the research design.

(2) Researcher-designed and managed trials, on-site, in local peoples' work and production sites, whether individual or shared space. Land users are consulted, their problems are addressed and they evaluate the results. There is little involvement of land users' management, since all labor and material inputs are planned and paid by the research institution.

(3) Researcher-designed and user-managed trials, on site. This is similar to case 2, above, with the difference that land users' resources and management are included in the trial, their evaluation and feedback are continuous, and land users' performance and judgment are part of the trial.
(4) Joint design and management of on-site trials by researcher and land users. Local people and outside researcher(s) collaborate in the design of the trial and confer on management decisions. Land users' management and decision-making are explicitly treated as experimental variables, their feedback and evaluation are high priorities in the research endeavor, and they consciously evaluate their own and researchers' decisions.
(5) Trials designed and managed by land users, with outside researcher(s) consulting. Outside researcher(s) enter into on-going trials as occasional consultants or regular collaborators, and document results and/or process. Researchers may or may not alter trial design.
(6) Trials designed and managed by land users. Outside researchers observe and document existing trials, experiments and on-going innovation. Outsiders also may produce documents for local review, revision and use.

The choice of trial types will also depend on the type of question, the variability of social, economic and ecological conditions in the region, and the time, space and precision required to produce useful answers to the questions. For example, farmers in southern Zimbabwe have participated in agroforestry projects that include species trials in community experimental plots, farm trials of tree establishment techniques, and resource management trials in shared and private lands. Trials have combined controlled experiments (types 2, 3, and 4) with more informal observation plots (types 5 and 6). In trials on-farm, on-station and in-the-commons farmers have participated in roles ranging from advisor to employee (Scoones, 1988; Wilson, 1987; Gumbo et al., 1988; Clarke, 1990; Matose, 1993; Seitz, 1993). In such cases the entire project can become an experiment. Researchers can carefully document the process and the effects of "work in progress," though they should take care not to turn the project into their own laboratory.

In contrast, some projects have emphasized farmer participation in formal research station experiments. Researchers in Rwanda have reported major advances in potato and bean varietal selection and breeding research through surveys of farmer knowledge (Haugerud, 1986) and involvement of farmer "experts" in on-station research (Sperling, 1994). This approach recognizes that farmers develop their knowledge and skill over long time periods and across a wide range of micro-environments. It couples the richness of farmer experience and judgment with the precision and control of research station experimental regimes.

4.3.3 Experience as Experiment

Beyond the terms of cooperation in specific experimental activities lies the question of how to reconcile our own experimental inclinations with the "science of survival." When outside researchers recognize rural people as independent innovators and co-researchers they also may reconstruct them in their own image. However, professional researchers need to recognize that their local collaborators may be scientific without being logical positivists. Their process of experimentation may be more like that of a concert pianist than an industrial chemist (Richards, 1989).

In his recent article on agriculture as a performance, Paul Richards (1989) notes that farmers must integrate all of their past experience at critical moments, such as drought, flood, pest outbreak or market fluctuations. They must make binding decisions that will affect the season's harvest or even their very survival as farmers. Long-term learning and innovation are likewise accelerated during times of rapid and dramatic change such as land tenure reform, large scale migration, an exodus of part of the work force, or the introduction of new technology such as plows or tractors.

Some types of research directly address the science of survival under uncertain conditions. Techniques from industrial psychology and marketing have proven useful to identify successful strategies for coping with change or for reducing vulnerability to economic and environmental stress (Jiggins, 1986 a, b, 1988). For example, aerospace industry researchers developed critical incident analysis to identify early indicators of trouble and to document the decisions taken by pilots that survived potentially fatal crashes. As Jiggins has demonstrated, this can be adapted to learn from rural people about crises that have been survived to better inform people from the place about their own options in similar future events. The technique consists of applying an experimental frame of mind to the documentation and interpretation of remembered events. It lends itself well to agroforestry studies within the context of wildlife management, watershed management, deforestation, reforestation and agricultural intensification.

Rural people can also reverse the time sequence to conduct "what if" simulations, applying an experimental mindset to visions of the future as well as to memories of the past. Extrapolating from current trends it is possible to imagine a range of possible futures (Rocheleau et al. 1988, 1996b; Davis-Case, 1989) based on a combination of conscious choices and chance occurrences. Choices could include: community resource management strategies, landscape design, household migration decisions, and introduction of trees as cash crops. Externally determined events might include: drought; plant and animal diseases; groundwater depletion; national wildlife, forest and land tenure policy; price fluctuation; or market collapse for particular commodities.

4.3.4 Sampling and Monitoring

Although scientists often make controlled laboratory conditions conform to the demands of particular statistical analyses, it is also possible to fit multivariate and time series analyses to complex field conditions. Oceanographers, meteorologists, geologists, and field ecologists are all accustomed to working with "experiments" designed by "nature." This differs from survey research in that researchers may gather detailed data on complex systems over time, within an experimental framework that embraces variability. They can monitor and test relationships between specific practices, processes, species, site characteristics, landholding types and land user groups. They can sample and monitor land use systems or trees in the landscape according to maps of existing variation in practice or in vegetation. Such "inferred" or "insinuated experiments" can explain the economic and ecological significance of these patterns in the landscape. Researchers also may use monitoring data to develop models that can simulate potential changes in land use systems. For example, researchers could use farmers' records and their own field observations to model an existing land use system and to predict the likely outcome of a prolonged drought or a new settlement program.

In most cases, rural people can be part of the system as well as active observers, recorders, analysts, evaluators and independent experimenters. They may monitor and evaluate projects and research process as well as specific technologies (Davis-Case, 1989). They can also conduct their own "perturbation experiments" with "real world" models to observe the response of local landscapes, livelihoods and ecosystems to specific changes. The "control" is in their memory. This type of agroforestry experiment may well prove more coherent from their point of view than a replicated experiment laid out according to a randomized block design.

4.3.5 Group Methods Applicable Throughout the Research Cycle

Farmer panels (Sperling, 1994), focus groups (Jiggins and Feldstein, 1994), group interviews (Buck, 1990; Rocheleau et al., 1988), group ranking exercises (Grandin, 1988; Scoones, 1989) and participant observation (Ashby, 1987; Ashby and Sperling, 1995) can all provide information about the substance and the distribution of knowledge, practice, resources, opinions and interests on a particular issue within a given community (Davis-Case, 1989). While individuals may provide detailed information in intensive interviews, their responses often represent one position on a larger spectrum that remains unknown or must be inferred by the researcher from a large random sample.

For any of these methods, researchers may choose a group at random, or select groups systematically to represent a range of characteristics present in the community. Alternatively, participant groups may be selected by the larger membership in a pre-existing group, they may volunteer according to researcher criteria, or key informants may nominate groups. Both the origin and composition of groups have strong implications for the substance, the style of interaction, and the locus of control within research activities. Researchers who collaborate with groups may work with: (1) pre-existing groups which take on research tasks as a group, (2) pre-existing groups which facilitate the participation of a sub-group of members as a special optional activity, or (3) groups created by and for researchers for the explicit purpose of research collaboration (Rocheleau, 1991a and 1994).

In the case of pre-existing groups, there are several key questions: whether they are formal or informal (as in legally registered vs. family and friends); whether membership and contributions are voluntary or coerced; whether the group represents the community as a whole or specific segments thereof (by gender, class, ethnicity, religion, occupation, location); whether they are traditional or recently initiated; whether they are internally or externally organized; and whether they are perennial, multi-purpose organizations or were formed to accomplish a specific task. If researchers misunderstand the purpose, composition or community context of a participating group it can bias the research results and distort the quality and distribution of participation within the community.

Pre-existing groups can apply their own usual terms of leadership and participation to research activities (Fernandez and Salvatierra, 1989). These may not be egalitarian nor are they likely to reward research aptitude and performance, but they have the advantages of familiarity, local control and credibility. Special research groups formed from pre-existing groups retain the advantage of group and community linkages and credibility, yet can provide freedom for group members and researchers to choose leaders and follow procedures that facilitate the research task. They can also remain accountable to the larger group and respect the spirit of its organization in research activities.

When researchers form new groups from the larger community, based on open enrollment, direct selection or "conscripts," then the control of group process and activities is far more likely to reside with the outside researchers. The form and substance of the research may differ substantially as a result. Such groups make convenient participants in quantitative research designs. They lend themselves readily to controlled experiments, test panels for pre-packaged products or as representative qualified informants (Norman et al., 1988). However, they are less likely to promote continuing local innovation, research and information exchange. This type of group may make significant contributions to formal research efforts (Sperling, 1994), but should not be confused with self-sustaining groups that participate in research efforts as part of their own long term agenda.

4.3.6 Individual, Household, or Group Data Collection and Record-Keeping

Record keeping, measurements, diaries, plant collections, oral histories, maps and sketches may be limited to "just the facts" or may include a substantial dose of rural people's judgment, skill and worldview. While all of these can be very labor intensive and are sometimes used in an exploitive fashion, these methods can allow for accurate data collection, analysis and interpretation by and for rural people.

Records might include only qualitative observations such as time and labor allocation within the household, or simple quantitative notes on nursery operation, such as records of seeds planted, plants germinated, seedling survival, number distributed, to whom, number planted, damage, growth, and survival (Davis-Case, 1989; Buck, 1989; Campbell, 1996). Records could also include measurements of tree growth, insect damage to leaves or stems of seedlings, and volume and/or weight of fuelwood harvested. Records might also provide substantial insight into seasonal and periodic price fluctuations of local products and purchased items.

Individual, household or group diaries provide scope for sharing judgments and reflections about changing conditions or new activities in a rural community. For example, a women's group might keep a long term narrative record of significant events, and peoples' interpretations of these occurrences, including suggested solutions to land use and environmental problems. A diary could also focus on a specific technical topic such as tree seedling condition and survival. Entries could emphasize pest attacks on planted seedlings, with comments about when, where and why pest attacks occur, and a description of the pests. A diary of tree use and management might also provide comments about which trees are harvested, whether there is adequate supply, the quality of products, the potential uses and users and the decisions about who will use what tree and what product. Alternatively, diaries could focus on sightings and observations of wild plants, including rare or endangered species sighted at home, at work sites, on regular trails or on extended journeys.

Individual, household or group plant or insect collections may serve several purposes, some a bit more "participatory" than others. A collection may be for the collector's own use; for outside researchers according to specific lists; or for shared use from jointly compiled lists of plant and insect types and sampling sites. The use may be as general as "basic research" or as practical as a reference guide to identify medicinal herbs for local preparation and use. Public participation in collecting work can also facilitate discussions of ecological history, ecosystem structure and function, and the future landscape. This, in turn, may affect planning, management and improvement of land use systems, including domestication of wild plants from forest or rangelands in agroforestry systems.

Oral history, while less tangible than some of the data collection methods described above may well prove crucial to subsequent planning efforts to shape the landscape and livelihood systems of the future. The objectives may range across a broad spectrum from: 1) the postmodern project of liberating subjugated knowledges (Stamp, 1989); to 2) empowerment of local communities and organizations (Fortmann, 1996; Rocheleau and Slocum, 1995); and 3) supplementing written and photographic records of soil erosion, water supply development, deforestation, reforestation, technology adoption or land use and land cover change (Tiffen et al. 1994; Rocheleau et al. 1996c). Oral history has recently gained formal recognition as a tool of environmental and land use research (Showers, 1989). The applications vary from personal life histories, which are now widely used by feminist scholars to portray the diversity and depth of experience among women and to illuminate the political and the sublime embedded within the personal and the everyday. Life history approaches can be extrapolated to discussions of community and regional history, and can be specifically focused on landscape, land use, land degradation, biodiversity and ecological processes (Rocheleau, 1995; Rocheleau and Slocum, 1995; Rocheleau et al., 1997b). Oral histories can be used also to construct matrices and diagrams describing inputs and outputs for a given area, or documenting energy flows and material cycles in a local ecosystem and its linkages to larger economic and ecological systems (Rocheleau and Ross, 1995; Shields et al. 1996).

Mapping of past, present and possible future landscapes is yet another way that research programs can collaborate with rural people to document, analyze and predict ecological and land use changes. Maps and sketches facilitate discussions of topics ranging from biodiversity to food production and water management, often under conditions of rapid land tenure change. The spatial configuration of landscapes is changing swiftly and dramatically in many agrarian, rangeland and forest landscapes and this method allows for an integrative and rapid portrayal of the range of micro-environments available to plants, animals and people in rural communities. The approach is particularly useful for botanical research as well as land use planning and resource management programs that transcend single plots and landholdings (Rocheleau et al. 1988, 1995, 1996b, 1997a and b; Chambers et al. 1989).

Visual aids for discussion and graphic representation can include a wide range of media and approaches. Researchers have recorded a number of techniques in recent field exercises: landscape drawings with pens and ink markers drawn by researchers and local residents (Rocheleau and Ross, 1995; Rocheleau et al. 1996b, 1997b); chalk sketches and maps on blackboards drawn by groups and individuals (McConnell, 1992); clay models, sand paintings, and stick-drawings on the ground (MacCracken et al. 1988; Chambers, 1993; Fortmann, 1996); felt board landscapes with plants, animals, people, landscape features and infrastructure for iterative construction of alternative scenarios by groups and individuals (Rocheleau and Ross, 1995); flow diagrams and systems diagrams of various types (Lightfoot et al. 1989) and computer mapping simulations of farms, watersheds and larger landscapes on portable computers (Eastman, 1997).

Beyond the local scale, some programs based in forest communities have also begun community-based mapping exercises to delineate established terrains of resource use and rights to particular places as territory, using survey maps (Peluso, 1995; Poole, 1985). Members of the Rubber Tappers Union and farmers' associations in the Brazilian Amazon have undertaken training to read and utilize aerial photographs and remotely sensed satellite imagery so as to locate their communities within larger regions and to conduct land use research in coordination with other communities. They have also mapped their lands at community and regional scale for use in surveys and legal proceedings against government and private sector encroachment on their resources (Anderson, 1990; Poole, 1995). The Land Care groups in Australia also depend on mapping and mapped information at multiple scales to facilitate participatory planning for resource management. The Arusha Diocesan Development Office (ADDO) has assisted Maasai communities in Tanzania to map their customary grazing lands, water sources and current settlements in response to increasing conflicts over land use and access (Fr. Ben Ole Nangore, personal communication, 1989). In Sri Lanka researchers have joined rural communities in the use of GIS for collaborative planning approaches with government and NGOs (Yapa, 1991). These are only a few examples of the increasing use of participatory mapping as a tool of agroforestry action research.

4.4 Finding Institutional Homes

Most national and international agricultural research institutions already provide some scope for participatory survey research, adaptive technology trials, and land user evaluation of new technologies and land use plans. Enterprising field researchers, from anthropologists to ecologists, have seized or created opportunities to inform mainstream technology research from local science and practice. They have incorporated rural peoples' contributions in formal and "informal" surveys, trials and research planning. Instances of such "injections" of local participation have been reported for several international agricultural research centers, as well as in several international environment and development research institutions (Zazueta et al. 1994; Thrupp, 1989; Kiriro and Juma, 1989; McCabe, 1990). Rapid rural appraisal is now widely used to identify problems, and to inform technology design and research planning. Several major international and national agroforestry and forestry research programs have incorporated participant observation, ethnobotanical surveys and "directed" participation, as in the case of farmer panels, focus groups, and key informants, both group and individual (Ashby and Sperling, 1995).

A parallel stream of reports have issued from the field through new organizational clusters explicitly created to give voice to participatory research experience in NGO and national government programs (Bunch, 1985; Guijt and Shah, 1997; ILEIA, 1988; Buck, 1989, 1990, 1993). The case studies, methodological summaries and field trial results include applications of rapid rural appraisal, agroecosystem analysis, participatory rural appraisal and a host of untitled but no less valid approaches. They all include rural peoples' skills, concerns and judgment in survey and monitoring activities.

The key message this conveys for sustainable development researchers and planners is simple. Several kinds of organizations, from local to international in scale, and from basic research to local empowerment in mandate, have successfully entered into participatory research activities in agroforestry and related fields (Korten, 1990; Uphoff, 1992). These programs have already accomplished much independently. However, they constitute a potentially more powerful mix of complementary skill, experience and institutional strengths. An expanded approach would join organizations with different sectoral expertise and mandates such as water, soil, crop production, forestry, wildlife,

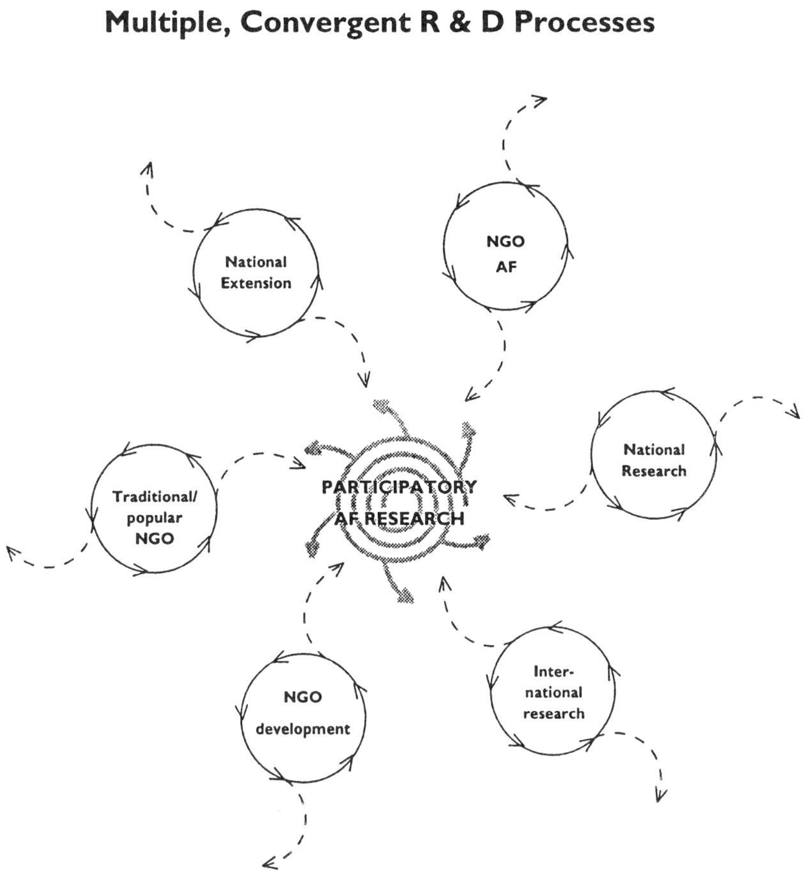

FIGURE 5. A "spinning wheel" model of institutional collaboration in participatory agroforestry research.

employment, education and culture. At the same time, the mix of participating organizations would combine distinct types of institutions: popular NGOs, technical NGOs, universities, national extension agencies, national research agencies, and international research, development, relief and environmental agencies.

The "spinning wheel" model of collaborating institutions, as depicted in Figure 5, suggests that each institution could spin on its own internal axis, yet contribute to a broader, shared circulation of participatory research for sustainable development. Each institution would also continue to contribute to specialized networks of like organizations. The work within the wheel would take place at a single site or a series of shared sites (Rocheleau, 1991a, 1994).

Depending on local and national circumstances, the institutions and their activities would vary substantially. In an agroforestry and rural livelihoods program the linked activities might include:

1. an ethnobotanical survey;
2. a series of formal trials to determine the best placement of a favorite local tree within a new settlement pattern;
3. exploratory trials with different types and sizes of tree nurseries;
4. management and tenure experiments to test alternative rules of use and access on-farm, along forest margins and in the forest;

5. ecological and social baseline surveys using local criteria to stratify the sample environments and social groups;
6. documentation of prior and on-going experiments by farmers;
7. participant observation by researchers in group agricultural and conservation work;
8. a seed evaluation, selection and collection program for favorite local and exotic plants;
9. a marketing study on existing and potential local tree products;
10. a local board to oversee all project work at the site, and
11. a farmer advisory group to collaborate on research station experiments of interest to the community.

A different agency might carry out each task, and the participants might include: an international church-based NGO; a national forestry research agency; a national agricultural extension program; local self-help groups; district officials; a national university research team of ecologists and social scientists; an association of local teachers; and an international conservation organization. Common interest in viable agroforestry systems and "livable landscapes" would drive the wheel, along with the cost savings of linking several activities in one place. The result could be one successful process, easy to multiply, rather than several specialized, incompatible environment and development successes that don't add up.

Throughout the world there are partial examples of the "spinning wheel" already turning. For example, the herbarium of the National Museums of Kenya has combined with Kenya Freedom From Hunger and Worldview International to conduct research and extension on indigenous wild food plants (Maundu Munyao, 1992). Their complementary skills have allowed them to link several distinct activities into a single coherent effort. They have been able to survey the use and knowledge of edible wild plants, to prepare and disseminate planting and cooking information on the best known plants, to promote domestication in community fruit and vegetable gardens, and to establish seed multiplication plots. In other cases "working groups" (networks that emphasize "work") have coordinated national research and extension agencies, economic planning, energy research, grassroots tree planting efforts and rural development agencies to promote agroforestry and social forestry within a single district, as in Indonesia (Mark Poffenberger and Fran Korten, personal communication, 1988) and Kenya (Bradley, 1991; Buck, 1990 and 1993; Kirkoff, 1990; Scherr, 1990 a and b; Barrow, 1992). The Total Catchment Management and Land Care groups in the Hunter Valley of Australia represent yet another example of multi-institutional and inter-disciplinary research and action on environment and development issues.

There is ample precedent within the international and grassroots NGO communities* for longstanding action research programs and collaboration with popular organizations in agroforestry technology innovation and development (Korten, 1990; Uphoff, 1992). These experienced groups are well-placed to inform and mediate the convergence of environment and development institutions with both research and action mandates to play complementary roles in participatory agroforestry research. We need not reinvent this particular wheel (Figure 2) from agriculture, forestry and conservation, but merely need to balance it for easy and effective use in a variety of agroforestry initiatives. Lest we become complacent about the ease of this task, we need only remind ourselves how a careless neglect of the relations of power can threaten such a carefully balanced process. This may well prove to be a necessary focus of participatory research over the coming decade (Cornwall et al. 1994; Marglin and Marglin, 1990; Rocheleau and Slocum, 1995).

The last two decades of experience in participatory agroforestry research methods has demonstrated what scientists and planners can and cannot do alone, working within national and interna-

* The best known examples include Oxfam, World Neighbors, International Technology Development Group (ITDG), Gandhian development organizations, International Institute for Rural Reconstruction (IIRR), IRED, International Organization of Consumer Unions (IOCU), Environment Development Action (ENDA), CARE International, SOS Sahel, and numerous church-based relief and development organizations.

tional institutions. Many have explored the soft edges of their own science and the regions of overlap with local science and practice in isolated rural communities throughout the world. Agroforestry researchers have also discovered a wealth of experience and information in the larger scientific and development community, often in institutions previously invisible to them, from their respective perches. Agroforestry scientists can use the decade ahead to employ their new-found collective skills in sustainable development and to demonstrate what is possible when they combine their efforts and insights with those of rural people. Participatory methods cannot guarantee socially just sustainable development. But they can facilitate democratic or self-determined programs to protect, create and maintain sustainable livelihoods and living landscapes for various possible futures. Neither participation nor environmental criteria automatically guarantee just, equitable and ecologically viable futures, but both constitute essential ingredients of a common future worth sharing.

5. CONCLUSION

Throughout the world people have experienced a profound shift in the division of land, labor, markets and organizational affiliation, as members of communities and households. Based on gender, class, age, ethnicity and other dimensions of difference rural people find themselves repositioned relative to each other, to the larger economy and to the ecosystems that support them, from their homesteads to the global environment. Agroforestry analysis and design in this nested context requires a telephoto lens to observe the interactions within and between ecological, economic and social systems from individual to planetary scale. The repositioning phenomenon also demands an agroforestry research framework that illuminates the complex patterns embedded in multiple and overlapping domains of resource use, access and control — a kind of kaleidoscopic view that both complicates and clarifies. Development of agroforestry for sustainable and just economies and ecologies will require a telephoto lens of observation across scales, with a kaleidoscopic analysis focused on the critical differences between the actors at each scale along the way.

The differences among people, both within and between households and communities, has crucial implications for agroforestry research and project design as well as for the development of the institutional frameworks to define and promote agroforestry development. The inclusion of diverse land user groups at the local level will require both planners and researchers to embrace complexity. Many agroforestry researchers may opt to forego simplistic research and design frameworks, for the sake of clarity, effectiveness and equity. The amount of variability within and between households and communities, in and of itself, will be a useful indicator of the range of environmental options and outcomes. The degree of differentiation and the distribution of power within and between communities also may be useful for framing agroforestry technologies and policies that are environmentally and socially effective.

Both the questions and the methods of agroforestry research may need to change to understand these complex social relations within households and communities and their articulation with larger systems. A new generation of "hybrid" agroforestry scientists may find themselves combining approaches such as participatory rural appraisal with household surveys, life histories and landscape/land use histories. Anthropologists and sociologists may find themselves using remote sensing imagery to locate the resources used by particular groups, while ecologists and land use planners may need to refer increasingly to detailed sketch maps and local taxonomies, and to combine these with local and individual environmental and social histories.

Agroforestry researchers and planners may begin their tasks with a very different set of questions:

How are individuals and households — or multiple and overlapping groups that cut across households and communities — differentiated with respect to resource use, access and control?

How do uneven relations of power affect differential access to: land; labor; commodities, markets, cash; and social organizations?

How will those differences affect the use of agroforestry technologies and how will technology changes affect existing and emerging relations of power and the interests of distinct groups?

Social science and agroforestry research can best address the present and future challenges of agroforestry by broadening the scope of "science" in general and by expanding the definition of social science and its "role" in agroforestry research. To fully realize the potential of AF, researchers will also need to explore and carefully negotiate the terms of collaboration between social and ecological domains of agroforestry as well as between local and outsider research, whether on social, ecological and economic dimensions of agroforestry.

What relevance does this have for agroforestry at the turn of the 21st century? If we accept the proposition that all science is local, then the task ahead would seem to be one of negotiation over the terms of joint exploration, across the borders of social and biophysical sciences and at the fertile boundaries of a multitude of local sciences and practices. The multi-faceted perspectives of the kaleidoscope and the spinning wheel model suggest practical ways to envision and to institutionalize such an approach. Values and visions, always implicitly at play in the scientific enterprise, perform an explicit and central role in this endeavor. The challenge for social and ecological science in agroforestry in the years ahead is to reconcile the power of general principles in international science to the opportunism and flexibility of everyday, place-based popular sciences and practices. The corollary of that challenge is the need to recognize and clarify the multiple objectives and values at play, the points of convergence and conflict among them, and the explicit terms of collaboration within shifting and uneven fields of power.

REFERENCES

Abel, N. O. J., Drinkwater, M. J., Ingram, J., Okafor, J., and Prinsley, R.T., 1989, *Guidelines for Training in Rapid Appraisal for Agroforestry Research and Extension*, Commonwealth Science Council and Zimbabwe Forestry Commission, London and Harare.

Anderson, A., Ed., 1990, *Alternatives to Deforestation: Steps toward Sustainable Use of Amazon Rainforest,* Columbia University Press, New York.

Anderson, A., May, P., and Balick, M.,1991, *Subsidy from Nature: Palm Forest Peasantry and Development on an Amazon Frontier.* Columbia University Press, New York.

Arizpe, L., Paz, F., and Velasquez, M., 1993, *Culture and Global Change: Social Perceptions of Deforestation in the Lacondon Rain Forest.* Miguel Angel Porrua, Mexico D.F. (Spanish or English).

Arnold, J. E. M. and DeWees, P., 1997, *Farms, trees and farmers: responses to agricultural intensification,* Earthscan, London.

Ashby, J. "A., 1987, The effects of different types of farmer participation on the management of on-farm trials," *Ag. Adm. Ext.* 24: 235–252.

Ashby, J. A. and Sperling, L., 1995, "Institutionalizing participatory, client-driven research and technology development in agriculture," *Development and Change* 26 (4): 753–770.

Atta-Krah A. N. and Francis, P. A., 1987, "The role of on-farm trials in the evaluation of composite technologies: the case of alley farming in southern Nigeria," *Agro. Syst.* 23: 133–152.

Balee, W.,1989, "The Culture of Amazonian Forests," in *Advances in Eco. Bot.*, Vol. 7, New York Botanical Garden.

Barrow, E., 1992, *Pastoralist Resource Management Systems in Kenya.* African Center for Technology Studies, Nairobi.

Batterbury, S., Forsyth, T., and Thomson, K., 1997, "Environmental transformations in developing countries: Hybrid research and democratic policy," *Geog. J.*, 163 (2): 126–132.

Batterbury, S. and Warren, A., 1997, "Approaches to environmental history and agricultural change in sub-Saharan Africa," in *Proceedings, Workshop on Environmental and Economic Change: Understanding Dynamic Processes in sub-Saharan Africa,* Corbett, J., Ed.,. January 1996, University College London/CSERGE, London.

Bebbington, T., 1990, "Farmer knowledge, institutional resources and sustainable agricultural strategies: a case study from the eastern slopes of the Peruvian Andes," *Bull. Lat. Am. Res.* 9(2).

Biggs, S. D., 1988, *Resource-poor farmer participation in research: a synthesis of experiences from nine national agricultural research systems.* ISNAR, The Hague, Netherlands.

Bradley, P., 1991, "Women Woodfuel, and Woodlots," Vol.1 *The Foundation of a Woodfuel Development Strategy for East Africa.* MacMillan Education, London.

Brokensha, D., Warren, D. M., and Werner, O., Eds., 1980, *Indigenous Systems of Knowledge and Development,* University Press of America, Lanham, MD.

Brookfield, H. and Paduch, C., 1995, "Appreciating agrodiversity: A look at the dynamism and diversity of indigenous farming practices," *Env.* 36 (5) 6–11 and 37–43.

Bruce, J., 1989, *Community Forestry: Rapid Appraisal of Tree and Land Tenure.* FAO Rome.

Bruce, J., Fortmann L., and Nhira, C., 1993, "Tenures in transition, tenures in conflict: Examples from the Zimbabwe social forest," *Rur. Soc.* 58,26–642.

Bryant, R., and Bailey, S.,1997, *International Political Ecology,* Routledge, London.

Buck, L. E., 1993, "Development of participatory approaches for promoting agroforestry collaboration between the Mazingira Institute, ICRAF, CARE-Kenya, KEFRI and the Forest Department, 1980–1991," in *Non-Governmental Organizations and the State in Africa,* Wellard, K. and Cobestake, J. G. Eds., Routledge, New York, 118–135.

Buck, L. E., 1990, "Planning agroforestry extension projects: The Care International approach in Kenya," in *Planning for Agroforestry,* Budd, W. W., Duchhart, I., Jardesty, L., and Stein, F., Eds.,. Elsevier, New York. 101–131.

Buck, L. E., 1989, *Agroforestry Extension Training Sourcebook,* CARE International, New York.

Bunch, R., 1985, *Two ears of corn: a guide to people-centered agricultural improvement.* World Neighbors, Oklahoma City, OK.

Campbell, C., 1996, "Out on the front lines but still struggling for voice," in *Feminist Political Ecology,* Rocheleau, D., Thomas-Slayter, B., and Wangari, E. Eds., Routledge, London and New York, 27–61.

Carney, J., 1988, "Struggles over land and crops in an irrigated rice scheme: the Gambia," in *Women and Land Tenure in Africa,* Davison, J., Ed., Westview, Boulder, CO.

Carter, J., 1996, *Recent Approaches to Participatory Forest Resource Assessment. Rural Development Forestry Network,* Overseas Development Institute, London.

Chambers, R. and Jiggins, J., 1986. *Agricultural research for resource-poor farmers: A parsimonious paradigm.* Discussion Paper 220. University of Sussex, IDS.

Chambers, R.and Ghidyal, B. P., 1985, "Agricultural research for resource-poor farmers: The farmer-first-and-last model," *Ag. Adm. Ext.* 20: 1–30.

Chambers, R., 1989, "Farmer-first: a practical paradigm for the third agriculture," *in Agroecology and Small Farm Development,* Alteri, M. and Hecht, S., Eds., CRC Press, Boca Raton, FL.

Chambers, R., 1983, *Rural development: Putting the last first,* Longman, London.

Chambers, R., 1981, "Rapid rural appraisal: Rationale and repetoire," *Pub. Adm. Dev.,* 1(2): 95–106.

Chambers, R., Pacey, A., and Thrupp, L., Eds., 1989, *Farmer First: Farmer Innovation and Agricultural Research,* Intermediate Technology Publications, London. 218

Clarke, J., 1990, *On-farm agroforestry research lessons and innovations from a pilot project in Zimbabwe,* paper presented at ICR agroforestry workshop on participatory methods for on-farm agroforestry research. Feb. 19–23. Nairobi, Kenya.

Cline-Cole, R., 1996, "African and Africanist biodiversity research in a Neo-Liberal context," *Africa.* 66 1: 145–159.

Colfer, C.,1989, "Two Complementary Approaches to Farmer Involvement: An Experience from Indonesia," in *Farmer First: Farmer Innovation and Agricultural Research,* Chambers, R., Pacey, A., and Thrupp, L. A., Eds., Intermediate Technology Publications, London, 151–157.

Collinson, M. P., 1981, "The exploratory survey: content, methods and detailed guidelines for discussions with farmers," *CIMMYT Eastern Africa Economics Program Farming Systems Newsletter,* April–June, Nairobi, Kenya.

Collinson, M.P., 1981, "A low-cost approach to understanding small farmers," *Ag. Adm. Ext.* 8(6): 433–50.

Conway, G. R., 1987, "The properties of agroecosystems," *Ag. Syst.* 24: 95–117.

Conway, G. R., 1985, "Agroecosystem analysis," *Ag. Admin.* 20: 31–55

Cornwall, A., Guijt, I., and Welborn, A., 1994," in *Beyond Farmer First*, Scoones, I. and Thompson, J., Eds., Intermediate Technology Publications, London.

Davis-Case, D., 1989, *Community Forestry, Participatory Assessment, Monitoring and Evaluation*. FAO, Rome.

Denevan, W., Treacy, J., Alcorn, J., Paduch, C., Denslow, J. and Flores Paitan, S., 1985, "Indigenous agroforestry in the Peruvian Amazon: Bora Indian management of swidden fallows," in *Change in the Amazon Basin, Volume 1: Man's Impact on Forests and Rivers*, Hemming, J., Ed., Manchester University Press, Manchester, U.K., 137–155.

Desloges, C. and Gauthier, M., 1997, *Community forestry and forest resource conflicts: An overview*, paper presented at the 11th World Forestry Congress. Antalya, Turkey October 13–22. FAO, Rome.

DeWees, P., 1989, "The woodfuel crisis reconsidered: observations on the dynamics of abundance and scarcity," *World Dev.* 17(8): 1159–1172.

Eastman, R., 1997, *Idrisi for Windows, Version 2*. Clark University, Worcester, MA.

Edwards, R. J. A., 1987a, *Farmers' knowledge: utilization of farmers' soil and land classification*, paper presented at the IDS Workshop on Farmers and Agricultural Research: Complementary Methods, Institute of Development Studies, University of Sussex, U.K. July 26–31, 1987,

Edwards, R. J. A., 1987b, *Farmers' groups and panels*, paper presented at the IDS Workshop on Farmers and Agricultural Research: Complementary Methods, Institute of Development Studies, University of Sussex, U.K., July 26–31, 1987,

Edwards, R. J. A.,1987c, *Mapping and informal experimentation by farmers: agronomic monitoring of farmers' cropping systems*, paper presented at the IDS Workshop on Farmers and Agricultural Research: Complementary Methods, Institute of Development Studies, University of Sussex, U.K., July 26–31, 1987,

Escobar, A., 1995, *Encountering Development: The Making and Unmaking of the Third World*, Princeton University Press, Princeton NJ.

Fairhead, J. and Leach, M., 1996, "Re-thinking the forest-savanna mosaic: colonial science and its relics in West Africa," in *The Lie of the Land: Challenging Received Wisdom on the African Environment*, Leach, M. and Mearns, R., Eds., Heinemann, Portsmouth, NH, 105–121.

Feldstein, H., Rocheleau, D., and Buck, L.,1990, "The Siaya agroforestry project in Kenya," in *Working Together: Gender Analysis in Agricultural Research, Vol. I*, Feldstein, H. and Poats, S. Eds., Kumarian Press, Hartford, CT, 167–209.

Fernandes, E. C. M., O'Kting'ati, A., and Maghembe, J., 1984. "The Chagga home gardens: A multistoreyed agroforestry cropping system on Mount Kilimanjaro (Northern Tanzania)," *Agro. Syst.* 2:73-86.

Fernandez, M. and Salvatierra, H., 1989, "Participatory technology validation in highland communities in Peru," in *Farmer First*, Chambers, R., Pacey, A. and Thrupp, L. A., Eds., Intermediate Technology Publications, 146–150.

Feuerstein M., 1986, *Partners in Evaluation: Evaluating Development and Community Programs with Participants*, MacMillan. London.

Ford, R., Kabutha, C., and Thomas-Slayter, B., 1993, "Participatory rural appraisal: a case study from Kenya," in *Rapid Appraisal Methods,* Kumar, K. Ed., The World Bank. Washington, DC.

Fortmann, L. and Nabane, N., 1992, *Fruits of their labors: gender, property and trees in Mhondoro district*, Natural Resource Management Occasional Paper 7, Centre for Applied Social Sciences, Zimbabwe, Harare.

Fortmann, L., 1996, "Gendered knowledge, rights and space in two Zimbabwe villages," in *Feminist Political Ecology*, Rocheleau, D., Thomas-Slayter, B., and Wangari, E. Eds., Routledge, London.

Fortmann, L., 1995, "Talking claims: discursive strategies in contesting property," *World Dev.* 23, 1053–1064.

Fortmann, L. and Rocheleau, D., 1984, "Women and agroforestry: Four myths and three case studies," *Agro. Syst.* 2: 253–272

Freire, P.,1972. *Pedagogy of the Oppressed*, Penguin, Hammondsworth, U.K.

Gomez-Pampa, A., Salvador Fores, J., and Sosa,V., 1987, "The 'Pet Kot,: A man-made forest of the Maya," *Interciencia*. 12: 10–15.

Guijt, I. and Shah, M. K., Eds., 1997, *The Myth of Community: Gender Issues in Participatory Development*, Intermediate Technology Publications, London.

Gumbo, D. J., Mukamuri, B. B., Muzondo, M. I., and Scoones, I., 1988, "Indigenous and exotic fruit trees: Why do people want to grow them?" In *Proceedings of the Conference on the Economics of Agroforestry*. Mbabane, Swaziland. April 23–27, Commonwealth Science Council, London.

Gupta, A. K., 1989, "Scientists' views of farmers' practices in India: barriers to effective interaction," in *Farmer First: Farmer Innovation and Agricultural Research*, Chambers, R., Pacey, A., and Thrupp, L. A., Eds., Intermediate Technology Publications, London.

Haraway, D. J., 1991, *Simians, Cyborgs, and Women: The Reinvention of Nature*, Routledge, New York.

Harding, S., 1986, *The Science Question in Feminism*, Cornell University Press, Ithaca, NY.

Haugerud, A., 1986, "An anthropologist in an African research institute: An informal essay," *Bull. Inst. Dev. Anthro.*, 4: 2.

Hildebrand, P. E., 1981, "Combining disciplines in rapid appraisal: the Sondeo approach," *Ag. Admin.* 8: 423–432.

Hildebrand, P. and Poey, F., 1985, *On-farm Agronomic Trials in Farming Systems Research and Extension*, Lynne Reiner Publishers, Boulder, CO.

Hildebrand, P., Ed., 1986, *Perspectives on Farming Systems Research and Extension*, Lynne Reiner Publishers, Boulder, CO.

Hope, A. and Timmel, S., 1984, *Training for Transformation* (Books 1,2,3), Mambo Press, Zimbabwe.

Hoskins, M., 1982, *Social forestry in West Africa: myths and realities*, paper presented at the annual meeting of the American Association for the Advancement of Science, Washington, DC.

Hoskins, M., 1996, *Opening statement*, E-mail conference on Conflict Management and Community Forestry, FAO, Rome.

I.L.E.I.A., 1988a, "Towards sustainable agriculture, Part One: Abstracts, Periodicals, Organizations, Part Two: Bibliography," *ILEIA Newsletter*, 4 (2) May.

I.L.E.I.A., 1988b, "Participative technology development," ILEIA Newsletter 4 (3) October.

Jiggins, J., 1988, *Farmer participatory research and technology development*, Occasional Papers in Rural Extension, No 5, Department of Rural Extension Studies, University of Guelph, Guelph, ON, Canada.

Jiggins, J., 1986a, "Problems of understanding and communication at the interface of knowledge systems," in *Gender Issues in Farming Systems Research and Extension*, Poats, S., Schmink, M., and Spring, A., Eds.,. Westview Press, Boulder, CO.

Jiggins, J., 1986b, "Women and seasonality: Coping with crisis and calamity," in "Seasonality and Poverty," Longhurst, R., Ed.,. IDS *Bulletin* 17 (3): 9–18.

Jiggins, J. and Feldstein, H., 1994, *Tools for the Field: Methodologies for Gender Analysis in Agriculture*, Kumarian Press, Hartford, CT.

Juma, C., 1989, *Biological Diversity and Innovation: Conserving and Utilizing Genetic Resources in Kenya*, African Centre for Technology Studies, Nairobi, Kenya.

Kabutha, C., Thomas-Slayter, B., and Ford, R., 1991, *Assessing Mbusyani: using participatory rural appraisal for sustainable resource management*, Clark University, Worcester, MA.

Kandeh, H. B. S. and Richards, P., 1996, "Rural people as conservationists: Querying neo-Malthusian assumptions about biodiversity in Sierra Leone," *Africa*, 66 1: 91–103.

Katz, E., 1993, "Forest food resources in the tropical mountains of the Mixtec Highlands, Mexico," in *Tropical Forests, People and Food: Bicultural Interactions and Applications to Development*, Hladik, C. M., Hladik, A., Linares, O. F., Pagezy, H., Semple, A., and Hadley, M., Eds., UNESCO, Paris.

Kean, S., 1987, "Developing a partnership between farmers and scientists: The example of Zambia's adaptive research planning team," *Exp. Agri.* 24 (part 3): 289–299.

Khon Kaen University, 1987, *Proceedings of the 1985 International Conference on Rapid Rural Appraisal. Rural Systems Research and Farming Systems Research Projects*, Khon Kaen University, Khon Kaen, Thailand.

Kiriro, A. and Juma, C., Eds., 1989, *Gaining Ground: Institutional Innovations in Land Use Management in Kenya*, African Centre for Technology Studies, Nairobi, Kenya.

Kirkoff, P., 1990, *Agroforestry in Africa: a Survey of Project Experience*, Panos, London.

Korten, D., 1990, *Getting to the 21st Century: Voluntary Action and the Global Agenda*, Kumarina Press, West Hartford, CT.

Langley, P., 1986, *Popular Participation as a Cargo Cult?*, ENDA African Environment Occasional Paper Series No. 95–86.

Leach, M. and Mearns, R., Eds., 1996, *The Lie of the Land: Challenging Received Wisdom on the African Environment*, Heinemann, Portsmouth, NH.

Lightfoot, C., de Guia O., Jr., and Ocado, F., 1988, "A participatory method for systems-problem research rehabilitating marginal uplands in the Philippines," *Exp. Agri.* 24 (part 3): 301–309.

Lightfoot, C., de Guia O., Jr., Aliman, A., and Ocado, F., 1989, "Systems diagrams to help farmers decide in on-farm research," in *Farmer First: Farmer Innovation and Agricultural Research*, Chambers, R., Pacey, A., and Thrupp, L. A. Eds., Intermediate Technology Publications. London.

Lundgren, B., 1987, "ICRAF's First Ten Years," *Agro. Syst.* 5,3: 197–217.

Marglin, F. and Marglin, S., Eds., 1990, *Dominating Knowledge: Development, Culture, Resistance*, Clarendon Press, Oxford. U.K.

Matose, F., 1993, *Gender and local knowledge of trees and crops in a pilot project in Zimbabwe*, paper presented to the Sixth International Forum of the Association for Women in Development, October 21–24, Washington, DC.

Maundu Munyao, P., 1992, *Biodiversity and wild foods*, manuscript, Indigenous Food Plants Program, National Museums of Kenya, Nairobi, Kenya.

Maurya, D. M., Bottrall, A., and Farrington, J., 1988, "Improved livelihoods, genetic diversity and farmer participation: A strategy for rice-breeding in rainfed areas of India," *Exp. Agri.* 24 (part 3): 311–320.

Mbogoh, S. G., 1991, "Crop production," In *Environmental change and dryland management in Machakos District, Kenya, 1930–1990: production profile*, Tiffen, M., Ed., 1–44. (ODI Working Paper 55). Overseas Development Institute, London.

McCabe, J. T., 1990, "Turkana pastoralism: A case against the tragedy of the commons," *Hum. Ecol.*, 18 (1): 81–104.

McConnell, W., 1992, *Local Ecological Knowledge and Environmental Management in the Republic of Mali*, M.A Thesis. Program in International Development and Social Change. Clark University, Worcester, MA.

McCracken, J. A., 1988, *Participatory rapid appraisal in Gujarat: a trial model for the Aga Khan Rural Support Program (India)*, November, International Institute for Environment and Development, London.

Michon, G., Mary, F., and Bompard, J., 1989, "Multistoreyed agroforestry garden system in West Sumatra, Indonesia," in *Agroforestry Systems in the Tropics*, Nair, P. K. R., Ed., 1989, Kluwer Academic Publishers, Boston, 243-268.

Molnar, A., 1989, *Community forestry rapid appraisal*, FAO, Rome.

Muller, E. U. and Scherr, S. J., 1990, "Planning technical interventions in agroforestry projects," *Agro. Syst.*, 12: 23–44.

Nair, P. K. R. Ed., 1989, *Agroforestry Systems in the Tropics*. Kluwer Academic Publishers, Boston.

National Environment Secretariat, Egerton University, Clark University, World Resources Institute (NES et al.), 1991, *Participatory Rural Appraisal Handbook: Conducting PRA's in Kenya*, World Resources Institute, Washington, DC.

Norman, D., Baker, D., Heinrich, G., and Worman, F., 1988, "Technology development and farmer groups: experiences from Botswana," *Exp. Agri.* 24 (part 3): 321–331.

Oakley, P., 1987, "State or process, means or end? The concept of participation in rural development ideas," *IDRC Bulletin*, March.

Oasa, E. and Jennings, B., 1982, "Science and authority in international agricultural research," *J. Conc. Asian Schol.*, 14 (4): 30–44.

Oldfield, M., and Alcorn, J. B., 1991, *Biodiversity: Culture, Conservation and Ecodevelopment*, Westview Press, Boulder, CO.

Peet, R. and Watts, M., 1997, *Liberation Ecologies*, Routledge. London.

Peluso, N., 1992, *Rich Forests, Poor People*, University of California Press, Berkeley, CA.

Peluso, N., 1995, "Whose woods are these? politics of mapping forests in Kalimantan," *Antipode*. Vol 27. No. 4 383–406.

Poats, S., Feldstein, H. and Rocheleau, D., 1989, "Gender and intra-household analysis in on-farm research and experimentation," in *The Household Economy: Reconsidering the Domestic Mode of Production*, Westview Press, Boulder, CO.

Poats, S., Schmink, M., and Spring, A., and Wilk, R., Eds., 1988, *Gender Issues in Farming Systems Research and Extension*, Westview Press, Boulder. CO

Poffenberger, M., 1990, *Keepers of the Forest: Land Management Alternatives in Southeast Asia*, Kumarian Press, West Hartford, CT.

Poole, P., Ed., 1985, "Geomatics: Who needs it?" Special issue of *Cultural Survival Quarterly*. 18 (4): 1–76.

Posey, D., 1985, "Indigenous management of tropical forest ecosystems: The case of the Kayapo Indians of the Brazilian Amazon," *Agro. Syst.* 3(2):139–158.

Raintree, J. B., 1983, "Strategies for enhancing the adoptibility of agroforestry innovations," *Agro. Syst.* 1(3): 173–188.

Raintree, J.B., 1987b, "The state of the art of agroforestry diagnosis and design," *Agro. Syst.* 5 (3): 219–250.

Raintree, J. B., 1987a, *D & D user's manual: an introduction to agroforestry diagnosis and design*, ICRAF, Nairobi, Kenya.

Rhoades, R., 1989, "The role of farmers in the creation of agricultural technology," in *Farmer First: Farmer Innovation and Agricultural Research*, Chambers, R., Pacey, A., and Thrupp, L.A., Eds., Intermediate Technology Publications. London.

Rhoades, R.,and Booth, R., 1982, "Farmer-back-to-farmer: a model for generating acceptable agricultural technology," *Ag. Admin.* 11:127–137.

Richards, H., 1985, *The Evaluation of Cultural Action*, International Development Research Centre, Ottawa, ON, Canada.

Richards, P., 1989, "Agriculture as performance," in *Farmer First: Farmer Innovation and Agricultural Research*, Chambers, R., Pacey, A., and Thrupp, L.A., Eds., Intermediate Technology Publications. London.

Richards, P., 1985, *Indigenous Agricultural Revolution: Ecology and Food Production in West Africa*, Hutchinson, London.

Rocheleau, D., 1997, "Sustaining what for whom? Differences of interest within and between households," in *Sustainable Resource Management in Central America*, Hatch, U. and Swisher, M., Eds., Oxford University Press. Oxford, U.K.

Rocheleau, D., 1995, "Maps, numbers, text and context: Mixing methods in feminist political ecology," *Professional Geographer.* 47 (4): 458–466.

Rocheleau, D., 1994, "Participatory research and the race to save the planet: Questions, critique, and lessons from the field," *Ag. Hum. Val.* 11 (2/3): 4–25.

Rocheleau, D., 1991a, "Participatory research in agroforestry: learning from experience and expanding our repertoire," *Agro. Syst.* 12 (2) 111–138.

Rocheleau, D., 1991b, "Gender, ecology and the science of survival: Stories and lessons from Kenya," *Ag. Hum. Val.* 8 (1): 156–65.

Rocheleau, D., 1988a, "Gender, resource management and the rural landscape: Implications for agroforestry and farming systems research," in *Gender Issues in Farming Systems Research and Extension*, Poats, S., Schmink, M., and Spring, A., Eds., Westview Press, Boulder. CO.

Rocheleau, D., 1988b, "Women, trees, and tenure: implications for agroforestry," in *Whose Trees? Proprietary Dimensions of Forestry*, Fortmann, L. and Bruce, J., Eds., Westview Press, Boulder, CO.

Rocheleau, D., 1987, "The user perspective and the agroforestry research and action agenda," in *Agroforestry: Realities, Possibilities and Potentials*, Gholz, H., Ed., Martinus Nijhoff, Doerdrecht, Netherlands.

Rocheleau, D., 1985, *Criteria for reappraisal and redesign: intra-household and between-household aspects of FSRE in three Kenyan agroforestry projects*, paper presented at the Annual Farming Systems Research and Extension Symposium, Kansas State University, Manhattan, KS, Oct. 7–10, 1984, ICR agroforestry Working Paper No. 37. ICRAF. Nairobi, Kenya.

Rocheleau, D., 1984, *An Ecological Analysis of Soil and Water Conservation in Hillslope Farming Systems: Plan Sierra, Dominican Republic.* Ph. D. Dissertation. Department of Geography, University of Florida, Gainseville, FL.

Rocheleau, D., Field-Juma, A., Benjamin, P., and Malaret, L. (tentative), *Forthcoming. Designs on Their Land: Social Relations and the Changing Ecologies of Ukambani, 1890–1990,* UNU Press, Tokyo.

Rocheleau, D. and Edmunds, D., 1997a, "Women, men and trees: Gender, power and property in forest and agrarian landscapes," *World Dev.* 25 (8) 1351–1371.

Rocheleau, D., Ross, L., and Morrobel, J., 1997b, *Forests, trees and fields of power: makings and takings of biodiversity in the regional agroforests of Zambrana-Chacuey, Dominican Republic*, paper presented to the Agrarian Studies Seminar, Yale University, New Haven, CT, 45

Rocheleau, D., Ross, L., and Morrobel, J. 1997c, "Gendered landscapes, gendered lives: maps and life histories from Zambrana-Chacuey, Dominican Republic," in *The Myth of Community: Gender Issues in Participatory Development,* Guijt, I. and Shah, K.M., Eds., Intermediate Technology Press, London.

Rocheleau, D., Ross, L., Morrobel, J., and Hernandez, R., 1996a, *Farming in the forest, gardening with trees: gendered landscapes and livelihoods in Zambrana-Chacuey, Dominican Republic. ECOGERN Case Study,* Clark University, Worcester, MA.

Rocheleau, D., Ross, L., and Morrobel, J., 1996b, "From forest gardens to tree farms: Women, men and timber in Zambrana-Chacuey, Dominican Republic," *in Feminist Political Ecology: Global Issues and Local Experiences*, Rocheleau, D., Thomas-Slayter, B., and Wangari, E., Eds., Routlege, London.

Rocheleau, D., Thomas-Slayter, B., and Edmunds, D., 1995, "Gendered Resource Mapping," *Cultural Survival Quarterly*, 18 (4): 62–68.

Rocheleau, D., Thomas-Slayter, B., and Wangari, E., Eds., 1996c, *Feminist Political Ecology: Global Perspectives and Local Experiences.* Routledge, London.

Rocheleau, D., Thomas-Slayter, B., and Edmunds, E., 1995, "Gendered Resource Mapping: Focusing on Women's Spaces in the Landscape," *Cultural Survival Quarterly.* 18 (4): 62–68.

Rocheleau, D. and Ross, L., 1995, "Trees as tools, trees as text: struggles over resources in Zambrana-Chacuey, Dominican Republic," *Antipode.* 27 (4): 407–428.

Rocheleau, D. E., and Slocum, R., 1995, "Participation in context: key questions," in *Power, Process and Participation: Tools for Change*, Slocum, R., Wichart, L., Rocheleau, D., and Thomas-Slayter, B., Eds., Intermediate Technology Press, London. 17–30.

Rocheleau, D., Steinberg, P., and Benjamin, P., 1995, "Environment, development, crisis and crusade: Ukambani, Kenya, 1890–1990," *World Dev.* 23 (6): 1037–1051.

Rocheleau, D., Wachira, K., Malaret, L., and Wanjohi, B., 1989, "Ethnoecological methods to complement local knowledge and farmer innovations in agroforestry," in *Farmer First: Farmer Innovation and Agricultural Research*, Chambers, R., Pacey, A., and Thrupp, L.A., Eds., Intermediate Technology Publications, London.

Rocheleau, D., Weber, F., and Field-Juma, A., 1988, *Agroforestry in Dryland Africa.* ICRAF, Nairobi, Kenya, 313.

Rocheleau, D., and Malaret, L., 1987, "Use of ethnoecology in agroforestry systems research," in University of Arkansas and Winrock International. How systems work," *Selected proceedings of the Farming Systems Research Symposium*, University of Arkansas and Winrock International, Fayetteville and Morrillton, A. R.

"RRA Notes, 1988–present," *IIED Newsletter*, Sustainable Agriculture Program. London.

Sachs, W., 1993, *Global Ecology: A new arena for international conflict*, Zed Press, London.

Sarin, M., 1996, *Case studies of gender, class and ethnic conflicts in joint forest management*, paper presented to the electronic conference on Conflict Resolution and Community Forestry, FAO Community Forestry Program, Rome.

Scherr, S. J., 1990, "The diagnosis and design approach to agroforestry project planning and implementation: examples from western Kenya," in *Planning for Agroforestry*, Budd, W.W., Duchhart, I., Hardesty, L. H., and Steiner, F. Eds., Elsevier, 132–160.

Scherr, S. J., Roger, J. H., and Odoul, P. A., 1990, "Surveying farmers' agroforestry plots: Experiences in evaluating alley-cropping and tree border technologies in western Kenya," *Agro. Syst.*, 11: 141–173.

Scherr, S. J., 1989, "Choosing priorities for agroforestry research," in *Social Science Perspectives in Managing Agricultural Technology*, Groenfeldt, D. and Moock, J., Eds., International Irrigation Management Institute and Rockefeller Foundation, New York.

Scherr, S. J., 1987, "Agroforestry research formulation: A case study of the initial stages from Cameroon," in *Multipurpose Trees: Selection and Testing for Agroforestry*, Huxley, P. and Westley, S. Eds., ICRAF, Nairobi, Kenya, 61–73.

Schmink, M. and Wood, C., 1992, *Contested Frontiers in Amazonia.* Columbia University Press. New York.

Schroeder, R. and Suryanata, K., 1997, "Case studies and class power in agroforestry systems," in *Liberation Ecologies*, Peet R., and Watts, M., Eds., Routledge, London.

Schroeder, R., 1993, "Shady practices: Gender and the political ecology of resource stabilization in Gambian garden/orchards," *Economic Geography.* 69 (4) 349–365.

Scoones, I. and Thompson, J., 1994, *Beyond Farmer First: Rural People's Knowledge, Agricultural Research and Extension Practice*, Intermediate Technology Publications, London.

Scoones, I., and McCracken, J., Eds., 1989, *Participatory rapid rural appraisal in Wollo: Peasant association planning for natural resource management*, Ethiopian Red Cross Society UMCC-DPP and IIED, London.

Scoones, I., 1988, *Community Management of Indigenous Woodland Project: Zvishavane and Chivi Districts*, ENDA, 59, Harare, Zimbabwe.

Scoones, I., Ed., 1989, *Participatory research for rural development in Zimbabwe: a report of a training workshop for ENDA-Zimbabwe Trees Project*, IIED, London.

Seitz, V., 1993, *Gender and rural organization in a pilot project in Zimbabwe*, paper presented at the Sixth International Forum of the Association of Women in Development, October 21–24. Washington, DC.

Shaner, W. W., Philipp, P. F., and Schmehl, W. R., 1982, *Farming Systems Research and Development: Guidelines for Developing Countries,* Westview Press, Boulder. CO

Shields, M. D., Butler, F., Thomas-Slayter, B., and Buenavista, G., 1996, "Developing and Dismantling Social Capital," in *Feminist Political Ecology,* Rocheleau, D. THomas-Slayter, B. and Wangari, E., Eds., Routledge, London. 155–179.

Shiva,V. 1988, *Staying Alive: Women, Ecology and Survival,* Zed Press. London.

Showers, K., 1989, "Soil erosion in the Kingdom of Lesotho: origins and colonial response, 1930s–1950s," *J. S. Afri. Stud.* 15, 263–86.

Slocum, R., Wichart,L., Rocheleau, D., and Thomas-Slayter, B., 1995, *Power, Process and Participation: Tools for Change,* Intermediate Technology, London.

Sperling, L., 1994, "Methods for integrating farmer experts into on-farm research," in *Tools for the Field: Methodologies for Gender Analysis in Agriculture*, Jiggins, J. and Feldstein, H. Eds., Kumarian Press, Hartford, CT.

Stamp, P., 1989, *Technology, Gender and Power in Africa*, International Development Research Centre and Kumarian Press. Ottawa, ON, Canada, and West Hartford, CT.

Stonich, Susan C., 1993, *"I Am Destroying the Land!": The Political Ecology of Poverty and Environmental Destruction in Honduras*, Westview Press, Boulder. CO.

Sumberg, J., and Okali, C., 1989, "Farmers, on-farm research, and new technology," in *Farmer First: Farmer Innovation and Agricultural Research*, Chambers, R., Pacey, A., and Thrupp, L.A., Eds., Intermediate Technology Publications, London.

Sutherland, A. J., 1987, *The benefits of adopting a community approach to farmer selection. Workshop on Household Issues in Farming Systems Research.* Lusaka, Zambia, April 27–30, CIMMYT.

Thomas-Slayter, B., Esser, A., and Shields, D., 1993, "Tools of Gender Analysis" *ECOGEN Working Paper*, Clark University. Worcester. MA.

Thrupp, L.A., 1989, "Legitimizing local knowledge: From displacement to empowerment for Third World people," *Ag. Hum. Val.* 6(3):13–24.

Tiffen, M., Mortimore, M., and Gichuki, F., 1994, *More People, Less Erosion: Environmental Recovery in Kenya*, John Wiley & Sons, Chichester, U.K.

Townsend, J., 1995, *Women's Voices from the Rainforest*, Routledge, London.

Townsend, J., 1993, "Housewifization in the Colombian rainforest," in *Different Places, Different Voices*, Momsen, J. and Kinnaired, V., Eds.,Routledge, London, 270–277.

Tripp, R., 1984, "On-farm research and applied nutrition: some suggestions for collaboration between national institutes of nutrition and agricultural research," *Food and Nutrition Bulletin.* 6 (3).

Uphoff, N., 1992, *Learning from Gal Oya: Possibilities for Participatory Development and Post-Newtonian Social Science*, Cornell University Press, Ithaca, NY.

Van Veldhuizen, L., Waters-Bayer, A., Ramirez,R., Johnson, D., and Thompson, J., Eds., 1997, *Farmers' Research in Practice: Lessons from the Field,* Intermediate Technology Publications, London.

Warren, D. M., Slikkerveer, L. J., and Brokensha. D., Eds., 1994, *Indigenous Knowledge Systems: The Cultural Dimension of Development*, Intermediate Technology Publications, London.

Wilken, G., 1988, *Good Farmers: Traditional Agricultural Resource Management in Mexico and Central America*, University of California Press, Berkeley, CA.

Wilson, K. B., 1987, *Research on trees in the Mazvihwa and surrounding areas*, unpublished report, ENDA, Harare, Zimbabwe.

Wilson, K. B., 1988, *Community management of indigenous woodland project: report to ENDA Zimbabwe.* ENDA, Harare, Zimbabwe.

Wilson, K. B., 1989, "Trees in fields in southern Zimbabwe," *J. So. Af. Stud.*, 15 (2): 369–383.

Wisner, B., 1990, *Power and Need in Africa*, Earthscan and Africa World Press, London and Trenton, NJ.

Woods, B., 1983, *Altering the present paradigm: a different path to sustainable development in the rural sector*, unpublished paper, The World Bank, Washington, DC.

Yapa, L., 1991, "Is GIS appropriate technology?" *Int. J. Geo. Info. Sys.* 5 (1): 41–58.

Zazueta, A., Cabarle, B., and Thrupp, L. A., 1994, "Linking participatory methods with policy dialogue and planning processes," in *Beyond Farmer First*, Scoones, I., and Thompson, J., Eds., Intermediate Techology Publications, London.

10 Contemporary Uses of Tree Tenure*

John W. Bruce and Louise Fortmann

1. INTRODUCTION

In his imagined account of the origins of property John Locke (1983) focuses initially on the creation of property rights in specific resources — acorns, apples, grass, turf, ore, and water — before he moves to consider rights in "the Earth itself," that is property rights to land. While Locke may or may not have intended this simply as a rhetorical device, it does in fact describe the separability of property rights found in practice. William Cronon (1983), for example, describes a system of resource-based property rights (clam banks, fishing ponds, berry picking areas, hunting lands) among the Native Americans of early colonial New England, noting

What the Indians owned — or, more precisely, what their villages gave them claim to — was not the land but the things that were on the land during the various seasons of the year.

In contrast, property rights in land are clearly privileged in the prevailing popular property discourse in the United States. That is, when people speak of "property," they tend to mean land. Anglo- American and European law assumes that a tree belongs to the owner of the land on which it stands and trees are treated as part of the "land."

In recent years property scholars have recognized the importance of separable rights in trees (Fortmann, 1985; Fortmann and Bruce, 1991). Attention to property in the form of tree rights reveals a remarkable variety in prevailing contemporary practices in which property rights in trees emerge as potential flexible and powerful policy instruments. Continuing from these insights, we have begun to explore the question of the circumstances under which rights in trees, as distinct from rights in land, emerge and disappear. In this chapter we examine cases ranging from peasant subsistence production to international corporate practice in which separable tree rights have significant practical importance with particular attention to their equity and policy implications. These cases fall into three general types: cases in which tree rights substitute for land rights, cases

* We are grateful to Russell Henly, California Department of Forestry and Fire Protection and Larry Ruth, College of Natural Resource, University of California at Berkeley, for their information on California forest practice regulations, to John Bliss, Auburn University, for providing Alabama forest statistics, and to Donald Moore for sharing his library.

1-56670-294-1/99/$0.00+$.50

in which tree rights are used to mobilize resources beyond those under the direct control of the land owner, and cases in which the state excises tree rights from the more general bundle of private land rights in order to protect the public interest. Our goal is to illuminate the ways in which tree tenure can or does provide a policy lever to encourage and facilitate the adoption and maintenance of socio-economic relations and technical practices that lead to sustainable agroforestry. Since some aspects of tree tenure are most clearly illustrated by examples from forestry, we have not limited our discussion to cases from agroforestry. The relevance of these examples to the more general question of sustainable agroforestry is summarized in the conclusions.

2. TREE RIGHTS SUBSTITUTE FOR LAND RIGHTS

Despite the general primacy of land rights, in at least three kinds of circumstances tree rights are more important than (or substitute for) land rights, namely when users don't need, can't get, or don't want property in land.

2.1 Users Do Not Need Land Rights

Tree rights are adequate and land rights unnecessary if the user only needs secure usufructuary rights to products from existing trees for harvesting. Such rights might be tied to general territorial rights of a group (e.g., Cronon, 1983) but rights to use trees are appropriated independently of any parcel rights to land. It is likely that property rights develop in trees first where land is plentiful and trees are the more scarce good. Contemporary examples include arrangements on Kalimantan where inheritance among children and cousins of usufructuary rights in fruit trees, especially durian, remains important in lieu of land ownership for the purpose of access to fruit and helps to maintain kin networks (Peluso, 1996). As such rights are distributed on the basis of kinship only, they may be of help to poorer relations within a kinship group. This safety net role of a particular tree product might be utilized on a broader scale, beyond a kinship group.

2.2 When Tree Producers Cannot Get Secure Access to Land

For disadvantaged groups or individuals who do not have access to land rights, such as wives, tenants, strangers and junior or former slave lineages, tenure in trees may be a substitute for rights to land. The separation of rights in trees from rights in land is also central to major community forestry programs developed in Asia, Africa and Latin America in the past decade.

In India, for example, the Joint Forest Management (JFM) Program seeks community participation in the afforestation of degraded state forest reserve land (Lindsay, 1994; Hobley, 1995). The program is sponsored by the central government but implemented by the states, and there are important variations from one state to another. The land is allocated (usually for the life of a tree crop) to a local village forestry committee. This often involves returning lands to local communities which utilized them before they were converted to forest reserves, and may indeed have continued to use them on an illegal basis in the years after reservation. The village forestry committee agrees on a plan of forest management with the Forestry Department. While the state retains the ownership of the land, the committee is given rights to manage the trees and take non-timber products from the land, subject to important restrictions such as prohibiting taking livestock into the forest.

Those who have sought to enhance community empowerment under the JFM program have not concentrated on a transfer of land ownership to communities. The state is reluctant to make such a transfer because the state's continuing ability to control use of the land allows it to enforce its requirement that the land be kept in trees, rather than agricultural crops which local people might find more profitable. It is also the threat which ensures community compliance with the forest management plan. Instead, there have been attempts to increase the level of community control over the planting and disposal of trees, and the objective is often stated as community "ownership" of the trees. These efforts are at very different stages in different states, but consistently it is tree rights which are being re-enforced to give local committees strong incentives, while land ownership provides the state with the security it requires. It is an approach which has made possible new access to forest land for the rural poor.*

2.3 When Tree-Producers Do Not Want Land Rights

Owning land can be an enormous practical and financial burden, so some clever resource users have found ways to do without it. Margaret FitzSimmons (1983) has traced the phenomenon of large agricultural capital freeing itself of "the costs and impediments of land ownership" in California's Salinas Valley. The effects of the resulting agricultural practices are significant in this case. Up to the end of World War II land owners were able to require their tenants (who had long leases) to practice crop rotation and green manuring. But as leases began to shift from shares of the crop to cash rents, the balance of power shifted, and with it environmentally friendly agricultural practices disappeared (FitzSimmons, 1983).

The parallel cases of capital freeing itself of the inconvenience of property in forestry are striking. Here too are investors who do not want the inconvenience of owning land and raising trees, but prefer to buy tree rights and harvest from both public and private land. Timber concessions on public land anywhere from Bangladesh to Indonesia to the Pacific Northwest are prime examples. The state bears most of the management burden — reforestation, pre-commercial thinning, fire suppression, and policing. The responsibility of the timber company varies from place to place but in practice may be as little as simply harvesting the logs from the land for sale (Hong, 1987; Broad, 1993).

U.S. forest products companies also rely on private smallholder timber land. For example, as timber companies have shifted back to the Southeast from the Pacific Northwest, they have moved into the poorest counties where their participation in the pulp and paper industry relies in part on the land of others. For example, in Alabama only 22% of the forest land is actually owned by the forest products industry with another 3% being leased. The forest products industry is heavily dependent on 70% of forest land owned by "other private" owners (Vissage and Miller, 1991).

These practices have clear equity implications. By leasing forest land for harvesting or buying standing timber off the land of others, forest products companies are able to shift the bulk of the risk of tree crop loss (from insect, disease, storm, drought and fire damage) and of ecological degradation (from poor harvesting practices and repeated monoculture rotations to private land owners and the public in the form of forest land-owning agencies, such as the U.S. Forest Service and the Bureau of Land Management.

3. RESOURCE MOBILIZATION

Land owners often need to mobilize more resources than they themselves control in order to grow trees (either the initial planting or on-going maintenance). This need arises in the case of trees because their planting involves a cost of obtaining the seedling, considerable opportunity costs, and maintenance costs, none of which can be recouped until the tree is grown and can be cut or begins to produce. By separating tree ownership from land ownership and assigning some of the rights from the ownership bundle to others, it is possible to mobilize their resources to meet the financial needs of this period. This may involve assigning them a share interest in current production, or all interests for a limited period of time, or a future, contingent interest in the trees.

Along the Nile in northern Sudan, in a practice still current but first reported in the early 20th century (Leach, 1919), date palms are owned separately from the land.** The practice makes it

* Extractive reserves in Latin America are another example of this approach although proposed reserves are sometimes more trendy than they are useful in actual practice (Hecht and Cockburn, 1990).

possible to create combinations of land in which to plant the seedling, the capital for the purchase of the seedling, and the water to nurture it. One individual is the owner of the seedling. It may be planted on the land of another. Water comes from a *sagia* (water lift) owned by a third party. If these are three different individuals, as is in fact often the case, the tree becomes owned in three equal shares by the parties. The difference between ownership of the tree and the land may further diverge as the result of the sale of an interest in the tree or in the process of inheritance of the land and tree. This shared ownership is reflected in the sharing-out of the crop of dates. This mechanism of joint tree ownership appears to ease the financial burden of planting and caring for the tree until it begins to produce.

Cocoa production provides a second example. The introduction of cocoa cultivation into West Africa created a dynamic market for the crop in the early part of the 20th century, and cultivation was spread inland by migrant farmers from coastal ethnic groups. This led to gradual changes in land tenure patterns and, in some cases, to the recognition of rights in trees separate from those in land. It is revealing to compare the cases of southern Ghana and southwestern Nigeria.

In southern Ghana, land under the traditional dispensation was controlled by sub-chiefs who sold the land to those who wished to establish cocoa farms. Sales of land had been practiced between sub-chiefs prior to the introduction of cocoa but the need for land for this commodity appears to have hastened the development of the land sale concept. Cocoa farmers did own their land, and mortgaged their farms when in need of capital for expansion (Hill, 1963). In this context of cocoa production, trees were treated as fixtures, objects whose ownership went with the ownership of the land.

In southwest Nigeria, land was, by contrast, controlled by lineages, not chiefs or sub-chiefs, and land acquisition by migrating cocoa farmers followed the model of "senior" (first arrived) landowning lineages making land available to recently arrived families which in time developed into "junior" lineages. The cocoa farmers were "land borrowers" and held their land initially subject to token payments to the senior lineages, acknowledging their ultimate control of the land; later these payments developed into more economic rents, and the relationships came to resemble tenancies. Here it was clear that the immigrants acquired no alienable interest in the land, and they could not borrow against the land. It became understood, however, that they did own their crops, their trees, and that these could be mortgaged. Pledging of personal property and slaves had existed earlier, and this model was now applied to cocoa trees as cocoa farmers sought the capital to expand their ventures (Adegboye, 1969; Berry, 1993).

Tree-ownership thus allowed a cocoa farmer to generate capital by taking the future interest "sticks" out of the "bundle of rights" constituting ownership of trees, and passing them to the lenders to secure the loans. Lack of a mortgageable title to land made it necessary to rely on trees, and sharpened what customarily may have a been a tradition that land-borrowers own their crops (an ethic of non-interference by the land-owning lineage) to a positive right of ownership which was not only secure but inalienable. Lack of secure land tenure led to reliance upon tree tenure rights to provide incentives for the cocoa planters, and this led to the use of the trees themselves as collateral for loans to expand production.

In upland Java in Indonesia, separating ownership of apple trees from that of land both eases the financial burden of getting the trees planted and allows the mobilization of finance invested in the orchards to meet urgent needs. In upland Java, fruit-based agroforestry has spread rapidly in the past two decades, and the tenure innovations noted here belong to the same period. One can now acquire rights in apple trees by either leasing or sharecropping of trees. Sharecropping occurs when a landowner who does not have the wherewithal to institute apple cultivation joins efforts with an apple producer, who plants the trees and provides the owner with a share of the product of the trees. Alternatively, apple trees may be leased, an arrangement in which capital-rich growers lease apple-trees from land-owning poor peasants. The peasants usually lease out the trees because

** Similar arrangements are also currently practiced with date trees in Morocco (personal communication Hsaïn Ilahiane).

of a pressing need for cash related to some expensive life event. This is preferable to leasing out the land with trees because it allows the peasant landowner to continue to make other subsistence uses of the land (Suryanata, 1994).

These arrangements are advantageous in that they ease the financial burdens of instituting and maintaining tree cultivation until production of fruit begins. They are particularly important when international or regional market forces are pushing the rapid expansion of new tree crops. But they also have proven to have important equity implications. In the Nigerian case of loans against cocoa trees, the borrower sometimes (unbeknownst to the family) has no intention of redeeming the trees and loses them to the creditor in the end (Berry, 1993). In the case from Java, landowners who have leased their trees often become paid laborers for the apple-producer, with the producer assuming a dominant role in the bargaining relationship, taking an increasing share in the revenue, and gradually suppressing subsistence production of other crops on the land.

4. PROTECTING THE PUBLIC INTEREST

One means the state uses to protect the public interest is excising tree rights from the more general bundle of private land rights. This may be deemed to be particularly necessary when forestland is held by several owners because practices on the holding of one owner may put trees on the holdings of other owners in danger.

State regulation of trees and forests in the U.S. state of California provide an excellent example. Because of the extreme danger fire poses in California's Mediterranean ecosystem, the California Public Resources Code PRC §4117 (State of California, 1997b) provides that local governments may

> ... adopt ordinances, rules, or regulations to provide fire prevention restrictions or regulations that are necessary to meet local conditions of weather, vegetation, or other fire hazards. Such ordinances, rules, or regulations may be more restrictive than state statutes in order to meet local fire hazard conditions.

Such measures include requiring private land owners to allow prescribed burns on their land "if the department has determined that burning of such vegetation is necessary for the prevention or suppression of forest fires" (State of California, 1997b §4118), prohibiting uncontrolled burning on private land adjacent to forest-covered land (State of California, 1997b §4422), requiring private land owners to allow access to their land for the purpose of controlling and extinguishing uncontrolled fires (State of California, 1997b §4170.5), and requiring a written permit for burning on private land adjacent to forest-covered land (State of California, 1997b §4423). That is, the state has removed the right to burn trees and forests from the bundle of tree tenure. Similarly, because of local historical dependence on the timber industry, the state requires private forest landowners and land managers to eradicate "pine beetles and other insect pests or plant diseases which are harmful, detrimental and injurious to timber and forest growth" on private land (State of California, 1997b §4713, §4714) and if such an infestation is "of such a character as to be a menace to the timber or timberlands of adjacent owners," state officials "may go upon ... private lands ... and shall cause the infestation or infection to be eradicated or controlled ..." (State of California, 1997b §4716). That is, freedom of choice in the management of trees which might pose a danger to other trees due to insects or disease is also removed from the tree tenure bundle.

In order to preserve the public interest in healthy forest ecosystems, specific habitats (old growth and riparian in particular), and soil and water conservation, the state of California has left only limited rights to harvest timber in the tree tenure bundle. The California Forest Practices Act requires a written Timber Harvesting Plan to be approved for most commercial timber harvesting (State of California, 1997a). The California Forest Practice Rules limit the size of allowable clear cuts, require restocking of harvested areas, regulate road construction, and restrict operations on steep slopes, in and near water courses, lakes and unstable soils, and in any area that would disturb, threaten or damage rare, threatened or endangered species of plants or animals (State of California, 1997a).

In addition to protecting forests in general, it is the policy of the State of California to preserve particular species and specific forest structures, in particular late succession/old growth, through regulation of timber harvesting on private land. In the event of timber harvesting, the maintenance of native species on private land is protected through three requirements:

> (1) Local seeds from state-specified seed zones must be used in restocking. (2) No high grading is allowed. Forest practices for natural regeneration require retaining good seed trees in any situation requiring natural regeneration. (3) Replanting has to retain a certain mix of species native to the site. The operator can't markedly increase the number of trees of a non-indigenous species out of its range — such as Giant Sequoia. (Menning et al., 1997).

In regard to forest structure, the state requires that for any late succession/old growth forest stand 20 acres or larger, long-term adverse effects on fish, wildlife and species listed under the Endangered Species Act "known to be associated with late succession/old growth forests *must* be identified and mitigated" (Menning et al., 1997). The state also removes timber harvesting from the bundle of rights on private land in order to protect the nesting sites of specific bird species, such as the Golden and Bald Eagles (Menning et al., 1997).

By such a strategy of excision of rights, it can be argued the state shifts risk to individuals and may remove from them the benefits of their investment of time, labor, and knowledge. But it should be noted that such policies serve not only the public interest but also the interests of individual forest land owners whose investments may be protected from the practices (or just bad luck) of up-stream, up-wind or bug-infested neighbors. Thus, the excision of particular parts of the tree tenure bundle is simultaneously driven by conservation interests and production interests, sometimes resulting in complementarity, sometimes resulting in bitter conflicts (as in recent conflicts over old-growth preservation and over salvage harvesting).

We would be remiss if we failed to note that rights in the chemical components of or genes from indigenous tree species is a hotly contested domain in intellectual property. While beyond the scope of this chapter, this is a component of tree rights that bears close watching.

5. CONCLUSIONS

This review proposes at least an outline of a solution to the perennial question of why tree rights exist separate from land in some contexts, but trees are simply treated as attachments to land in other contexts. It also makes a point with practical implications. The separation of rights in trees from rights in land is not an archaic mode of ownership associated primarily with subsistence in the age of hunter-gatherers. It continues after ownership rights in land are well developed as the primary means by which society allocates rights to use and transfer land, even land with trees. It is an important mode by which those who have land reach accommodations with those who wish to use land to grow trees, meeting the minimum expectations of the growers. It is a means by which a landowner can tap the resources of others to cover the costs of tree planting and maintenance, trading shares in the ownership of the trees for their investments. The separating out of some of the rights from the bundle of ownership rights in trees and appropriation of them by the state is simultaneously an important contemporary conservation strategy and an important production strategy. And in recent years timber companies have increasingly moved away from growing trees on their own land. Already having used timber cutting concessions on federal and state land, they are moving to similar arrangements on private land, buying rights to cut the trees grown by land owners, avoiding the taxation and maintenance costs involved in ownership of forest land.

The separation of tree rights from land rights is an important tool for tree management. It can be used to attain either public or private purposes. It can undermine property rights in land or enhance them. How it is used has important equity implications. In the case of the apple lords in Indonesia, the separation conferred such economic clout on the lessees that they, rather than the lessors, the landlords, eventually came to control the relationships. But it can be used to advance equity objectives as well. Federal court decisions affirming Native American treaty rights to use trees on federally and privately owned land create use rights in sugar maples, recognizing that certain rights from the ownership bundle were retained by the tribes when they lost the ownership of the land. Near the White Earth Reservation in Minnesota, this is allowing reconstitution of traditional sugar mapling, and in 1995, the White Earth Land Reclamation Project put forward to county, state and federal foresters a plan under which the White Earth Chippewa would take over the management of forest on extensive reserve land, land which had once formed part of their reservation but which had been lost through takings for tax defaults during the allotment period. The shifting of tree rights is an important tool for transferring resources without interfering with the underlying land rights, incurring the costs implied in a land market solution or a taking for public purposes.

We have ranged broadly over forestry and agroforestry situations, and it may be useful in conclusion to recap the uses of tree tenure with particular significance for adoption and sustainability of agroforestry. First, tree rights can be used to provide tenure security for the tree investment in land over which would-be tree-planters do not enjoy tenure security. By providing an alternative to land tenure security, it can remove the need for norms which attribute ownership of land to those who plant trees on it, and so allow tree-planting in ambiguous land tenure situations. In so doing, it can open new land areas to agroforestry. Second, it can serve as a means by which resources beyond those of the tree-planter are mobilized to meet the cash and labor costs of planting and maintaining trees in the farming system. While these potentials are stated here as tools in agroforestry programming, they sometimes exist under indigenous agroforestry and those working with such systems must be aware of them and their potential value. Finally, a caution: the separation of rights in trees from rights in land can also impede agroforestry. Legal "protection" of species with agroforestry potential can undermine any incentives for the household to plant those species within its farming system.

REFERENCES

Adegboye, R. O., 1969, "Procuring loan through pledging of cocoa trees," *J. Geo. Ass. Nigeria,* 12: 68.

Berry, S., 1993, *No condition is permanent: the social dynamics of agrarian change in Sub-Saharan Africa,* University of Wisconsin Press, Madison, WI, 17.

Broad, R. with Cavanagh, J., 1993, *Plundering paradise: the struggle for the environment in the Philippines.* University of California Press, Berkeley, CA.

Cronon, W., 1983, *Changes in the Land: Indians, Colonists and the Ecology of New England,* Hill and Wang, New York, 65.

FitzSimmons, M. I., 1983, *Consequences of agricultural industrialization: environmental and social change in the Salinas Valley, California 1945–1978,* unpublished Ph.D. Dissertation, University of California at Los Angeles, 259–261, 320.

Fortmann, L., 1985, "The tree tenure factor in agroforestry with particular reference to Africa. *Agro. Syst.* 2: 229.

Fortmann, L. and Bruce, J., 1991, "Property and forestry," *J. Bus. Adm.* 20: 471.

Hecht, S. and Cockburn, A., 1990, *Fate of the forest: developers, destroyers and defenders of the Amazon,* Harper Perennial, New York.

Hill, P., 1963, *The migrant cocoa-farmers of Southern Ghana: A study in rural capitalism,* Cambridge University Press, Cambridge, U.K., 11, 186.

Hobley, M., 1995, *From participatory principles to pragmatic practice*, paper prepared for "Making forest policy work," OF/ICED/GS Seminars, Oxford University, Rural Development Forestry Program. Overseas Development Institute, London.

Hong, E., 1987, *Natives of Sarawak: survival in Borneo's vanishing forest.* Institut Masyarakat, Pulau Pinang, Malaysia.

Leach, T. A., 1919, "Date-trees in Halfa Province," *Sudan Notes and Records* 2: 98.

Lindsay, J. M., 1994, *Law and community in the management of India's state forests*, Working Paper, Lincoln Institute for Land Policy. Lincoln Institute for Land Policy, Cambridge, MA.

Lock, J., 1983, "Of property," in *Property: Mainstream and Critical Positions*, University of Toronto Press, Toronto, 1704, 17–27.

Menning, K. M., Johnson, K. N., and Ruth, L., 1997, "A review of current non-federal policies on non-federal lands in the Sierra Nevada that affect aquatic, riparian, upland and late-successional biological diversity," in *Sierra Nevada Ecosystem Project, Final Report to Congress, Addendum,* University of California Centers for Water and Wildland Resources, Davis, CA., 6, 16, 22.

Peluso, N., 1996, "Fruit trees and family trees in a forest: property, ethics of access and environmental change in Indonesia," *Comparative Studies in Society and History* 38: 510.

State of California, 1997a, *California Forest Practice Rules, Title 14, Chapters 4 and 4.5, California Code of Regulations,* January 1997 version, Compiled by the California Department of Forestry and Fire Protection, Sacramento, CA.

State of California, 1997b, "Public resources code," *West's Annotated California Codes*, West Publishing Co., St. Paul, MN.

Suryanata, K., 1994, "Fruit trees under contract: Tenure and land use change in the uplands of Java, Indonesia," *World Dev.,* 22: 1567.

Vissage, J. S. and Miller, P. E., 1991, "Forest statistics for Alabama counties-1990," *Resource Bulletin* SO-158, USDA Forest Service Southern Forest Experiment Station, New Orleans, LA.

11 A Utilitarian Approach to the Incorporation of Local Knowledge in Agroforestry Research and Extension

Fergus L. Sinclair and Daniel H. Walker

1. INTRODUCTION

The assertion that local knowledge plays a central role in the successful execution of agroforestry research programs has been transformed over the last two decades from a marginal strand of opinion to an increasingly central tenet of development programs. Few agroforestry professionals would now be likely to contest an assertion that agroforestry development programs are likely to be more

1-56670-294-1/99/$0.00+$.50

effective if designed, implemented and evaluated with due regard to the knowledge already held by the farmers or forest dwellers involved – a stark contrast to historically dominant views of the development process. While this new ethos is rapidly becoming the received wisdom, practice has tended to lag well behind the rhetoric. This lag can, in part, be ascribed to inevitable inertia in institutional structures, skill development and training (Pretty and Chambers, 1994). We suggest here, however, that this lag is also caused by the relatively slow pace of the development and deployment of methods that explicitly deal with people's knowledge, as opposed to the development of overall methodologies for participatory rural development, within which local knowledge may be considered important but remains largely implicit. Reluctance to use explicit methods to acquire local knowledge stems from conflation of the practically useful knowledge that people have about agroecology, which we argue can be usefully abstracted, and the social and cultural context in which this knowledge is used by various actors in the development process.

We propose that a utilitarian approach to making effective use of local ecological knowledge in generating timely solutions to pressing resource problems is required. In doing so, we do not underestimate the ethical, political and philosophical complexities of shifting paradigms in relation to the role of local people in development, but we do set out to debunk some myths and false dichotomies that have polarized views on local knowledge and hampered the development of practical methods for working with rural people's knowledge (Agrawal, 1995; Heyd et al., 1996). We propose that any agroforestry research and extension initiative that aims to improve the productivity and sustainability of existing farming systems should be founded on a rigorous and reportable analysis of the knowledge held by practitioners that underpins the management of those practices. We argue, contrary to some recent trends in thinking about agricultural development (Scoones and Thompson, 1994), that explicit treatment of rural people's ecological knowledge is both possible and useful, and that it complements and augments the adoption of participatory approaches to development.

The approach we propose is compatible with, and required by, a growing consensus that agroforestry programs are more successful where based on incremental change of existing systems and practices than where based on the extension of prefabricated technology packages that seek abrupt transformation. There is considerable evidence that although farmers may be interested in elements of technology packages (a tree species, for example), they rarely adopt whole packages in the form in which they are disseminated (Buck, 1990; Kerkhof 1990). Agroforestry researchers have not, in general, been able to design practices which farmers adopt, which is perhaps not all that surprising given the complexity involved. This is reflected in the current shift in the emphasis of agroforestry activity in which concentration on the central development of one or two agroforestry practices (such as alley cropping, contour hedgerows or improved fallows) for widespread adaptation and dissemination is being superseded by the facilitation of local development of tree resources at field, farm and landscape scales. This is a major shift in thinking which has led to redefinition of agroforestry as an incremental activity leading to increases in tree cover within farms and farming landscapes over time (Leakey, 1996). It implies support for local development initiatives that involve continuous integration of trees within farming systems, in various productive niches, over a considerable time period.

In this chapter, we address the roles and relationships between systems thinking, participatory methodologies and local knowledge. We then provide a pragmatic definition of local knowledge as being understanding of the world that can be articulated by an informant. We explore and explain differences inherent in alternative treatments of the nature of local knowledge and justify a utilitarian approach in terms of the practical gains in improving the effectiveness of research and extension that it makes possible. We then describe methods designed to make these opportunities a reality. The application of these methods is then illustrated through consideration of some case studies before we sum up the argument for adopting an explicit approach to incorporating local knowledge in agroforestry development.

2. CONCEPTS, DEFINITIONS, AND ISSUES

2.1 Systems Thinking, Participatory Approaches, and Local Knowledge

Agroforestry is an interdisciplinary subject that gained international prominence during the 1980s largely as a development imperative in the tropics. It combines resource conservation at low input levels with sustainable production of food, fuel and income, and therefore is likely to be particularly relevant to resource-poor farmers who have not been able to benefit from green revolution technologies (Pretty, 1995). As such, a systems perspective (Jones and Street, 1990) is important in agroforestry research and extension because:

- as an interdisciplinary science, agroforestry requires a theoretical framework for integrating disciplines;
- as a development imperative at the household level, resource-poor farmers face real problems that are by their nature interdisciplinary and, because their farms involve many interdependent resources and components, require an holistic approach to problem solving if improvements are to be reintegrated into farms, and
- as a development imperative at the level of public policy, a means is required of integrating action across hierarchical levels, principally aggregating household activity up to resultant effects at the scale of the landscape or catchment.

While the word *system* frequently is encountered in the agroforestry literature, the application of systems methods (Simmonds 1985) is far less common. While there are examples of exemplary use of a systems perspective (Rocheleau, 1987), much agroforestry research has been essentially agronomic, with particular practices, such as alley cropping, substituting for a commodity focus. This is characterized by research, even some at an adaptive, on-farm level, focusing on the plot rather than the farm household or community. In the case of alley cropping, this approach often did not result in improved practices that could be integrated back into whole farms (Carter, 1995). The view that much agroforestry activity in the past has been essentially about trees as a commodity focus is strengthened by the development of a particular variant of farming systems research and extension methods for agroforestry — D&D (Raintree, 1990), rather than agroforestry acting as a focal point for the unification of diagnostic procedures. A lack of a systems perspective is also evident elsewhere. There is no identifiable systems focus to the journal *Agroforestry Systems*, for example, which juxtaposes research on components, that may have no explicit relation to whole systems at all with system descriptions that may have very little analytical content and may not employ systems concepts or terminology consistently (Nair, 1989).

To make progress, we start from the premise that agroforestry is generally seeking to address problems experienced by farmers in a holistic manner. This is consistent with agroforestry increasingly being viewed as a broad, interdisciplinary activity that embraces a progressive view of the development of tree resources within the context of sustainable agroecosystem and community development (Sinclair, 1997), such that a systems perspective is implicit. So, while institutional barriers to marshalling information across disciplines may remain at an implementational level, and there may be methodological constraints in doing so in an effective and rigorous manner, systems thinking needs to be a central orthodoxy in agroforestry.

However, a research and extension program which is based on a systems framework generated from the perspective of external natural resource professionals may be just as prone to irrelevance to the rural people intended as beneficiaries as a more traditional program of reductionist science and technology transfer which has made no attempt to take system context into account. System descriptions are a human construct that abstract, simplify and explain reality. Being a simplification and abstraction, they are based on assumptions and objectives — different assumptions and objec-

Box 1. Different Perspectives of Catchment Management Among Farmers and Natural Resource Scientists in Queensland, Australia

Integrated Catchment Management (ICM) is a policy and program aimed at fostering natural resource management in Queensland, Australia, at a catchment scale from the bottom up where the bottom is defined as the local community. However, interviews and workshops with farmer groups in the Herbert Catchment of north Queensland have demonstrated that while farmers knew precisely what the word "catchment" meant, they did not naturally view the landscape that they live and work in as being part of a catchment and, therefore, did not see the catchment as a natural unit of management. By contrast, for natural resource professionals who were the authors and champions of ICM, a model of the landscape as a set of catchments that provided natural units for rational management was both intuitive and powerful. The limitations of this simple but external model of the landscape makes the expected "meeting of minds" between the two groups more demanding than would otherwise have been anticipated.

tives leading to different descriptions. Where there is a human agent within the system (which occurs wherever there is management of the system) the nature of the abstraction and approach to management of the system are inextricably entwined. So, where the model of systems function that underpins research is based on a completely external perspective, the operational relevance to targeted beneficiaries may be seriously compromised. This can be illustrated with reference to a simple example (Box 1).

The qualitative descriptions by professionals of the vastly more complex systems characteristic of the agroforestry domain (see for example Ranasinghe and Newman, 1993) are likely to differ far more from local perspectives – to the extent that they may seem at times to be closer to professional tourism than a sound basis for a rational development program.

One reaction to these partial failings of the systems approach has led to a significant attack on the systems ethos from an unexpected corner and the creation of a false dichotomy between hard and soft systems methods.* As the pendulum swings away from an interventionist external paradigm for development to an increasingly participatory** one, it is being argued that development professionals should leave the systems thinking to the target communities — if provided with the range of technological solutions available, local communities can undertake their own systems synthesis to generate solutions (Blaike et al., 1997). It would appear to follow from this that farmers do their own analyses, research and information exchange and there is no need for a systems perspective on the part of development professionals. It is increasingly fashionable to put this view forward in terms of contrasting hard and soft systems approaches (Scoones and Thompson, 1994) with the implication that since agricultural development is inherently a process played out by various people making decisions (referred to as actors), only soft systems methods (Checkland and Scholes, 1990) that concentrate on the social organization of knowledge (Engel, 1995) are appropriate. In this frame of analysis, attempting to obtain a systematic understanding of what local people know about the ecology of their environment and farming practices is criticized as being

* Hard systems are characterized by having easy-to-define objectives, clearly defined decision-making procedures and quantitative measures of performance. They are often mechanical, although biological systems are also often of this kind. The more intimately that people are involved in a system, the less appropriate this view becomes. Soft systems, in contrast, are those in which objectives are hard to define, decision-making is uncertain and measures of performance are, at best, qualitative (Spedding, 1988).

** Participatory approaches have been variously defined, and different degrees of participation can be envisioned (Primbert and Pretty, 1997). We use the word here to denote "active participation" where local people are involved in making decisions about what happens rather than merely being consulted in a decision-making process controlled by others (Cornwall et al., 1994).

inherently extractive, because it is assumed that it must entail the imposition of an external rationality (Scoones and Thompson, 1994). This view questions the validity of systematic study of local knowledge by development professionals as a part of the development process but, notwithstanding its current popularity, unnecessarily polarizes the debate on relevant methodology for working with rural people and their knowledge and raises concerns that on further examination are invalid for reasons which are summarized below and elaborated in the rest of the chapter.

First, in the context of agroforestry research and extension, hard and soft systems methods are not mutually exclusive but, in fact, play different roles and are appropriate for different purposes and hence for application at different hierarchical levels (Figure 1). The "hardness" of a system is determined by how well-defined, stable and quantitative the system mechanism and purposes are. It is, therefore, useful to conceive of agricultural activity as consisting of harder ecological systems at the organism, field and farm level that are managed by people within household, village and other social structures, thereby creating increasingly soft systems as more of the social structure is included. Farmers, therefore, albeit within a complex soft systems framework, are managing the ecology of their farming practices, which can be tractably understood, at least to some extent, using a blend of hard and soft systems methods. Thus, partial views of the relatively "hard" ecology of a farming system are a useful and valid tool, providing their partial nature is fully understood, inasmuch as what farmers actually do in their farming practice is determined by other social and cultural factors as well as their knowledge of agroecology. Indeed, a systematic appreciation of farmers' ecological knowledge can sometimes help to distinguish the action of socio-economic determinants.

Second, although it is a danger that has often been apparent when agricultural scientists have analyzed local knowledge, seeking to obtain a systematic view of local knowledge does not necessarily imply the imposition of an external rationality. It is possible to seek local explanations and to represent them. The western scientific tradition does not have an exclusive claim to systematic understanding of agroecology, and it would be somewhat patronizing to posit that farmers do not have locally derived understanding that may be systematic. There is, in fact, increasing evidence from various agrarian societies from studies using a variety of different methods that farmers often possess a sophisticated, locally derived and general understanding of agroecology comparable with, although different from, scientific formulations (Berlin, 1992; Fairhead and Leach, 1994; Richards, 1994; Thapa et al., 1995). It is important then not to confuse the development and use of formal methods with the imposition of a particular rationality. Quite different explanations for particular ecological phenomena or the efficacy of farming practices can be codified using formal notation. Indeed, dealing with local knowledge explicitly may assist in revealing where external interpretation occurs, as well as common ground between different knowledge systems. The use of formal methods to represent local knowledge (a process of abstraction rather than extraction) can, therefore, be seen as according rural people's knowledge a status on a par with scientific knowledge rather than as an *extractive* process that somehow diminishes or distorts it.

Third, empowerment of local people in a participatory development process does not necessarily preclude the need for strategic research and extension by development professionals. In fact, the existence of sophisticated ecological knowledge among farmers, based on their own observation and experimentation, improves the prospects for research to be done by supporting institutions that addresses fundamental questions that farmers have difficulty tackling because of resource constraints or limitations in their ability to make observations. In fact, while in many situations it is unlikely that professional researchers can improve on farmers' adaptive research (Millar, 1994), they may well be in a position to contribute by doing strategic research and extending information about ecological mechanisms that

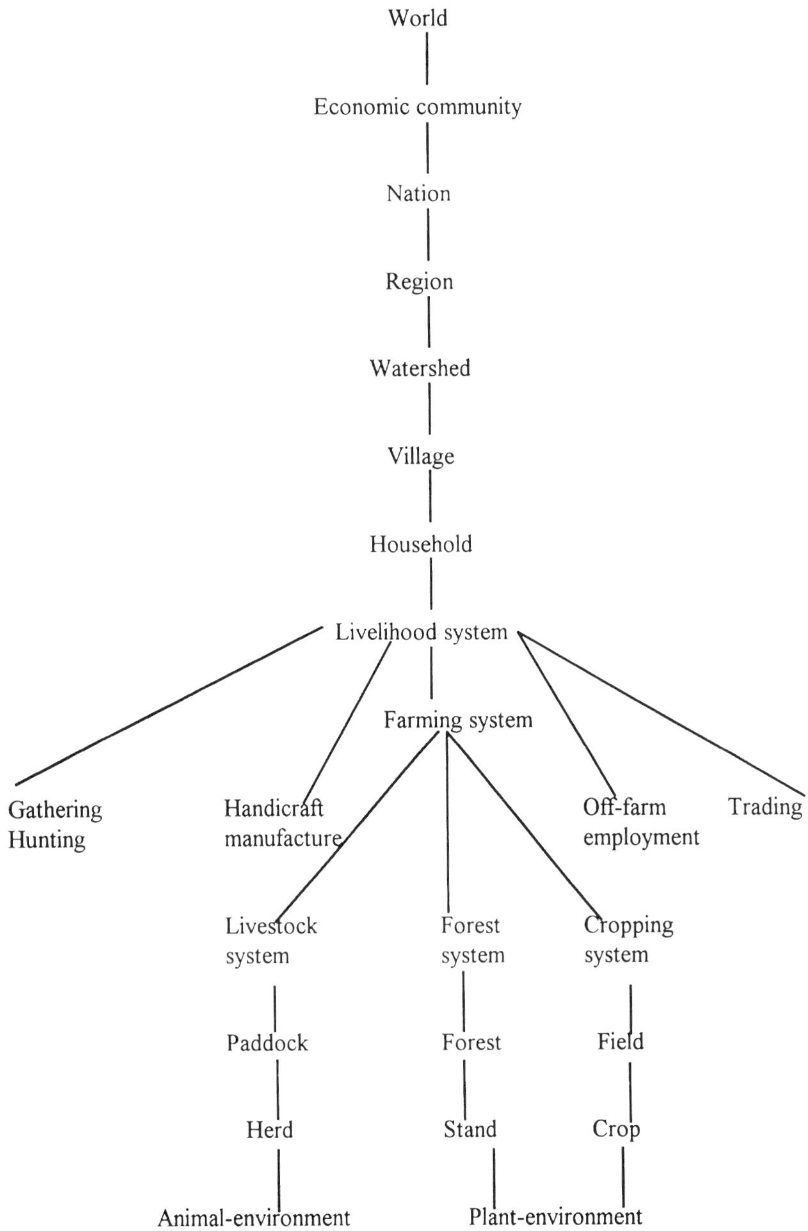

FIGURE 1. Schematic view of different hierarchical levels at which agroecosystems can be identified. (Adapted from Conway, G.R., in *Systems Theory applied to Agriculture and the Food Chain*, Jones, J.G.W. and Street, P.R., Eds., Elsevier Applied Science, London, 1990, 205.)

farmers find difficult to observe (Bentley, 1994). To design such strategic research and extension activity on the basis of what farmers require and to be able to communicate effectively about agroecolgy, development professionals need to be aware of what farmers already know.

We suggest, therefore, that a valid reaction to the failings of a systems approach based on external perspectives is to seek an internal (local) construction of a model of the system that development professionals can access and use. Furthermore, we consider that this approach is both compatible with and complementary to participatory approaches to research and extension.

In summary, it is reasonable to propose that initiatives that are aimed at impacting on management (implying that people change their behavior) are likely to be much more efficient where premised on the conception of the system used in management than on some external conception or none at all. This does not mean that the models of system functioning held by farmers are more correct (when compared to reality) than the models that might be generated by external professionals or that they are even adequate. However, the assertion that the farmer's model of the system is an important starting point forms the fundamental premise of this paper. In short, we argue that a representation of a local community's view of the systems and practices under development is a prerequisite to relevant research and extension.

2.2 The Nature of Local Knowledge

Local knowledge has been viewed and defined in a number of contrasting and, perhaps for different purposes, equally valid ways. Our purpose here is to define the nature of our present focus and point to trends in the way in which local knowledge is perceived as it becomes a more frequent focus of study in rural development. Our central argument here is that increasing recognition that local agroecological knowledge is often a contemporary and dynamic explanation of how the world works with potentially universal application, which is primarily derived from observation and experimentation by farmers, is more useful than concentrating on unique and idiosyncratic elements of traditional knowledge that are culturally bound, predominantly heuristic and preserved through generations.

2.2.1 Knowledge as Understanding That Can Be Articulated

For the purposes of this chapter, we make use of the following definitions. Information is a continuum that has data and knowledge as two extremes. Data is a recorded set of observations (which may be quantitative or qualitative). Knowledge is the outcome, independently of the interpreter, of the interpretation of data. Whether a piece of information is considered to be data or knowledge depends on the abstractness of the question under consideration — data becomes knowledge by asking a more specific question, knowledge becomes data by asking a more general question. However, both data and knowledge (and, therefore, information) can, by this definition, be quite clearly distinguished from understanding. Understanding is the outcome, specific to the interpreter, of the interpretation of data or knowledge: the comprehension that the interpreter achieves. So the interpretation of observation may advance an individual's understanding. Some part of this advance can be articulated and communicated as information. Whether this information is considered to be data or knowledge is a moot point. The key issue is that it can be articulated and communicated. Nevertheless, because in the current context we are interested in information about the world that is of more general rather than more specific applicability, we will use the term knowledge throughout. We believe that the distinction between what can and cannot be articulated is of paramount importance in taking a practical view of local knowledge and its contribution to research and extension. Furthermore, it is compatible with the knowledge engineering paradigm that underpins many powerful theoretical, methodological and implementational advances in cognitive science and artificial intelligence.

Knowledge is thus seen as a central aspect of culture, derived from education and experience, that may be used, in conjunction with consideration of a value system and competing priorities and possibilities, to make decisions and, therefore, lead to actions (such as farming practice). Actions also may be intuitive, short circuiting the consideration of knowledge in their genesis. What is perceived about the natural world, however, is influenced by what can be observed and the way in which perceptions are transformed into knowledge is influenced by the mode of learning, which may differ from one society to another. It is clear then that the way in which knowledge is acquired and transformed into decisions depends on the cultural context, but knowledge is distinguishable from other aspects of a person's, or a community's, culture.

This definition of knowledge as being something that can be separated from the person who knows it, essentially a position that conforms to a Popperian philosophy of science (Popper, 1972), is quite different from the more widespread sociological definition used within the recent farming systems literature which views the two as inseparable. So, for example, Scoones and Thompson (1994) challenge "the assumption of a positivist view of investigation that sees knowledge as a tangible stock or store to be tapped" and see "the process of knowing" as "engaged, value-bonded and context determined." They argue that the "human mind is not simply a mirror that accurately reflects a reality," but that "interpretation, translation and representation are social acts that cannot be assumed to be neutral and objective." Taking this perspective further, Röling (1996) states, "Reality is not assumed to exist independently of the human observer or to project itself on the mind through the senses ..." Blaike et al.(1997) put these theoretical considerations into context by stating, "These processes [of advancing ideas], and hence knowledge, are inextricably linked to the social, environmental and institutional contexts within which they are found. For example, a partly deforested hillside in the tropics may be viewed very differently by a local subsistence farmer ..., a cattle rancher, and an NR scientist ..." In addressing perceptions of landscape, Colquhoun (1997) takes an even more extreme view that knowledge includes "intangible deep feeling of what meets us in a landscape." These definitions make local knowledge essentially indistinguishable from local people. We argue that they conflate knowledge, values, perceptions and even emotions. While these definitions may in many contexts be useful, they serve to seriously constrain how local knowledge can be gathered and used. Defining local knowledge as statements (or assertions) about the way the world works that local people can articulate clearly and precisely allows us to explore effective means of accessing that knowledge.

2.2.2 Local Knowledge, Indigenous Knowledge, and Technical Knowledge

The terms local knowledge and indigenous knowledge have often been used interchangeably. However, we use local knowledge to denote locally derived understanding which is based more on experience and real world observation than indigenous knowledge, which may reflect cultural beliefs and values to a greater extent. This distinction has important practical implications. So, investigation of indigenous knowledge (principally the realm of anthropological and ethnographic research, which are particularly concerned with mapping and explaining cultural differences), has tended to focus on knowledge that is idiosyncratic and culturally bound, often being traditional and handed down orally. Such investigation has practical as well as academic application – of 242 published papers retrieved from a search on indigenous knowledge of CAB Abstracts for 1996/7, 85 are concerned with medicinal aspects of rare ethno-botanical knowledge or the ethics associated with the search for such knowledge. Some indigenous knowledge, and hence research into it, has profound implications for resource management and development. Historically, for example, it has been suggested that the cultural beliefs held by Mayan Indians that people were literally made of maize (thus making it a sacred plant) precluded commercial exploitation of this crop by indigenous people, whereas Hispanic settlers were not so restricted (Asturias, 1949). Such cultural beliefs can be pervasive, the sacred nature of maize in the previous example was bound up with notions of soil fertility and its decline through continuous cropping. Sallas (1994), working today in the Peruvian Andes, reports similar culturally derived constraints to the adoption of modern potato varieties, that although higher yielding than local varieties, are less resistant to some pests and diseases and require inputs of chemicals for yields to be achieved. The modern potato varieties and the chemicals applied to them are considered by older people in the community to "bring more diseases" and to be "poisonous" both to "mother earth" and to the people who eat the potatoes.

However, while providing important insights and methods relevant to investigation and use of local knowledge in resource management, we argue that this focus on "indigenous knowledge" is less useful in the current context than a more recent and applied interest in local knowledge as a dynamic resource based on contemporary experience and observation (Howes and Chambers, 1980; Bunch, 1989; Fairhead and Leach, 1994; Richards, 1994; Rhoades and Bebbington, 1995; Thapa et al., 1995). Furthermore, overemphasis of cultural belief can mask important aspects of practice. So, for example, the Kantu of Kalimantan, Indonesia use the presence or absence of particular species of "omen birds" to select sites for swidden plots (a practice for which various ecological rationales can be postulated). However, in practice, farmers are quite willing to ignore omen birds where these would otherwise preclude what they saw as an appropriate site (Dove, 1981). This illustrates the interplay between culturally derived indigenous knowledge and local knowledge based on practice, observation and personal experience.

However, there is also a danger in relying too heavily on practice in the investigation of local knowledge. While indigenous technical knowledge (ITK) has been the focus of much research (IDS, 1979) we argue that there is a fundamental distinction between the decision rules that constitute management and the knowledge underlying them. Furthermore, we contend that many management actions undertaken by farmers have an explanatory basis, although such explanations may differ from a scientific rationale and may require careful interaction with farmers to elicit.

Unfortunately, farmers' explanations for agroecological phenomena have not received adequate attention. In the present enthusiasm for moving from considering what people do, as in studies of ITK — or farmer's performance, to the cultural context in which knowledge is generated — or the process of knowledge generation and development (Scoones and Thompson, 1994), there is a risk that the gains from explicit representation of farmer's explanatory understanding of agroecology will continue to be overlooked. The key issue here, once again, boils down to whether it is valid to abstract farmers' knowledge about ecology from the broader set of considerations that govern agricultural practice and whether this can be done without overly distorting it. A major confusion has arisen here between the limitations of using external scientific ("etic") explanations of local practice and seeking the farmer's own ("emic") explanation. So, for example, Richards (1989) in an influential article, explains how complex intercropping layouts in a field, that scientists may interpret as a carefully planned arrangement based on consideration of inter-specific competition and weed and pest control may, in fact, represent a set of contingent responses by the farmer to various events throughout the season so that what is observed at the end of the season is the result of a completed performance (influenced by a range of unfolding ecological and socio-economic factors), rather than the result of a specific plan in the mind of the farmer at the beginning of the season. However, while it is clear in this case that the farmer may not have planned the design from the outset, and that what transpired over the season and why can only be interpreted by considering the sequence of events that actually occurred in time (which may be impractical if the dialogue begins at the end of the season), knowledge of plant ecology on the part of the farmer is not precluded. It has not, in fact, been adequately sought.

A major justification for abstracting farmers' ecological knowledge from other social and cultural aspects determining agricultural practice is that it can be postulated that core knowledge about natural processes is likely to have some applicability across contexts rather than being entirely unique to the situation and society in which it was derived. This is a somewhat contentious view that flies in the face of postmodernist approaches to anthropology (Tyler, 1987). There is, however, some compelling evidence to support this likelihood. First, Berlin's seminal work on ethnobiological classification (Berlin, 1992) reveals widespread regularity in the way in which people classify and name plants and animals in quite different non-literate societies. Thus, he asserts that when people function as ethnobiologists, they "discern" rather than "construct" order from their observation of a biological reality, in which groups of plants and animals appear as a series of discontinuities, whose structure and content are seen by all human beings in essentially the same way. He contrasts this with other areas of human experience which are culturally constructed and to which anthropologists have devoted much study, such as social organization, ritual, religious beliefs and notions of beauty, for which no such universal regularity may be expected. Second, anthropologists researching local knowledge in the context of agricultural development are increasingly revealing widespread explanatory knowledge of agroecology. Thus Richards (1994), in an article about knowledge of rice production held by Mende-speaking people in Sierra Leone, argues that local knowledge "is in conformity with general scientific principles" and that farmers have "detailed knowledge of the way in which crops and soils, or crops and pests, interact, under a variety of local climatic conditions." Fairhead and Leach (1994) use examples of knowledge of agroecological processes related to the management of trees and watercourses held by Kouranko farmers in Guinea to challenge conventional perceptions that ecological knowledge is inevitably socially differentiated within communities. They found that although there were social and cultural reasons why different groups of people, such as men and women, were differentially inclined to discuss various aspects of their knowledge, this did not imply they actually had different knowledge. Thus they state "even if the capacity to express ecological knowledge is socially bounded, the knowledge itself might not be." This accords with studies, discussed later in this chapter, in which formal methods have been used to represent farmers' ecological knowledge, most notably in Nepal, where sophisticated local concepts and terminology about tree-crop interactions and the nutritive value of tree fodder were found to be widely held among rural people both within villages and across the whole of the eastern mid-hills (Thapa et al., 1995; Thapa et al., 1997; Joshi and Sinclair, 1997).

2.2.3 Combining Local and Scientific Knowledge

Given that farmers have much more intimate experience of their production practices than external professionals, it is reasonable to assume that the people who have been operating them have developed an understanding of them. Indeed, such understanding and its utility has been increasingly widely documented in recent years as discussed above. We argue that the knowledge held by farmers can provide a resource for science irrespective of the participation of the farmers in the use of that knowledge. Investigating local knowledge may be a powerful and efficient means of rapidly filling gaps in scientific understanding about agroforestry. So that amalgamating specific local knowledge and general scientific knowledge may be more powerful in designing appropriate research and extension than the use of either alone. Indeed, it has been claimed that agroforestry professionals are in a unique position to learn from traditional knowledge and practice and combine with indigenous experimental initiatives and potentials (Rochleau, 1987). At a minimum, local knowledge can provide a useful basis for preliminary formulation of hypotheses which can then be referred to scientists (Howes and Chambers, 1980).

Productive synthesis of local and scientific knowledge, however, depends fundamentally on whether the potential complementarity of local and scientific knowledge is matched by compatibility. In principle, scientific knowledge attempts to explain causality and engender that causality in universally applicable and predictive principles. Opinion is divided as to whether indigenous or local knowledge can also be said to provide systematic explanation of causality.

Niamir (1990) suggests, for example, that local knowledge is entirely based on what people think it is necessary to know — they may see correlations but feel it is unnecessary to explain causality. She states that local knowledge does not devise general principles and absolutes but allows an understanding of the heterogeneity of local conditions, while formal science seeks general applicability but, particularly in natural science, has difficulty in attempting to cope with the variability that is found in the real world. By operating at different scales, this view regards the two systems as complementary but of a fundamentally different nature and held in different conceptual frameworks.

By contrast, there is evidence to suggest that local knowledge is not directly utilitarian (Berlin, 1973; Howes and Chambers, 1980). Howes and Chambers (1980) state that "ITK, like scientific

TABLE 1
Practical differences between indigenous and scientific knowledge (after Brokensha, personal communication).

	Indigenous knowledge	Scientific knowledge
Scale	local	universal
Leaders	secular	professional
Transmission	informal	formal
Access	generally open	generally open
Experiments	limited	vast
Orthodoxy	present	present
Supernatural	present	absent

knowledge should be regarded in the first instance as something which became possible as a result of a more general intellectual process of creating order out of disorder, and not simply as a response to practical human needs, such as health and sustenance." Knight (1980) argues that "ethnoscience and formal science share a common quest for explanatory theory in accounting for the apparent diversity, complexity, disorder and regularity in the environment." The resulting theories produce a wider causal context than simple common sense although they may be complementary in day-to-day activities.

In addition to a common motivation, we argue that there is no fundamental distinction between the mechanisms by which indigenous and scientific knowledge are advanced. It is claimed that logical analysis and experimentation underpin advances in both (Howes and Chambers, 1980). Indeed, there is clear evidence for deliberate experimentation among local people, and it seems probable that active experimentation is widespread (Richards, 1939; Richards, 1985; Bunch, 1989; Millar, 1994; Rhoades and Bebbington, 1995).

While motivations and mechanisms may be comparable in a fundamental sense there are, nevertheless, important practical differences between scientific and local knowledge. Table 1 summarizes some of these differences.

It is clear, then, that scientific and local knowledge differ in four aspects.

- **Methodological structure** Science has, theoretically, a formal structure of hypothesis, proof or disproof (through experimental evaluation rather than logical deduction) and acceptance or rejection. It seems likely that such rigorous procedures are rare beyond the western scientific tradition.
- **Institutional framework** The international structure of scientific education, dissemination of findings, peer review and debate facilitate progress in science. While comparable procedures exist within some culturally defined communities, for example the Poro and Sande secret societies of the Kpelle of Liberia, (Murphy, 1980), the network of individuals involved in support, interaction and innovation and the breadth of experience that can be drawn on will be very small in comparison with the scientific network.
- **Technical facilities and ability** Science depends heavily on advances in instrumentation and methodology to advance understanding. Such instrumentation expands the range of sensory perception open to the scientist. Means of augmenting the basic senses for local communities are much more limited. Richards (1980) points out that "the farmer will not know that which he cannot observe fully and completely" and that "qualitative judgments and farm decisions based on quantification cannot be better than the level of accuracy inherent in the quantitative procedure used by the farmer." The level of accuracy attained by farmers will normally be significantly lower than that attainable by scientific research.

- **Scale of perspective** Science seeks, in principle, universal explanation. For the practical purpose of managing a land use system, farmers normally only seek understanding with regard to their immediate surroundings.

In summary, we suggest that there is no reason to believe that scientific and indigenous knowledge about agroforestry are fundamentally different in origin and motivation. While this does not necessarily mean that they are held in comparable conceptual frameworks, it does suggest some level of compatibility. Science is well equipped to generate robust and widely applicable theoretical frameworks. Indigenous knowledge may provide an effective source of the detailed and context-specific information needed to test and apply predictive theory. This suggests a complementarity.

2.3 Local Knowledge as a Basis for Defining Researchable Constraints and Priorities

Having established that local knowledge may represent a valuable synthesis of farmers' experience, which may complement what research and extension workers know, as well as the scientific and professional literature, it is self-evident that it should be included in the review process when planning and prioritizing research. Simply educating natural resource professionals about farmer knowledge and practice may change their perception of what is important, the terminology they use and the type of research they do. Where an explicit model of local knowledge is available, however, it is possible to take this process much further by exploring the local knowledge base and formally comparing it against both farmer practice and scientific knowledge to identify researchable constraints and opportunities for extending existing knowledge that is not available locally.

2.3.1 Educating Professionals

In many circumstances, research and extension staff have only a limited appreciation of local knowledge and, in some cases, of local practice. This clearly constrains their ability to understand the nature of the problems facing farmers, the constraints they are under and the opportunities for improvement. It also affects their ability to communicate with farmers, affecting both their ability to obtain information from farmers about their system during diagnosis and then their ability to disseminate information and other products of research (germplasm, tools, techniques) to farmers in the extension phase. Farmers' knowledge also may be more developed in some aspects, particularly those that are locally important, than that of researchers and the scientific literature. There are some stark examples of where ignorance of farmer knowledge and practice has led to inappropriate research and extension efforts and how systematic documentation of such local knowledge has led to researchers changing their behavior (Box 2).

It is worth noting, however, that there may be considerable institutional resistance to researchers learning from farmers because it challenges conventional perceptions of people's roles and power relations (Pretty and Chambers, 1994). For example, prior to the systematic documentation of farmers' knowledge described in Box 1, an expatriate tree fodder scientist had criticized the proposal to acquire local knowledge because his experience had been that farmers were inconsistent in ascribing fodder value to tree species, and so he did not think their knowledge was worth collecting. His experience in the field had been that "farmers would point to a particular tree and say it produced high value fodder, and then later on in the same day point to another example of the same species and ascribe it a lower fodder value." However, when farmers' knowledge was subsequently analyzed, far from revealing inconsistency, this was found to represent a stable understanding of intraspecies variability. Thus, it was actually found that farmers classified six out of the 90 woody species found on farms to a sub-species level not yet botanically recognized and differentiated the fodder value of the variants (Thapa et al., 1997). Recent laboratory assessment of nutritive value corroborates these differences in some cases (Thorne et al., 1997). In a similar vein, the director of a local research center had to be taken into the field and shown the number of naturally

Box 2. Researchers in Nepal Design Different Types of Research When They Learn About What Farmers Already Know

In the Eastern mid-hills of Nepal it was assumed during the 1980s by the forest service that tree planting on farmland was constrained by lack of appropriate planting material and so nurseries were set up and seedlings offered to farmers. Take up of seedlings by farmers was low, and it was later discovered through acquiring local knowledge that farmers were, in fact, already managing abundant natural regeneration on their crop terrace risers (Thapa et al., 1995). Basically there were plenty of naturally regenerating seedlings — farmers cut back those they did not want to develop into fodder trees. Furthermore, they chose which species they did allow to grow on the basis of a sophisticated understanding of the seasonal feeding value of the fodder they produced (Thapa et al., 1997) and the extent to which they affected crop yield and soil erosion (Thapa et al., 1995) — aspects that research and extension staff had not adequately considered in choosing nursery stock. Farmers also used terminology not ubiquitously understood by research and extension staff to describe tree-crop interactions. Perhaps most notably, farmers were concerned about canopy modification of rainfall drop size, because they thought that larger drops caused higher rates of soil erosion. The process of water droplets falling from leaves was locally known as *tapkan*.

Not only had research and extension staff been unaware of this farmer knowledge, but also it was actually contradicted in the scientific literature which held until recently that drop size was independent of canopy morphology (Brandt, 1989; Thornes, 1989). Scientific understanding of how leaves of different types affected drop size was revised, and brought in line with that of the Nepalese farmers, in 1993 when new instrumentation allowed more reliable measurement of drop size (Hall and Calder, 1993). The key point here, though, is that once researchers were made aware that tree-crop interactions were important to farmers and that farmers had a cogent interest in minimizing negative impacts of trees on soil and crops they could see the relevance and importance to farmers of research in this area and had the terminology to communicate with farmers about it. In the last few years researchers at frontline agricultural research institutions serving the eastern mid-hills have done work directly on tree-crop interactions (Joshi and Devkota, 1996) and there are now plans for tree-crop interactions to form a central basis of agroforestry research in the Western Development Region (Paudel et al., 1997).

regenerating seedlings on crop terrace risers before he would believe that they were there. Similar gains from researchers understanding more about farmer knowledge have been found in a range of agroecological, and socio-economic situations and institutional settings (Sallas, 1994; Richards, 1994; Millar, 1994; Bentley, 1994).

In summary, there is evidence that researchers can learn from local knowledge both about farmer practice and the ecological processes operating in farmer's fields. Where researchers do so they may change their view of what research is useful to farmers both because of the information they gain about the operative processes and the confidence they have that research of a particular type is relevant to and can be communicated to farmers. There also may be considerable barriers to some researchers learning from farmers.

2.3.2 Local Perception of Inadequacy of Knowledge

In certain circumstances farmers may be aware of aspects of their knowledge systems which are less well developed than they would like, which then constrains their ability to manage their farming practice. This represents articulation of research requirements by farmers. Examples abound of situations where farmers recognize deficiencies and adopt their own experimentation to counteract this. In the quite different settings of Ghana and Peru (Millar, 1994; Rhoades and Bebbington, 1995), farmers' motivation for experimentation has been found to include:

Box 3. Farmers' Recognition of Deficiencies in Their Knowledge About Below-Ground Tree-Crop Interactions

While farmers recognized six tree attributes (leaf size, texture and inclination angle, crown diameter and density and tree height) that affected *tapkan* (see Box 2) and shade, and described causal mechanisms for how each attribute affected them, their knowledge of below-ground competition was restricted to a rough classification of 40 out of the 90 tree species found on farms as being either *malilo* — enhancing soil fertility and less competitive with crops, or *rukho* — competitive with crops (Thapa, 1994). Causative knowledge about why trees were classified in these ways included only two elements: a gross classification of root systems as predominantly shallow or deep and some knowledge of the speed of decomposition of leaf litter (which occurred above-ground and so could be observed). Given that trees were regularly lopped for fodder, a number of issues pertinent to practical management arose with respect to species differences in root system characteristics and the effects of different lopping strategies on root development and competitiveness, that farmers were unable to address.

- **Curiosity** For example, can cocoyam grow well in mixed cropping under mango? Do potato cultivars with apical dominance yield fewer but larger tubers than cultivars without apical dominance? Can new potato varieties be generated by planting true seed?
- **Attempts to find solutions to problems** For example, can the parasitic cereal weed striga (*Striga hermonthica*) be controlled by rotating crops? Does storing potatoes in diffused light conditions increase the incidence of tuber moth pest (*Phthorimaea operculella*)? Are aphids (*Myzus persicae*) attracted to green but not red sprouts on potatoes?
- **Attempts to adapt technologies or components to local conditions** In this case, farmers are concerned with whether the technology works or the component will grow in their environment, and whether and how it can be integrated into their farming system (for example, in Ghana, a recommendation to plant a new cassava variety vertically was adapted to horizontal planting because it took the people planting too long to identify which way up to insert the cutting vertically to ensure the aerial part was pointing upwards).

Clearly, where farmers set up their own experimentation, their knowledge system may be incrementally augmented and deficiencies addressed. But in some cases, and particularly with agroforestry where trees as large and long-lived components are involved, farmer observation and experimentation may be fundamentally constrained, thereby constraining the farmers' ability to obtain the knowledge they require to optimize their management. Returning to the example of Nepali farmers outlined in the previous section in Box 2, it was found that while the farmers possessed a sophisticated understanding of above-ground tree-crop interactions that they could observe, their knowledge of below-ground interactions was relatively sparse, despite the fact that they considered them important (Box 3).

In agroforestry, both below-ground interactions and considerations of sustainability, represent frequent areas where farmers' knowledge may be constrained by their ability to observe processes, because they involve observations over long time periods and because soil is an impenetrable medium. Organisms and processes which are too small to observe with the naked eye also represent a potential area where farmer knowledge is constrained, particularly with regard to biological pest and disease control (Bentley, 1994). Areas where farmers identify that they lack knowledge may represent appropriate foci for supportive research and extension by institutional providers because the farmers have identified a need for knowledge, clearly have difficulty in obtaining it through their own efforts and are, therefore, likely to be receptive to the results.

Box 4. Comparative Analysis of Scientists' and Farmers' Knowledge About the Tannin Content of Tree Fodder and Its Implications for Feeding Farm Animals

The existence of formally documented records of farmers' and researchers' knowledge about the nutritive value of tree fodder (Thapa, 1994; Thapa et al., 1997) made it possible to compare the equivalence of terms used by farmers and scientists. This was done using automated reasoning procedures on computer (Kendon et al., 1995) and it was found that there was some equivalence between the way in which farmers used the term "leaf bitterness" and scientists used the term "tannin content" — put simply, fodder that scientists described as having a high tannin content tended to be described by farmers as bitter. However, while scientists had some detailed knowledge about the role of tannins in protein digestion by ruminants and decomposition of leaf litter, they knew very little about the actual tannin contents of the 90 native species used by farmers and how this varied seasonally. In contrast, farmers did not possess detailed knowledge about the mechanism of action of tannins in ruminant digestion, although they did associate leaf bitterness with lower palatability and nutritive value (Thapa et al., 1997), and their local classification of fodder appears to implicitly encompass effects of tannins on protein supply to the duodenum in cattle (Thorne et al., 1997). Farmers did, however, articulate detailed knowledge about how leaf bitterness varied in a large number of tree species throughout the season. This demonstrates complementarity between farmers' and scientists' knowledge that could be exploited in designing appropriate research (farmers' understanding of intraspecies variability has already led researchers to revise strategies for sampling tree material for analysis of its nutritive value). Clearly, because of the complementarity, the combination of what farmers and scientists know represents a more powerful resource than either knowledge system alone.

2.3.3 Comparison of Local and Scientific Knowledge

It is clear from the example in the previous section (Box 3), where farmers themselves identified a lack of knowledge about below-ground interactions, that a comparison of farmers' knowledge with scientific knowledge of ecological processes could also have highlighted this gap. While scientific knowledge is also skewed in favor of atmospheric interactions, a basic framework for understanding root interactions in agroforestry (Schroth, 1995) and the contributions that tree leaf litter can make to soil fertility and the nutrient requirement of crops (Palm, 1995) is available, and advances are being made in methods to observe below-ground interactions (van Noordwijk and Ong, 1996). Conversely, in the example in Box 2, as already discussed, there was evidence of scientists being in a position to learn from farmers about canopy modification of rainfall drop size through similar comparative analysis of the two knowledge systems. The possession of sophisticated understanding of processes by farmers is more widespread than may be expected. Fulani pastoralists in northern Nigeria, for example, developed a home-made cow vaccine against a contagious bacterial disease (they put infected lung tissue under the skin of healthy cows) and well before scientists understood that the causative virus in foot-and-mouth disease could be transmitted through the air, avoided transmission of the disease by passing with healthy cattle upwind of infected herds (Bunders et al., 1996). As intimated in the previous section, it was Latin American farmers who first observed that tuber moth infection of potatoes in store was increased by diffused light conditions, which was later scientifically verified and that aphids were attracted to green rather than red shoots (Rhoades and Bebbington, 1995). It is this potential for mutual exchange of information that makes comparative analysis of farmer and scientific knowledge systems so powerful. This can be illustrated with respect, once again, to the knowledge of Nepalese hill farmers (Box 4).

Comparison of farmers' knowledge with science can either help to identify research objectives more clearly or identify areas where existing knowledge available to researchers can be directly extended to farmers. With respect to multilayered tree gardens in Sri Lanka, for example, comparison of farmers' knowledge about tree pruning with scientific understanding, revealed that farmers did not recognize suppressed trees nor the extent of the scope for removing lower branches of timber trees to improve timber quality without large reductions in increment (Hitinayake, 1996). Since the upper canopy trees have a large influence, through the shade they cast, on the potential productivity of the understory spice and beverage crops, introducing scientific knowledge of tree pruning to farmers has the potential to make a large impact on garden productivity. Similarly, in Honduras, farmers were found to be unaware of the existence of parasitoids (small, solitary wasps and flies whose larval stages are parasitic on insects that damage crops) and entomopathogens (fungal, bacterial and viral diseases that infect and kill insects) because of difficulties in observing them, but when taught about them, were able to manipulate natural enemy populations and enhance biological pest control (Bentley, 1994).

In summary, comparison of local knowledge and scientific knowledge can be powerful and facilitate development of both the farmers' knowledge system and that of researchers, leading to more relevant research and identification of opportunities for extension of existing knowledge. Comparison of knowledge systems is assisted by their formal documentation but it is important to stress that comparative analysis does not imply attempting to validate one system against another but to seek:

- common ground to facilitate communication among farmers and researchers, and
- to identify areas of complementarity where knowledge can be usefully exchanged and combined.

2.3.4 Comparison of Local Knowledge and Practice

Another revealing form of comparison is where farming practice is compared with what farmers articulate that they know about agroecology. As has been pointed out previously, knowledge and practice are not synonymous, not least because what farmers actually do represents a complex set of compromises involving consideration of ecological, economic and social factors occurring through the season, and where trees are concerned, over many seasons. Where farming practice appears to compromise productivity or sustainability, it is useful to distinguish deliberate trade-offs that farmers make, where they are cognizant of the consequences of their actions, from situations where they are unaware that aspects of their practice may be sub-optimal or have negative impacts. Such analyses, as are shown in Box 5, can reveal leverage points where research may be particularly effective.

Clearly, formal comparisons of knowledge and practice require documentation of both what farmers know and what they do. In the past, there has generally been much more information collected about practice than knowledge, and in many situations there may be sufficient descriptive information available about agroforestry practice to embark on this form of analysis if comparable information on what farmers know can be gathered. In this respect knowledge acquisition may be focused on areas where there appear to be productivity or sustainability problems in the system, specifically to ascertain by comparative analysis whether these represent knowledge gaps or a conscious trade-off.

2.3.5 Setting Priorities

Much has been written about setting priorities in natural resources research (Alston, Norton and Pardey, 1995) and the relevance of various methodologies that may be used depends very much on the scale of operation. There is neither space nor is it appropriate to address this whole subject here, but it is relevant to consider briefly how an explicit treatment of local knowledge may change views of the relative merits of alternative research strategies and impact on the adoptability of research results at the level of frontline research and development institutions working directly with client farmers. Assuming that the people who are to benefit from a research effort have been identified, prioritization of competing research proposals requires consideration of:

Box 5. Identification of Leverage Points by Comparison of What Farmers Do With What They Know

Continuing the unfolding example related to Nepalese hill farming used in Boxes 2–4, in addition to documenting farmers' and researchers' knowledge, a detailed tree inventory was conducted, permitting comparison of what farmers said about trees and how they actually incorporated them into their farming systems (Thapa, 1994). Statistical analysis of the location of trees that farmers classified as *malilo* and *rukho*, showed that farmers had a higher proportion of their *malilo* trees growing in association with crops than their *rukho* trees, consistent with *malilo* trees being considered less competitive with crops and to enhance soil fertility. In contrast, nearly half of the trees that farmers had which they classified as causing heavy *tapkan*, were grown on crop terrace risers where, according to local knowledge, they would reduce crop yield and promote soil erosion. Thus, despite having a clear understanding that large-leaved tree species were competitive with crops and promoted soil erosion, farmers still planted them in association with crops. The explanation for this was that farmers were trading off the negative impacts of these trees on crops and soil against their high fodder value at key times in the dry winter season. Subsequent analysis of formally documented local knowledge (Joshi, 1998), indicates a positive relationship between leaf size and palatability of tree fodder among the species used by farmers (>70% of trees classified by farmers as large-leaved were also classified as having high palatability, whereas <20% of trees classified as small-leaved were highly palatable). This represents a key constraint in the system where farmers are having to sacrifice crop yield and tolerate soil erosion in order to obtain fodder at key times in the season. Hence, this identifies a leverage point where research, to for example, introduce or breed a smaller-leaved tree with the same fodder characteristics as the large-leaved species that farmers are currently using, may represent an adoptable advance that addresses current constraints.

- the importance of particular areas of research to the people who are intended as beneficiaries and the likely adoptability of results,
- the likelihood of success in a particular research endeavor, the size of the potential gain and the time it is likely to take to get there, and
- the cost (the resources that will be consumed by the research effort).

Balancing these aspects, which are not unrelated from one another, clearly requires dialogue between rural people and researchers. It is the first of these areas where consideration of local knowledge can be expected to have the largest impact, and an area where agroforestry research has not always performed well (Kerkhof, 1990). There has been a lot of criticism in this respect of the prominence of the international research effort on the development of alley cropping with fast growing hedgerow species over the last decade, largely because adoption rates of the technology have been low outside a fairly restricted domain characterized by adequate rainfall and initial soil fertility and where labor is relatively abundant and land scarce (Carter, 1995; Sanchez, 1995).

One of the reasons that alley cropping received so much research attention, even in areas where it may not have been an appropriate candidate technology, was the lack of alternative agroforestry solutions for researchers to work with. This can be seen as a natural consequence of the way in which diagnostic procedures in an adaptive research context end up with researchers trying to match technology models they have available to problems encountered in the field (Raintree, 1990). The fact that in many cases farmers, who may not have adopted alley cropping *per se*, nevertheless adapted aspects or components of the technology into their farming systems (Buck. 1990; Scherr, 1990; Sturmheit, 1990), indicates that closer attention to farmer knowledge from the outset may have yielded avenues of research that would have produced more directly adoptable research results. It is evident from the examples already given in this chapter that explicit consideration of local knowledge is likely to widen the horizon that researchers consider. Where farmers have sufficiently sophisticated agroecological knowledge, it is likely to lead to research that addresses particular constraints at the level of components, rather than whole technologies, which, because they represent incremental improvements, are inherently more adoptable because they are less disruptive of the existing farming system (Scherr, 1990).

In evaluating competing proposals, it may be necessary to compare the likely adoptabiltiy of results from different research thrusts. While this remains an inherently complex task that will involve some element of subjective judgment, as well as the use of participatory methods, careful analysis of local knowledge can provide useful information to assist in decision making. For example, where local knowledge is well documented, it is possible to consider both how large a change particular innovations may be likely to cause to current knowledge and practice and to trace the likely consequences of proposed changes (which may include unintended effects). This sort of analysis essentially enhances what has been referred to as prescreening in the farming systems research literature (Collinson, 1987). However, any formal analyses of documented knowledge in this context is not intended to replace dialogue with farmers, or circumvent a participatory process, but help to identify important questions and areas that may form the basis of interaction.

2.4 Measuring the Impact of Research

Funding for natural resources research is in general tightening. The CGIAR system is faced with shrinking real budgets that hardly match their enlarged mandates, while many national research systems in developing countries are in a more restricted position (Anderson, 1997). In this climate, requirements to evaluate research impact, often against more explicit criteria imposed by donors in terms of poverty alleviation and sustainable use of natural resources, is acting to raise the profile of the monitoring and evaluation of research outcomes (Alston et al., 1995). In overall terms, as discussed previously, investment in research may have different forms of immediate outputs:

- materials (such as an improved crop variety or tool),
- knowledge increments (such as a new crop husbandry method), and
- increments to human capital (referring to enhancement of the capacity of people to innovate in the future).

Clearly these items are not unrelated. For example, it may require advances in knowledge to produce a new crop cultivar, and the possession of new knowledge may increase people's capacity to innovate. Ultimately these immediate outputs are useful only if they result in some desired goal in terms of people's living standards, principally in agroforestry, an overall increase in the productivity and sustainability of farming systems. The consideration of impact in overall terms is beyond the scope of the present discussion, but measuring the generation of immediate outputs from research and their uptake by farmers is a more tangible goal and of direct relevance to how local knowledge is viewed.

It is undoubtedly easier to measure the production of material outputs from research and their subsequent uptake by farmers than to discern and trace the impact of a knowledge increment because it is easier to identify and count material things than it is to do the same for what people do or know. Furthermore, more fundamental knowledge increments (which may have far reaching consequences for the way in which people farm, and hence ultimately a large impact) will be far more difficult to trace than the extension of a simple methodological advance, such as a specific crop rotation that controls pest or weed burdens, that can be fairly easily recognized. This has potentially serious implications, since if there is a premium on being able to demonstrate impact, then a negative feedback may be created where, because the most useful outcomes are the hardest to measure, they become more difficult to justify. This is particularly pertinent in an agroforestry context because it is a recent area of scientific endeavor and, therefore, many fundamental research questions remain (Sinclair, 1997). The need to be able to measure knowledge increments more readily is brought into sharp focus by recent though controversial suggestions that developments in how crops are managed are likely to be more important in addressing "second generation" problems in the post Green Revolution era than more conventional research aimed at material outcomes (Byerlee, 1997).

Explicit documentation of local knowledge is clearly relevant to the problem of assessing incremental change, since it is necessary to have baseline data against which to evaluate progress. Furthermore, if research institutions maintain an explicit corporate record of what is known about key areas of their research (which may represent problem foci experienced by farmers), such as tree-crop interactions or nutritive value of tree fodder, then as new knowledge is generated through the research process, it can be added to the developing knowledge base and an explicit record of knowledge increment is possible (Walker et al., 1997). Since a major output of the research process is new knowledge, then explicit definition of what knowledge research aims to produce, followed by an explicit process of recording what knowledge actually emerges from the research is a prerequisite for evaluating success. This can increasingly be achieved using formal methods that employ knowledge engineering techniques implemented on computer (described in the next section), rather than relying on paper as the principal medium to record, store and access knowledge and bibliographic techniques to assess the contributions that are made to what is known.

As with priority setting discussed in the previous section, consideration of impact depends on the scale of operation being assessed. Large scale evaluations of investment in agricultural research and extension at international and national levels by, for example, the World Bank (Purcell and Anderson, 1997) suggest a positive impact, but this generally relates to research infrastructure, staff development and operational efficiency and the delivery of select, timely messages in extension. It is notable that improvements in the human resource base for research have been highlighted by the World Bank while mixed results are acknowledged in attempts to improve research-extension-farmer linkages (Purcell and Anderson, 1997). It is precisely these linkages that formal methods of documenting local knowledge may be able to improve.

3. METHODS FOR ACQUIRING LOCAL KNOWLEDGE

3.1 Representing Local Knowledge

In order to take advantage of the opportunities for the integration of local knowledge into research planning and evaluation (as outlined above) we clearly require efficient and effective collection and collation of knowledge from local communities. Furthermore, to be useful, this knowledge needs to be recorded in some form that allows it to be effectively stored, accessed, analyzed, synthesized and, thereby, to be available for future use. So, making effective use of local knowledge requires methods of representation (the process of recording the knowledge articulated by informants) that:

- are sufficiently rigorous to provide a repeatable means of knowledge capture from disparate sources,
- result in comparable sets of knowledge from these various sources, and
- are sufficiently flexible to cater for a diverse and uncertain domain.

The approach taken to representing knowledge defines both how the knowledge can be subsequently used and how much it is transformed from the manner in which it was articulated by the source. Typically, the knowledge that someone articulates is expressed in a form of relevance to a particular discourse — in other words as a part of an argument. This does not mean, however, that the knowledge articulated could not be equally useful in some quite different context. So, the key requirement in representing the articulated knowledge is to be able to record it in a way that makes it as widely and advantageously available for use in other contexts. This means that knowledge needs to be abstracted from the particular argument in which it was articulated. We call this disaggregation. However, the process of disaggregation may cause the loss of important information about the way in which a piece of knowledge could be combined with others in reasoning about a new situation. As a result, contextual information about that knowledge must also be represented. Furthermore, having disaggregated knowledge into sets of statements there is also a requirement to manage these sets so that relationships between statements are maintained and a coherent whole results. Hence, any approach to representation must tackle three fundamental tasks, disaggregation, capturing context and managing sets of statements.

Development and evaluation of alternative means of addressing these requirements is presented in Sinclair and Walker (1998). Application of these methods demonstrated that simple and intuitive approaches to knowledge representation can provide a basis for generating coherent sets of statements in natural language that are precise enough in their meaning to be useful in developing an abstraction of knowledge held by a defined community about a defined subject. Nevertheless, the nature of language and the complexity of the agroforestry practices under consideration resulted in significant impediments to the development of useful abstractions of knowledge and, thereby, "models" of the behavior of the systems in question. Statements often do not have a single unambiguous meaning, but can be interpreted in a number of different ways, often as a result of the use of unexplained value judgments (such as poor, good or greater), the use of undefined or inadequately defined terms or inadequate description of context, assumptions or other caveats. Even where these issues are overcome, the volume of information that goes into making up a detailed account of, for example, the ecology of an agroforestry practice can become overwhelming. The inevitable uncertainty, inconsistency and disagreement that accompany a description of such a system by an individual, a community or a set of communities (that may include farmers and researchers) poses further challenges.

Many of these issues can be tackled by adopting a formal approach to the representation of knowledge. This involves ensuring that individual statements are syntactically precise and capturing the relationships that exist between statements. This can be achieved by applying a formal grammar that restricts the syntax of the representation. These techniques, borrowed from knowledge-based systems technologies, a field of artificial intelligence, ensure rigorous consideration of each statement and enable automated reasoning with statements, thus opening up various means of appraising the consistency and coherence of the knowledge base. The basic assumption underlying the application of such an approach is that the majority of articulated local knowledge about agroforestry can be adequately represented using a simple definite clause grammar (Sterling & Shapiro, 1986) that restricts the syntax of the statements. While any practical grammar does restrict expressiveness as compared to natural language, this assumption is a natural extension of our basic argument that we should make particular use of articulated local knowledge. Application of one such formal grammar is described in Walker and Sinclair (1998) while the case studies presented below relied on its use as implemented in a software toolkit called AKT (the Agroforestry Knowledge Toolkit) (Walker et al., 1995).

Diagrammatic representation of statements and the linkages between statements provides another approach to managing the complexity of systems descriptions. Diagramming is a familiar approach to communicating complex abstracts of system functioning (Spedding, 1988) and has been used successfully in arriving at consensus among people in eliciting local knowledge (Lightfoot et al., 1989). A diagramming syntax that complements the formal grammar described previously is outlined by Walker and Sinclair (1998) and is implemented as an interface to potentially large and complex knowledge bases in the AKT software (Walker et al., 1995).

3.2 Eliciting Local Knowledge

Developing a representative abstraction of local explanation of the behavior of an agroforestry practice or system is a significant undertaking and might typically involve hundreds of interviews with 50 or so informants. Approaches to interviewing are well developed and documented (see, e.g., Cordingley and Betsy, 1989, Diaper, 1989, Cooke, 1994); however, it is worth emphasizing the relationship between the processes of knowledge elicitation and representation that together constitute knowledge acquisition. Techniques for formal representation such as those outlined above provide a powerful mechanism for incremental evaluation and modification of a growing knowledge base in response to new material and provide a means of focusing further interviews. Where implementation of the formal representation enables formal reasoning, such techniques can provide a powerful mechanism for augmenting this iterative evaluation.

3.3 Strategies for Knowledge Acquisition

Development of a coherent, comprehensive and consistent knowledge base that captures what is known about the behavior of an agroecosystem in a useful way demands more than simply cycling through an iterative process of knowledge elicitation, representation and evaluation. A structured sequence of activities with differing objectives and sampling strategies at each stage maybe be required. The case studies presented in this paper used a four-stage strategy as follows (Walker and Sinclair, 1998).

1. *Scoping* – an introduction to the source community which may use methods familiar from approaches to rapid rural appraisal and participatory rural appraisal.
2. *Definition* – development of an overall understanding of the domain, defining boundaries and identifying terminology.
3. *Compilation* – recording detailed knowledge on the domain by repeated interviews with a small, purposively selected set of key informants.
4. *Generalization* – testing the representativeness of the elicited knowledge of the community as a whole in order to: validate the knowledge base as representative of the knowledge held by the community; explore the distribution of knowledge among people within the community; and augment the knowledge base with detail not recorded in the compilation phase.

3.4 Summary

In this section we have argued that formal approaches to knowledge elicitation and representation provide a powerful means of developing concise and coherent abstractions of the knowledge held by local people about the practices that they pursue. While we have illustrated this point with reference to one particular set of formal approaches, our fundamental position is that some formal approach to representation is required if testable and, thereby, defensible, abstractions of local knowledge systems are to be generated and used in informing development professionals, the planning and prioritizing of research and extension, and the evaluation of potential or actual impact of research.

4. CASE STUDIES

In this section brief details of the outcome of applying formal methods of knowledge acquisition in three contrasting agroecological environments in Nepal, Sri Lanka and Kenya are presented and the usefulness of the approach in these different circumstances compared. The comparability of these case studies lies in the common methodology used and the fact that they all relate to some aspect of agroforestry. In each case local researchers, all graduate natural resource scientists, were involved in developing the AKT knowledge-based systems approach to agroforestry research and extension (Walker et al., 1995), an outline of which has already been given in the previous section. The AKT software and approach evolved iteratively in response to the needs and knowledge encountered at five field sites, including the three discussed below, and two others, involving rangeland degradation in Tanzania (Kilahama, 1994) and the sustainability of jungle tea in hill evergreen forest in northern Thailand (Preechapanya, 1996).

4.1 Tree Fodder in Nepal

Initially the context for this work was a single hill farming community in Nepal in which a detailed study of what farmers knew about their on-farm tree fodder resources was conducted (Thapa, 1994; Thapa et al., 1995). This revealed a widespread and sophisticated understanding of tree-crop interactions, the feeding value of tree fodder (Thapa et al., 1997) and how these were affected by tree management actions.

Boxes 2–5 illustrate general points made in previous sections. Subsequent research has investigated, on the one hand, the use of local knowledge in planning research and extension at an institutional level at Pakhribas Agricultural Centre, a frontline research centre with a mandate for agricultural research throughout the eastern mid hills of Nepal (Walker et al., 1997; Joshi and Sinclair, 1997; Joshi, 1998) and, on the other hand, the comparability and complementarity of farmers' detailed knowledge of fodder quality with laboratory assessments of nutritive value (Thorne et al., 1997).

Contrary to initial expectations, conceptual understanding of interactive processes such as shading and splash erosion caused by water droplets falling from tree leaves (*tapkan*) were consistently held by people across gender, ethnic, and attitudinal divides. This conceptual knowledge, based on contemporary experience and observations of farmers, was found to be systematic (Figure 2) and to underpin decision-making about agroforestry practice (Boxes 2 and 5). The knowledge was found to be both comparable with scientific knowledge and to complement it. While farmers' knowledge was actually in advance of conventional scientific wisdom until 1993 with respect to canopy modification of rainfall drop size (Box 2), the more important complementarity lay in the fact that farmers knew a lot about how attributes of trees and tree fodder varied among the 90 tree species that were cultivated on farm land and within species, how attributes varied through the season. While the overall conceptual knowledge referred to earlier was held by all members of the farming community, detailed knowledge of particular tree attributes and how they varied, was distributed according to the experience that particular farmers had of growing particular tree species, and this varied markedly across different locations and within locations according to gender, with women in general having more detailed knowledge about fodder value than men. The non-uniformity of the distribution of detailed knowledge resulted in considerable utility in simply gathering local knowledge from different people to create an encyclopedic resource.

With respect to the comparability of farmers' knowledge with that of researchers, while farmers' conceptual understanding was certainly systematic, it generally involved a greater level of aggregation. Thus, with respect to the impacts of trees on soil fertility, farmers combined competitiveness (in terms of resource capture) with fertility enhancement (in terms of amount and quality of leaf litter fall) in classifying trees as either *malilo* or *rukho* (Box 3). Similarly, farmers classified tree fodder using two independent systems of categorization which were related to the overall practical effects on animals of feeding particular tree fodder as a dietary supplement. Thus, the *posilopan* of a fodder was considered to affect animal productivity and health, and appears to bear some relation to protein supply to the duodenum in ruminants, while *obanopan* (literally referring to the hotness and dryness of a fodder) was said to determine the extent to which particular fodders satisfied appetite (affecting animal behavior) and influenced the consistency of dung (of practical relevance in a system where the dung is manually collected for composting), and appears to have some correspondence to overall dry matter digestibility. (See Pell, Chapter 2.) Thus, while some simple attributes map quite closely from the local to the scientific knowledge system, such as leaf bitterness to tannin content (Box 4), farmers also have broad classificatory concepts that aggregate a number of aspects that scientists tend to see as distinct because these aggregated concepts are of direct practical relevance to farmers.

The most important revelation from this work relates to the way in which it confirms a multiplicity of views about local knowledge, while dispensing with the need for them to be seen to conflict. Thus, farmers' conceptual understanding was commonly held in keeping with authors who have found similarly widespread understanding of general agroecological processes in, for example, Guinea (Fairhead and Leach, 1994), Ghana (Millar, 1994), Sierra Leone (Richards, 1994), Honduras (Bentley, 1994), Guatemala and Peru (Rhoades and Bebbington, 1995). But detailed species-specific knowledge was socially differentiated in keeping with other authors who have focused attention on differences in knowledge according to, for example, gender (Rusten and Gold 1991), socio-economic status (Winarto, 1994) and age (Brokensha and Riley, 1980). Farmer knowledge was found to be systematic and comparable with scientific knowledge (Berlin, 1992; Richards, 1994) but also different from it in many important respects, some of these conferring complementarity (Richards, 1994) others asserting different utilitarian knowledge structures and classifications that are locally situated, as has been found elsewhere (van der Ploeg, 1989). Thus in Nepal we have a situation where an appraisal of local knowledge conforms both to assertions about the importance of multiple knowledges associated with different actors in the development process and the existence of widely held systematic knowledge of agroecology comparable with science that can be usefully abstracted. This clarity, demonstrating how aspects of farmers' knowledge in Nepal simultaneously accords with various views of local knowledge that are often presented as mutually exclusive alternatives, comes from having an explicit and durable record of the local knowledge system.

4.2 Multilayered Tree Gardens in Sri Lanka

The context for this work was the complex and species-diverse multilayered tree gardens in upcountry Sri Lanka, often referred to as Kandy Forest Gardens (Perera and Rajapakse, 1991). The trees, however, generally are planted rather than natural forest vegetation being involved and similar species associations are found throughout Sri Lanka rather than being confined to Kandy district. Key aspects of the gardens are their traditional nature and long term sustainability. Work on local knowledge in conjunction with the universities of Peridenya and Sri Jayawardenepura was stimulated by frequent though unsupported assertions by agricultural scientists that, despite the seemingly random arrangement of plants in the gardens, farmers must have a systematic understanding of species interactions and arrange plants accordingly (Jacob and Alles, 1987). Initial research, therefore, set out to acquire local knowledge of species interactions (Southern, 1994) and plant selection and siting (Jinadasa, 1995) and was followed by a comparative study of the use of conventional Farming Systems Research and Extension (FSRE) methods with knowledge-based systems techniques (Hitinayake, 1996; Hitinayake and Sinclair, 1996).

In stark contrast to the widespread and systematic local knowledge about tree-crop interactions found in Nepal, knowledge acquisition about plant interactions in Sri Lanka did not reveal a coherent system of agroecologcial knowledge related to how plants were arranged. Farmers did articulate knowledge of the shade tolerance and requirements of some species for light, water and nutrients and employed a fairly detailed classification of soil based on color, texture and fertility to describe microsites within their gardens. However, the actual distribution and arrangement of plants in the gardens was not well explained by what farmers knew about resource requirements of different species. This concurs with Richard's (1989) explanation of local intercropping practices as the result of a performance, in which farmers make incremental decisions through time about plant siting and management, so that the resultant plant arrangement does not conform to a planned design. Ecological considerations about species interactions and requirements are one aspect which may be considered by farmers at each point in time that a management decision is made, but other

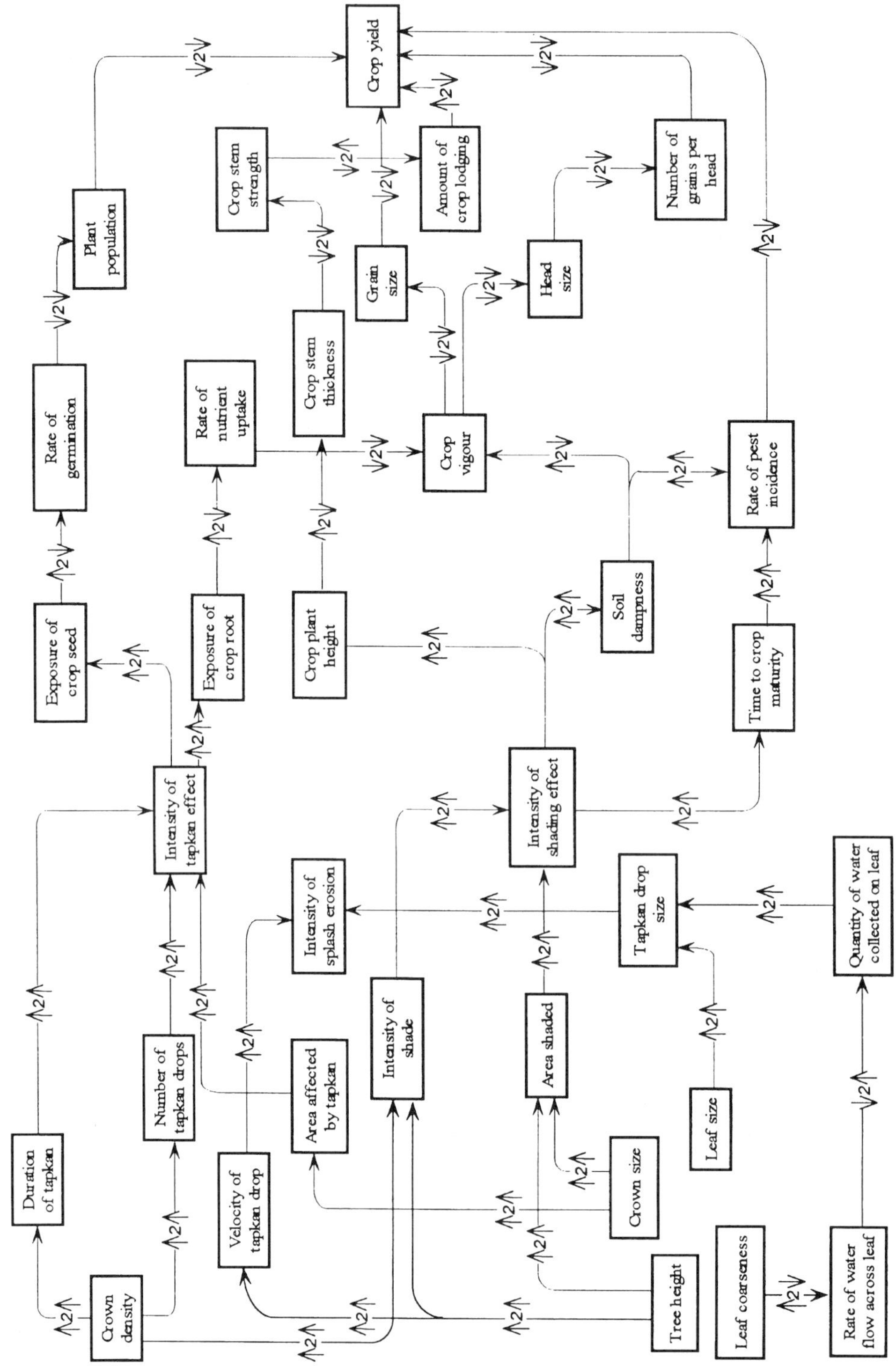
Crop yield
Plant population
Crop stem strength
Amount of crop lodging
Number of grains per head
Rate of germination
Rate of nutrient uptake
Crop stem thickness
Grain size
Head size
Crop vigour
Exposure of crop seed
Exposure of crop root
Crop plant height
Soil dampness
Rate of pest incidence
Time to crop maturity
Intensity of tapkan effect
Intensity of shading effect
Intensity of splash erosion
Tapkan drop size
Quantity of water collected on leaf
Duration of tapkan
Number of tapkan drops
Area affected by tapkan
Intensity of shade
Area shaded
Leaf size
Crown size
Velocity of tapkan drop
Crown density
Tree height
Leaf coarseness
Rate of water flow across leaf

FIGURE 2.

socio-economic aspects and convenience will also be involved. Clearly in a perennial system with long-lived components and year-round planting and harvesting, the extent of the aggregation of these small incremental decisions in terms of the resultant vegetation structure will be far greater than in seasonal cropping patterns. There is also, however, a second level of contrast with the Nepalese case study in that the farmers in Sri Lanka were far less dependent on the outcome of species interactions both because of the relatively benign environmental conditions that made specific decisions about what to grow less critical and because the gardens were a supplementary activity to staple production of paddy rice elsewhere. In Nepal, farmers in marginal conditions were increasingly having to grow fodder trees on staple crop land (because of declining fodder supply from adjacent common property forest areas), thus knowledge of tree-crop interactions was critical to their overall farm productivity and livelihood whereas, in Sri Lanka, garden productivity was far less critical to farmers. This relationship between dependence on resources and the extent and coherence of local agroecological knowledge about them was also borne out within Nepal when knowledge from different communities was compared (Joshi and Sinclair, 1997).

The supplementary nature of the gardens actually leads to a great variability in garden structure, which appears to be largely based on idiosyncratic objectives and management by individual farmers mitigating against a widespread local conceptual framework that can be used as a basis for researchers to understand the system. This has major implications for what research may be relevant and how this can be identified. Given the supplementary nature of gardens, their complexity and variability and the idiosyncrasy of farmers' objectives, it is not surprising that attempts by researchers to produce model garden designs which improve on traditional practice have not been successful (Sinclair, 1996). There are prospects, however, discernible from consideration of local knowledge, for interventions at a more basic level, dealing not with whole garden structures but with specific components which occur frequently, and their management. Nearly all gardens, for example, have upper canopy timber trees as important components, and so selection and breeding of associative timber tree ideotypes for multispecies associations (rather than even-aged, monocultural plantations) and development of tree pruning strategies that optimize timber increment and light transmission to lower stories could have a large impact on the productivity of gardens generally despite their variable nature. This illustrates the general finding that analysis of local knowledge can often reveal opportunities for improving parts of complex and variable farming systems that farmers can then incorporate into their practice.

4.3 Fruit Tree Management in Kenya

In Kenya, the knowledge-based systems techniques, that had developed through work in other contexts and countries (see above) to a stable methodology, were applied in conjunction with KEFRI (the Kenyan Forestry Research Institute) to the consideration of fruit trees grown on farms in a seasonally arid area of Machakos district (Kiptot, 1996). This work revealed knowledge about tree-crop interactions that was roughly intermediate in its sophistication between that encountered in Nepal and Sri Lanka. The main areas of knowledge that were documented related to fruit quality,

---

FIGURE 2. Diagrammatic representation of farmers' causal knowledge of tree-crop interactions in Solma village in the mid hills of eastern Nepal. Arrows connecting one box to another show the direction of causal influence, breaking each of these arrows is a symbolic explanation of the nature of the influence with: upward arrows referring to increases and downward arrows to decreases; the first arrow refers to whether the causal factor is increasing or decreasing and the second arrow conveys the same information with respect to the effect factor; the number between the arrows refers to whether causation is one-way (1) or two-way (2). Two-way causation is where a decrease in x causing an increase in y also implies that an increase in x will cause a decrease in y, this reversibility does not hold in one-way causal relationships. (Adapted from Thapa, B., *Farmers ecological knowledge about the management and use of farmland tree fodder resources in the mid-hills of eastern Nepal*, Ph.D. Thesis, University of Wales, Bangor, 1994.)

pest and disease interactions and effects of trees on soil and crop production. The aim here is to point to aspects of this third case study that confirm some general patterns encountered in other knowledge systems that have been studied.

Taking, as an example, farmers' knowledge about effects of trees on soils and crop yield, it was found that they generally understood a number of processes associated with the way in which tree leaf litter contributed to soil fertility (Figure 3). This knowledge was based on observation and deliberate experimentation. Thus, farmers could classify the litter of different tree species in terms of its usefulness as a fertilizer, so that while at one end of the scale mango (*Mangifera indica*) leaf litter was not considered to be a good green manure, that of *Balanites aegyptiaca* was highly valued as fertilizer. Their knowledge about the comparative fertilizing power of leaf litters was derived from deliberate trials. Farmers generally applied fertilizer to patches of crop fields where yields were lower than elsewhere in the field and assessed fertility effects by the extent to which this improved yield on this area in subsequent years. By applying different leaf litters separately, they were able to assess their relative performance. This conforms with the sort of farmer experimentation and knowledge generation found in a range of other contexts as previously discussed, but it is also important to note that the farmers' conception of fertility represents a local construct that combines what researchers consider various independent aspects of soil fertility. This is directly comparable to the way in which the Nepali farmers' practically-based classifications of fodder value contrasted with scientific knowledge. In the present case, farmers recognized three principal effects of leaf litter — their nutrient contribution (referred to by farmers as manure), effects of decaying litter on surface soil characteristics and hence infiltration of water and the retention of soil moisture by leaf mulch. It is interesting here that the farmers' description of soil softness, although not immediately transferable to any scientifically defined attribute of soil structure, compares with how farmers describe soils in completely different contexts. For example, Andean farmers describe soils as *suavecita* (soft) or *dura* (hard) referring to surface structure but related in this context to the extent to which soil has been tilled in previous years (van der Ploeg, 1989). One of the most ubiquitous conceptualizations that farmers use, which does not map easily to scientific knowledge is a hot/cold or heating/cooling conceptual pair, which may be variously applied. Thus, in Nepal, farmers use *obano/chiso* to refer to the extent to which tree fodder satisfies animal appetite; in Sri Lanka farmers describe plants as *seraiy* or *sitelay* when referring to their competitiveness (cooling species are thought to conserve moisture and so be less competitive to neighboring plants); in Thailand certain ground flora species such as *Eupatorium adenophorum*, are considered beneficial in jungle tea gardens because they keep soil cool and moist (*din yen*), while others such as *Imperata cylindrica* are thought disadvantageous because they promote hot soil (*din ron*) (Preechapanya, 1996); Andean farmers use *fria/caliente* to refer to soil fertility associated, though not directly, to subsoil nutrient and humus content (van der Ploeg, 1989).

The significance of these observations is that they suggest general patterns in the way in which farmers in quite different cultures and agroecological conditions conceive and describe interactions among trees, crops and soil, as well as revealing particularities associated with differing practical imperatives. Explicit representation of farmers' explanations of their agroecological knowledge is clearly a useful first step in appreciating what farmers know and in considering its proximity to the scientific knowledge being used and generated in the formal agricultural research sector.

5. CONCLUSIONS

While the central role of local knowledge in the development process is increasingly widely acknowledged, the approaches to realizing the roles that have been applied to date are not sufficiently rigorous to make full use of it. Participatory methods in development are a significant revolution and provide important opportunities for local people to collaborate in guiding development agendas that are appropriate to local values, aspirations and needs. However, we suggest that participatory methods alone will either be hampered by the use of inappropriate, externally generated "models"

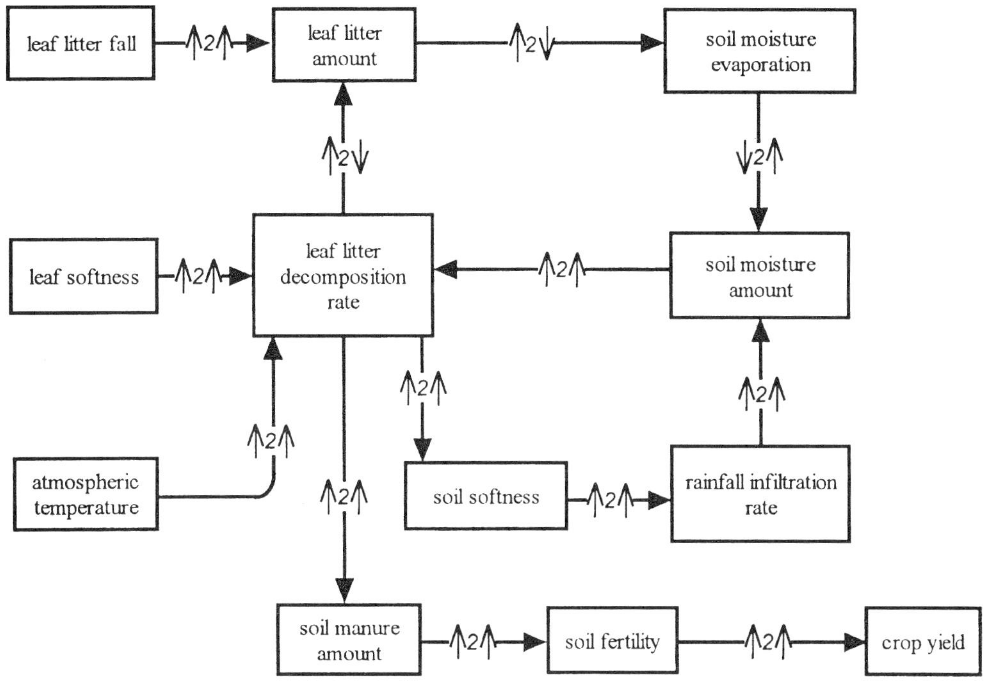

FIGURE 3. Diagrammatic representation of farmers' causal knowledge of effects of tree leaf litter on soil and crop yield in Katangi, Kenya. For an explanation of symbols used, see caption to Figure 2. (Adapted from Kiptot, E.C., *An investigation of farmers' ecological knowledge about fruit trees grown on farms in South Yatta, Kenya*, MPhil thesis, University of Wales, Bangor, 1996.)

of the system (whether formal or implicit) held by natural resource professionals or be hampered by an inadequate systems perspective on the part of those professionals — resulting in programs which may be hard to evaluate, inefficient and miss significant opportunities for development. We have argued that it is important that program design and execution involve a formal and explicit consideration of local explanation of system structure and function and that this is best achieved through the application of formal methods that result in testable representations of local knowledge. However, the methods for achieving these objectives need significant further development and the implications associated with their adoption need to be carefully evaluated.

REFERENCES

Agrawal, A., 1995, "Indigenous and scientific knowledge: some critical comments," *Indigenous Knowledge and Development Monitor*, 3(3), 3–6.

Alston, J. M., Norton, G. W. and Pardey, P. G., 1995, *Science under Scarcity: Principles and Practice for Agricultural Research Evaluation and Priority Setting*, Cornell University Press, Ithaca, NY.

Anderson, J. R., 1997, *Millennial challenges in international agricultural research: Striving for international public goods in an era of donor fatigue*, Symposium paper for IAAE Conference, August, Sacramento, CA.

Asturias, 1949, *Men of Maize*, Editorial Losada, S.A., Buenos Aires, Argentina, 335.

Bentley, J. W., 1994, "Stimulating peasant farmer experiments in non-chemical pest control in Central America," in *Beyond Farmer First*, Scoones, I. and Thompson, J., Eds., Intermediate Technology Publications, London, 147–150.

Berlin, B., 1992, *Ethnobiological classification: Principles of categorization of plants and animals in traditional societies.* Princeton University Press, Princeton, NJ, 335.

Berlin, B., 1973, "Folk Systematics in Relation to Biological Classification and Nomenclature," *Ann. Rev. Ecology Systematics,* 4, 259.

Blaike, P., Brown, K., Stocking, M., Tang, L., Dixon, P., and Sillitoe, P., 1997, "Knowledge in action: local knowledge as a development resource and barriers to its incorporation in natural resource research and development," *Ag. Syst.*, 55(2), 217.

Brandt, C. J., 1989, "The size distribution of throughfall drops under vegetation canopies," *Catena*, 16, 507.

Brokensha, D. and Riley, B. W., 1980, "Mbeere knowledge of their vegetation, and its relevance for development (Kenya)," in *Indigenous Knowledge Systems and Development,* Brokensha, D., Warren, D. M., and Werner, O., Eds., University Press of America, Ch.6.

Buck, L., 1990, "Planning agroforestry extension projects, the CARE international approach in Kenya," in *Planning for Agroforestry,* Budd; W. W., Duchhart, I.; Hardesty, L. H., and Steiner, F., Eds., *Developments in landscape management and urban planning, 6c,* Elsevier Science Publishing, Amsterdam, 321–335.

Bunch, R., 1989, "Encouraging farmers' experiments," in *Farmer First,* Chambers, E., Pacey, A., and Thrupp, L. A., Eds., Intermediate Technology Publications, London, 55.

Bunders, J., Haverkort, B., and Hiemstra, W., 1996, *Biotechnology: building on farmers knowledge,* MacMillan Education, U.K., 256.

Byerlee, D., 1997, "Knowledge-intensive crop management technologies: Concepts, impacts and prospects in Asian Agriculture," in *IRRI conference proceedings, mimeo,* World Bank, Washington, DC.

Carter, J., 1995, "Alley farming: have resource-poor farmers benefited?" *Nat. Res. Perspect,* 3. Overseas Development Institute, London, 10.

Checkland, P. and Scholes, J., 1990, *Soft systems methodology in action,* John Wiley & Sons, Chichester, U.K.

Collinson, M., 1987, "Farming systems research: Procedures for technology development," *Exp. Agri.*, 23, 365.

Colquhoun, M., 1997, "An exploration into the use of Goethean science as a methodology for landscape assessment: The Pishwanton Project," *Agri. Ecosys. Env.* 63, 145.

Cooke, N. J., 1994, "Varieties of knowledge elicitation techniques," *Int. J. Human-Computer Stud.*, 41, 801.

Cordingley, E. and Betsy, S., 1989, "Knowledge elicitation techniques for knowledge based systems," in *Knowledge elicitation: Principles, techniques and applications,* Diaper, D., Ed., Ellis Horwood Ltd., Chichester, U.K.

Cornwall, A, Guijt, I., and Welbourn, A., 1994, "Acknowledging process: methodological challenges for agricultural research and extension," in *Beyond Farmer First,* Scoones I. and Thompson, J., Eds., Intermediate Technology Publications, London, 98–117.

Diaper, D., 1989, *Knowledge elicitation: principles, techniques and applications,* Ellis Horwood Ltd., Chichester, U.K.

Dove, M. R., 1981, *Swidden agriculture in Indonesia: The subsistence strategies of the Kalimantan Kantu,* Mounton Publishers, Berlin.

Engel, P. G. H., 1995, *Facilitating innovation,* Agricultural University Press, Wageningen, Germany.

Fairhead, J. and Leach, M., 1994, "Declarations of Difference," in *Beyond Farmer First,* Scoones I. and Thompson, J., Eds., Intermediate Technology Publications, London, 75.

Hall, R. L. and Calder, I. R., 1993, "Drop size modification by forest canopies: measurements using a disdrometer," *J. Geophys. Res.* 98,18, 465.

Heyd, T., Warren, D. M., Showers, K. B., Serrano, R. C., Semali, L., Köhler-Rollefson, I. U., van Hooft, K., Hess, C. G., Haverkort, B., Dialla, B. E., and Brokensha, D. W., 1996, "Comments on article by Arun Agrawal," *Indigenous Knowledge and Development Monitor,* 4(1), 12–19.

Hitinayake, G., 1996, *Farmers' knowledge and the development of complex agroforestry practices in Sri Lanka,* Ph.D. thesis, University of Wales, Bangor, U.K., 275.

Hitinayake, H.M.G.S.B. and Sinclair, F.L., 1996, "Categorising complex agroforestry practices for farming systems research and extension (FSR/E): The case of forest gardens in Kandy district, Sri Lanka," *Proceedings of the 14th international symposium on sustainable farming systems, Colombo, November, 1996: Thematic Workshop 4: methodological issues: systemic questions to basic disciplines,* 45-69.

Howes, M. and Chambers, R., 1980, "Indigenous technical knowledge: analysis, implications and issues," in *Indigenous knowledge systems and development,* Brokensha, D., Warren, D. M., and Werner, O., Eds., University Press of America.

IDS, 1979, "Rural development: Whose knowledge counts?" *IDS Bulletin* 10(2). Institute of Development Studies, University of Sussex.

Jacob, V. J. and Alles, W. S., 1987, "The Kandyan gardens of Sri Lanka," *Agro. Syst.*, 5.

Jinadasa, P. N. M., 1995, *Indigenous ecological knowledge about mother plant selection and plant siting in Kandy homegradens of Sri Lanka,* M.Phil. thesis, University of Wales, Bangor, U.K., 123.

Jones, J. G. W. and Street, P. R., 1990, *Systems Theory applied to agriculture and the food chain*, Elsevier Science Publishers, London, 365.

Joshi, L., 1998, *Incorporating farmers knowledge in the planning of interdisciplinary research and extension,* Ph.D. thesis, University of Wales, Bangor, U.K.

Joshi, L. and Sinclair, F. L., 1997, *Knowledge acquisition from multiple communities*, Report, School of Agricultural and Forest Sciences, University of Wales, Bangor, U.K., 34.

Joshi, L. and Devkota, N., 1996, "Effect of *Ficus auriculata* on maize and millet crops at mid altitude *Bari*." *PAC Working Paper* 153. Pakhribas Agricultural Centre, Dhankuta, 34.

Kendon, G., Walker, D. H., Robertson, D., Haggith, M., Sinclair, F. L., and Muetzelfeldt, R. I., 1995, "Supporting customised reasoning in the agroforestry domain," *New Rev. App. Expert Sys.,* 1, 179–192.

Kerkhof, P., 1990, *Agroforestry in Africa. A Survey of Project Experience,* Panos Publication Ltd., London, 216.

Kilahama, F. B., 1994, *Trees and indigenous ecological knowledge about agroforestry practices in the rangelands of Shinyanga region, Tanzania*, Ph.D. thesis, University of Wales, Bangor.

Kiptot, E.C., 1996, *An investigation of farmers' ecological knowledge about fruit trees grown on farms in South Yatta, Kenya*, MPhil thesis, University of Wales, Bangor, U.K.

Knight, G. G., 1980, "Ethnoscience and the African farmer: rational and strategy (Tanzania)," in *Indigenous knowledge systems and development*, Brokensha, D., Warren, D. M. and Werner, O. Eds., University Press of America.

Leakey, R. R. B., 1996, "Definition of agroforestry revisited." *Agrofor. Today,* 8(1), 5.

Lightfoot, C., de Guia., Aliman, A., and Ocada, F., 1989, "Systems diagrams to help farmers decide in on-farm research," in *Farmers first: Farmer innovation and agricultural research*, Chambers, R., Pacey, A. and Thrupp, L. A., Eds., Intermediate Technology Publication Limited, London, 93.

Millar, D., 1994, "Experimenting farmers in northern Ghana," in *Beyond Farmer First*, Scoones, I. and Thompson, J., Eds., Intermediate Technology Publications, London, 160.

Murphy, W. P., 1980, "Secret knowledge as property and power in Kpelle society," *Africa* 50(2), 193.

Nair, P. K. R., 1989, *Agroforestry systems in the tropics.* Kluwer Academic Publishers, London.

Niamir, M., 1990, *Community Forestry, decision making in natural resource management in arid and semi-arid Africa,* Community Forestry Note 4. Food and Agriculture Organization of the United Nations.

Palm, C. A., 1995, "Contribution of agroforestry trees to nutrient requirements of intercropped plants," in *Agroforestry: Science, Policy and Practice*, Sinclair, F. L., Ed., Kluwer Academic Publishers, Netherlands, 105–124.

Paudel, K. C., Pandit, R., Amatya, L. K., Gurung, D. B., Bhandari, H. S., Harding, A. H., and Bhattarai, S. P., 1997, *Agroforestry research strategy for Lumle Agricultural Research Centre, 1997–2001*, 37.

Perera, A. H. and Rajapakse, R. M. N., 1991, "A baseline study of Kandyan Forest Gardens of Sri Lanka: Structure, composition and utilization," *Forest, Ecology and Management*, 45, 269, 1991.

Popper, K. R., 1972. *Objective knowledge: An evolutionary approach*, Clarendon Press, Oxford, U.K., 380.

Preechapanya, P., 1996, *Indigenous ecological knowledge about the sustainability of tea gardens in the hill evergreen forest of northern Thailand*, Ph.D. Thesis, University of Wales, Bangor, U.K., 286.

Pretty, J. N., 1995, *Regenerating Agriculture: polices and practice for sustainability and self-reliance*. Earthscan Publications, London, 320.

Pretty, J. N. and Chambers, R., 1994, "Towards a learning paradigm: New professionalism and institutions for a sustainable agriculture," in *Beyond Farmer First*, Scoones, I. and Thompson, J., Eds., Intermediate Technology Publications, London, 182–202.

Primbert, M. and Pretty, J., 1997, *Diversity and sustainability in community based conservation*, paper presented at the UNESCO-IIPA regional workshop on Community-based conservation, India, February 9–12.

Purcell, D. L. and Anderson, J. R., 1997, *Agricultural Extension and Research: Achievements and Problems in National Systems*, The World Bank, Washington, DC, 282.

Raintree, J. B., 1990, "Theory and practice of agroforestry development and design," in *Agroforestry Classification and Management,* MacDicken, K. G. and Vergara, N. T., Eds, John Wiley & Sons, New York, 58.

Ranasinghe, D. M. S. H. K. and Newman, S. M., 1997, "Agroforestry research and practice in Sri Lanka," *Agro. Syst.* 22(2), 119.

Rhoades, R. and Bebbington, A., 1995, "Farmers who experiment: An untapped resource for agricultural research and development," in *The cultural dimension of development,* Warren, D. M., Slikkerveerm L. J., and Brokensha, D., Eds., Intermediate Technology Publications, London, Ch.22.

Richards, A. I., 1939. *Land, labor and diet in northern Rhodesia.* Oxford O.U.P. for the International Institute of African Languages and Culture.

Richards, P., 1980, "Community environmental knowledge," in *African rural development, Indigenous knowledge systems and development,* Brokensha, D., Warren, D. M., and Werner, O., Eds., University Press of America.

Richards, P., 1994, "Local knowledge formation and validation: the case of rice in central Sierra Leone," in *Beyond Farmer First,* Scoones, I. and Thompson, J., Eds., Intermediate Technology Publications, London, 165.

Richards, P., 1989, "Agriculture as a performance," in *Farmer First,* Chambers, R., Pacey, A., and Thrupp, L. A., Eds, Intermediate Technology Publications, London, 39.

Richards, P., 1985, *Indigenous Agricultural Revolution.* Hutchinson & Co.

Rocheleau, D. E., 1987, "The user perspective and the agroforestry research and action agenda," in *Agroforestry: Realities, possibilities and potentials,* Gholz, H. L., Ed, Martinus Nijhof Publishers.

Röling, N., 1996, Creating human platforms to manage natural resources: first results of a research program, in *Agricultural R & D at the crossroads: Merging systems research and social actor approaches,* Budelman A., Ed., Royal Tropical Institute, Netherlands, 149.

Rusten, E. P. and Gold, M. A., 1991, "Understanding an indigenous knowledge system for tree fodder via a multi-method on-farm research approach," *Agro. Syst.,* 15, 139–165.

Sallas, M. A, 1994, "'The technicians only believe in science and cannot read the sky': The cultural dimension of the knowledge conflict in the Andes," in *Beyond Farmer First,* Scoones, I. and Thompson, J., Eds., Intermediate Technology Publications, London, 57–68.

Sanchez, P. A., 1995, "Science in Agroforestry," in *Agroforestry: Science, Policy and Practice,* Sinclair, F. L., Ed., Kluwer Academic Publishers, Netherlands, 5–55.

Scherr, S. J., 1990, "The diagnosis and design approach to agroforestry project planning and implementation: examples from western Kenya," in *Planning for Agroforestry,* Budd, W. W., Duchhart, I., Hardesty, L. H., and Steiner, F., Eds., Elsevier, Netherlands, Ch. 6.

Schroth, G., 1995, "Tree root characteristics as criteria for species selection and systems design in agroforestry,: in *Agroforestry: Science, Policy and Practice,* Sinclair, F. L., Ed., Kluwer Academic Publishers, Netherlands, 125–144.

Scoones, I. and Thompson, J., 1994, "Knowledge, power and agriculture — towards a theoretical understanding," in *Beyond Farmer First,* Scoones, I. and Thompson, J., Eds., Intermediate Technology Publications, London, 16.

Simmonds, N. W., 1985, "Farming Systems Research: A Review," *World Bank Technical Paper,* 43.

Sinclair F. L., 1997, *Professional update on agroforestry.* Department for International Development — Natural Resources Advisors Conference, 13.

Sinclair, F. L., Walker, D. H., 1998, "Acquiring qualitative knowledge about complex agroecosystems. Part 1: Representation as natural language," *Ag. Syst.,* 56(3), 341–363.

Southern, A. 1994, *Acquisition of indigenous ecological knowledge about forest gardens in Kandy district, Sri Lanka,* M.Phil. thesis, University of Wales, Bangor, U.K.

Spedding, C. R. W., 1988. *An introduction to Agricultural Systems,* Elsevier Applied Science, U.K., 189.

Sterling, L., and Shapiro, E., 1986. *The Art of Prolog,* MIT Press.

Sturmheit, P., 1990, "Agroforestry and soil conservation needs of smallholders in southern Zambia," *Agro. Syst.,* 10(3), 265.

Thapa, B., 1994, *Farmers ecological knowledge about the management and use of farmland tree fodder resources in the mid-hills of eastern Nepal,* Ph.D. thesis, University of Wales, Bangor, U.K., 271.

Thapa, B., Sinclair, F. L., and Walker, D. H., 1995, "Incorporation of indigenous knowledge and perspectives in agroforestry development. Part 2: Case-study on the impact of explicit representation of farmers' knowledge," *Agro. Syst.,* 30, 249.

Thapa, B., Walker, D. H., and Sinclair, F. L., 1997, "Indigenous knowledge of the feeding value of tree fodder," *An. Feed Sci. Tech.,* 67, 97–114.

Thornes, J., 1989, Solutions to soil erosion. *New Scientist* (3 June), 45.

Thorne, P. J., Sinclair, F. L., and Walker, D. H., 1997, "Using local knowledge of the feeding value of tree fodder to predict the outcomes of different supplementation strategies," *Agro. Forum*, 8(2), 45.

Tyler, S., 1987, *The Unspeakable: Discourse, Dialogue, and Rhetoric in the Postmodern World*, University of Wisconsin Press, Madison, WI.

Van der Ploeg, J. D., 1989, "Knowledge systems, metaphor and interface: the case of potatoes in the Peruvian highlands," in *Encounters at the Interface: a Perspective of Social Discontinuities in Rural Development*, Long, N, Ed., Wageningse Sociologishce Studies, 27.

Van Noordwijk, M., Ong, C., and Buresh, R., 1996, "Current root research issues at ICRAF," *Agro. Forum*, 7(1), 12.

Walker, D. H., Sinclair, F. L., 1998, "Acquiring qualitative knowledge about complex agroecosystems. Part 2: Formal Representation," *Ag. Syst.*, 56(3), 365–386.

Walker, D. H., Sinclair, F. L., Joshi, L., and Ambrose, B., 1997, "Prospects for the use of corporate knowledge bases in the generation, management and communication of knowledge at a frontline agricultural research centre," *Ag. Syst.* 54(3), 291.

Walker, D. H., Sinclair, F. L., and Kendon, G., 1995, "Development of a knowledge based systems approach to agroforestry and extension," *AI applications* 9 (3), 61.

Winarto, Y. T., 1994, "Encouraging knowledge exchange: integrated pest management in Indonesia," in *Beyond Farmer First*, Scoones, I. and Thompson, J., Eds., Intermediate Technology Publications, London, 150.

12 Trees in Managed Landscapes: Factors in Farmer Decision Making

J. E. Michael Arnold and Peter A. Dewees

1. INTRODUCTION

The presence of trees in part of contemporary farming systems has its origins in two attributes of trees. One is their role in sustaining crop production and their impact on the physical environment, most notably through the restoration of soil nutrients and energy, and protection against damage from wind and water. The other is the role various tree products play in the household economy. This includes products used directly by the rural households as food, fuel, construction materials, etc.; inputs to agriculture such as fodder, mulch and raw materials for making agricultural implements and storage structures; and products or activities that provide household members with employment and income. The presence or absence of trees also may have a role in securing or

1-56670-294-1/99/$0.00+$.50

maintaining rights of use or tenure. Simple visual observation discloses that today there are few farming systems which do not incorporate trees in some fashion or another.

Until recently — in aggregate large but widely dispersed — the tree resource present in the rural landscape has been neglected. In comparison with what is known about the crop and livestock components of agriculture, very little is known about existing tree management practices, about farmers' perceptions of the value of trees and of different tree outputs in meeting their needs and production objectives, and about the constraints farmers face that limit their potential to develop tree resources within their farming system. Programs to stimulate tree management at this level have been severely hindered by this lack of knowledge.

This paper presents results of recent research and analysis undertaken to address some of these gaps in knowledge. The analysis moves away from the needs-based approach that dominated much earlier work on the subject, and instead examines tree management in terms of farmer livelihood strategies and of the dynamics of rural change. It draws in particular on a number of detailed case studies of situations in eastern Africa and south Asia*.

In the next part of the paper we examine the patterns of tree management that farmers adopt in different agroecological and economic situations, and the farm household strategies that influence farmer decisions about tree growing. The subsequent section looks at on-farm tree management from the perspective of recent development thinking and strategies, and the final section identifies policy implications related to the identification and design of interventions in support of tree growing by farmers.

2. TREE MANAGEMENT AND FARM HOUSEHOLD STRATEGIES

Households use the resources available to them to pursue a strategy of livelihood security that includes food sufficiency, social security, risk management, and income generation, among its objectives. As objectives and resource availability change over time, the strategy the household pursues will be modified. Similarly, as changes in socio-economic conditions alter the costs and returns faced by farmers, the choices they make between different courses of action in order to pursue their strategy will also change.

Using these concepts as a starting point, tree cultivation and management may be explained as being one or more of four categories of response by farmers to change:

- to maintain supplies of tree products as production from off-farm tree stocks declines due to deforestation or loss of access;
- to meet growing demands for tree products as populations grow, as new uses for tree outputs emerge, or as external markets develop;
- to help maintain agricultural productivity in the face of declining soil quality or increasing damage from exposure to sun, wind or water run off;
- to contribute to risk reduction and risk management in the face of needs to secure rights of land tenure and use, to even out peaks and troughs in seasonal flows of produce and income, and in seasonal labor demands, or to provide a reserve of biomass products and capital available for use as a buffer in times of stress or emergency.

---

* This paper is based on a study (Arnold and Dewees 1995, 1997) that draws on the results of detailed original research studies, by Peter A. Dewees (Kenya), Michael R. Dove (Pakistan), Don Gilmour (Nepal), Narpat S. Jodha (India), N.C. Saxena (India), Sara Scherr (Kenya) and Katherine Warner (eastern Africa). The analysis, and a substantial part of the research, was supported by research funds from the Rockefeller Foundation and the Ford Foundation, and a grant to the Oxford Forestry Institute from the Economic and Social Committee for Overseas Research of the UK Overseas Development Administration. Work on the present paper also benefited from additional support provided by the EPAT/MUCIA project.

TABLE 1
Patterns of planted trees on farms

- *Trees maintained on non-arable or fallow land.* Low intensity management of naturally regenerated trees on uncultivated land is likely to occur in more extensive farming and grazing systems.
- *Trees grown around the house.* This often emerges even when there is still plentiful tree cover, in order to introduce fruit and other valued species. Where protection against livestock or burning is difficult, the fenced area around the house can be the only niche where trees can be grown.
- *Tree growing along boundaries and in other interstitial sites.* Found where trees need to be separated from crops in areas of intensive land-use, or where trees are the dominant means of boundary demarcation, or where lines of trees serve a protective purpose (e.g., windbreaks and contour planting).
- *Intercropping on arable land.* Generally takes the form of trees scattered, or in clumps or rows (alley cropping), as part of sometimes complex agricultural crop production. Occurs where trees provide benefits to agricultural crops through shade, shelter or soil improvement, or intercropping is mutually beneficial to both trees and crops because of shared water, soil, nutrient, and light resources. In its most highly developed forms, as in multi-storied multiple species compound farms and 'home gardens', tree/crop mixtures can represent important components of the overall farm system.
- *Monocropping on arable land* (farm woodlots). This is usually associated with the growing of trees to produce cash crops, such as poles, pulpwood, bark or for fruits such as cashew nuts, and is most commonly found in the more advanced market-oriented agricultural areas. Tree crops are also employed as a low cost means of using poor sites.

Source: Arnold and Dewees 1995

Trees are located within farmland in the patterns summarized in Table 1. In most farm systems trees are present for a combination of more than one of the four purposes listed above. Home gardens, for example, incorporate a variety of woody perennials that contribute to nutrient recycling and soil protection, that yield produce that supplements output from other parts of the farm system, and that help spread farm work, output and income more evenly throughout the year (Ninez 1984).

Within this broad overall framework, farmer decisions about tree management are likely to be influenced by a number of factors, including a decline in access to off-farm tree resources, land and tree tenure and control, agrarian transformation and growth in market transactions, factor availability and allocation, and management of risk.

In practice there are wide variations in responses among farmers even within a single community. Individual farm households have different resource endowments, needs and objectives. Women farmers often confront a situation that differs from that faced by men farmers. Within the household, men and women, and young and old, frequently have different agendas involving trees and tree products. This diversity needs to be kept in mind in interpreting the review of general patterns, trends and inter-relationships between factor allocation processes and farmer decision making that follows.

2.1 Decline in Access to Off-Farm Supplies

In many agricultural systems rural households fill gaps in the resource and income flows from their farm and household resources by drawing on off-farm resources. For many, the complementary inputs of fodder, fuel, green mulch, food and income that are often critical to the continued functioning of their agricultural systems have come from nearby areas of forest, woodland or scrubland that are collectively used as common property resources.

In the past, such resources were usually subject to some form of local control by those who used them, in order to prevent their overuse. In recent times, both the resource and the management system which evolved have nearly everywhere come under increasing pressure from growing populations, and from the effects of economic and political changes at the level of both the nation and the community.

The imposition of State control first over the forest resource and then over the land has widely reduced access and rights of usage to public resources in most countries. Distributions of public land to small farmers, and encroachment by larger farmers, have also increasingly taken these resources out of common use. Concurrently, many traditional methods of access control, usufruct allocation, and conflict resolution have become ineffective or have disappeared. Among the factors undermining local management have been the replacement of traditional village leadership with elected village councils, greater commercialization of the products of the resource, technological changes that encourage alternative uses of the land, and demographic changes resulting from growing populations and in-migration of outsiders. Increased differentiation within communities, as their composition changes, reduces communal cohesion and uniformity of interest about how to manage collective resources. As a consequence, much use has deteriorated into unregulated 'open access' use, and remaining resources have generally been heavily reduced in size and degraded in quality (Jodha 1990, Arnold and Stewart 1991, Shepherd 1992, Davis and Wali 1993, Lynch and Talbott 1995).

At its simplest the growth in establishment of trees on-farm can be seen as a response to this decline in access to tree resources off-farm. However, some of the changes that accompany depletion of common pool resources also may alter demand for the tree products previously supplied from these off-farm resources. Irrigation of dry land, for example, is likely to reduce the need for draft animals, and hence for fodder, and is also likely to create new and more productive sources of the latter than could be provided by fodder trees. Availability of fertilizer at an affordable price could remove the need for green mulch, or for intercropping with nitrogen-fixing tree species (though there is no substitute for the soil-conserving outcome which results from the addition of organic matter). Alternatives may be available that present a lower opportunity cost to the farmer than creating supplies of tree products — hence the widespread use of dung and crop residues in place of fuelwood. Other economic options available to the farm household — off the farm as well as on it — may offer a better use of its resources than adding or intensifying tree management. Finally, improved food security programs are likely to reduce the need for maintaining tree stocks as a 'buffer' against poor years or hard times. It cannot be assumed, therefore, that the tree products that were most in demand from off-farm sources will necessarily determine what trees farmers wish to establish on-farm (Arnold 1996).

2.2 Agroecological and Land Use Characteristics

Patterns of tree growing also vary with the agroecological characteristics of the area, and the land use practices that are shaped by climate and soil. Burning of pasture and grazing of livestock, for example, make it impossible to grow trees except in protected areas. Competition with crops for light, water and nutrients on intensively used crop land, and practices such as site preparation with tractors (that would be impeded in their operation by the presence of trees), cause trees to be removed from fields to boundaries and other niches not used for crop cultivation.

This suggests that on-farm tree growing is most likely to form a significant component of the farm household system in higher rainfall, more arable areas without large numbers of free-ranging livestock, and where higher site productivity enables needed tree products to be produced efficiently on-farm. In dry regions, with extensive agricultural and livestock systems that are dependent on biomass products and woody perennials, and exhibiting low site productivity, woodland management usually offers greater biological potential than tree planting as a means of maintaining supplies of outputs such as fodder and grazing.

This is borne out by the comparisons across agroecological zones and land use systems. Tree growing generally increases as one moves towards more intensive agriculture and land use, and as access to natural tree stocks declines (Shepherd 1992, Warner 1993). There is also a general progression over time within most systems towards more *planted* trees, as agriculture intensifies and existing stocks diminish (Arnold and Dewees 1995, 1997).

2.3 Land and Tree Tenure

The shift in tree resources from public to private control reflects the greater security of access provided by the latter. However, the evidence does not support the argument that only private ownership achieves sufficient security. Individual households plant trees in a wide variety of tenurial contexts other than private property. In the customary systems of collective control of land prevailing over much of Africa, with individual households having rights of cultivation and use on an area of land, systems have long been in existence that distinguish security of access to trees from security of access to land. Persons who plant trees are assured of continued rights to the produce even after they have relinquished control of the land on which the trees are located (Fortmann 1987).

The most important factor affecting tree growing in such systems appears to be the existence or absence of rights of exclusion — in particular, exclusion of grazing on the household's fallow fields. Where this is discouraged, because livestock management is important, or where it cannot be enforced, tree growing is unlikely to take place. Where farmers can exercise this degree of control, economic factors are probably more important than land tenure in determining decisions about tree growing (Lawry 1989, Shepherd 1992, Godoy 1992).

Nevertheless, there are situations where tenure or control restrictions do affect tree growing decisions more directly. Annual contracts for sharecroppers, for instance, exclude multi-year crops such as trees (Dove 1995). When land-use rights can be established by planting trees on the property, this may encourage or discourage people to plant trees. Where the State is empowered to appropriate forest or woodland areas, people are frequently unsure about their rights to grow trees, and are likely to be reluctant to do so.

2.4 Agrarian Change and Growth in Market Transactions

Although some of the instances where farm households have adopted trees solely or primarily as a cash crop are striking, as in the north India situation described in Box 1, self-sufficiency in particular tree products is the primary objective of most households in planting trees (Arnold and Dewees 1995, 1997). However, as agriculture shifts from a predominantly subsistence basis to greater involvement in market transactions, tree growing at the farm level becomes exposed to a number of economic influences. Access to market outlets for tree products can extend the range of the farm household's income generating options, and the market place provides opportunities to substitute purchased inputs such as fertilizer for inputs previously supplied by growing trees (Dewees and Scherr 1996).

Box 1. Eucalypt Growing as a Cash Crop in Uttar Pradesh, India

When the government farm forestry program started in the late 1970s, many farmers in Uttar Pradesh adopted eucalyptus as a farm crop, in an area where farm trees had not been grown before, but then gave up its planting when the first production and marketing cycle was completed. A survey showed that eucalypt planting was taken up more by wealthier farmers who had more land, had more assets, faced shortages of labor and problems of supervision, and had diversified sources of incomes. Tree growing also took place more in the commercialized and agriculturally prosperous region.

The results suggested that eucalypt growing was discontinued due to higher than anticipated costs, lower crop yields in the vicinity of the planted trees, low output prices, and uncertainties over yields and markets. Farmer access to markets was adversely affected by government controls on private production and transport of wood products, by government sales of pulpwood at administered prices, and price controls on domestic fuels (kerosene and gas).

Source: Saxena 1991, 1992

Box 2. Black Wattle Farm Woodlots in Murang'a District, Kenya

Between 20 and 30% of smallholder farming land in the transitional high potential, coffee/tea zone of Murang'a District has been planted with woodlots of black wattle (*Acacia mearnsii*), grown for its bark (yielding tanning extracts), charcoal, fuelwood and building poles. Black wattle became important during the 1930s as a cash crop that enabled male labor to go off to the cities and plantations in search of wage employment. Its adoption was also favored because planting of permanent crops could more firmly establish land use rights.

Though the markets for tannin bark and charcoal have become restricted, and much of the wattle sits on land that could produce much more lucrative crops, such as tea, the woodlots are still widely maintained, and new woodlots continue to be established. A detailed study in the early 1990s indicated that this is still largely due to constraints on availability of labor for farm activities, and of capital with which to hire labor or invest in more input-intensive crops such as tea. The study results suggest that woodlots are more likely to be established as households age and as labor becomes scarce, and that woodlot clearance takes place when labor is more available to cultivate the holding. Patterns of resource allocation suggest that woodlot growing households are more risk averse. Woodlot growing parcels tend to be on more steeply sloping sites than other parcels, and therefore harder to work and more prone to erosion if cleared.

Source: Dewees 1993

A distinction can usefully be made between local rural markets, and the growth of urban and industrial markets for tree products. Local markets for fruits, fuel, poles and other tree products emerge as an outcome of the need for specialization and exchange: often first as barter trade as a response to shortages, as demands on the time of women (and other household members) increase leaving them less time for gathering what is needed to meet household needs, and as cash incomes rise and allow some the option of purchasing rather than gathering or growing. Households that are managing tree stocks in order to provide themselves with such products will sell what is surplus to their needs, in order to exploit the opportunity to generate additional income. As this happens, the distinction between production for subsistence or sale has progressively less meaning. Not only will producers sell what is surplus to their subsistence needs, but they will sell a commodity needed in the household if the opportunity cost of doing so is advantageous. For example, even when a household is short of firewood, it may engage in its collection and sale, because cash income (for example to purchase food) may be in even shorter supply.

Production for urban and industrial markets for wood products is more likely to be practiced by farmers in areas where the process of agrarian change has brought about greater involvement in commodity markets and in an intensification of agriculture based on cash crop production. In some instances, farmers may enter the market for tree products where they lack other income opportunities. In central Kenya, for example, the rapid and widespread adoption of black wattle as a cash crop in the 1940s was an outcome of colonial restrictions on the growing of alternative cash crops (Box 2). In other cases, returns from tree crops may appear to be more attractive or stable than from alternative crops, as was the case during the phase when farmers were adopting eucalypts in north India (Saxena 1992).

The policy environment may effectively discourage farmers from participating in these markets and, where supplies can be harvested from lower cost alternatives, it becomes difficult for sustainable tree cultivation and management to compete. Most urban wood fuel markets, for example, are still supplied by the mining of natural tree stocks, with producers paying little, if anything, for the raw material, so that the cost of fuelwood delivered to the market consists mainly of labor and transport costs. In most countries, much fuelwood — and poles and other categories of wood —

also comes from State forests and plantations, and is sold at administered prices. Private producers may be subjected to controls on harvesting, transport and sale, designed to protect against illegal felling for sale from state forests. Resulting cumbersome and costly bureaucratic procedures lead to producers having to depend on intermediaries to market their produce.

A combination of these factors helps explain the limited occurrence of private production of fuelwood and poles for urban and industrial markets. This can be illustrated from the information in Table 2, which compares three different situations. In the Sudan, production by farmers is simply not viable in competition with the low cost supplies from wood generated through agricultural land clearing. In both central Kenya and northern India the cultivation of trees as cash crops did emerge but was curtailed in the face of competitive and policy constraints (Dewees and Saxena 1995a). Unless such impediments disappear, urban and industrial markets are likely to be less important than local rural markets in most tree growing situations.

2.5 Factor Availability and Allocation

The varying role of trees in different situations often reflects differences in the availability of the different factors of production — land, labor and capital. Where the amount of arable land is the limiting resource, trees, as a land use that produces low returns per unit of area, generally become restricted to homesteads, boundaries and other niches where they do not compete with the agricultural crops. However, there can be exceptions to this where tree outputs complement or supplement crop outputs, thereby increasing total returns per unit of area. Home gardens, with their vertically layered structure of trees, shrubs and ground cover crops making effective use of space above and below the soil surface, provide notable examples of this. The labor-intensive technique of alley cropping (involving the planting of rows of leguminous tree-shrubs in fields of annual crops to provide soil nutrients) can, under the specific conditions where capital scarcity restricts the use of fertilizer, maintain the productivity of land-constrained farms.

As farm households have increasingly to depend on income earned from employment off-farm, labor rather than land often becomes the main resource constraint which determines farmer options. As is illustrated in Table 3, shifts in the ratio of labor to land may encourage the growing of trees as labor to undertake other land uses becomes constrained. In central Kenya, black wattle woodlots are more likely to be established as household labor seeks to cultivate other more lucrative crops such as tea (Dewees 1991). In north India, the earliest adopters of eucalyptus as a cash crop were asset-rich households which had significant sources of off-farm income, and which were seeking to minimize labor supervision requirements (Saxena 1992). The last of these characteristics demonstrates another attraction of trees to some — their suitability as a crop for absentee owners or farmers, as an alternative to selling or leasing the land, or as a way to establish continuing rights to idle land. In the middle hills of Nepal (Box 3), the advantages of having tree stocks closer to the users, as the amount of labor available to gather fodder, mulch, livestock bedding and fuelwood declined, appeared to be an additional labor-related factor in increased tree growing on farms.

The north India experience also demonstrates that, although trees require less capital than most crops to establish and maintain, they can lock this capital up (and the underlying land) for periods of several years, with little if any intermediate return. Tree crops may therefore be an option mainly for those who do not rely on that land for household self-sufficiency — such as larger farmers or those with sufficient off-farm income.

As trees are often grown because of shortages of inputs, such as fertilizer or labor, and of capital with which to obtain these, it is likely that increased access to capital would trigger a move away from extensive and site-enhancing uses of tree cover, towards adoption of more valuable crops and intensive land uses. Box 4 recounts the sequence of changes in the role of trees in an area in south India where farm households were subjected first to a major outflow of labor, and subsequently to a large inflow of capital. It needs to be kept in mind, therefore, that increased

TABLE 2
Comparison of wood product markets, and their operation and impact in providing incentives for farmers to cultivate and manage trees

Characteristics of markets and systems of production	Country case		
	Sudan	**Kenya**	**India**
Type of wood and tree product markets	charcoal	wattle bark, charcoal, fuelwood, building poles	building poles, fuelwood, pulpwood
Sources of supply which meet market demands	agricultural land clearance	farmer-grown *Acacia mearnsii* woodlots	farmer grown eucalyptus woodlots
Competing sources of supply	none	plantation-grown wattle for bark and charcoal, land clearance for charcoal, eucalyptus woodlots for building poles and fuelwood	government grown timber supplies
Time-scale and impact of markets on farmer tree growing	never provided any incentive for tree growing	long term farmer response allowing scope for the emergence of new markets for tree products	short term farmer response resulting in overproduction and market collapse
Other incentives not directly linked to markets	none of relevance; some irrigation is made available for eucalyptus and *Acacia nilotica,* but it is not economic to grow these species for charcoal on any scale	seed packets and technical advice provided by Colonial government and tanning extract companies; before Independence, scope for cultivation of other cash crops was limited by legislation	subsidized seedlings and technical advice provided
Earliest adopters of tree growing innovations	no adopters	earliest adopters were landed and wealthy, though later adopters included small-scale farmers as well	earliest adopters were landed and wealthy, later spread to other farmers; some planting financed by urban capital
Comparative advantage of sustainable production systems	only comparative advantage would be in terms of proximity to urban markets; high costs of production mean that it is unlikely sustainable systems of production could be competitive	land tenure constraints before Independence encouraged households to plant trees to establish land rights; trees also favored by households with labor supervision problems or with low demands for income	low costs of labor and supervision vis-a-vis annual crops; low profitability of other crops; markets for other crops thought to be more unstable than markets for tree products.
Bioeconomic constraints to the adoption of tree growing innovations	low productivity and high costs of growing trees in an arid environment	high opportunity costs for land (tea and coffee more lucrative than trees)	planting on field boundaries caused competition with other crops
Policy constraints to the adoption of tree growing innovations	no clear policy constraints; bioeconomic constraints are more critical	movement controls on charcoal; price controls; paraffin subsidies	movement controls; large-scale and heavily subsidized government production for the most lucrative markets limited farmer access to them
Characteristics of the ways markets operate	single entrepreneurs control most aspects of the market production and delivery system; heavily integrated and sophisticated markets	more segmented market operation, with transporters providing the crucial link between rural producers and urban markets	movement controls favor the emergence of village agents who have the dominant role in linking producers with centers of demand

Source: Dewees and Saxena 1995a

TABLE 3
Trees and land and labor allocation

- As tree planting and husbandry requires less inputs of labor than most other crops, it may be seen to be a feasible land-use option when the opportunity costs of labor are high because there are good wage opportunities in other labor markets.
- Problems with supervising and hiring-in labor can act as incentives for households to plant or to maintain trees instead of other more labor-intensive crops.
- Older households, having a smaller resident active labor force on which to draw, may adopt less labor intensive forms of land-use such as tree growing.
- Trees may be planted by households with access to sufficient income from non-farm sources, which consequently have less need to cultivate their land intensively.
- The quality of land within a holding, as well as across holdings in a given agroecological zone, may vary greatly. Trees may be planted in those areas which would require most labor to cultivate in order to even out labor demands.
- Trees may be planted and maintained as an alternative to sale of land that is surplus to the household's immediate needs in order to retain resources which can be passed on to the next generation. Tree growing also may be preferred to renting out of surplus land because the latter might jeopardize the tenure holder's long-term rights of ownership.

Source: Derived from Dewees and Saxena 1995

Box 3. The Increase in Trees on Farmland in the Middle Hills, Nepal

Recent studies have indicated that the density of trees on some categories of farmlands in the Middle Hills zone has increased substantially in recent decades, despite increasing population pressures on the land. Comparison of aerial photo cover from 1964 and 1988 for selected plots in two districts in central Nepal showed a more than fourfold increase in tree density. Field investigations in the same areas suggest that farmers pursue a strategy of natural regeneration and planting first on stream beds and banks and other uncultivated land, then on the walls of rainfed terraces and then on the walls of irrigated terraces. Over time the density of planted trees in the total increases.

The increase in trees on private lands seems to be greatest where scarcity of forest products from natural forests is most pronounced. Other evidence suggests that changes in labor availability, as more farmers seek work off-farm, makes it difficult to find the time to collect leaf material for compost and fodder for animals from the forest. Consequently, more trees are cultivated close to the home to make collection easier. This shift is facilitated by changes in livestock management practices that reduce the quantities of tree produce needed. Increased access to markets has increased the value of planted trees such as **utis** (*Alnus nepalensis*), so that trees are increasingly regarded as a crop.

Source: Gilmour 1995

wealth, or improved functioning of land, labor and capital markets, that would enable farmers to respond to imbalances in factor availability, would probably be accompanied by a reversal of some of the shifts towards more tree cover that are happening at present.

2.6 Management of Risk

In agricultural systems where other forms of accumulating and holding capital, such as livestock herds, are not available, trees may be managed for this purpose. As they can be harvested and sold whenever the owner chooses, they are particularly appropriate as a flexible reserve that can be called down in times of emergency, or to fund exceptional financial outlays such as weddings or the purchase of land (Chambers and Leach 1987). Farmers in arid areas also exploit the resistance

Box 4. Changes in the Tree Component in Home Gardens in Kerala, India

A study of tree cropping in an area in Kerala, India, records the following changes over recent years. Home gardens — in which perennial crops such as coconut, arecanut, rubber, and pepper, are intercropped with seasonal and annual crops such as pulses, bananas, tubers and vegetables — have long existed as part-systems of farms with more extensively managed areas. As rising population pressures on the land leads to decreasing land-holding size, uncultivated land is first brought into use, resulting in the removal of natural tree cover on these areas. This is followed by more active management of the home gardens, to reduce the range of cultivated trees to those with multiple uses, with priority being given to those species valued for fruit, fodder and mulch and suitable as supporting structures for cultivation of pepper, betel vine and various climbers. In the process, the density of trees and the intensity of their cultivation increase.

As pressure on the land increases further, and size of land-holding continues to decline, the farm population increasingly has to seek off-farm employment. The reduction in labor available for work on-farm results in a cut back in crops requiring intensive cultivation, and the vegetation reverts towards a forest condition. Subsequently, the inflow of capital from remittances sent by the migrant workers, has enabled some farmers to intensify land use again, using purchased inputs of fertilizer and herbicide. This reduces the importance of trees in soil-nutrient maintenance and weed suppression, and they tend to be removed as an impediment to the more intensive agricultural practices. Removal of trees has been accelerated by rapid rises in the prices of timber and land, leading to a shift in land use to more valuable cash crops. Trees are then cultivated only where they are competitive as cash crops. For example, *Ailanthus triphysa* is grown to supply wood stock to the match industry.

Source: Nair and Krishnankutty 1984

of certain trees and woody shrubs to drought to create a buffer against periods of low rainfall. In western Rajasthan, India, for instance, farmers have traditionally coped with an environment subject to repeated periods of low rainfall, and the resulting reduced crop yields, by adopting a strategy of mixing extensive and intensive uses of land through crop-fallow rotation, and the complementary use of annuals and woody perennials — with the latter providing reserves of fodder and other biomass in years of poor rainfall (Box 5).

Tree components are also widely included in a farm system because they help even out seasonal peaks and troughs in flows of produce and income and in demands on farm labor. Arboreal fodder, for example, is the principal dry season source of animal feed in both the semi-arid areas in western Rajasthan, and in the middle hills of Nepal. In systems where trees form a substantial part of farm output, e.g., where home gardens are well developed, tree products are also important in diversifying farm output — reducing the exposure of the farm household to failure or price falls on the part of individual crops (Ninez 1984).

Farmers allocate land for tree planting in ways which complement or supplement existing crops in preference to configurations which cause problems of crop competition. The cultivation of *Prosopis cineraria* as an intercrop in Rajasthan reflects its value in enriching the soil and protecting the adjacent pearl millet against wind, as well as its value as a fodder crop (Jodha 1995). In contrast, farmers in Uttar Pradesh have discontinued the cultivation of eucalyptus trees on field boundaries around cropland because they caused substantial reductions in the yields of nearby crops (Saxena 1991).

It is also likely that farmers use lower implicit discount rates in making decisions about activities that can contribute to risk minimization — or to meeting self-sufficiency needs — than they do to income generating activities. Low yielding long gestation tree options that appear to produce low financial returns may be seen to be viable components of the system once their role in relation to risk or food production is understood.

Box 5. Trends in Tree Management in Western Rajasthan, India

In this arid region, people have historically based their livelihood systems on production of grain in association with nitrogen-fixing trees, on livestock management, and on retaining a substantial part of the lands as common property to ensure a reserve of biomass products for use in low rainfall years. During the past few decades, land reform, heightened population pressures on the land, and progressive commercialization of agriculture in response to growing access to markets, have required major changes in these biomass-centered strategies.

Depletion of the area of common lands and overexploitation of the resource that remained has forced greater reliance on private tree management, but the adoption of tractor cultivation has hindered growth in the latter, because trees impede the use of tractors and tractor drawn ploughs destroy the roots of the ber (*Ziziphus numularia*) bushes important as a source of fodder and fuel. Private tree management has also been discouraged by the difficulty of protecting trees in areas where traditionally there has been open access seasonally to fields. With improved access to markets, many farmers have increasingly cut trees for sale as timber and fuelwood. Concern to strengthen the stock of biomass products as insurance against drought period shortages, and the potential for private reward, have recently led to an increase in the intercropping of one locally important tree species, kejri (*Prosopis cineraria*), which can be adapted to cultivation with tractors. However, studies indicate that the system as a whole is being seriously threatened by the large net reduction in tree resources.

Source: Jodha 1995

3. FARMER TREE MANAGEMENT AND DEVELOPMENTAL STRATEGIES

The upsurge in interest, among development agencies, in tree management on farms stemmed from perceptions that this could have a number of important positive impacts on development. One perceived impact was that tree planting could offset deforestation, and mitigate the environmental damage caused by the excessive removal of tree cover. A second was the perception that tree planting could help meet people's fuel and other basic self-sufficiency needs at minimal cost. A third was the view that trees could be a potential tool for resource-poor farmers to help them stabilize and improve their farm system. Tree crops could help them to increase output and generate income, and secure a greater degree of self-sufficiency, with low inputs of capital and labor. In this section we examine the extent to which these developmental concepts and strategies have proved to be consistent with the information about tree management decision making by farmers discussed in the previous section.

3.1 Environmental Impacts

In the 1970s, promotion of tree growing on farms in areas suffering deforestation and environmental damage was seen as a means of creating new wood stocks where they were readily accessible to the main body of users, thereby reducing pressure on remaining forests. Planting was also encouraged as a way of reestablishing a protective tree cover in environmentally fragile landscapes.

In practice, tree resources which have been established on farms in practice serve particular quite narrowly defined purposes — production of fruit or fodder, shelter from the wind, restoration of soil nutrients, boundary demarcation, etc. The patterns of tree stocks and tree cover that emerge on farm lands are therefore quite different from those which are found in natural forests. Farmers establish shelterbelts on windswept drylands, and plant trees along terrace risers or field boundaries in hilly land, in order to reduce soil erosion due to wind and water at the micro level of the field or farm. But they are very unlikely to establish the large contiguous areas of tree cover needed, for example, to influence hydrological flows in mountain areas, or to create the species diversity found in a forest. Even a farm woodlot is not a functioning ecosystem in the sense that a forest is.

Thus, while tree growing by farmers may be, indirectly or directly, a response to deforestation, and can create additional supplies of wood and other forest products, it does not recreate forests. Trees in farming systems are therefore more usefully seen not as part of the forest resource, but in the context of farm household livelihood needs and strategies. These strategies may provide important farm level environmental services, such as maintaining soil nutrients and limiting soil erosion, to which trees contribute. Environmental benefits that may accrue from farmer tree management should consequently be seen as a by-product of farmers' pursuit of their livelihood goals. It is rare for farmers to decide to plant trees for environmental reasons if they are not facing serious soil loss or site deterioration.

3.2 Meeting Basic Needs of the Poor

Another development that focused attention on farmer tree growing in the 1970s was growing recognition of the extent to which rural households depended on fuelwood for cooking their food. Mobilizing farm households to grow more trees was identified as the most effective way for the rural poor to avert or reverse shortages of fuelwood, and of other essential tree products, at little cost to themselves.

In many parts of the developing world, much of the fuelwood used by rural households does come from the trees and other woody plants on the farm. However, there was a failure to relate the target of more tree growing to farmers' other objectives, and to alternative uses of their resources. The needs approach to analysis of the place of farm trees, with its very restricted focus on particular needs and products, tended to obscure the dynamics of farmers' economic responses to changes in demand and supply and scarcity and abundance. Solutions to real or apparent scarcities of tree products were framed in terms of tree growing options without adequate recognition of the alternatives, or of the adaptations to scarcity already practiced by farmers.

Fuelwood "gap" analyses extrapolated present consumption and supply patterns without recognizing the various ways in which people actually adjust to decreases in fuelwood supplies, or the fact that fuel shortages are often due to constraints other than shortages of wood (e.g., shortages of labor can limit a household's ability to collect fuelwood). Also, the growing of trees always involves some cost in terms of land, labor and capital. The produce of trees therefore has a real value to the farm household. There will nearly always be alternative fuel supplies (such as crop residues and dung) that are less costly for resource-constrained farmers to use than home-grown fuelwood, and there will nearly always be uses to which tree products can be put that have higher value than as fuel (Dewees 1989, 1995).

The often very large programs that were set in place to encourage and support tree growing by farmers, in order to increase local fuelwood supplies, consequently often had disappointing results. Many resulted in very little additional planting. Where there was planting it was of trees for fruit, fodder, protection, construction timbers or products for sale. The narrow focus on fuelwood in program planning delayed recognition of the fact that most spontaneous farm level tree management involved a variety of different trees, grown for a range of different products, with fuelwood being produced as a by-product or co-product. There are few situations where farmers have been found to be growing trees to use solely for fuel.

3.3 Household Livelihood Security

In the 1980s attention switched increasingly to the income generating potential of tree growing, as most farm households became increasingly dependent on earning income with which to meet some of their livelihood needs. Though the potential for planting trees to produce wood products for the market is often limited, sale of part of the output from trees grown primarily to meet household needs is increasingly practiced. Markets for tree products can be important to the poor if the costs and risks involved are low: there may be low costs and early returns to market entry;

Box 6. Farmer Tree Growing Strategies in Western Kenya

In Siaya and South Nyanza Districts in western Kenya, on-farm tree planting and management have become progressively more intensive with the transition to permanent cropping, the disappearance of communal tree resources, and the rise of local cash markets for wood fuel, poles, seedlings and fruit. Planting of trees was historically in underused parts of the farm such as around homesteads and along field pathways and borders. With progressive increase in the intensification of pressure on land, farmers employed a large and growing number of different tree species and management practices, and cropland became the dominant site for tree planting. Building poles have replaced fuelwood as the principal use, with green manure, fruit, shade, medicine, timber and stakes as other uses.

The predominant reasons why farmers have been increasing the numbers and land area in trees under conditions of increasing land scarcity appear to have been to obtain critical consumption goods which would otherwise have to be purchased, to diversify their sources of cash income, and to protect food security in the face of declining crop yields. While the initial focus was on self-sufficiency objectives, interest quickly turned to commercial opportunities, with consequent demand for greater assistance with marketing. A high degree of inter- and intra-household variability in the demand for specific tree products and services was observed. These are influenced by differences in resource access, opportunity costs, and varying strategies for achieving short- and long-term food and income security.

Source: Scherr 1995

market channels may effectively serve small-scale as well as large-scale producers, and production systems that can be developed incrementally and do not put other parts of the farm system at risk. For example, farmers in an area studied in western Kenya, who had been increasingly planting trees to meet their household needs for poles, fruit and fuelwood, expanded their production of these products when local markets for them developed, in order to sell some part of their output. Sales within communities and between neighbors had become an important part of many households' income generating strategies (Box 6).

However, some of the initiatives to support "farm forestry" have encouraged commercial tree growing on farms on a scale, or in situations, that encountered market or marketing problems of the kind discussed earlier — as happened in the area of north India described in Box 1. Issues have also arisen about the compatibility of some tree cash crops with other aspects of farm household livelihood systems. In some situations where trees have become important cash crops, concerns have been expressed (notably in India) that the spread of eucalyptus-based farm forestry on agricultural sites has diverted land from the production of essential foods, and has been reducing rural employment. It is true that, as a low input land use, the replacement of food crops by the growing of trees can result in a reduction in employment. However, other underlying factors have been central to the conversion of agricultural land to farm forestry, such as poor and unstable prices for food crops, rising labor costs and the difficulties of supervision, and the need for more stable, remunerative and less labor intensive forms of land use. Tree growing is in many instances a logical response to these agrarian changes. It also may result in increased employment off-farm, in small enterprise activities that process, trade in, or use tree products. Where there are sizeable areas of eucalypts in parts of India, for example, small enterprises have been established to process the oils from their leaves. Similarly, the black wattle woodlots on farms in central Kenya have been important as a source of raw material for those who make charcoal for the Nairobi market.

The allocation of farmland to grow tree monocrops such as eucalypts, however, could have other adverse effects. The narrow markets for the polewood they produce, and the danger of substantial fluctuations in prices in these markets, exposes the household to the risk of uncertain income — as was the case in north India. Income flows from such tree crops are also 'lumpy', coming at intervals of several years, with little if any income being generated in the intervening years (Longhurst 1987). Multi-purpose trees and multi-species systems that are found in most farming situations, on the other hand, are more likely to contribute to a sound mixed subsistence/cash crop household economy. Fruit or fodder products, and shade, protection, green manure or soil amelioration from the presence of trees being grown as cash crops, all contribute to stable food supplies as well as to income generation. Virtually all trees will also produce some fuelwood for use in the household. Tree species that can be intercropped with early yielding crops reduce the problems of delayed returns usually associated with longer rotation tree production.

Incremental adaptive changes of this kind are likely to pose less of a risk to small farmers than changes that involve major alterations to existing land use or product flows. These features — together with the constraints on competitive access to wood markets — help explain the limited occurrence of trees as field crops, and the prevalence of land use patterns in which a variety of trees are fitted into different niches in the farm landscape.

4. POLICY IMPLICATIONS

Policy and project interventions in support of tree growing by farmers have often been poorly matched to the role of trees and tree products. Partly because of the selective thrusts of the prevailing developmental strategies, many of the interventions to date have been overly narrowly focused on just one tree-related issue — often on fuelwood supplies. As a result, policy measures have often been introduced that inadvertently encouraged tree growing where trees are not an appropriate component of the farm household economy, or that induced growing of inappropriate trees. Others have pursued solutions that would require changes in the institutional or social framework that could not realistically be achieved in connection with tree growing, or have failed to focus on the critical areas where change could be brought about.

Recently more holistic approaches have been adopted that recognize farmers' multiple objectives, and the need to balance tree-based solutions against alternative courses of action. These experiences suggest that there is less need to stimulate farmers to grow trees by subsidizing tree establishment, and more need for interventions that help match production to demand. In particular, higher priority should be given to changing policies and practices that presently constrain farmers' access to markets, and that depress market prices for their tree products (Dewees and Scherr 1996). These commonly include lack of market information, poorly functioning trading systems serving small producers, competition from subsidized supplies from State forests and plantations, fuelwood prices which are depressed by subsidies to alternative fuels, and restrictions on private harvesting and trading of wood products. There is a danger that, by hindering farmer access to tree product markets, governments may inadvertently be interfering with the shift from a subsistence to a market economy. These issues, and a number of other policy-related points, are discussed further in this final section.

4.1 Subsidies and Related Fiscal Measures

4.1.1 Subsidized Tree Planting

In the past, project interventions have centered on stimulating more tree planting through provision of subsidized planting stock, and/or cash payments to offset establishment and maintenance costs. However, it is difficult to identify evidence that indicates that this type of support is needed, or that it is effective (cf. Dewees 1995a). As tree establishment requires only low inputs of capital, it is not clear that the cost of doing so constrains many farmers who wish to grow trees.

Subsidizing planting stock can also have negative impacts on the emergence of sustainable tree growing. Free or below-cost supplies from public sources are likely to discourage the emergence of local production of tree seedlings, so that tree planting is likely to cease or sharply decline if the support is withdrawn. Cash subsidies can also encourage farmers to establish tree crops in order to benefit from the short term cash payments rather than the longer term investment returns from trees. This can lead to trees being established where they are not viable, and at the expense of better uses of the land such as grazing or food crops (cf. Arnold et al. 1990).

Government help as a provider of planting stock can play an important pump-priming role in stimulating tree planting, and in providing species that are not locally available, or that are difficult to raise. But, where the constraint on tree growing is the costs of establishment or husbandry, rather than lack of knowledge or expertise, use of credit, and measures to reduce costs, such as staggered planting, and interplanting with crops that produce intermediate yields, would avoid the distortions arising from use of subsidies.

The analyst considering the role of trees in a farming system needs to be aware of the impact of agricultural policy measures on the viability of tree growing. These can include policies that influence adoption of new agricultural technologies such as use of tractors, and policies that affect shifts in livestock management and hence demand for grazing and fodder, as well as price policies. Where it is not practical to remove these distortions, it could be argued that a countervailing fiscal intervention may be needed in order to restore the true competitive position of the latter. But, this is not necessarily best achieved by subsidizing the establishment phase; it may be more appropriately achieved by improving the demand and market prospects of the tree products.

4.1.2 Price Controls and Subsidies to Competing Suppliers

Governments' preference for subsidies for tree planting as a way of stimulating supplies of wood fuels (charcoal and fuelwood) can be because of political difficulties in the way of allowing prices of the latter to rise (Dewees 1995b). Prices of wood fuels in urban markets are frequently kept low, as are food prices, in order to ensure the political support of an urban electorate. This is achieved by price controls on wood fuels, or by subsidizing alternative fuels such as kerosene or bottled gas.

In many countries the government also intervenes in the market as a producer from State forests. Some products, notably fuelwood, are made available at deliberately subsidized prices, because of their importance to the urban, and rural, poor. Others are effectively sold at below-cost prices because the process of setting and collecting harvesting royalties is inefficient. While providing support to production by farmers through one part of its forestry program, many governments thus compete with them through the industrial forestry component of their forestry programs. This fundamental internal inconsistency within forest policies is found in many countries.

4.2 Legislation and Regulations

4.2.1 Regulations Controlling Private Production and Sale

In the short term, the scope for improving the position of the farmer producers probably lies mainly in removing or relaxing regulatory constraints that reinforce the structural and scale advantages that the State, through its forestry administration, possesses as a producer of many forest products. Regulations requiring private producers or traders to obtain permits to harvest, transport and sell roundwood, were usually introduced in order to help protect against illegal felling for sale from State forests. Controls of this kind may be reduced and simplified by excluding from the controls planted species, such as eucalypts, that are not present in the natural forests that forest services seek to protect.

4.2.2 Security of Tenure

As was noted earlier, the arguments that prevailing tenurial regimes fail to provide the security necessary to induce people to grow trees often seems to be overstated. Where the present tenure situation does seem to pose a constraint to tree growing, it is likely to be more realistic to seek solutions that can be effected within the existing legal and tenurial framework, than to try and alter the latter. Changes in both formal and customary tenure are usually difficult to accomplish. Moreover, past changes have often engendered a strong distrust of government intervention in this area, because they can disenfranchise large segments of the local population. The prospect of change can thus itself introduce uncertainty, and so may inhibit investments in long term activities such as tree growing. Lesser changes, such as clarification of linkages between the presence or absence of trees and control of the land could often provide the necessary assurance that farmers need.

4.3 Public Investment in Support Services

4.3.1 Research and Extension

Many tree planting support programs have been characterized by poor technical prescriptions and practices. Implementation has been forced to run ahead of capability to provide adequate extension and technical packages, and without sufficient regard being paid to institutional constraints and possibilities. There is need for a greater content of applied research that responds to the needs, opportunities and constraints actually faced by farmers.

This should focus *inter alia* on helping to identify changing demands that farmers could exploit — including investment in provision in market intelligence if it is needed. This needs to be matched by a much broader based approach to extension, with farmers being able to choose from a menu of options, in recognition that there is a high degree of variability between, and even within, households for specific tree products and services. A greater use of pilot activities in the initial phase of support programs, and a more measured build-up, should enhance the likelihood that support services are able to respond to local conditions.

4.3.2 Strengthening the Planning Base

Analysis and planning for support of tree management at the farm level have been hampered by a poor database. The historical data on changes in production and use that normally provide the starting point for policy analysis have seldom been assembled for tree resources within agricultural systems. However, there are a number of ways in which existing information can be used in order to improve the information base.

Archival research can often yield important pointers to past change in the presence of trees within land and resource use, and the reasons for change. Aerial photographic coverage and satellite imagery from different periods can provide more direct and detailed evidence of the nature and extent of past changes in tree cover (and provide a basis for designing follow up field studies). Secondary data may be reorganized in ways that permit comparison of patterns of tree occurrence and management across different agroecological regions, land use systems, and conditions of wealth and market access. Careful monitoring and evaluation of projects, and of experiments, is another valuable source of information of use in analysis. Use of participatory planning processes, in which local people identify their own priorities for tree management, is another very important means of generating more reliable project agendas.

5. CONCLUSIONS

As access to forest resources and products decreases, trees managed by farmers are often of increasing importance in meeting multiple household objectives. Tree management practices reflect the variety of different ways in which trees can contribute to such strategies. They also reflect different patterns of availability of land, labor and capital to the farm household, and the alternatives available to the latter for employing these resources. The role and adoption of trees will consequently change as these dimensions of the farm situation change.

Analysis of the policy and technical interventions needed to support viable tree management by farmers needs to compare tree solutions with alternatives, and to recognize the adaptations to scarcity and opportunities already practiced by farmers. At the policy level, interventions to reduce market and demand constraints to tree growing appear to be more important than providing incentives to plant trees. At the technical level, farmers are likely to need access to a menu of tree options that correspond to their different needs and opportunities.

REFERENCES

Arnold, J. E. M., 1996, "Economic factors in farmer adoption of forest product activities," in *Domestication and Commercialization of Non-Timber Forest Products in Agroforestry Systems*, ICRAF, Non-Wood Forest Products Paper 9, FAO, Rome.

Arnold, J. E. M., Alsop, R., and Bergman, A., 1990, *Evaluation of the SIDA supported Bihar social forestry project for Chotanagpur and Santhal Parganas, India*, Swedish International Development Authority, Stockholm.

Arnold, J. E. M. and Stewart, W. C., 1991, "Common property resource management in India," *Tropical Forestry Papers* 24, Oxford Forestry Institute, University of Oxford, U.K.

Arnold, J. E. M. and Dewees, P. A., Ed., 1995, *Tree Management in Farmer Strategies: Responses to Agricultural Intensification*, Oxford University Press, Oxford, U.K. (Paperback edition, 1997, *Farms, Trees and Farmers: Responses to Agricultural Intensification*, Earthscan, London.)

Chambers, R. and Leach, M., 1987, "Trees to meet contingencies: savings and security for the rural poor," *IDS Discussion Paper* 228. Institute of Development Studies at the University of Sussex.

Davis, S. H. and Alaka, W., 1993, "Indigenous territories and tropical forest management in Latin America," *Policy Research Working Paper Series No. 1100*, World Bank, Washington, DC.

Dewees, P. A., 1989, "The Woodfuel Crisis Reconsidered: Observations on the Dynamics of Abundance and Scarcity," *World Dev.*, 17, 8, 1159–72.

Dewees, P. A., 1991, "The Impact of Capital and Labor Availability on Smallholder Tree Growing in Kenya," D.Phil. dissertation, University of Oxford, U.K.

Dewees, P. A., 1995, "Farmer responses to tree product scarcity: the case of fuelwood. In Tree management in farmer strategies: responses to agricultural intensification," Arnold, J. E. M.and Dewees, P. A., Eds., Oxford University Press, Oxford, U.K.

Dewees, P. A., 1995a, "Trees on farms in Malawi: private investment, public policy and farmer choice." *World Dev.*, 23, 7, 1085–1102.

Dewees, P. A., 1995b, "Forestry policy and woodfuel markets in Malawi," *Nat. Res. Forum*, 19, 2, 143–152.

Dewees, P. A. and Saxena, N. C., 1995a, "Wood product markets as incentives for farmer tree growing," in *Tree management in farmer strategies: responses to agricultural intensification*, Arnold, J. E. M. and Dewees, P. A., Eds., Oxford University Press, Oxford, U.K.

Dewees, P. A. and Saxena, N. C., 1995b, "Tree growing and household land and labor allocation: case studies from Kenya and India," in *Tree management in farmer strategies: responses to agricultural intensification*," Arnold, J. E. M. and Dewees, P. A., Eds., Oxford University Press, Oxford, U.K.

Dewees, P. A. and Scherr, S. A., 1996, "Policies and markets for non-timber tree products, " *EPTD Discussion Paper 16*, International Food Policy Research Institute, Washington, DC.

Dove, M. R., 1995, "The shift of tree cover from forests to farms in Pakistan," in *Tree management in farmer strategies: responses to agricultural intensification*, Arnold, J. E. M. and Dewees, P. A., Eds., Oxford University Press, Oxford, U.K.

Fortmann, L., 1984, "The tree tenure factor in agroforestry with particular reference to Africa," *Agro. Syst.*, 2: 231–248.

Gilmour, D. A., 1995, "Rearranging trees in the landscape in the middle hills of Nepal" in *Tree management in farmer strategies: responses to agricultural intensification*, Arnold, J. E. M. and Dewees, P. A., Eds., Oxford University Press, Oxford, U.K.

Godoy, R. A., 1992, "Determinants of smallholder commercial tree cultivation," *World Dev.*, 20, 5, 713–725.

Jodha, N. S., 1990, "Rural common property resources: Contributions and crisis," *Ec. Pol. Weekly*, Quarterly Review of Agriculture, 25, 26.

Jodha, N. S., 1995, "Trends in tree management in arid land use in western Rajasthan," in *Tree management in farmer strategies: responses to agricultural intensification*, Arnold, J. E. M. and Dewees, P. A., Eds., Oxford University Press, Oxford, U.K.

Lawry, S. W., 1989, "Tenure policy in natural resource management in Sahelian West Africa," *LTC Paper 130*, Land Tenure Center, University of Wisconsin, Madison, WI.

Longhurst, R., 1987, "Household food security, tree planting and the poor: the case of Gujarat," *IDS/ODI Workshop on Commons, Wastelands, Trees and the Poor: Finding the Right Fit*, Institute of Development Studies at the University of Sussex, June 1987.

Lynch, O. J. and Talbott, K., 1995, *Balancing Acts: Community-Based Forest Management and National Law in Asia and the Pacific.* World Resources Institute, Washington, DC.

Nair, C. T. S. and Krishnankutty, C. N., 1984, "Socio-economic factors influencing farm forestry: a case study of tree cropping in the homesteads in Kerala, India," in *Community forestry: socio-economic aspects*, FAO/East-West Center, Bangkok, Thailand.

Ninez, V. K., 1984, "Household gardens: theoretical considerations on an old survival strategy," *Potatoes in Food Systems Research Series, Report No. 1*, International Potato Center, Lima, Peru.

Raintree, J. B., 1991, "Socioeconomic attributes of trees and tree planting practices," *Community Forestry Note 9*, FAO, Rome.

Saxena, N. C., 1991, "Crop losses and their economic implications," *Agro. Syst.* 16(3).

Saxena, N. C., 1992, *Adoption and Rejection of Eucalyptus on Farms in North-West India.* D.Phil. dissertation, University of Oxford. (Published, 1994, as *India's Eucalyptus Craze: The God that Failed*, Sage Publications, New Delhi, India).

Scherr, S. A., 1995, "Tree growing to meet household needs: farmer strategies in western Kenya," in *Tree management in farmer strategies: responses to agricultural intensification*, Arnold, J. E. M. and Dewees, P. A., Eds., Oxford University Press, Oxford, U.K.

Shepherd, G., 1992, "Managing Africa's tropical dry forests: a review of indigenous methods," *ODI Agricultural Occasional Paper 14*, Overseas Development Institute, London. (Also in Messerschmidt, 1993).

Warner, K., 1993, "Patterns of Tree Growing by Farmers in Eastern Africa," *Tropical Forestry Papers 27*, Oxford Forestry Institute and International Centre for Research in Agroforestry, Oxford. U.K., and Nairobi, Kenya.

13 Economic and Environmental Benefits of Agroforestry in Food and Fuelwood Production

David Pimentel and Aaron Wightman

1. INTRODUCTION

When agriculture first began more than 10,000 years ago, the world population was only about five to ten million, or about the population of New York City. It took more than a million years for the population to reach this level due to a population growth rate of less than 0.0001%. Today the rate of increase is 15,000 times faster. Currently the population is nearly six billion and an additional quarter million people are added every 24 hours. With the current rate of increase, the world population will double to 12 billion in less than 50 years (PRB, 1995).

More than 99% of human food comes from the terrestrial environment and the remaining less than 1% comes from the oceans, lakes and other aquatic ecosystems (FAO, 1991). To provide a diverse, nutritious diet of plant and animal products about 0.5 ha of cropland per capita is needed

1-56670-294-1/99/$0.00+$.50

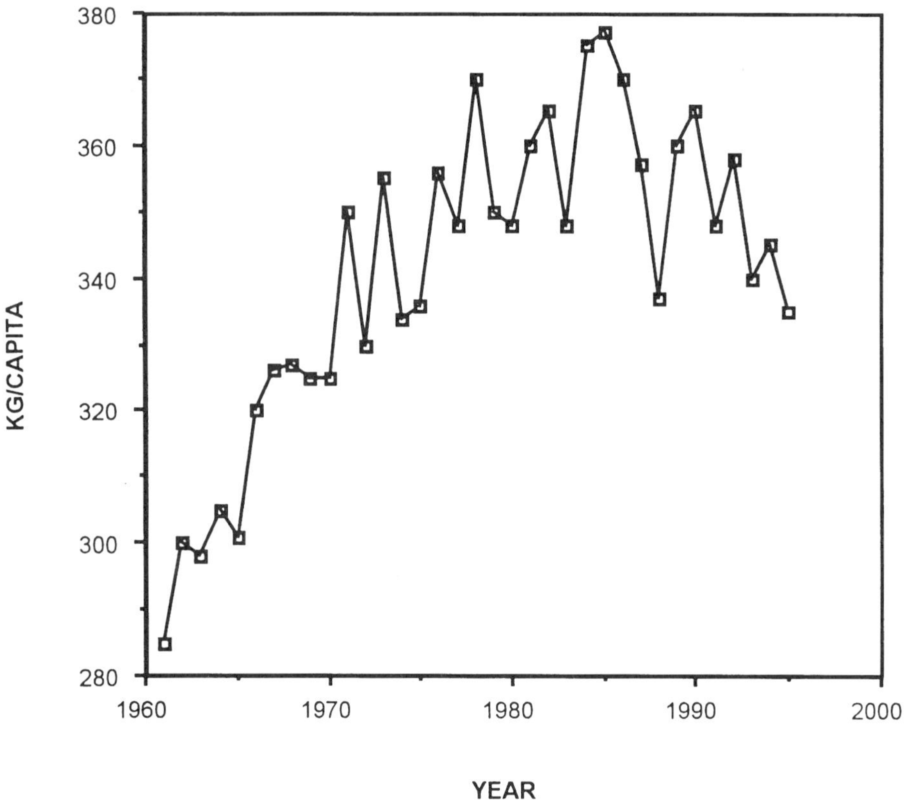

FIGURE 1. World cereal grain production per capita. (Adapted from Harris, 1996; WATI database system compiled by the Economic Research Division, USDA and FAO Production Yearbooks).

(Lal, 1989). At present, the world cropland average is slightly more than 0.25 ha, or nearly half this needed amount.

Not only is the arable land decreasing because of rapid population growth, bur it also is declining because of soil erosion. Worldwide, about 12 million ha of arable land are destroyed and abandoned each year because of nonsustainable agricultural land practices (Lal and Stewart, 1990; WRI, 1994; Pimentel et al., 1995). Shortage of cropland is, in part, the cause of the food shortages and poverty that many humans are experiencing today.

The food situation worldwide is becoming critical. Per capita food production increased rapidly during the latter half of the 20th century due to technological innovation. Such a phenomenon is unlikely to solve the agricultural problems that loom in the future of developing nations (Kidd and Pimentel, 1992). Furthermore, this revolution in food production has failed to reach many of the rural poor in developing nations. At present, more than two billion humans are malnourished and experience unhealthy living conditions (Maberly, 1994; Bouis, 1995; WHO, 1995). The number of humans who also are diseased is the largest number ever, and about 40,000 children die each day from disease and malnutrition (Kutzner, 1991; Tribe, 1994). Since 1985 per capita food production, measured by the production of cereal grains which make up 80% of human food, has been declining (Figure 1) (FAO, 1990; FAO, 1994).

In this chapter, we provide a global overview of the forces of soil degradation that threaten the productivity of arable land. We then examine the economic and environmental benefits of agroforestry in food and fuelwood production, especially crop and forestry production in tropical regions. Also from a macro perspective, we assess the role of agroforestry in conserving soil, water, nutrient

TABLE 1
Average soil erosion from cropland (tons/ha/year) worldwide. (Adapted from Pimentel et al., 1997c)

Country	Rate	Country	Rate
Argentina, Paraguay, Brazil	19	Kenya	5–47
Australia	1–50	Madagascar	25–250
Belgium	10–25	Nepal	40
China	30	Niger	35–70
England	1–44	Papua New Guinea	6–320
Ethiopia	42	Peru (Andes)	20–70
India	30	Poland	1–4
Italy	0.2–47	Thailand (North)	128
Ivory Coast	60–570	USA	17
Japan	1–28	Zimbabwe	50
Indonesia, Java	60–120		

resources, increasing food production, and providing a better source of fuelwood. We conclude by advocating stronger profit incentives and institutional reforms as well as education and training programs to enable more widespread and effective agroforestry practice.

2. EFFECTS OF SOIL EROSION ON CROPLANDS

Soil erosion is a major environmental and agricultural problem worldwide (Pimentel et al., 1995). Although erosion has occurred throughout the history of agriculture, it has intensified in recent years. Soil is being lost about 30 times faster than the sustainability rate worldwide (Pimentel et al., 1995).

About 80% of world agricultural land suffers moderate to severe erosion and 10% suffers slight to moderate erosion (Speth, 1994). Of the world's agricultural land, about one-third is devoted to crops and the remaining two-thirds devoted to pastures for livestock grazing (USDA, 1989; WRI, 1992). Croplands and pastures are susceptible to erosion, but croplands are more vulnerable because the soil is repeatedly tilled and usually left without a protective cover of vegetation. Still, erosion may exceed 100t/ha/yr in severely overgrazed pastures (Lal, 1993). It is estimated that more than half of the world's rangelands are overgrazed and are subject to erosive degradation (Worldwatch Institute, 1988). Moreover, natural erosion occurs throughout the world, especially on steep slopes, barren lands, and stream beds (Carson, 1985; Troeh and Thompson, 1993).

Even though the problem is widespread on cropland, erosion rates are generally highest in Asia, Africa, and South America — ranging from one to 570 t/ha/yr — and lowest in the U.S. and Europe — ranging from one to 47 t/ha/yr (Table 1). Although the sustainable soil-loss tolerance differs for soil types, a soil-loss rate of about one t/ha/yr is approximately sustainable (Hudson, 1981; Lal, 1984). A comparison of erosion rates by country is often misleading because of the high degree of variation in erosion rates between regions. In Africa, for example, an average of 20 to 40 t/ha are lost each year, yet in many East African locations as much as 225 t/ha/yr may be lost annually (Lal and Stewart, 1990). Similarly, the U.S. currently has an average combined wind and water erosion rate of about 13 t/ha/yr, but in some states, such as Iowa, erosion rates average 30 t/ha/yr (USDA, 1994).

Even these relatively low rates greatly exceed the sustainable rate of soil formation of approximately one t/ha/yr (Troeh and Thompson, 1993). The tolerable soil loss values (T) range from two to 11 t/ha/yr but as Troeh and Thompson indicate, these T values are much higher than the actual rate of soil formation and conversion of parent material into soil.

Throughout the world erosion rates are lowest in areas of undisturbed forests, ranging from 0.004 to 0.05 t/ha/yr (Bennet, 1939). However, once forest lands are converted to conventional agriculture, erosion rates soar to as much as 750 t/ha/yr because of tilling, vegetation removal, and overgrazing (Roose, 1988). As the world population continues to grow, the rates of deforestation and agricultural soil erosion are expected to increase (Kendall and Pimentel, 1994).

Although modern agricultural practices are contributing to the soil erosion problem, the failure of farmers and governments to recognize and address the soil erosion problem is equally important for the current soil crisis. Erosion is ignored by some farmers because it is difficult to measure visually in one storm or even in one season (Troeh et al., 1991; Stocking, 1995). Generally the damage goes unnoticed when a storm erodes 15 t/ha and removes slightly more than 1 mm of soil by sheet erosion. Governments also ignore erosion because of its insidious nature; that is, there are not major crises because the soil is slowly and quietly lost year after year.

3. EROSION PROCESSES

Erosive processes are set in motion by the energy transmitted from either rainfall, wind or a combination of these forces. Although the effects of erosion are not easily observed on a daily basis, water and wind are both capable of quickly damaging the soil. Raindrops hit exposed soil with an explosive effect, launching soil particles into the air. On sloping land more than half of the soil contained in the splashes is carried downhill. In most areas, raindrop splash and sheet erosion are the dominant forms of erosion (Allison, 1973; Foster et al., 1985). When rainfall is intense and rapid runoff occurs, gullies form ranging from one to 100 meters deep and large volumes of water and soil are swept away (Lal and Stewart, 1990).

When wind speeds reach 25 mph or more, the wind detaches soil particles from unprotected soil. Airborne soils can be transported thousands of miles. For instance, soil particles from eroded lands in Africa are transported as far as Brazil and Florida (Simons, 1992), and Chinese soil has been detected in Hawaii (Parrington et al., 1983).

These examples illustrate the power of erosive forces. Erosion is often thought of as a gradual process with effects that only become visible over an extended period of time. Yet it is truly a dynamic force with tremendous capability to degrade arable land. In the worst cases, a productive topsoil that is the product of centuries of soil forming processes can be lost in the matter of a few years.

4. INFLUENCE OF VEGETATIVE COVER ON EROSION AND SOIL LOSS

Vegetative cover included in agroforestry reduces erosion in many ways and determines how vulnerable land is to erosion. Living and dead plant biomass reduce soil erosion by intercepting and dissipating raindrop and wind energy. Above-ground foliage slows the velocity of water running over the soil and decreases the volume of water and the soil lost in surface runoff (Langdale et al., 1992). In addition, plant roots physically bind soil particles, thus stabilizing the soil and increasing its resistance to erosion (Gray and Leiser, 1989).

Over the last 20 years, more than 200 million ha of tree cover worldwide have been removed, exposing the land to rain and wind (Brown, 1990). Grass cover has also been reduced. When vegetative cover is reduced, soil erosion rates increase dramatically. In Missouri, for example, barren land lost soil 123 times faster than land which was covered with sod (<0.1 t/ha/yr erosion) (USFS, 1936). Similarly, in Oklahoma, areas with ryegrass or wheat cover lost 2.5–4.8 times less water than land without cover (Sharpley and Smith, 1991).

Loss of vegetative cover is extremely widespread in many developing countries where people collect leaves, roots, wood, and crop residues to provide household fuel, thus leaving the soil barren and most susceptible to erosion (Pimentel et al., 1986). About 60% of crop residues in China and

90% in Bangladesh are removed and burned for fuel each year (Wen, 1993). In areas where fuel is extremely scarce, even the roots of grasses and shrubs are collected (L. McLaughlin, pers. comm., 1991). Without the protection of crop residues and roots, soil erosion rates can increase ten to 100 fold. Tree and crop vegetation and their residues are key to preventing erosion.

4.1 Loss of Productivity

Due to the erosion-associated loss of productivity and other factors, the per capita food supply has been reduced over the past ten years and continues to fall (Kendall and Pimentel, 1994). The Food and Agricultural Organization reports that per capita production of grains, which make up 80% of the world food supply, has been declining since 1985 (Figure 1).

Crop yields on severely eroded soil are much lower than those on protected soils because erosion reduces soil fertility and water availability. Maize yields on severely eroded soils have been reduced by 12 to 21% in Kentucky, 0 to 24% in Illinois, 25 to 65% in the Southern Piedmont of Georgia, and 21% in Michigan (Frye et al., 1982; Olsen and Nizeyimana, 1988; Mokma and Sietz, 1992). In parts of the Philippines, erosion has been responsible for maize productivity declines as high as 80% over the last 15 years (Dregne, 1992).

The majority (about 60%) of soil erosion on U.S. croplands is due to rainfall, but in the arid states, wind erosion is more important (WRI, 1992). For example, Texas cropland is affected mostly by wind erosion (average soil loss of 33 t/ha/yr), while Tennessee cropland suffers primarily from water erosion (average soil loss of 32 t/ha/yr) (Crosson, 1985).

Erosion by water and wind adversely affects soil quality and productivity by reducing infiltration, water-holding capacity, nutrients, organic matter, soil biota, and soil depth (Troeh et al., 1991; OTA, 1982; El-Swaify et al., 1985). Each of these factors influences soil productivity individually and interacts with the other factors, thus making the assessment of soil erosion impacts on productivity extremely difficult.

4.2 Water Resources

All crops require enormous quantities of water for their growth and the production of fruit (NSESPRPC, 1981; Follett and Stewart, 1985; Falkenmark, 1989). For example, during a single growing season a hectare of maize (7,500 kg/ha yield) transpires more than four million liters of water (Leyton, 1983), and an additional two million liters/ha concurrently evaporate from the soil (Waldren, 1983; Donahue et al., 1990).

As erosion decreases the depth of soil and organic matter content, its ability to retain water significantly declines (Buntley and Bell, 1976). Frye et al. (1982) reported than an uneroded soil 60 cm deep could hold 1500 m^3/ha of water whereas eroded soil 20 cm deep had a water-holding capacity of only 500m^3/ha. In the tropics, Lal (1976) reported that erosion may reduce infiltration by up to 93% and thus dramatically increases water runoff. When erosion increases the amount of water runoff, less water enters the soil matrix and is available for the crop.

Because crops require such large quantities of water, it is vital that water soaks into the soil instead of running off. When erosion occurs, soils absorb from ten to 300 mm/ha/yr less water, or between 7% and 44% of total rainfall (Van Doren et al., 1950; Wendt and Burwell, 1985; Wendt et al., 1986; Kramer, 1986; Edwards et al., 1990; Hauser and Jones, 1991; Murphee and McGregor, 1991). This loss of water can severely reduce crop productivity; even a runoff rate of 20% to 30% of total rainfall results in significant water shortages for crops (Elwell, 1985).

Water runoff and soil loss can be reduced by using cropping methods, such as intercropping, that increase soil organic matter and ground cover (Reid, 1985). For example, when maize is interplanted with shrubs and trees, water runoff can be reduced two-fold or more and soil loss can be reduced 10- to 20-fold compared with maize grown without shrubs and trees (Kidd and Pimentel, 1992).

4.3 Soil Nutrients

In addition to causing water deficiencies, soil erosion causes shortages of basic plant nutrients such as nitrogen, phosphorus, potassium, and calcium which are essential for crop production. A fertile agricultural topsoil typically contains one to 6 kg of nitrogen, one to 3 kg of phosphorus, and two to 30 kg of potassium per ton, whereas an eroded soil may have nitrogen levels of only 0.1 to 0.5 kg per ton (Troeh et al., 1991; Alexander, 1977; Van Dijk, 1980; Lal, 1980; Foth and Ellis, 1988). When erosion occurs these nutrients are lost because the wind and water selectively remove the fine particles leaving behind large particles and stones. Soil removed by erosion typically contains about three times more nutrients (range 1.6 to 10) than the soil left behind (Lal, 1980; Bhatt, 1977; Young, 1989).

When nutrient reserves are depleted by erosion, plant growth is stunted and crop yields decline. Areas that suffer severe erosion may produce 15% to 30% lower maize yields than uneroded areas (Olsen and Nizeyimana, 1988; Mokma and Sietz, 1992; Dregne, 1992; Crosson, 1985; OTA, 1982; El-Swaify et al., 1985; NSESPRPC, 1981; Follett and Stewart, 1985). Under the current average soil erosion rates (13 t/ha/yr), the loss of nitrogen, phosphorus, and potassium could result in a long-term drop in maize yields of 800 kg/ha, or roughly 12% of the average maize yield (Tisdale et al., 1985).

4.4 Soil Organic Matter

Organic matter is a necessary component of soil because it facilitates the formation of soil aggregates, increases soil porosity, and thereby improves soil structure, water infiltration, and ultimately overall productivity (Langdale et al., 1992; Greenland and Hays, 1981; Tisdall and Oades, 1982; Chaney and Swift, 1984). In addition, organic matter increases water infiltration, facilitates cation exchange, enhances root growth, and stimulates the proliferation of important soil biota (Allison, 1973). Also, it provides an essential source of both macro-nutrients and micro-nutrients for plant growth (Allison, 1973; Volk and Loeppert, 1982). Approximately 95% of the nitrogen, 25–50% of the phosphorus, and over 70% of the sulfur in soil is contained in its organic matter (Allison, 1973).

Fertile topsoils typically contain about 100 tons of organic matter (or 4% of total soil weight) per hectare (Follett et al., 1987; Young, 1990). Because the majority of organic matter is concentrated near the soil surface in the form of decaying leaves and stems, erosion of topsoil results in a rapid decrease in soil organic matter levels. Several studies have demonstrated that the soil removed by either wind or water erosion is 1.3 to five times richer in organic matter than the soil left behind (Allison, 1973; Barrows and Kilmer, 1963). Thus, the loss of 13 t/ha of soil by rainfall removes nearly two t/ha of organic matter (Young, 1990). Since farming in the United States began, an estimated one-half of soil organic matter has been lost from most cultivated soils (Curry-Lindahl, 1972). About 2 to 5% of organic matter in the soil disappears each year because of mineralization (Troeh and Thompson, 1993).

Once the organic layer is depleted, soil productivity and crop yields decline because of the degraded soil structure and depletion of nutrients. The reduction of soil organic matter from 4.3% to 1.7%, for example, lowered the yield potential for maize by 25% (Lucas et al., 1977). In addition, studies in Uzbekistan have demonstrated that reducing the soil humus level from 1.43% to 0.92% reduced grain yields by about 50% (Libert, 1996). In parts of Rosto Oblast it was reported that reducing soil humus from 500 to 250 t/ha reduced grain yields by about 70% during dry years (Libert, 1996).

Many farmers apply inorganic fertilizers to their soils to replace the nutrients lost when organic matter is swept away. However, the continuous use of commercial fertilizers alone cannot sustain high crop yields. In tropical Africa, a soil that had lost over 60% of its organic matter produced maize yields at only 20% of its initial level despite the continuous application of 100 kg of nitrogen, 20 kg of phosphorus and 50 kg of potassium per hectare each season (Agbooloa, 1990).

4.5 Soil Biota

Although soil biota are often ignored in assessing the impact of erosion, they are a critical component of the soil and constitute a large portion of the soil biomass. Indeed, a large portion of the soil is living. One square meter may support populations of about 200,000 arthropods and enchytraeids, and billions of microbes (Wood, 1989; Lee and Foster, 1991). A hectare of good quality soil contains an average of 1,000 kg of arthropods; 150 kg of protozoa; 150 kg algae; 1,700 kg bacteria; and 2,700 kg of fungi (Pimentel et al., 1980).

Soil biota have a profound effect on crop productivity because they recycle the basic nutrients required for plants in the ecosystem (Pimentel et al., 1980; Van Rhee, 1965). Earthworms and other soil biota also enhance productivity by increasing water infiltration. Through tunneling and burrowing activities, soil biota create thousands of pores and channels within the soil matrix which allow water to percolate efficiently into the soil.

Soil biota further contribute to soil productivity by mixing the soil profile, enhancing aggregate stability and preventing soil crusting. Earthworms commonly bring 10–250 t/ha of soil from underground to the soil surface per year (Edwards, 1981; Lee, 1985). The churning and mixing of soil redistributes nutrients, aerates the soil, and increases infiltration rates, thus making the soil overall more favorable for plant growth.

The erosion that accompanies conventional agriculture may decrease the diversity and abundance of soil organisms (Atlavinyte, 1964; Atlavinyte, 1965). On the other hand, agricultural practices that maintain the soil organic matter content at optimum levels favor the proliferation of soil biota (Reid, 1985). Thus, the simple practice of straw-mulching may increase biota three-fold (Teotia et al., 1950), while the application of organic matter or manure may increase earthworm and microorganism biomass as much as five-fold (Ricou, 1979).

5. BENEFITS OF AGROFORESTRY FOR SOIL AND WATER CONSERVATION AND CROP PRODUCTION

To assess how and to what extent erosion decreases crop productivity, it is necessary to consider the multiple factors that influence erosion rates, as well as the soil components that affect productivity. An empirical model is developed that demonstrates how soil erosion causes the loss of soil nutrients, depth, biota, organic matter, and water resources and how these losses translate into reduced crop productivity. The model is based on the following set of assumptions: 1000 mm of rainfall, soil depth 15 cm, slope of 5%, loamy soil, 4% organic matter, and a soil erosion rate of 30 t/ha/yr. This model provides a perspective on the interdependence of the various factors associated with the ecological effects of erosion.

On the basis of empirical evidence, it appears that when soil erosion by water and wind occurs at a rate of approximately 30 t/ha/yr, an average of 75 mm of water, four tons of organic matter, and 30 kg of available nitrogen are lost from each hectare each year. In addition, soil depth is reduced by 2.6 mm, the water-holding capacity is decreased by about 0.2 mm, and soil biota populations are diminished. When combined, these losses translate into about a 15% reduction in crop productivity over the short term (one year). The loss of water and nutrients account for nearly 90% of the loss in productivity. This model assumes that nutrients and water are not applied in an effort to offset the degradation.

If this soil erosion rate were allowed to continue for 30 years, the soil at the end of this period would no longer be productive and the land would have to be abandoned (Lal and Stewart, 1990; Pimentel et al., 1995).

Agroforestry practices include several vegetation management techniques that, in many cases, can provide continuous vegetative cover to help conserve soil and water resources. Agroforestry especially helps increase the total biomass per unit land area which aids in not only conserving soil and water resources but also in increasing crop yields as well as fuel biomass. In addition, this

TABLE 2
Increase in crop yields because of the benefits of shelter belts (includes area occupied by the tree shelter belts). (Adapted from Pimentel et al., 1997a).

Crop	Location	% Increase in Crop Yield
Corn	China	16
Corn	Egypt	17–74
Cotton	Egypt	35
Millet	Niger	23
Millet	China	44
Rice	China	25
Rice	Egypt	10
Sorghum	China	43
Soybeans	China	36
Wheat	Egypt	38

approach to land use can benefit natural biota which are essential to agricultural and forestry productivity. In the following sections, several agroforestry practices and the benefits of each practice are considered.

5.1 Shelter Belts and Trees

Trees are effective as shelter belts and windbreaks, helping to control water and wind erosion as well as increasing crop yields (Young, 1993). For example, tree and shrub shelter belts help reduce wind energy from 65% to 87% and in turn decrease erosion by as much as 50% (Stocking, 1995). The use of shelter belts in developing countries has increased the yields of various crops from 10% to 74% (Table 2). Note, the increased yields per hectare include the area occupied by the shelterbelts.

Similarly, the establishment of windbreaks (predominantly *Azadirachta indica*) in the Maija valley of Niger resulted in an initial net increase of millet ranging from 854 to 1,102 kg/ha, after accounting for the land used for the shelterbelt (Le Houerou, 1987). Although millet yields later decreased somewhat as the trees grew and competition between the trees and crops intensified, the yields were still 15% above those of the monocultured millet. In the long term, additional benefits are expected to accrue because soil erosion was controlled on the tree-protected land.

5.2 Conserving Water and Soil Resources

Alternating three rows of a N_2-fixing leguminous fuelwood tree like *Leucaena* with three rows of maize perpendicular to the slope of 10% to 15% will control erosion if all the maize residues of 2,000 kg plus 2,500 kg of small branches and leaves of the leguminous tree are left on the surface of the land (Kidd and Pimentel, 1992). The erosion rate under these conditions with 1,000 mm of rainfall or wind would be about one t/ha/yr (Kidd and Pimentel, 1992). A similar study with a maize *Leucaena* intercrop on a steep slope of 44% produced similar results. Plots protected by *Leucaena* hedgerows lost two t/ha/yr topsoil compared with 80 t/ha/yr for an unprotected plot (Banda et al., 1994). Furthermore, the hedgerows caused the formation of microterraces with improved soil characteristics. Some reports indicate that this type of alley cropping has some disadvantages (Jeanes et al., 1996), but others report significant increases in yield employing this technology (Kidd and Pimentel, 1992). One of the reported problems with alley cropping is the competition between the crop and the leguminous tree for water and nutrients.

Another technique that has been employed is to culture the crop, such as maize, separately from the leguminous tree. For example, instead of alternating three rows of maize and three rows of trees, the farmer might plant 25 rows of maize and then 25 rows of trees again perpendicular to the slope. In this case, the farmer would have the additional task of transporting the small twigs and leaves from the leguminous trees and placing them in the maize area. However, the reward would be a higher maize yield than that produced by a monoculture without the benefit of leguminous trees. The amount of maize and leguminous tree material covering the land would be the same as in the alley cropping system. This should be sufficient to protect the land from water and/or wind erosion. In addition, with the strips of trees planted perpendicular to the slope of the land, this strip should trap most soil and water moving down the slope.

5.3 Trees and Insect Pests

Trees in agroforestry systems provide the benefit of protecting non-tree crops from insect pests. Just as many polyculture systems suffer less damage from insect pests, agroforestry systems are prone to less intense pest outbreaks than monocultures (Rathore, 1995). Trees protect crops from insect damage in a variety of ways. The most direct influence of trees on insect herbivory is that trees act as a physical barrier to insect movement. In addition, trees mask odors and interfere with orientation as well as providing habitat for natural enemies of pest species (Rathore, 1995).

6. BENEFITS OF MAIZE PRODUCTION IN AN AGROFORESTRY SYSTEM

Agroforestry techniques can be used effectively in the tropics to produce maize grain and fuelwood for rural families. When properly designed, agroforestry systems conserve soil and water while producing fuelwood as well as crops (Kidd and Pimentel, 1992).

Although studies comparing the costs and benefits of agroforestry systems are scarce (Scherr, 1987), a reasonable cost-benefit analysis can be extrapolated from several proposed models. To demonstrate the advantages of a food and fuelwood agroforestry system, consider the following models based on data from studies in Central America: Pimentel et al. (1976; 1986; 1987; 1995); Torres (1983); Rachie (1983); Nair (1984); Lal (1989); Pimentel and Pimentel (1996); and unpublished research by Pimentel. The first model system is an agroforestry system involving combined maize and *Leucaena* production (Brewbaker, 1987). For comparison, the second model system is a low-fertilizer maize system with no added nitrogen fertilizer utilizing conventional production methods.

Each model system covers one ha of land, which is the typical amount owned by a rural Central American family. Environmental conditions were assumed to be 1500 mm rainfall/yr and a temperature range during the growing season of 22 to 32°C. The slope of the land was assumed to be about 5%.

Using the agroforestry system, one half of the hectare is planted to maize and the other half to *Leucaena*. This design includes two rows of maize alternated with two rows of Leucaena. The maize in this agroforestry system is planted at twice the density of the low-input maize. Thus, the same number of maize plants are present in this agroforestry system as in the low-fertilizer system. Competition between Leucaena and maize is reduced in this alley cropping system by cutting the *Leucaena* back to a stump of about 8 cm just before the maize is planted.

Other assumptions are embodied in the model. For instance, the soil is assumed to be suitable for *Leucaena* production though no single tree species is well suited to every possible site. *Leucaena* grows poorly on reddish oxidized tropical soils in particular (Brewbaker, 1989). Also, equal or nearly zero damage from pests is assumed for each system. This may well represent the actual scenario in the field. Though it is true that monocultures are more prone to large scale pest damage than polyculture, *Leucaena* has a number of pests, including larger browsing mammals such as wild pigs, deer, porcupines, and at least one insect (Brewbaker, 1989). Therefore, damage to each system due to pests may roughly average out.

Short-term outcome is of primary importance to poor tropical farmers since they can tolerate only small risks and rely heavily on their crops for survival. Thus, we have set the time frame for a cost-benefit analysis at one year. However, the year under scrutiny cannot be the first year that the agroforestry system is in place because it takes three or more seasons for the *Leucaena* to become established and to be coppiced before the maize can be planted and benefit from the intercrop. The short-term analysis will stem from the outcome of each system four years after the implementation of the agroforestry system for that reason.

6.1 Costs

Production costs for the year in question include labor, machinery and seeds. Draft ox-power of about 200 hr/ha and human labor of 400 hr/ha provide the labor for each system. These labor inputs are typical of maize production using draft animals and human labor (Pimentel and Pimentel, 1996). Only a small amount of machinery (2.5 kg/ha) and seeds (15 to 21 kg/ha) augmented the systems. The total energy input for the low fertilizer system was 1.8 million kcal. The agroforestry system required a total energy input of 1.9 million kcal.

Translating production costs into a common metric is useful for comparing the costs of the two model systems. In terms of U.S. dollars, the total production cost of the low-fertilizer system is \$163. Due primarily to the cost of purchasing *Leucaena* seeds in addition to maize seed, the total production cost of the agroforestry system is \$188 (Table 3).

6.2 Benefits

In the fourth year the total maize production of the agroforestry system is 1,800 kg/ha/yr compared with 1,000 kg/ha/yr for the low fertilizer system. A yield of 1,000 kg/ha/yr is common for low fertilizer maize production (Torres, 1983; E. Perez, pers. comm., 1997). The agroforestry system is more productive with a yield 80% higher than that of the traditional system. The agroforestry system produces about 5,000 kg/ha of *Leucaena* biomass (Lulandala et al., 1995). Of this amount 3,000 kg of small twigs and leaves are coppiced and worked into the soil as mulch. The practice of leaving 3,000 kg of leaves and small twigs on the land is highly effective in controlling soil erosion. One ton of topsoil eroded from good agricultural land contains about 4 kg of N, 1 kg of P, and 20 kg of K. Thus, preventing 20 to 30 t/ha/yr of soil erosion has major benefits in preventing the loss of soil nutrients from the land. Nitrogen, organic matter, and erosion protection added by the coppiced *Leucaena* are responsible for the increase in productivity (Rachie, 1983; Torres, 1983; Liebhardt, 1983). Mureithi et al. (1994) reported 30% reduction in maize and stover yields in an *Leucaena* intercrop when the tree was not coppiced and mulched. When the *Leucaena* was used as mulch the productivity of the maize increased to 44% over a sole maize crop. The quantity of N applied to the land via the 3,000 kg of *Leucaena* is more than 60 kg/ha (J. Halliday, pers. comm., 1987). Each kilogram of N should increase maize yields about 60 kg (Liebhardt, 1983); however, we assume only half of the potential is actually available in this model system.

In 120 days the N released from the *Leucaena* cuttings ranges from about 46% to 96% (Yamoah et al., 1986). Thus, most of the N stored in the cuttings of *Leucaena* are available to the crop during the growing season. The remaining N is slowly released to the ecosystem, and some will be available to the crop planted later (Troeh and Thompson, 1993).

Furthermore, the agroforestry system produces 2,000 kg of *Leucaena* stems that are harvestable as fuelwood. The small stems of two to 5 cm in diameter are preferred as a cooking fuel. The 2,000 kg of fuelwood are sufficient to provide approximately three persons with fuel for one year (Pimentel et al., 1986). However, the 1,800 kg of maize grain produced on the same hectare are sufficient to feed five to six persons for a year.

TABLE 3
Inputs, Maize Yields, and Fuelwood Harvested per Hectare by Two Model Production Systems. (Adapted from Kidd and Pimentel 1992)

	With low fertilizer			With legume tree		
Parameter	Quantity	10^3 kcal	Econ	Quantity	10^3 kcal	Econ
Labor (hr)	400	210,000	$100.00	500	262,500	$125.00
Draft animal (hr)	200			200		
Concentrate (kg)	150	525,000	30.00	150	525,000	30.00
Stover and *Leucaena*	295	885,000	10.00	295	885,000	10.00
Machinery (kg)	2.5	67,500	4.00	2.5	67,500	4.00
N (kg)	0	0	0	0	0	0
P (kg)	10	63,000	5.00	10	63,000	5.00
K (kg)	15	37,500	3.00	15	37,500	3.00
Ca (kg)	20	6000	2.00	20	6000	2.00
Seeds (kg)	15	60,000	9.00	15	60,000	9.00
Total costs		1,838,000	$163.00		1,906,500	$188.00
Maize grain yield (kg)	1000	4,000,000	200.00	1800	6,000,000	360.00
Maize stover yield (kg)	1000	4,000,000	8.00	1800	6,000,000	14.40
Residue harvested	0	0		0	0	
Wood biomass yield dry (kg)	0	0		4500	18,000,000	
Fuelwood harvested	0	0		2000	8,000,000	40.00
Biological N added (kg)	0	0		60	1,260,000	31.80
Gross Income			$208.00			$452.20
Remarks	Low input costs plus high rates of soil erosion and water runoff. System is *not* sustainable and has low productivity.			Low input costs with low rates of soil erosion and water runoff. System is sustainable and produced 80% more food than the low fertilizer system. This system also produced 2000 kg of fuelwood. Because fuelwood was produced on the farms, the labor used to collect fuelwood was saved which allowed the opportunity to employ this labor for other profitable operations.		

The *Leucaena* trees and maize planted on the contour of sloping land act to conserve soil and water resources. Planting on the contour, leaving the maize residues, and applying the *Leucaena* mulch limit soil erosion to less than one t/ha/yr (Nair, 1984; Lal, 1989). This is in contrast to the low fertilizer system, which has a soil erosion rate of 30 t/ha/yr.

With the value of maize at $0.20/kg the gross income of the low fertilizer system is $200/ha. After deducting production costs the system yields a net profit of $37. The agroforestry system has a gross income of $360 from the 1,800 kg of maize. In addition, with a value of just under $0.05/kg for *Leucaena* fuelwood, the gross income from the stems cut for fuelwood is $92.20. The total gross income for the agroforestry system is $452.20. After deducting the production cost of $188 this yields a net profit of $264.20 per hectare.

6.3 Risk

Like most economic evaluations of agroforestry systems, the above analysis does not include risk as a factor. In order to quantify a particular agroforestry system's proneness to risk, a sensitivity test is needed. The most common approach is to randomly change the costs, benefits and discount rates and examine the effect on profitablity (Hoekstra, 1983). Though a sensitivity test is beyond the scope of this cost-benefit analysis, a few things can still be said concerning risk.

By ignoring risk the model embraces the assumption that a fixed input will result in some predictable level of output. In actuality, the productivity of an agricultural system, given a certain amount of input, is a stochastic phenomenon (Hoekstra, 1983). Events and conditions such as pest outbreaks, weather, market conditions and political change can alter the expected productivity of a crop or the value of the product. Resource poor farmers want an agricultural system that involves relatively low risk and a short term return for their efforts (ICRAF, 1993). Furthermore, poor farmers cannot easily divert resources to tree production which produces yields or augments crop yields several years in the future (Mol, 1989).

Still, agroforestry may appeal to risk averse, resource poor farmers. Agroforestry increases food and economic security (Pimentel et al., 1997a) albeit within the constraints of culture and economy (Belsky, 1993). Many farmers grow trees as a capital reserve that can be harvested when added income or tree resources are needed (Mol, 1989). Trees in an agroforestry system can be utilized in the same way. For instance, stems can be thinned or selectively cut from a shelterbelt in years of economic strain or hedgerows can be planted in a time of economic prosperity in anticipation of less profitable seasons when the stems may be harvested. A study undertaken in Senegal by Caveness and Kurtz (Caveness and Kurtz, 1993) indicated that the main reason for adopting agroforestry was for the production of wood and other tree products rather than the addition of nutrients to the soil.

The model *Leucaena*-maize agroforestry system has a degree of security for the poor farmer because, other than his salary, his total investment is only $188 (Table 3). Also, because of the environmental sustainability built into the design of this system, the farmer is less likely to be affected by future droughts, pests, and other problems (Shannon and Vogel, 1994). At maximum, we assume that poor farmers can invest only a small portion of their income for crop production. These farmers often depend on wood as their prime source of fuel; hence, under such conditions agroforestry technology can be helpful to rural people because it increases their supplies of both food and fuel with minimal inputs.

An agroforestry system can lessen peak seasonal demands for labor by lengthening the ripening period (Filius, 1988). An increased amount of time available to harvest crops increases the probability that all crop products will be properly collected. This is important because of increasing migration of the rural poor, especially youths, to urban areas. In tropical Africa, experience has shown that labor shortages occur during the peak labor periods of planting and harvest (La Anyane, 1985).

Trees reduce dependency on markets by providing products that would otherwise need to be purchased such as fuel and building materials (Mol, 1989). These products can also be provided by an agroforestry system if it allows the trees to be harvested. Alley cropping is such a practice where harvesting the trees does not disrupt the system and can provide a significant amount of needed products. Reducing market dependency protects farmers from price fluctuations and high inflationary rates.

Of course, the issue of risk is not entirely addressed by this added security. A certain amount of risk will always be felt or perceived by farmers when adopting a new technology. In fact, agroforestry systems must be carefully tailored to local conditions and needs. So, adopting a new method must involve a certain amount of uncertainty. Indeed, the system may be entirely unsuccessful due to unforeseen or unknown factors. For example, an effort to promote alley cropping in Nigeria failed due to decreased yields and increased rodent damage (Bayliss-Smith, 1994). The failure of this effort was due, in part, to the acidic soils of the region and unanticipated pests. Such problems can be difficult to predict. Since each agroforestry system will not be thoroughly tested for each piece of land where the system will be adopted, implementing new agroforestry techniques will usually involve a certain amount of risk. However, in the context of severe land degradation and erosion, the risks involved in trying something new are a problem for resource poor farmers.

Where agroforestry has been adopted farmers have lowered risks by adopting new techniques incrementally (Scherr, 1995). A slow, step-wise conversion to new techniques allows the farmer to identify problems without losing an entire crop. Techniques that present problems are modified or abandoned and beneficial techniques are eventually applied to a larger area of the farm. Thus, risks are minimized and beneficial techniques are applied on a larger scale with a greater degree of security.

6.4 Social and Economic Considerations

Farmers make decisions based on more complex criteria than a simple cost-benefit analysis. Agricultural methods are judged on the basis of how well they meet their basic needs including food, shelter, and cash income. Concurrently, the potential of agricultural methods is constrained by factors such as infrastructure, social obligations, available resources, and the availability of information. There is significant evidence that agroforestry techniques can help meet the needs of rural farmers (Rocheleau, 1987). In addition, agroforestry is a flexible technique that can be tailored to local environments and the needs of local people.

For resource poor farmers, economic security is dependent upon environmental health and stability. Environmental degradation, such as the soil erosion occurring over much of the world's arable land, contributes to a more desperate type of poverty and malnutrition. Thus, any activity that conserves or improves soil productivity and the environment without restricting usage improves economic security in general. Agroforestry techniques can protect soil and water resources and other aspects of the environment and maintain or increase the potential usage of environmental resources.

As previously mentioned, one major crisis facing many rural poor in developing nations is malnutrition. Land shortages caused by increasing populations and degradation of available arable land are the primary causes of food shortages. The agroforestry system described in the cost-benefit analysis mitigates both problems. Currently the world cropland average is 0.25 ha per capita, about half of the area necessary for a diverse plant and animal diet similar to that in the United States and Europe. Agroforestry can increase the productivity of existing land thus enabling the rural poor to meet their nutritional needs on the limited amount of land. Also, agroforestry systems protect cultivated soils from degradation thereby maintaining the quality of the arable land that is available to the rural poor.

Disseminating information on agroforestry techniques is a sensitive issue. Researchers and agricultural scientists need to be sensitive to the role of the farmer and aware of the conditions surrounding the current mode of agriculture when promoting agroforestry techniques. Complicating this endeavor is the low level of literacy among many of the rural poor. Furthermore, an insufficient number of professionals are educated in agroforestry to train others and oversee agroforestry projects (York, 1990). Agroforestry is in its academic infancy at the moment. Therefore, there is no widely accepted framework for agroforestry education and current education programs do not optimize cooperation and collaboration on projects (Krishnamurthy et al., 1994).

Though agroforestry can provide added security to food and the household economy, risk is the greatest obstacle to farmer adoption of agroforestry systems. Often, it is better to build upon existing techniques than to introduce totally new techniques. Agroforestry has the advantage of having been practiced throughout the history of agriculture. Many agroforestry systems have been in place since before agroforestry's surge in popularity in the academic community (Budowski, 1993). Traditional practices are similar to those described by science. For example, a traditional system in India uses agroforestry to maintain sustainable crop production during periods of normal rainfall.

During droughts, forage for livestock is harvested from the trees (Chabeda, 1981). Farmers may be more willing to adopt agroforestry if they are familiar with the use of trees in agriculture and know that integrating tree and crop production has been successful for other people (Nielsen,

1994). Indeed, the perceived risk will be greater to a farmer with "new scientific" techniques than with "traditional" techniques. Furthermore, the study and understanding of traditional agroforestry techniques will give scientists new insights that will allow them to develop more effective agroforestry systems for specific sites. Thus, utilizing traditional agroforestry techniques can help lower perceived risks and real risks.

Surveys conducted by Scherr (1995) in Kenya indicated that adoption of agroforestry is most likely when clear incentives for new land use practices come about. In addition to benefits from sustainability and improved yield, incentives provided by government and the private sector will also be crucial in the adoption of agroforestry. The pressures of economy and immediate needs push farmers toward less sustainable practices with greater short term gains. This needs to be countered with incentives that help meet needs in the short term and encourage the implementation of more beneficial and sustainable practices in the long term.

7. AGROFORESTRY AND LIVESTOCK PRODUCTION

Agroforestry systems that include livestock provide potential increased productivity and economic security. A nutritious diet for the livestock is provided by some tree species, or livestock can graze beneath the trees once they have reached a certain height, with minimal damage to the tree (ASPAC/FFTC, 1995). In Burundi, field trials demonstrated that trees used in reforestation projects provide feed resources for sheep (Branckaert and Mbayahaga, 1993). Tree foliage is an important source of feed for livestock in India because trees provide a reliable source of feed during conditions of floods and drought (Solanki, 1991). Arthur and Ahunu (1992) report that sheep can be raised in pastures under the canopy of plantation trees with minimal damage to the trees. Furthermore, the sheep raised in these pastures were similar in quality to those raised in normal open pastures. The potential of integrating trees and livestock is especially pressing due to the increasing paucity of grazing lands due to reforestation, desertification and the establishment of parks. By integrating livestock into an agroforestry system, farmers can maximize the potential of their land and maintain healthy livestock.

Cattle can be integrated into an agroforestry system such as the *Leucaena*/maize system previously described either on a rotational basis or indirectly by using tree foliage as feed. Livestock can be grazed in the alley crop in between crop plantings. The livestock will graze on crop residues (ASPAC/FFTC, 1995) and the *Leucaena* foliage. The nutritional value of a tree varies between tree species and type of livestock grazed. For this reason, the tree has to be chosen for both its suitability for the crop grown and the livestock to be grazed (Solanki, 1991). Tree fodder can provide significant nutritional benefits to livestock. Reynolds and Jabbar (Reynolds and Jabbar, 1994) reported that tree fodder increases livestock survival in West Africa. They also reported that supplementing the diets of dairy livestock with *Leucaena* fodder in East Africa increased milk production. Goats used for dairying in an alley crop system with a leguminous tree showed high productivity when fed branches from the trees and other sources of fodder (Madany, 1991). (See Chapter 2.)

To retain some of the advantages formerly provided by the trees in the mulch, the livestock manure and urine must be applied to the ground where the mulch would have been applied. Benefits to the crop from the livestock waste will be less than the benefits from mulch because much of the nitrogen in the manure and urine is lost to volatilization, leaching and surface run-off (Steenvorden, 1988). Manure should be turned under the soil to minimize loss to runoff and the atmosphere. Urine is more difficult to maintain in the soil, so much of the nutrients in the urine will be lost.

Livestock are an important source of food and economic security for the rural poor. They provide food and capital reserves to be used in the event of a poor crop harvest such as droughts. Furthermore, livestock convert plant matter that is unfit for human consumption, such as crop residues and tree fodder, into useful products including meat, milk and wool.

TABLE 4.
Biomass burned as fuel in developing countries.
(Adapted from Pimentel et al., 1997a)

Fuel	10^6 tons	tons/person/yr	$/ton	Total $ × 10^9	10^9 kcal
Fuelwood	3000	0.7	35	105	8000
Crop residues	1300	0.3	10	13	4600
Dung	800	0.2	35	28	2400
Total	4100	1.2		146	15,000

Cattle production on pasture with leguminous trees can be beneficial both economically and environmentally. For example, pastures with 200 leguminous trees per hectare are estimated to increase the production of forage from 3000 kg/ha to 5000 kg/ha per year. This includes some added forage provided by the trees themselves. The gross weight production of cattle on the pasture alone was calculated to be 60 kg/ha per year. The pasture with the leguminous trees was calculated to produce approximately 100 kg/ha per year, or 60% more than the pasture system. Assuming $1 per kilogram of beef, the yield for the pasture system would be $60/ha and the pasture-tree system $100/ha per year. The environmental benefits of the silvopastoral system would be reduced soil erosion and water runoff.

8. FUELWOOD NEEDS AND BENEFITS OF AGROFORESTRY

Biomass fuels are used for cooking in developing countries and for heating during the cool evenings. Currently, various kinds of plant biomass provide up to 90% of the energy needs of the poor in developing countries (Chatterji, 1981; Peskin et al., 1992). An estimated 4.1 billion dry tons of biomass is burned as fuel annually, yielding 15 quadrillion (= 15 × 10^{15}) kcal of heat (Pimentel et al., 1994) (Table 4). Of this, about one-half (2.0 billion tons) is fuelwood, 33% is crop residues, and about 17% dung (Pimentel et al., 1994).

More than one half of the people who depend on fuelwood have inadequate supplies (de Montamlebert and Clement, 1983). In some countries, such as Brazil, where forest areas are fairly abundant at present, the rural poor burn mostly wood and charcoal. However, elsewhere in developing countries crop residues account for most of the biomass fuel — e.g., 55% in China, 77% in Egypt, and 90% in Bangladesh (Pimentel et al., 1986; Wen and Pimentel, 1992; Agarwal, 1992). FAO (FAO, 1978) estimates that the poor spend 15% to 25% of their income for fuel. In the Philippines, agroforestry is currently being practiced to meet fuelwood needs (Bensel, 1994). The demand for fuelwood in urban areas is met by agroforestry systems in which *Leucaena* and *Gliricidia sepium* are the main tree species grown.

Cooking of food is essential for preventing disease, improving nutrition, and increasing the palatability of many foods (Pimentel and Pimentel, 1996). Under usual cooking conditions, approximately two kcal are required to cook one kcal of food (Pimentel et al., 1986). Foley and van Buern (1982) and Smith et al. (1983) estimate that a person in a developing country will use nearly 700 kg of dry wood or one m^3 per year for cooking.

For both cooking and heating in Nepal, an average of 846 kg/person/yr of wood was used (Metz, 1994). Others report that from 912 kg to 1,200 kg/yr are required per person for cooking and heating (Tesoro, 1983; Appasamy, 1993). Fuelwood for cooking and heating may cost almost as much as food in some developing countries.

Collecting biomass for fuel requires a substantial amount of time and effort. For example, in Indonesia, India, Ghana, Mozambique, and Peru families spend from 1.5 to five hours each day collecting biomass to use as fuel (Peskin et al., 1992).

9. AGROFORESTRY AND THE CONSERVATION OF BIODIVERSITY

Beyond the diverse food and other products that forests provide for humans, they enhance food yields because they help conserve the biodiversity that is essential to human survival and the integrity of the environment (Wilson, 1988; Ehrlich and Wilson, 1991; Sayer and Whitmore, 1991; Pimentel et al., 1992). For instance, although tropical forests cover only 7% of the earth's surface, they contain more than one-half of the world's total plant and animal species (Wilson, 1988). An estimated 30,000 species of insects live in tropical forests and a hugely disproportionate amount of other flora and fauna live in this small region of the tropics (Boehnert, 1988). Natural biota not only help degrade wastes, recycle vital nutrients, and pollinate crops and natural vegetation but also provide natural enemies of pests that attack crops and natural vegetation.

Unfortunately, as many as two million species of plants and animals are projected to be exterminated throughout the world during the next 50-year period (Reid and Miller, 1989). This enormous loss is primarily caused by the removal of vast forest areas, soil degradation, and the heavy use of agricultural chemicals such as pesticides (Pimentel et al., 1992; Pimentel et al., 1997b).

To date, biological conservation has focused mainly on protecting national parks that cover only 3.2% of the world's land area (Reid and Miller, 1989). Certainly parks which include forests, grasslands, and other habitats are valuable resources. Often overlooked or taken for granted, however, is the protection of the biological diversity that exists in our vast, managed agricultural and forest ecosystems and human settlements. These combined areas cover approximately 95% of the terrestrial environment and require greater efforts to conserve the extensive biodiversity that exists there (Pimentel et al., 1992; Western, 1989).

One way to increase and conserve biological diversity is to increase plant diversity in managed ecosystems of forestry/agriculture and even in small household gardens (De Foresta and Michon, 1992). Multispecies gardens support a diverse group of natural biota and help the small farmer produce an abundant and varied food crop. When the garden contains trees and/or shrubs, the farmer benefits not only from a more effective use of soil nutrients but also has a source of fuelwood. For example, Michon and de Foresta (1991) describe a two ha home garden in Malaysia that contained 260 useful trees and other plant species out of 350 species found in the garden. Indeed, the diversity of species in home gardens may rival that of a natural forest system (Padock and De Jong, 1991). Examples of such multispecies gardens are found in Java where small holder farmers cultivate 607 crop species, including fruit trees, in their gardens (Dover and Talbot, 1987).

This overall species diversity is comparable to that found in deciduous subtropical forests. In Sumatra, the Damar people have developed profitable and productive agroforestry land use systems. These gardens simultaneously preserve hundreds of forest plants including rare epiphytes, herbs, and shrubs and provide habitats for at least 46 mammal species (17 protected in Indonesia), 92 bird species, plus most of the mesofauna (De Foresta and Michon, 1992).

Tree and shrub shelter belts as well as hedgerows maintained along the edges of cropland and pastureland also increase biodiversity (Elton, 1958). As mentioned, they reduce soil erosion and moisture loss in addition to increasing the biomass present in the managed ecosystems. Shelter belts and hedgerows frequently are a refuge for many forest species, including beneficial parasites and predators, which can invade crops and pastures to help control pest insects and weeds (Altieri et al., 1987; Paoletti et al., 1989). For example, in Indonesia a coccinelid beetle introduced to control an insect pest on the *Leucaena* tree that was interplanted in coffee plantations moved onto the coffee trees and helped to control a cottony-cushion scale pest of coffee (I.N. Oka, pers. comm., 1993).

The potential of agroforestry to reduce erosion and the use of agricultural chemicals has the potential to benefit aquatic organisms. Agricultural run-off and eroded soils eventually end up in streams, rivers and lakes. These chemicals and sediments are harmful to many aquatic organisms. Wide spread chemical usage and erosion on agricultural lands has led to the virtual destruction of many aquatic ecosystems. In turn, aquatic organisms are among the most endangered in the world.

In the United States for example, 364 species and subspecies (about 34% of all North American fish) are listed as "endangered," "threatened" or "of special concern" under the Endangered Species Act. In addition, 213 of 297 mussel species are endangered, as well as 162 of 338 crayfish and 204 of 10,000 aquatic insects (Allan and Flecker, 1993). This decline is attributed to habitat change and destruction, introduced species, overfishing and chemical pollution.

Agriculture is a common cause of both habitat change and chemical pollution. Sediment from deforested and naked agricultural soil ends up in stream and river bottoms, eliminating microhabitat for benthic invertebrates and spawning gravel for fish. Sediment also hinders filter feeders like the mussels which are essentially buried in sediment. The disruption of soil structure encourages overland flow which can result in flash floods with flow high in suspended sediment that can scour stream and river bottoms. Similarly, agricultural chemicals including pesticides and fertilizers end up in streams and rivers. As a result, chemically sensitive species are eliminated. Also, fertilizers can cause cultural eutrophication in which dense phytoplankton populations out-compete large aquatic plants. The decomposition of phytoplankton cells can lead to low oxygen levels that kill invertebrates and fish.

Agroforestry may reduce erosion by significant amounts, reducing sedimentation in streams and rivers. Nutrients provided by coppiced leaves and branches may diminish the need for chemical fertilizers, depending on the species of tree and nature of the nutrients. Also, the nutrients provided by tree matter are less volatile than chemical fertilizers and less likely to be leached from the soil. Finally, because agroforestry techniques can increase the retention of water in soils the need for irrigation is lessened. Though the ecologically destructive dam building spree occurring world-wide is in large part due to energy needs, demand for irrigation water is also a factor.

10. CONCLUSION

Population growth and the degradation of arable land have contributed to the world food crisis. Rural poor people are left with a decreasing supply of land and food. What land is available is quickly being degraded by erosion, leading to a decline in crop and livestock productivity. For instance, during the past 40 years approximately 30% of the world's arable land resources have had to be abandoned because of erosion. This serious loss continues and is one of the factors related to the worldwide malnutrition problem.

Agroforestry techniques can provide the means to increase biomass and, in turn, food crops and livestock while at the same time improving the productivity of degraded soils. In some situations with some crops employing agroforestry practices, grain yields may be nearly doubled. With livestock in some agroforestry systems, the yield in animal products may be increased approximately 60% while protecting the soil from erosion.

Since agricultural productivity is unlikely to keep up with the rate of population growth, only global population control can solve the crisis. Yet even if this came to be, in the immediate present the human population would still be six billion with more than two billion malnourished and many of the rural poor trying to make use of marginal land. Therefore, the use of agroforestry to increase food supplies and protect environmental resources could be one of the long-term solutions to hunger. In addition, agroforestry provides fuelwood to help meet cooking and heating needs. Cooking and heating aid in preventing diseases, further easing the maladies of the rural poor.

Resource poor farmers are likely to adopt agroforestry techniques when they are profitable and can help protect the farmer from risk and uncertainty. Agroforestry techniques encompass a wide variety of techniques and a diverse array of crop, livestock, and tree species. Systems can be designed to provide for many farmers' needs and can be tailored to suit local conditions. In fact, the greatest obstacles to the success of agroforestry systems do not originate with the agroforestry systems themselves, but the lack of education and knowledge concerning agroforestry practices. In some cases, the lack of information increases the risks and problems in implementing agroforestry techniques.

REFERENCES

Agarwal, A., 1992, *The Price of Forests*, Center for Science and Environment, New Delhi.

Agbooloa, A. A., 1990, "Organic matter and soil fertility management in the humid tropics of Africa," in *Organic Matter Management and Tillage in Humid and Subhumid Africa, Proceedings of the Third Regional Workshop of the Africaland Program*, Pushparajah, E., Latham, M., and Elliot, C. R., Eds., ISBRAM, Antananarivo, Madagascar, 231.

Alexander, M., 1977, *Introduction to Soil Microbiology*, second edition, John Wiley, New York.

Allan, J. D. and Flecker, A. S., 1993, "Biodiversity conservation in running waters," *Biosci.*, 43, 32.

Allison, F. E., 1973, *Soil Organic Matter and its Role in Crop Production*, Elsevier, New York.

Altieri, M. A., Trujuillo, F. J., and Farrell, J., 1987, "Plant-insect interactions and soil fertility in agroforestry systems implications for the design of sustainable agroecosystems," in *Agroforestry: Realities, Possibilities and Potentials*, Gholz, H. L., Ed., M. Nijhoff, Dordrecht, the Netherlands, 89.

Appasamy, P. P., 1993, "Role of non-timber forest products in a subsistence economy: the case of a joint forestry project in India," *Eco. Bot.,* 47, 258.

Arthur, P. F. and Ahunu, B. K., 1992, "Integration of sheep with plantation tree crops in the humid tropics," *World Rev. Anim. Prod.*, 27, 75.

ASPAC/FFTC, 1995, "Integrated crop-livestock production in slopeland areas," *ASPAC/FFTC*, 107, 5.

Atlavinyte, O., 1964, "Distribution of earthworms (Lumbricidae) and larval insects in the eroded soil under cultivated crops," *Pedobiologia*, 4, 245.

Atlavinyte, O., 1965, "The effect of erosion on the population of earthworms (Lumbricidae) in soils under different crops," *Pedobiologia*, 5, 178.

Banda, A. Z., Maghembe, J. A., Ngugi, D. N., and Chome, V. A., 1994, "Effect of intercropping maize and closely spaced *Leucaena* hedgerows on soil conservation and maize yield on a steep slope at Ntcheu, Malawi," *Agro. Syst.*, 27, 31.

Barrows, H. L. and Kilmer, V. J., 1963, "Plant nutrient losses from soils by water erosion," *Adv. Agron.*, 15, 303.

Bayliss-Smith, T., 1994, "The adoption of alley cropping by tropical rain forest farmers: top-down enthusiasm but bottom-up rejection?" *Res. Sem. Rur. Dev. Stud. Inst. Soc. Stud.*

Belsky, J. M., 1993, "Household food security, farm trees, and agroforestry: a comparative study in Indonesia and the Philippines," *Human Org.*, 52, 130.

Bennet, H. H., 1939, *Soil Conservation*, McGraw-Hill Book Co., New York.

Bensel, T., 1994, "Fuelwood, agroforestry and the environment in the uplands of Cebu, Philippines," *APA News*, 6–7, 21.

Bhatt, P. N., 1977, "Losses of plant nutrients through erosion process — a review," *Soil Cons. Dig.*, 5, 37.

Boehnert, J., 1988, *Agroforestry in Agricultural Education*, Josef Margraf, Gaimersheim, Germany.

Bouis, H. E., 1995, "Breeding for nutrition," *J. Fed. Am. Sci.*, 48, 4, 1.

Branckaert, R. and Mbayahaga, J., 1993, "Silvopastoralism: an appropriate solution for land use in Burundi," *Revue-Mondiale-de-Zootechnie*, 76, 35.

Brewbaker, J. L., 1987, "*Leucaena*: a multipurpose tree genus for tropical agroforestry," in *Agroforestry: a decade of development*, Steppler, H. A. and Nair, R. K. R., Eds., International Council for Research in Agroforestry, Nairobi, Kenya, 289.

Brewbaker, J. L., 1989, "*Leucaena*: Can there be such a thing as a perfect tree?" *Agrofor. Today*, 1, 4.

Brown, L. R., 1990, "Assessing the planet's condition," *EPA J.*, 16, 2.

Budowski, G., 1993, "Agroforestry: a discipline based on traditional knowledge," *Revista-Forestal-Centroamericana*, 2, 14.

Buntley, G. J. and Bell, F. F., 1976, "Yield estimates for the major crops grown on the soils of west Tennessee," *Tenn. Ag. Exp. Sta. Bull.*, 561.

Carson, B., 1985, "Erosion and Sedimentation Processes in the Nepalese Himalaya," *Occasional Paper No. 1.*, International Centre for Integrated Mountain Development, Kathmandu, Nepal.

Caveness, F. A. and Kurtz, W. B., 1993, "Agroforestry adoption and risk perception by farmers in Senegal," *Agro. Syst.*, 21, 11.

Chabeda, A. E. O., 1981, "The role of leguminous and non-leguminous fodder trees and shrubs in arid and semi-arid environments in Kenya," in *Kenya National Seminar in Agroforestry*, Buck, L., Ed., ICRAF, Nairobi, Kenya, 397.

Chaney, K. and Swift, R. S., 1984, "The influence of organic matter on aggregate stability in some British soils," *J. Soil Sci.*, 35, 223.

Chatterji, M., 1981, "Energy and environment in the developing countries: an overall perspective and plans for action," in *Energy and Environment in Developing Countries*, Chatterji, M., Ed., John Wiley, New York, 3.

Crosson, P., 1985, "National cost of erosion on productivity," in *Erosion and Soil Productivity*, American Society of Agricultural Engineers, St. Joseph, MI, 254.

Curry-Lindahl, K., 1972, *Conservation for Survival*, William Morrow & Co., New York.

De Foresta, H. and Michon, G., 1992, "Complex agroforestry systems and conservation of biological diversity (II)," *Malayan Nat. J.*, 45, 488.

deMontamlebert, M. and Clement, J., 1983, *Fuelwood Supplies in Developing Countries*, FAO Forestry Paper No. 42, Food and Agricultural Organization of the United Nations, Rome.

Donahue, R. H., Follett, R. H., and Tulloch, R. N., 1990, *Our Soils and Their Management*, Interstate Pub. Inc., Danville, IL.

Dover, M. and Talbot, L. M., 1987, *To Feed the Earth: Agroecology for Sustainable Development*, World Resources Institute, Washington, DC.

Dregne, H. E., 1992, "Erosion and soil productivity in Asia," *J. Soil Water Conserv.*, 47, 13.

Edwards, C. A., 1981, "Earthworms, soil fertility and plant growth," in *Workshop on the Role of Earthworms in the Stabilization of Organic Residues*, Beech Leaf Press, Lansing, MI, 61.

Edwards, C. A., Lal, R., Madden, P., Millier, R. H., and House, G., 1990, *Sustainable Agricultural Systems*, Soil and Water Conservation Society, Ankeny, IA.

Ehrlich, P. R. and Wilson, E. O., 1991, "Biodiversity studies: science and policy," *Science*, 253, 758.

El-Swaify, S. A., Moldenhauer, W. C., and Lo, A., 1985, *Soil Erosion and Conservation*, Soil Conservation Society of America, Ankeny, IA.

Elton, C. S., 1958, *The Ecology of Invasions by Animals and Plants*, Methuen, London.

Elwell, H. A., 1985, "An assessment of soil erosion in Zimbabwe," *Zimbabwe Sci. News*, 19, 27.

Falkenmark, M., 1989, "Water scarcity and food production," in *Food and Natural Resources*, Pimentel, D. and Hall, C. W., Eds., Academic Press, San Diego, CA, 164.

Filius, A. M., 1988, "Economic aspects of agroforestry," in *Viewpoints on Agroforestry*, Wieisum, K. F., Ed., Agricultural University, Wageningen, Netherlands, 169.

Foley, G. and van Buern, A., 1982, "Energy in transition from rural agriculture," in *Energy in the Transition from Rural Subsistence*, Wionczek, M. S., Foley, G. and van Buren, A., Eds., Westview Press, Boulder, CO, 1.

Follett, R. F. and Stewart, B. A., 1985, *Soil Erosion and Crop Productivity*, Am. Soc. Of Agron., Crop Sci. Soc. Of Am., Madison, WI.

Follett, R. F., Gupta, S. C., and Hunt, P. G., 1987, "Conservation practices: Relation to the management of plant nutrients for crop production," in *Soil Fertility and Organic Matter as Critical Components of Production Systems*, Soil Science Society of America and American Society of Agronomy, Madison, WI.

Food and Agricultural Organization of the United Nations, 1978, *Forestry for Local Community Development*, Food and Agricultural Organization of the United Nations, Rome.

Food and Agriculture Organization, 1990, *FAO Quarterly Bulletin of Statistics*, 3, 2, 1.

Food and Agriculture Organization, 1991, *The State of Food and Agriculture*, Food and Agriculture Organization of the United Nations, Rome.

Food and Agriculture Organization, 1994, *FAO Quarterly Bulletin of Statistics*, 7, 2, 1.

Foster, G. R., Young, R. A., Ronkens, M. J. M., and Onstad, C. A., 1985, "Processes of soil erosion by water," in *Soil Erosion and Crop Productivity*, Follet, R. F., and Stewart, B. A., Eds., Am. Soc. Of Agron., Crop Sci. Soc. Of Am., Madison, WI, 137.

Foth, B. D. and Ellis, B. G., 1988, *Soil Fertility*, John Wiley & Sons, New York.

Frye, W. W., Ebelhar, S. A., Murdock, L. W., and Bevins, R. L., 1982, "Soil erosion effects on properties and productivity of two Kentucky soils," *Soil Sci. Soc. Am. J.*, 46, 105.

Gray, D. M. and Leiser, A. T., 1989, *Biotechnical Slope Protection and Erosion Control*, Kreiger Publishing Co., Malabar, FL.

Greenland, D. J. and Hays, M. H.,1981, *The Chemistry of Soil Processes*, John Wiley, New York.

Harris, J. M., 1996, "World agricultural futures: Regional sustainability and ecological limits," *Eco. Eco.*, 17, 2, 95.

Hauser, V. L. and Jones, O. R., 1991, "Runoff curve numbers for the Southern High Plains," *Trans. ASAE*, 34, 142.

Hoekstra, D. A., 1983, *The Use of Economics in Agroforestry*, Working Paper No. 2, ICRAF, Nairobi, Kenya.

Hudson, N. W., 1981, *Soil Conservation*, second edition, Cornell University Press, Ithaca, NY.

International Council for Research in Agroforestry, 1993, *ICRAF Annual Report 1993*, Van Houten, H., Ed., ICRAF, London.

Jeanes, K. W., Gutteridge, R. C., and Shelton, H. H.,1996, "Competition effects between Leucaena and maize grown simultaneously in alley cropping system in subtropical Australia," *Exp. Agri.*, 32, 1, 49.

Kendall, H.W. and Pimentel, D., 1994, "Constraints on the expansion of the global food supply," *Ambio*, 23, 198.

Kidd, C. and Pimentel, D., 1992, *Integrated Resource Management: Agroforestry for Development*, Academic Press, San Diego, CA.

Kramer, L. A., 1986, "Runoff and soil loss by cropstage from conventional and conservation tilled corn," *Trans. Am. Soc. Agric. Engineers*, 29, 774.

Krishnamurthy, L., Nair, P. K. R and Latt, C. R., 1994, "Agroforestry education and training: The Latin American perspective," *Agro. Syst.*, 28, 1, 1.

Kutzner, P. L., 1991, *World Hunger. A Reference Handbook*, ABC-Clio, Santa Barbara, CA.

La Anyane, S., 1985, *Economics of Agricultural Development in Tropical Africa*, John Wiley & Sons, New York.

Lal, R., 1976, *Soil Erosion Problems on an Alfisol in Western Nigeria and their Control*, IITA, Lagos, Nigeria.

Lal, R., 1980, "Losses of plant nutrients in runoff and eroded soil," in *Nitrogen Cycling in West African Ecosystems*, Rosswall, T., Ed., UNEP, Royal Swedish Academy of Sciences, Stockholm, 31.

Lal, R.,1984, "Soil erosion from tropical arable lands and its control," *Adv. Agron.*, 37, 183.

Lal, R., 1989, "Land degradation and its impact on food and other resources," in *Food and Natural Resources*, Pimentel, D., Ed., Academic Press, San Diego, 85.

Lal, R., 1993, "Soil erosion and conservation in West Africa," in *World Soil Erosion and Conservation*, Pimentel, D., Ed., Cambridge University Press, Cambridge, U.K., 7.

Lal, R. and Stewart, B. A., 1990, *Soil Degradation.*, Springer-Verlag, New York.

Langdale, G. W., West, L. T., Bruce, R. R., Miller, W. P., and Thomas, A. W., 1992, "Restoration of eroded soil with conservation tillage," *Soil Tech.*, 5, 81.

Le Houerou, H., 1987, "Indigenous shrubs and trees in silvopastoral systems of Africa," in *Agroforestry: A Decade of Development*, Steppler, H. and Nair, P., Eds., ICRAF, Nairobi, Kenya.

Lee, E. and Foster, R. C., 1991, "Soil fauna and soil structure," *Aust. J. Soil Res.*, 29, 745.

Lee, K. E., 1985, *Earthworms: Their Ecology and Relationships with Soils and Land Use*, Academic Press, Orlando, FL.

Leyton, L., 1983, "Crop water use: principles and some considerations for agroforestry," in *Plant Research and Agroforestry*, Huxley, P. A., Ed., International Council for Research in Agroforestry, Nairobi, Kenya, 379.

Libert, B., 1996, *The Environmental Heritage of Soviet Agriculture*, CAB International, Oxon, UK.

Liebhardt, W. C., 1983, "Variability of fertilizer recommendations by soil testing laboratories in the United States," in *Environmentally Sound Agriculture*, Lockeretz, W., Ed., Praeger, New York, 167.

Lucas, R. E., Holtman, J. B., and Conner, L. J., 1977, "Soil carbon dynamics and cropping practices," in *Agriculture and Energy*, Lockeretz, W., Ed., Academic Press, New York, 333.

Lulandala, L. L. L., Munishi, P. K. T., and Maliondo, S. M., 1995, "Effect of spacing and management regime on *Leucaena* fodder yield and nutrient removal in alley cropping systems in Tanzania," *Int. Tree Crops J.*, 8, 129.

Maberly, G. F., 1994, "Iodine deficiency disorders: contemporary scientific issues," *J. Nutrit.*, 124, supplement 8, 1473s.

Madany, M. H., 1991, "Intercropping fodder trees: a case study from Somalia, *Rangelands*," 14, 4, 208.

Metz, J. J., 1994, "Forest product use at an upper elevation village in Nepal," *Environ. Manag.*, 18, 371.

Michon, G. and de Foresta, H., 1991, "Complex agroforestry systems and the conservation of biological diversity," *Malayan Nat. J.*, 45, 488.

Mokma, D. L. and Sietz, M. A., 1992, "Effects of soil erosion on corn yields on Marlette soils in South-Central Michigan," *J. Soil Water Conser.*, 47, 325.

Mol, S., 1989, *Agroforestry and Small Farmers: Some Issues on Risk and Uncertainty*, Working Paper No. 56, ICRAF, Nairobi, Kenya.

Mureithi, J. G, Tayler, R. S., and Thorpe, W., 1994, "The effects of alley cropping with *Leucaena leucocephala* and of different management practices on the productivity of maize and soil chemical properties in lowland coastal Kenya," *Agro. Sys.*, 27, 31.

Murphee, C. E. and McGregor, K. C., 1991, "Runoff and sediment yield from a flatland water shed in soybeans," *Trans. Am. Soc. Agric. Engineers*, 34, 407.

Nair, P. K. R., 1984, *Soil Productivity Aspects of Agroforestry*, ICRAF, Nairobi, Kenya.

Nat. Soil Erosion-Soil Prod. Res. Planning Committee, 1981, "Soil crosion effects on soil productivity: a research perspective," *J. Soil Water Conser.*, 32, 82.

Nielsen, F., 1994, "Indigenous agroforestry systems in southern Luwero District Uganda and their potential in agricultural development," *CDR Working Paper Centre for Development Research*, 16, 97.

Office of Technology Assessment, 1982, *Impacts of Technology on U.S. Cropland and Rangeland Productivity*, Office of Technology, U.S. Congress, Washington, DC.

Olsen, K. R. and Nizeyimana, E., 1988, "Effects of soil erosion on corn yields of seven Illinois soils," *J. Production Agric.*, 1, 13.

Padock, C. and De Jong, W., 1991, "The house gardens of Santa Rosa: diversity and variability in an Amazonian agricultural system," *Econ. Bot.*, 45, 166.

Paoletti, M. G., Stinner, B. R., and Lorenzoni, G. C., 1989, "Agricultural ecology and the environment," *Agr. Ecosyst. and Environ.*, 27, 1.

Parrington, J. R., Zoller, W. H., and Aras, N. K., 1983, "Asian dust: Seasonal transport to the Hawaiian Islands," *Science*, 246, 195.

Peskin, H. M., Floor, W., and Barnes, D. F., 1992, *Accounting for Traditional Fuel Production: The Household Energy Sector and Its Implications for the Development Process*, The World Bank, Washington, DC.

Pimentel, D. and Pimentel, M., 1996, *Food, Energy and Society*, University Press of Colorado, Niwot, CO.

Pimentel, D., Terhune, E. C., Dyson-Hudson, R., Rochereau, S., Samis, R., Smith, E., Denman, D., Reifschneider, D., and Shepard, M., 1976, "Land degradation: effects on food and energy resources," *Science*, 194, 149.

Pimentel, D., Garnick, E., Berkowitz, A., Jacobson, S., Napolitano, S., Black, P., Valdes-Cogliano, S., Vinzant, B., Hudes, E., and Littman, S., 1980, "Environmental quality and natural biota," *Biosci.*, 30, 750.

Pimentel, D., Wen, D., Eigenbrode, S., Lang, H., Emerson, D., and Karasik, M., 1986, "Deforestation: interdependency of fuelwood and agriculture," *Oikos*, 46, 404.

Pimentel, D., Allen, J., Beers, A., Guinand, L., Linder, R., McLaughlin, P., Meer, B., Musonda, D., Perdue, D.,Poisson, S., Siebert, S., Stoner, K., Salazar, R., and Hawkins, A., 1987, "World agriculture and soil erosion," *Biosci.*, 37, 277.

Pimentel, D., Stachow, U., Takacs, D. A., Brubaker, H. W., Duman, A. R., Meaney, J. J., O'Neil, J., Onsi, D. E., and Corzilius, D. B., 1992, "Conserving biological diversity in agricultural/forestry systems, "*Biosci.*, 42, 354.

Pimentel, D., Herdendorf, M., Eisenfeld, S., Olander, L., Carroquino, M., Corson, C., McDade, J., Chung, Y., Cannon, W., Roberts, J., Bluman, L., and Gregg, J., 1994, "Achieving a secure energy future: environmental and economic issues," *Ecol. Eco.*, 9, 3, 201.

Pimentel, D., Harvey, C., Resosudarmo, P., Sinclair, K., Kurz, D., McNair, M., Crist, S., Shpritz, L., Fitton, L., Saffouri, R., and Blair, R, 1995, "Environmental and economic costs of soil erosion and conservation benefits," *Science*, 267, 1117.

Pimentel, D., McNair, M., Buck, L., Pimentel, M., and Kamil, J., 1997a, "The value of forests to world food security," *Hum. Ecol.*, 25, 1, 91.

Pimentel, D., Wilson, C., McCullum, C., Huang, R., Dwen, P., Flack, J., Tran, Q., Saltman, T., and Cliff, B., 1997b, "Economic and environmental benefits of biodiversity," *Biosci.*

Pimentel, D., Harvey, C., Resosudarmo, P., Sinclair, K., Kurz, D., McNair, M., Crist, S., Shpritz, L., Fitton, L., Saffouri, R., and Blair, R., 1997c, "Land use, erosion, and water resources," in *Water Resources: Environmental Planning, Management, and Development*, Biswas, A. K., Ed., McGraw-Hill, New York, 37.

Population Reference Bureau, 1995, *World Population Data Sheet*, Population Reference Bureau, Washington, DC.

Rachie, K. O., 1983, "Intercropping tree legumes with annual crops," in *Plant Research and Agroforestry*, Huxley, P. A., Ed., ICRAF, Nairobi, Kenya, 103.

Rathore, M. P. S., 1995, *Insect Pests in Agroforestry*, Working Paper No. 70, ICRAF, Nairobi, Kenya.

Reid, W. S., 1985, "Regional effects of soil erosion on crop productivity — northeast," in *Soil Erosion and Crop Productivity*, Follet, R. F. and Stewart, B. A., Eds., Am. Soc. of Agron., Crop Sci. Soc. Of Am., Madison, WI, 235.

Reid, W. V. and Miller, K. R., 1989, *Keeping Options Alive: The Scientific Basis for Conserving Biodiversity*, World Resources Institute, Washington, DC.

Reynolds, L. and Jabbar, M., 1994, "The role of alley farming in African livestock production," *Outlook on Agriculture*, 23, 2, 105.

Ricou, G. A. E., 1979, "Consumers in meadows and pastures," in *Grassland Ecosystems of the World: Analysis of Grasslands and their Uses*, Coupland, R. T., Ed., Cambridge University Press, Cambridge, U.K., 147.

Rocheleau, D. E., 1987, "The user perspective and the agroforestry research agenda in *Agroforestry: Realities and Potentials*, Gholz, H. L., Ed., M. Nijhoff, Dordrecht, the Netherlands.

Roose, E., 1988, "Soil and water conservation lessons from steep-slope farming in French speaking countries of Africa," in *Conservation Farming on Steep Lands*, Soil and Water Conservation Society, Ankeny, IA, 130.

Sayer, J. A. and Whitmore, T. C., 1991, "Tropical moist forests destruction and species extinction," *Biol. Conserv.*, 55, 199.

Scherr, S. J., 1987, *Setting priorities for agroforestry research and development: Potential contributions of economics research*, Working Paper No. 51, ICRAF, Nairobi, Kenya.

Scherr, S. J., 1995, "Economic factors in farmer adoption of agroforestry: Patterns observed in western Kenya," *World Development*, 23, 787.

Shannon, P. A. and Vogel, W. O., 1994, "The effects of alley cropping and fertilizer application on continuously cropped maize," *Trop. Agri.*, 71, 3, 163.

Sharpley, A. N. and Smith, S. J., 1991, "Effects of cover crops on surface water quality," in *Cover Crops for Clean Water*, Hargrove, W. L., Ed., Soil and Water Conservation Society, Ankeny, IA, 41.

Simons, M., 1992, "Winds toss Africa's soil, feeding lands far away," *New York Times*, Oct. 29, 1992, A1.

Smith, K. R., Aggarwal, A. L., and Dave, R. M., 1983, "Air pollution and rural biomass fuels in developing countries: a pilot study in India and implications for research and policy," *Atmosphere and Environment*, 17, 2343.

Solanki, S. S., 1991, "Utilization of trees for livestock feeding," *Asian Livestock*, 16, 153.

Speth, J. G., 1994, *Towards an Effective and Operational International Convention of Desertification,* United Nations, International Convention of Desertification, Int. Negotiating Comm., New York.

Steenvorden, J. H. A. M., 1988, "How to reduce nitrogen losses in intensive grassland management," *Ecological Bulletins*, 39, 126.

Stocking, M., 1995, "Soil erosion and land degradation," in *Environmental Science for Env. Mgt.*, O'Riordan, T., Ed., Longman Scientific & Technical, Essex, England, copublished with John Wiley & Sons, New York, 223.

Teotia, J. P., Duky, F. L., and McCalla, T. M., 1950, *Effect of Stubble Mulch on Number and Activity of Earthworms,* Bulletin 165, Nebraska Experimental Research Station, Lincoln, NE.

Tesoro, F. O., 1983, *Utilization of selected non-timber forest products in the Philippines, First ASEAN Forestry Congress Proceedings*, Book II, Volume 4, ASEAN, Manila.

Tisdale, S. L., Nelson, W. L., and Beaton, J. D., 1985, *Soil Fertility and Fertilizers,* MacMillan Publishing Co., New York.

Tisdall, J. M. and Oades, J. M., 1982, "Organic matter and water-stable aggregates in soils," *J. Soil Sci.*, 33, 141.

Torres, F., 1983, "Potential contribution of *Leucaena* hedgerows intercropped with maize to the production of organic nitrogen and fuelwood in the lowland tropics," *Agro. Syst.*, 1, 4, 323.

Tribe, D., 1994, *Feeding and Greening the World*, CAB International, Oxon, UK.

Troeh, F. R. and Thompson, L. M., 1993, *Soils and Soil Fertility*, fifth edition, Oxford University Press, New York.

Troeh, F. R., Hobbs, J. A., and Donahue, R. L., 1991, *Soil and Water Conservation*, second edition, Prentice Hall, Englewood Cliffs, NJ.

U.S. Department of Agriculture, 1989, *The Second RCA Appraisal: Soil, Water, and Related Resources on Nonfederal Land in the United States, Analysis of Conditions and Trends*, Soil Conservation Service, USDA, Washington, DC.

U.S. Department of Agriculture, 1994, *Summary Report 1992 Natural Resources Inventory*, Soil Conservation Service, USDA, Washington, DC.

U.S. Forest Service, 1936, *The Major Range Problems and Their Solution*, U.S. Senate Document No. 199 — Separate No. 1, U.S. Forest Service, Washington, DC.

Van Dijk, H., 1980, "Survey of Dutch soil organic matter research with regard to humification and degradation rates in arable land," in *Soil Degradation: Proc. of the Land Use Seminar on Soil Degradation*, Boels, D., Davies, D. B., and Johnston, A. E., Eds., Wageningen, Netherlands, 133.

Van Doren, C. A., Stauffer, R. S., and Kidder, E. H., 1950, "Effect of contour farming on soil loss and runoff," *SSSAP*, 15, 413.

Van Rhee, J. A., 1965, "Earthworm activity and plant growth in artificial cultures," *Plants and Soil*, 22, 43.

Volk, B. G. and Loeppert, R. H., 1982, "Soil organic matter," in *Handbook of Soils and Climate in Agriculture*, Kilmer, V. J., Ed., CRC Press, Boca Raton, FL, 211.

Waldren, R. P., 1983, "Corn," in *Crop-Water Relations*, John Wiley & Sons, New York, 187.

Wen, D., 1993, "Soil erosion and conservation in China," in *Soil Erosion and Conservation*, Pimentel, D., Ed., Cambridge University Press, New York, 63.

Wen, D. and Pimentel, D., 1992, "Ecological resource management to achieve a productive, sustainable agricultural system in northeast China," *Agric. Ecosys. Env,.*, 41, 215.

Wendt, R. C. and Burwell, R. E., 1985, "Runoff and soil losses from conventional, reduced and no-till corn," *J. Soil Water Conser.*, 40, 450.

Wendt, R. C., Alberts, E. E., and Helmers, A. T., 1986, "Variability of runoff and soil loss from fallow experimental plots," *SSSAJ*, 50, 730.

Western, D., 1989, "Conservation without parks: Wildlife in rural landscapes," in *Conservation for the Tewnty-First Century*, Western D. and Pearl, M. C., Eds., Oxford University Press, New York, 158.

Wilson, E. O., 1988, *Biodiversity*, National Academy of Sciences, Washington, DC.

Wood, M., 1989, *Soil Biology*, Chapman & Hall, New York.

World Health Organization, 1995, *Bridging the Gaps*, World Health Organization, Geneva.

World Resources Institute, 1992, *World Resources*, Oxford University Press, New York.

World Resources Institute, 1994, *World Resources 1994–95*, World Resources Institute, Washington, DC.

Worldwatch Institute, 1988, *State of the World*, Worldwatch Institute, Washington, DC.

Yamoah, C. F., Agboola, A. A., and Wilson, G. F., 1986, "Nutrient contribution and maize performance in alley cropping systems," *Agro. Syst.*, 4, 3, 247.

York, E. T., 1990, "The importance of agroforestry education and training," *Forest Sci.*, 35, 7.

Young, A., 1989, *Agroforestry for Soil Conservation*, CAB International, Wallingford, U.K.

Young, A., 1990, "Agroforestry, environment and sustainability," *Outlook Agric.*, 19, 155.

Young, A., 1993, "Agroforestry as a viable alternative for soil conservation," *Entwicklung und Landlicher Raum*, 27, 3.

14 Domestication of Tropical Trees: From Biology to Economics and Policy

Roger R. B. Leakey and Thomas P. Tomich

1. INTRODUCTION

Simmonds (1976) in his book the "Evolution of Crop Plants" lists about one thousand plant species associated with the domestication of 105 food and 30 non-food crops (fibre, fodder, oil, latex, *etc.*, excluding timber). A total of 193 of these 1000 plants are tree species of which about 50 are domesticated, mostly providing fruits. Trees are also grown for timber. Around the world about 200 timber species are commonly planted (Faulkner, 1976; Evans, 1992), but few of these can really be considered to be more than semi-domesticated, although for some of them breeding programs are in progress. Together these food and non-food crops originate from all corners of the globe (see Simmonds, 1976 for maps). Frequently, through cross-breeding, there are genes from many related, but non-sympatric, species brought together to create the many cultivars and varieties of the different crops. Thus, over time, mankind has transferred genes geographically and frequently grown crops in areas far removed from the natural range of the original species.

In addition to the plants domesticated as crops, a further group is domesticated for amenity, for growing in gardens and along streets to beautify our homes and towns. Ornamental garden plants number about 15000 (Brickell,1996) and like crops include herbaceous and tree species. Some of these garden plants, like roses, are highly domesticated, while others are raised from seed collections made in the wild. There is relatively little overlap between garden and crop plants, although a few tree species of the temperate zone e.g., *Betula pubescens, B. verrucosa* and *Fagus sylvatica* feature in both lists.

1-56670-294-1/99/$0.00+$.50

Like plants, animals have also been domesticated for agriculture and for pleasure giving us livestock and pets. In both plants and animals the proportion that have been domesticated for both purposes out of the large number available is small. For the plant kingdom the proportion domesticated is about 0.3% in agriculture and 4.1% if garden plants are included. However, if the potential for domestication is limited to higher plants (angiosperms, gymnosperms and pteridophytes) of which there are 250,000 species (Wilson, 1994), the proportions go up to 0.5 and 6.5% respectively. What has determined the proportion of species that we have domesticated for agriculture? This question is particularly striking when one considers that hunter gatherers have made use of very large numbers of species, some 20000 are edible (Kunin and Lawton, 1996). Compendia of plants that have been and are still used by mankind leave many areas unrepresented, but in Ghana for example, most tree species have a medicinal or other use in addition to their timber (Irvine, 1961); as do many herbaceous plants (Abbiw, 1990). Agriculture is therefore not making full use of the diversity of species available. What are the reasons why so few plants species have been domesticated?

Homma (1994; 1996) has emphasized that domestication follows extractivism when either markets expand beyond the capacity of the forest to supply the products, or when the resource is over exploited so that again the market cannot be fulfilled. Another consideration, is that through domestication, sought after products can be produced on-farm through the practice of agroforestry, close to the market, making "extraction" cheaper and easier. Typically, domestication lowers the price of the product, although potentially this can also be offset by higher yields, enhanced quality or an expanded period of production. Based on the experience of already cultivated crops, domestication also typically results in the spread of the species from their natural range to be introduced elsewhere around the world. Contrary to this expansion, the products from domesticated species are sometimes replaced on the market by synthetics, resulting in their disappearance from agriculture. To the extent that domestication lowers production costs, hence prices, incentives to develop synthetic substitutes are reduced.

The domestication and commercialization of trees producing timber and non-timber forest products is seen by Sanchez and Leakey (1997) as an essential component of developing agroforestry as a means of balancing food security with the utilization of natural resources, within an ecological framework akin to the normal dynamics of natural ecosystems (Leakey, 1996). Previous papers have focused on biological issues associated with the domestication of indigenous trees (e.g., Leakey and Jaenicke, 1995; Leakey and Simons, 1997). Within this biological context, this paper examines some of the social, economic and policy issues.

2. WHAT IS DOMESTICATION?

The domestication of agroforestry trees, as in other species, involves accelerated and human-induced evolution to bring species into wider cultivation through a farmer-driven and often market-led process (ICRAF, 1997 a). It is an iterative procedure involving the identification, production, management and adoption of desirable germplasm. Strategies for individual species vary according to their functional use, biology and target environments. Domestication can occur at any point along the continuum from the wild to the genetically-transformed state.

3. THE ROLE OF DOMESTICATION IN AGROFORESTRY

Cash generation from tree products has the potential to reduce farmers' reliance on subsistence food production and to improve their welfare (Tomich *et al.,* 1994). This trend is also seen with valuable timbers which are again increasingly becoming important cash-earners for small-scale producers in agroforestry systems. This is clearly seen in the Philippines and Costa Rica where farmers are growing *Gmelina arborea.* In Costa Rica these farmers are also finding that the market for tree seeds is lucrative.

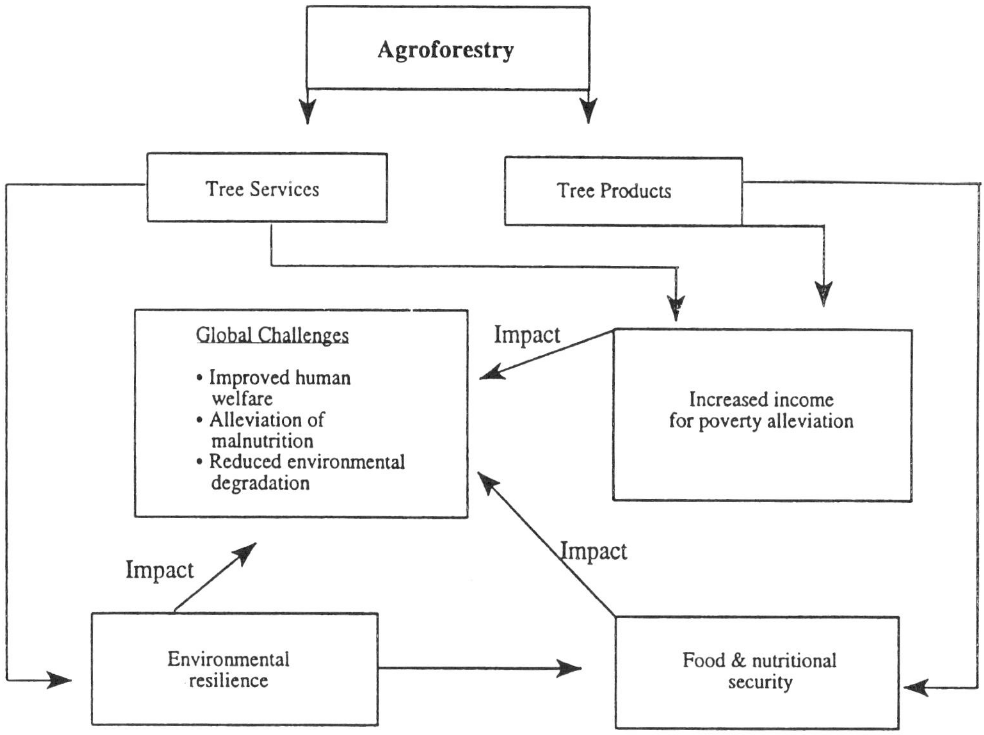

FIGURE 1. The relationship between the two functions of trees, the three goals of ICRAF and the impact required to mitigate two global challenges (modified from ICRAF 1997a).

At ICRAF it is the relationships between trees, people and the environment that are seen as the crux of agroforestry and the object of tree domestication. Tree services and tree products impinge on human welfare, and environmental degradation through their beneficial effects on food and nutritional security, poverty alleviation and environmental resilience (Figure 1); the three goals of agroforestry. Hence, the domestication of agroforestry trees, one of the five pillars of ICRAF's program (ICRAF 1997 a), is aimed at enhancing the services and products from trees. Currently, specific emphasis is being placed on the products from trees, as these are seen as having potentially great impacts on farmers' income and the health and well-being of rural households; as well as indirect effects on the urban poor, many of whom are dependent on tree products for their essential needs. A recent estimate suggests that 1.5 billion people in the tropics may depend on tree products for many of their needs (Leakey and Sanchez, 1997).

4. INCOME FROM NON-TIMBER FOREST PRODUCTS

Is the income from selling non-timber forest products sufficient to lift poor farmers out of the poverty trap?

There is a growing body of data on the value of these so-called "minor forest products." Some of this information is at the level of regional or global trade, some at the level of a market place and some at the level of the household. There is certainly a need for much more data, particularly at the household level before we can really answer the above question. However, Watson (1990) has reported that in monetary terms the output of tree-based compound gardens in Nigeria are

TABLE 1
Comparison of net present values and internal rates of returns of production systems in 1991 prices, using a 15-year time horizon, Yurimaguas, Peru (After: ICRAF, 1995)

Production option	Net present value (U.S. $ ha^{-1})	Internal rate of return
Multi-strata system	6727	831
Peach palm plantation	2061	64
Shifting cultivation	218	19

five–10 times greater than crop fields, with returns to labor of four–eight times greater. These orders of magnitude are supported by economic data from multi-strata systems in Yurimaguas, Peru (Table 1).

At the global level, the pharmaceutical products extracted from the bark of the *Prunus africana* tree and used to treat prostate hypertrophy and cancers were reported in 1993 to sell for U.S. $150 million yr^{-1} (Cunningham and Mbenkum, 1993). This figure is now thought to have gone up to $220 million in 1997 (Cunningham, pers. comm.) The raw bark in Cameroon sells at U.S. $0.5 per kg. The yield from a tree can be 55 kg tree^{-1}, but sustainable harvesting cutting a panel from the trunk every five years is preferable and can yield more over a period of 30–40 years (Cunningham *et al.,* In press). The bark extract sells in Cameroon for $966 kg^{-1}. Similarly, the trade in Peach palm fruits and heart of palm which are bottled and canned for the gourmet food market are another U.S. $100 million (Clement and Villachica, 1994).

Low input and high input continuous cultivation both had negative net present values.

At the regional or market level, the street value of chewing sticks *Garcinia* sp. have been reported to be sold from Kumasi market for an estimated U.S. $9 million per annum (Falconer, 1990), while the damar gum from *Shorea javanica* trees grown in agroforests around Pesisir, near Krui in south-west Sumatra, brings U.S. $7.55 million to the community from about 10,000 ha of these agroforests. These damar agroforests contain about 160 *S. javanica* trees ha^{-1} in a mixture with 85 other trees of 35–40 other species. The resin is used in the manufacture of paints, varnish and incense (Michon *et al.,* in press). These damar agroforests also produce fruits and timber that are marketed regionally. The Cinnamon agroforests around Kerinci Seblat National Park in Sumatra, are equally lucrative to their farmers, and were reported to produce $7.7 million worth of cinnamon bark in 1970. Since then the area under production in Kerinci has risen from 23336 ha to 42600 ha (Aumeeruddy, 1994) and the price has risen 8 fold (5 fold by comparison with the price of rice) to 2700 Rupiah kg. Another important agroforestry product from Indonesia is rubber from the jungle rubber agroforests that now produce almost 75% of Indonesia's production; the value of this trade was almost $.9 billion in 1995 (International Rubber Study Group, 1997). Calculations by Barlow *et al.* (1994) indicated that Indonesian smallholders are the lowest cost (i.e., most efficient) natural rubber producers in the world.

An extrapolation from market surveys in Cameroon suggests that the trade of four indigenous fruits (*Ricinodendron heudelottii, Irvingia gabonensis, Dacryodes edulis* and *Cola acuminata*) from the humid forest zone over the six-month period January–July 1995 was valued at $1.2 million (Table 2). A proportion of this trade was to neighboring countries of Gabon, Nigeria, Equatorial Guinea and Central African Republic (Ndoye, 1995); 30% of this trade goes to Gabon and Equatorial Guinea. In this case, the reason given by individual farmers for selling non-timber forest products was to acquire cash for basic needs of their household (74%), to pay for school fees (9%), and for other needs (17%).

TABLE 2
Market details of four non-timber forest products in Cameroon (After: Ndoye, 1995)

	US $ Price/kg	Wt traded (tonnes)	Value of trade U.S. $
R. heudelottii	2.7	172	460,200
I. gabonensis	2.2	140	301,590
D. edulis	0.4	587	244,480
C. acuminata	1.0	212	211,970
others	—	401	122,580
Total for non-timber forest products		1512	1,340,820

In South Africa, the fruits of the Marula tree (*Sclerocarya birrea*) are being used to make a jelly, a beer and also a liqueur called "Amarula." Amarula, which is produced by the Distillers Corporation, is breaking onto the international market being available in Zimbabwe, Kenya, Australia, U.K. and elsewhere. In addition, a Marula beer "Afreeka," produced by Mitsubushi Corporation, has recently been undergoing market trials in England. Wines from *S. birrea* "Marulam" and *Uapaca kirkiana* "Masau" are also bottled and marketed in Zambia. In support of this commercial activity, Shackleton (1996) has reported for the Transvaal a conservative estimate of fruit yield as 50 kg tree^{-1} and a density of 3 trees ha^{-1} at a value of \$1.4 kg^{-1} (i.e., \$210 ha^{-1}).

The previous examples from southern Africa indicate the opportunities to open new markets for formerly wild indigenous fruits. The value and quantities of indigenous fruit currently used throughout southern Africa are not known, although in excess of 700 tons were marketed for various uses in 1987 (van Wyk, quoted by Cunningham and Davies (in press). Shackleton *et al.* (1997) report that 46% of all households surveyed cultivate indigenous fruits mostly for domestic use, although trade accounted for income of about $27 per month, while 68% grew commercial fruit trees (citrus, mango, avocado, *etc.*). *S. birrea* was grown on more than half the farms growing indigenous fruits. The highest income ($822 per month) from indigenous fruits was achieved by a household involved in selling beer made from fruits. This indicates the potential of these fruits to increase farm income.

In the Sahel the sale of Néré fruits (*Parkia biglobosa*) can increase household income by about $270 per year, nearly half the average annual income (ICRAF, 1997 b). Néré fruits, however, can vary in price from 17 to 30 cents — they are cheapest in the south and most expensive in the north (Lamien *et al.*, 1996). However, soumbala, the fermented product from Néré fruits sells at nearly \$1.0 kg^{-1}. Despite the reduction in crop production due to the trees in the traditional Sahelian Parklands, the overall returns per hectare are increased by \$55 with eight *Vitellaria paradoxa* (Karité/sheanut) and *Parkia biglobosa* (Néré) trees ha^{-1} (Bonkoungou, 1995).

In Kenya, the Australian *Grevillea robusta* is commonly grown on field boundaries for fuel and timber. A study found that planting 37 *G. robusta* trees down one side of a 1 ha maize plot was profitable, even at a real discount rate of 20% (Tyndall unpublished, quoted by Cooper *et al.*, 1996). This study measured the value of timber and fuelwood, but did not value other benefits, such as control of erosion, the role of trees as windbreaks, boundary demarcation, *etc.*

These few examples, serve to illustrate that tree products do represent significant sources of income for resource-poor, subsistence farmers, and that a change towards specialist tree-based land uses, such as the SE Asian agroforests, can raise income still further.

5. THE UNEVEN PACE OF DOMESTICATION

The domestication of cereals, started as far back as 8000 BC in the Near East, at the same time as cultivation began. Barley, the most abundant and cheapest grain, was the standard fare of the poor, the ration of the soldier, serf and slave, and the staff of life for the Greek peasantry (Harlan, 1976). The domestication of similar basic food stuffs has progressed ever since.

There are important differences among the missions of the CGIAR (Consultative Group for International Agricultural Research) and other contemporary public, private sector and colonial era research institutions. The Green Revolution, led by the CGIAR since the 1960s has made enormous progress in the domestication of the staple food crops, raising food security through increased crop yields. The best-known example of the impact of the Green Revolution was in India where in 1968 the wheat harvest leaped by 45% to 16.5 million tons (Tribe, 1994). This was the result of breeding short-strawed, high-yield varieties and the use of fertilizers, irrigation and pest control — i.e., domestication through genetic improvement and better management. It was calculated 12 years ago that the research of wheat and rice breeders had together resulted in an annual increase of more than 40 million tons of grain. The process continues in these and some 15–20 other crops (Tribe, 1994). The agricultural revolution has boosted other areas of economic growth, leading to the suggestion that "agriculture is the engine of economic growth" (Mellor, quoted by Tribe, 1994). Thus the Green Revolution contributed to two important public goods. The first is increased food security (at both the household and the national level) resulting from dramatic increases in output of the major staple cereals (rice, wheat and maize). The second is significantly reduced poverty in the developing countries where smallholders have been the main producers of these staples. This reduced poverty arose through the direct effects on farm income arising from the increased productivity of these staples, as well as indirectly through multiplier effects that boosted economic growth and employment (Tomich *et al.*, 1995).

Although many were public institutions, colonial research organizations tended to focus on domestication of a few crops, which were judged to have significant export potential and that were suited to large-scale production by colonial investors. This is an important reason why, among trees, domestication has been limited to a relatively few species with established world markets and that suited large-scale production (rubber, cocoa, coffee, tea, oilpalm, *etc.*). Smallholders' needs were largely ignored regarding identification, propagation and management of improved germplasm. Barlow and Jayasuriya (1994) have documented the bias against smallholders in rubber research in Malaysia. Likewise, until independence, Kenyan smallholders were prohibited from planting lucrative export crops like coffee and tea.

The dominance of the private sector in agricultural research has been accelerated by the advent of genetic engineering techniques, which have now been widely applied to food crops. While there are many positive aspects of this trend, it is important to be aware of two attendant shortcomings. First, since profit to the seed companies is the driving force, these private firms will underinvest in research on technology that produces significant public goods, such as food security, poverty alleviation and environmental conservation. Second it is much easier for private firms to reap profits from research in some types of germplasm than from others. Hybrid maize (at least in high-income countries like the USA) is one of the best examples where private incentives created appropriate outcomes regarding the development and distribution of improved germplasm (Griliches, 1957). The market-driven development and dissemination of hybrid maize rests, in no small part, on a fundamental feature of the technology: farmers cannot produce their own hybrid seed so they must buy seed each season from the company that holds the patent. This is the mechanism that enables companies to earn a return on their investments in research. It is different in the case of most tree species, in which domestication has often involved vegetative propagation (Table 3), because the long generation interval makes breeding a slow process and the outbreeding nature of these trees leads to heterogeneity in progenies. Vegetative propagation allows the capture and multiplication of any one genotype within its array of variation, and clonal propagules so formed are genetically uniform. By contrast to hybrid seed, producers using clonal planting stock can mass produce their own planting stock. (See Chapter 8.) In the absence of dependable mechanisms to enforce intellectual property rights (patents and plant breeders rights) and low-cost means of collecting fees for use of new tree germplasm, it will be difficult for private sector firms to earn a return from their

TABLE 3
Examples of the domestication history of some cash crops (Source: Simmonds, 1976)

Cash crop (Reference)	Area of Origin	Local use	Dissemination	Commercial interest	Mode of domestication
Rubber *Hevea brasiliensis* (Wycherley, 1976)	Amazon Basin	Edible seeds, Latex	1876: collected near Boim, Rio Tapajoz in Brazil. 2800 seedlings raised at Kew Gardens, England. 2397 seedlings dispatched to Sri Lanka, Malaysia, Singapore and Indonesia	1890s commercial plantings. 1937 Ford Motor Co. selected for resistance to South American Leaf Blight. Firestone Rubber Company breeding program in Guatemala.	1917 bud grafting started to form clones — 100% gain. 1930 Seed orchards planted for breeding, led to first generation clones with another 100% gain. Multiple clone — rootstock, stem and crown budding
Tea *Camellia sinensis* (Visser, 1976)	Highlands of SE Asia (Nepal to Japan)	Beverage	1878 — plantations started in Java and Malawi. Later also transferred to Kenya, Congo, Mozambique, Russia, Turkey and Argentina	1835 Plantations became possible with the end of monopoly of East India Company	1930 — Breeding program in Java. 1940 — Clonal selection in Sri Lanka and Java and started later elsewhere.
Cloves *Eugenia caryophyllus* (Wit, 1976)	Moluccas and New Guinea	Spice	Cultivated in China for over 2000 years French took cloves to Réunion, Mauritius in 1770–72. 1818–20 introduced to Madagascar and Zanzibar	Dutch East India Company took over from Portuguese in 17th Century — the spice trade was one of the driving forces of the exploration of the East in the 16th Century.	Breeding for disease resistance in Zanzibar and Indonesia between 1918 and 1935. Hybridization. Need for vegetative propagation techniques

investments. Prospects for a "Woody Plant Revolution" for smallholder farmers, heralded by Leakey and Newton (1994 c), thus rest on publicly-funded tree domestication and the ability of farmers to propagate their selected clones. This use of public funds is however fully justified as many of the Cinderella trees suitable for domestication have promise to provide significant public goods, including poverty alleviation and environmental benefits, as argued in the previous section. There are, however, some interesting developments in other directions, in which synthetic products, or those from genetic modifications to existing crops, are being produced that might be substituted for natural non-timber forest products. Trends in this direction are, however, also being matched by others in which industry (e.g., Daimler Benz) are promoting the small-scale production of natural products as alternatives for synthetic products, such as fiberglass (Panik, in press).

A possible reason why so few tropical trees have been domesticated as clonal varieties is the view of foresters that horticultural approaches were inappropriate for forest trees and that expensive, slow and long-term breeding programs were the way forward. Fortunately these concepts are now changing in favor of the clonal option (Ahuja and Libby, 1993).

Tracing the evolutionary history and geography of fruit trees is often complex: See *Ficus carica* (Storey, 1976); *Malus* sp. and *Pyrus* sp. (Watkins, 1976) and *Citrus* sp. (Cameron and Soost, 1976). What is clear, however, is that vegetative propagation of rare genotypes with superior characteristics arising either naturally by chance or from breeding programs, has often been the most effective domestication strategy for fruit trees, and is now one also applied to timber trees (Leakey, 1987). This ensures the uniformity required by the trade. Originally, at least in Europe, the domestication process of important fruits, and incidentally of garden flowers, was often in the hands of the elite and hobbyists whose activities enhanced their social standing at their own dinner tables, through country fairs and simply by word of mouth within the various strata of rural life. Now commercial interests, which have really expanded over the last 50 years, dominate and the horticultural trade is big business. It has flourished, and is diversifying, with the arrival of the Supermarket (e.g., the kiwi fruit). This diversification has triggered some interest in the semi-domesticated tropical fruits, starting with mangoes, avocados, macadamia nuts and progressing to litchi, rambutan, mangosteen, peach palm, *etc*. A recent review of the potential of agroforestry trees to provide novel food products (Leakey, In press a), has identified their diversity in chemical composition and the opportunity for their use as food ingredients.

Little is know about smallholders' contributions to the ongoing processes of domestication of other tropical perennials, who have sometimes taken advantage of "sports," the spontaneous, random changes that have occasionally led to significant improvements. Tomich *et al.* (in press) documented a case in the highlands of Sumatra, where a smallholder identified a high-yielding robusta coffee variety in his own field, which was subsequently adopted by other small-scale producers in surrounding districts. In addition, smallholders have also initiated domestication in response to depletion of wild trees. Perhaps the most noteworthy example also comes from Sumatra, where smallholders in Lampung province successfully domesticated damar (*Shorea javanica*) through cultivation (Michon and de Foresta, 1996), despite the substantial propagation difficulties of the Dipterocarps. While these examples demonstrate that there is 'indigenous technical knowledge' into which science can tap its domestication efforts, it is also clear that the pace of domestication of woody species will continue to be slow if smallholders are left to the task on their own.

6. RECENT INITIATIVES IN DOMESTICATION

In recent years, starting with the work of Okafor (1978, 1980), Okafor and Lamb, (1994), a new movement has arisen to domesticate some of the totally wild trees of tropical forests and woodlands for their fruits, timbers, and other products (Leakey *et al.*, 1982; Leakey and Newton, 1994 a,b; Leakey *et al.*, 1996). The trees, which have been overlooked by science have been labelled as "Cinderella" species (Leakey and Newton, 1994 c). They are species whose products have been traditionally collected, gathered and utilized by man, and are still of enormous importance to many people around the tropics for food and nutritional security, and so also for welfare. Although these species are mostly producers of non-timber forest products, some of the commercially-important tropical hardwoods like Mahogany, Meranti, Obeche, Iroko, Laurel and others, can also be seen as Cinderella species. They have been exploited from the wild, but in most cases not replanted. In the case of Mahogany, some planting has occurred since colonial times in areas outside its natural range and where the shoot borers *Hypsipyla grandella* and *H. robusta* were absent. Within their natural range, however, shoot borers have severely limited planting and science has tried and basically failed to overcome the problem of how to grow it commercially, although there is hope for the future (Newton *et al.*, 1994). Now, with time, the shoot borer also is a severe constraint to new plantings outside the natural range.

7. ICRAF'S TREE DOMESTICATION PROGRAM

Activities to domesticate some of the indigenous trees producing high-value products started in 1994 and now are in progress in the humid and semi-arid lowlands of West Africa, the bimodal highlands (Miombo ecozone) of southern Africa and western Amazonia (Leakey and Jaenicke, 1995; Ladipo *et al.*, 1996; Leakey and Simons, 1997).

TABLE 4
Priority tree species selected for domestication by implementation of farmer preference surveys and priority setting guidelines (Franzel *et al.*, 1996) by ICRAF and partners

	West Africa		Latin America
	Humid Lowlands	Semi-arid* Lowlands	Peruvian Amazon*
Priority order			
1	*Irvingia gabonenis/I.wombolu*	*Adansonia digitata*	*Bactris gasipaes*
2	*Dacryodes edulis/D. klaineana*	*Vitellaria paradoxa*	*Cedrelinga catenaeformis*
3	*Ricinodendron heudelottii*	*Parkia biglobosa*	*Inga edulis*
4	*Chrysophyllum albidum*	*Tamarindus indica*	*Calycophyllum spruceanum*
5	*Garcinia kola/G. afzelii*	*Zizyphus mauritiana*	*Guazuma crinita*

* Preliminary, awaiting full-economic evaluation

TABLE 5
Priority fruit tree species selected by ICRAF and partners for domestication by market and ethnobotanical surveys.

	Southern Africa Plateau
Priority order	
1	*Uapaca kirkiana*
2	*Sclerocarya birrea*
3	*Zizyphus mauritiana*
4	*Vangueria infausta*
5	*Azanza garckeana*

In each of these eco-regions the domestication activities started with a priority setting process that assessed the preferences of farmers, the markets, research in progress, researchability and future potential of identified species (Jaenicke *et al.,* 1995; Franzel *et al.,* 1996). In all regions this led to the identification of 5–10 species (Tables 4 and 5) for the initial focus of ICRAF's work, which relies heavily on collaboration with National Agricultural Research Systems (NARS). In addition to these farmer-identified and market-led priority species, a few species with industrial priority have also been identified for domestication. They are *Prunus africana, Pausinystalia johimbe,* and *Pterocarpus erinaceous.*

So far, in this new initiative, rangewide germplasm collections have been made for *Irvingia* sp., *Uapaca kirkiana, Sclerocarya birrea, Calycophyllum spruceanum* and *Guazuma crinata.* In addition, follow-up collections have been made to existing germplasm collections of *Inga edulis* and *Bactris gasipaes.* New collections have also started for *Prunus africana* and *Pterocarpus erinaceous.* Prior to this recent priority setting process, germplasm collections had also been made for: *Prosopis africana, Combretum aculeatum, Bauhinia rufescens, Faidherbia albida, Sesbania sesban, Markhamia lutea,* and *Grevillea robusta* landraces.

Strategies for domestication vary depending on the value, type of product, duration of productivity, duration of reproductive cycle (Leakey and Simons, in press: Simons, 1996 a). Strategies also have to include germplasm dissemination pathways (Simons 1996 b) and the need to minimize risks of narrowing the genetic base when developing clonal cultivars (Leakey, 1991).

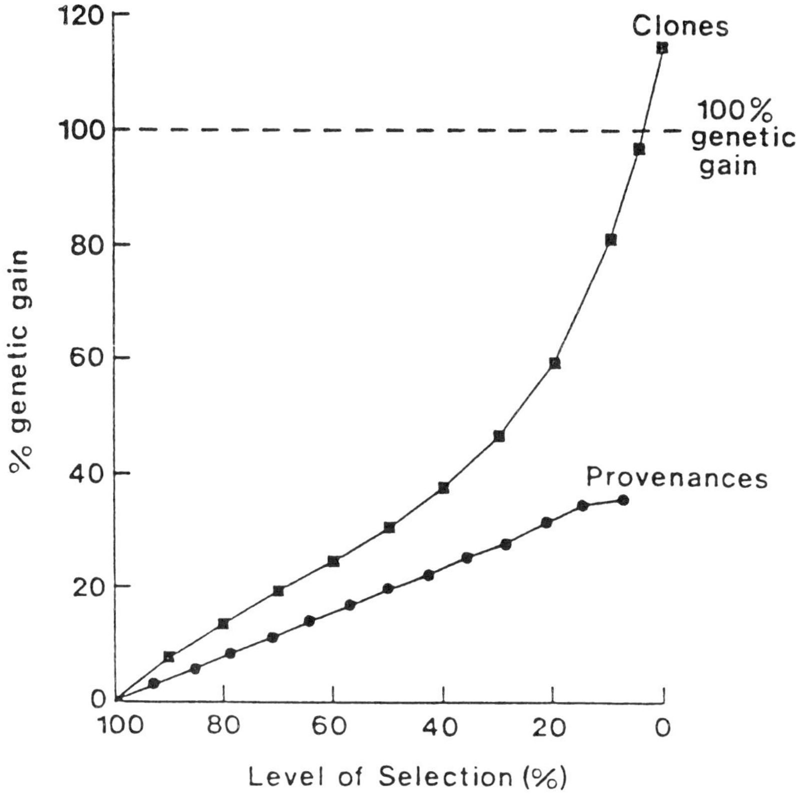

FIGURE 2. The potential genetic gains in volume production (year 5) by different levels of provenance and clonal selection, as determined by field trials with 98 clones of *T. scleroxylon* originating from 14 provenances (after Leakey, 1990).

Genetic characterization of the collected germplasm of *Irvingia gabonensis* and *I. wombolu; Prunus africana; Calyrophyllum spruceanum; Sesbania sesban; S. guetzei* and *S. bispinosa* is in progress, using molecular procedures based on the polymerase chain reaction and randomly amplified polymorphic DNA (RAPDs).

8. EXPECTED GAINS FROM DOMESTICATION

The use by smallholders of tree crops domesticated for plantation agriculture (coffee, rubber, *etc.*) gives confidence that the domestication of new tree crops for timber and non-timber forest products can substantially improve yields and also the quality of the products. In new initiatives it is important to realize two things: first, that the more intensive the level of genetic selection, the greater the expected returns (see Figure 2) and second, that domestication is not a one-shot event. It is, and has to be, a continuing process in which new germplasm is continuously being introduced into the domestication process, so that the genetic base of the selected cultivars is broadened and so that with each successive generation of selection more superior traits are being included into the cultivars. Despite the apparent contradiction between gains from selection and the need to maintain genetic diversity, these two concepts have to be part of a wise risk-averse strategy for domestication (Leakey, 1991).

A few examples will illustrate these principles. In the West African timber tree *Triplochiton scleroxylon* (Obeche), clones developed from a random selection of 100 seedlings originating from 15 different provenances varied eight–nine fold in the volume of their basal log after only four years growth (Ladipo *et al.,* 1991). Selection of the best 33% resulted in expected gains in volume of about 30%, selection of the best 10% would raise this to 80% gains, while the best 4% of the clones were more than 100% better than average (Figure 2). Theoretically therefore, to have a clonal population of 100 clones capable of yields in excess of 100% above average would require the testing of 2500 clones — a very major task. Hence, for this species a Predictive Test that can be applied in the nursery was developed (Ladipo et al., 1991), so that even larger numbers of seedlings could be screened on an annual basis.

Much larger clonal screening programs have been developed for *Eucalyptus* hybrids (Zobel, 1993). Genetic gains of 200% in yield have been achieved, and integrated with selection for 15 other valuable traits (see review by Leakey, 1987), so that quality and pest-resistant high yielding clones are already in commercial production.

In Rubber (*Hevea brasiliensis*), the first generation of clonal tests resulted in clones yielding twice the average quantity of latex. In the subsequent generation of clones developed from a breeding program between the first generation clones, a further doubling of yield was achieved (Wycherley, 1976). Now in the 2nd and 3rd bred generations high yields are combined with other traits such as pest resistance. Similar results were obtained in Avocado (*Persea americana*), and now breeding programs are seeking tolerance to a wider range of environments (Bergh, 1976).

Among the lesser known indigenous fruits of West Africa, a market survey has illustrated five-fold variation in fruit size, in pulp:seed ratio and in the price per kilogram of pulp (Leakey and Ladipo, 1996). This again illustrates the potential for cultivar development. Interestingly in this case, high prices were not only associated with large fruit size or high pulp:seed ratio, suggesting that even in wild, undomesticated fruits the market recognizes other traits, presumably flavor (Figure 3).

In India, the yields of fruit from *Zizyphus mauritiana* (Ber) have been increased five fold, while in Hawaii new clones of Macadamia nuts (*Macadamia integgrifolia* and *M. tetraphylla*) have four times the yield of the earlier clones, with 10% more first grade kernels (Cannell, 1989).

Examples of tree domestication for medicinal products are less common, but perhaps the best example are the hybrid clones of *Duboisa,* an Australian species, whose scopolamine content almost has been doubled. Scopolamine is a naturally occurring alkaloid that induces amnesia and is used as a prophylactic for motion sickness. The tree was converted to a crop with improved growth characteristics, predictable yield and quality, and lower susceptibility to nematodes and stem borers (Ohlendorf, 1996).

The achievement of the sorts of success quoted above is possible with almost any tree species; it just requires investment in a dedicated program of domestication. Rapid progress is possible with relatively small inputs, thanks to two common characteristics of tree species:

- they are outbreeding, and thus strongly heterogeneous. The resultant large tree-to-tree variation makes the selection of superior individuals easy
- most trees are easily propagated vegetatively, while in the juvenile state (Leakey *et al.,* 1982, 1990; 1994)

The latter is essential to take advantage of the former. Vegetative propagation can be used to mass produce genetically identical copies of the originally selected tree. This is the basis of horticulture worldwide, over many centuries (Hartmann *et al.,* 1990). (Also see Chapter 8.)

9. WHAT IS BIOLOGICALLY POSSIBLE?

To answer this question we only have to look at examples from our everyday dinner table. Most of the staple food crops, rice, maize, wheat, etc. have been domesticated through generations of

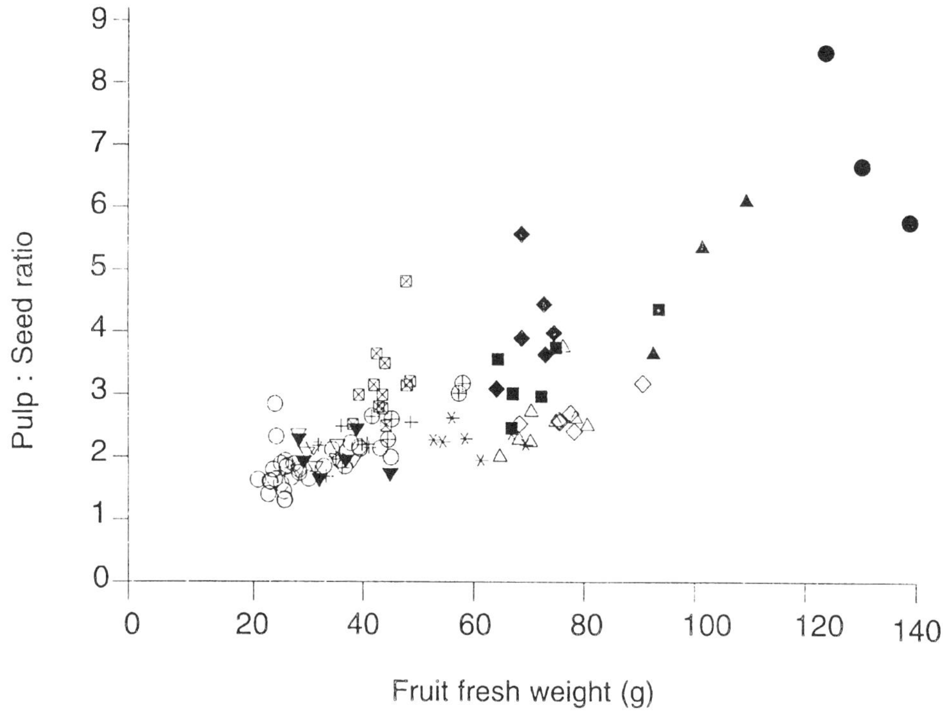

FIGURE 3. Variation in *Dacryodes edulis* fruit characteristics. Each symbol represents a different fruitlot bought in a market in Yaoundé, Cameroon (filled symbols are fruitlots that sold for more than 500 CFA per kg of pulp).

breeding. These are species most easily domesticated that way, as they are not out-breeding. Contrastingly, the clonal option has been pursued, in tandem with breeding, for apples, pears, figs, oranges, *etc*. But the first great leap into domestication was by the route described earlier:

- identify a superior tree
- propagate it vegetatively

This was the origin of the Granny Smith apple, found in the back-garden of Mrs. Smith's house in Bristol in England. Cox's Orange Pippin had a similar origin.

Nowadays, apple cultivars are produced by complex breeding programs between cultivars and the progeny are grafted onto dwarfing rootstocks, so that what was once a large tree that didn't fruit for many years has now become little more than a stick one meter tall, covered in fruits within two–three years. In this way 70% of the dry matter production of four–five-year-old Golden Delicious apples is in the fruits. Similar improvements in the Harvest Index have been made in other tree crops — e.g., Peach from 0.35 to 0.7; and coffee from 0.44 to 0.75 (Cannell, 1989). However, again this level of sophistication is not needed from the start of the domestication process since horticulture has another trick up its sleeve to take advantage of the biology of trees. The maturation process in trees occurs progressively as trees grow bigger so that after a number of years the crown is mature and capable of flowering and fruiting, while the base of the trunk is juvenile. Horticulturists graft mature crown shoots or buds onto juvenile seedling rootstocks and the resulting tree is capable of fruiting early while still small, and hence has an increased Harvest Index. Once again, therefore, rapid progress can be made in reducing the size of fruiting trees and in obtaining a crop after a shorter period of time — both important aspects of domesticating fruit trees.

10. DOMESTICATION PRIORITIES AND PITFALLS

Priorities for agricultural research investments should go to those commodities and land use systems with the highest expected returns. For the CGIAR and other public research institutions, the analysis of priorities must include the provision of public goods — including food security, poverty alleviation and environmental benefits — and should not be restricted to pecuniary returns. In this respect, it is useful to compare and contrast the types and likely scale of impact of additional investments in research on staple cereals with alternative investments in agroforestry systems.

The absolute number of poor, hungry people continues to increase, especially in Africa and some parts of Asia (Tomich *et al.*, 1995). The major cereals (rice, maize and wheat) and certain other starchy staples have a very large share of production in these rural economies and of consumption expenditures of poor households. For 'Cinderella trees' and other agroforestry species, the number of species concerned is much greater and the extent of utilization of each species is much narrower.

From this global perspective, the staple cereals often look like a better research investment than agroforestry because returns to investment in developing new technology are proportional to the extent of its extrapolation domain, both in terms of area and of the target population. In assessing alternatives for a \$10 million grant, for example, work likely to lead to a net gain worth \$10/yr on the cereal plots of each of 100 million poor people is clearly a better investment than a program to domesticate an agroforestry tree that only affects 1 million poor people, even if the increase is worth \$100/yr to each of them. But is this a fair comparison, as the impact of agroforestry research need not be restricted to a single species. Strategic research on methods that are applicable to domestication of many 'Cinderella tree' species has potential for impact across the tropics which may have indirect benefit for cereal production. Below we take a comparative perspective on opportunities and pitfalls related to marketing high-value products *vis à vis*, poverty alleviation, food security, and environmental benefits from research investments.

10.1 Marketing High-Value Tree Products

Some tree species have the potential to become high-value cash crops for large numbers of farmers and are capable of producing significant export earnings as inputs for international pharmaceutical or food processing industries. The bark of *Prunus africana*, hearts of peach palm (*Bactris gasipaes*) and gum arabic (*Acacia senegal*) each account for hundreds of millions of dollars *per annum* in international trade, each of which could be enhanced by domestication, especially if dialogue with potential commercial interests, could ensure that the product characteristics of importance to the industry are included in the genetic selection program. (Leakey, In press a)

The same is true for countless other tree species that face a narrower market with uncertain prospects for growth in demand. Because of the impact of increases in supply on grain prices, a large share of the benefits of technologies accrued to consumers (Scobie and Posada, 1984). The situation often is different for cash crop exports, since the bulk of consumers are in high-income countries. If a burst in productivity causes a collapse in price, it is hard to argue that the benefits accruing to well-off consumers justify public research investments. Nevertheless, whether or not prices collapse depends on demand prospects and features of the markets for the commodity in question. This is not the place for a comprehensive discussion of marketing, but two major issues deserve mention:

First, although increases in quantity supplied usually cause a more than proportionate fall in prices for many primary products, this static relationship can be either offset or enhanced by changes in other economic and demographic determinants of demand. Key questions are: whether demand for the product (at the local, regional or international level) is growing or declining. Demand in national markets for product can be boosted by rising prices for substitutes and rising incomes and urbanization can increase demand for some products. If a product is exportable, the same economic and demographic forces affect market prospects at the international level. Indeed, rising incomes together with population growth have helped to offset downward price pressure from productivity gains in many primary commodities (Tomich *et al.*, 1994). These forces supported cereal prices despite the output increases arising from the Green Revolution, and also may benefit some tree products, such as those that are marketed as "organic" products or natural alternatives to synthetic food additives.

The second issue concerns market share: will productivity increases have a significant effect on overall supply? The relevant market includes substitutes; as a result many markets are wider than is commonly believed. For instance, natural rubber is less than half the world's elastomer market, where it must compete with synthetics; palm oil has an even smaller share of the world's complex edible oils market. Paradoxically, this competition from other commodities is a blessing in this context because the smaller the market share of a particular commodity, the smaller the fall in price if its supply increases. Significant productivity increases carry little risk of price collapse for internationally traded commodities that have a small share of a large, established (and expanding) market.

Because of the above factors, market prospects must be analyzed at the earliest stage of identification of candidates for domestication, well before significant scientific resources have been committed. This must be done on a commodity-by-commodity basis. It should be clear, however, that when urbanization and income growth are rapid, domestic market prospects are not good for products that are considered 'inferior' to available substitutes by higher-income, urban consumers. There is also significant risk of a price collapse if domestic markets are limited and the commodity is not exportable because of technical, economic, or policy constraints. These marketing pitfalls can, and must, be anticipated through careful assessment of demand and market prospects. The question is more complex if the commodity faces price competition from a substitute. Although this might seem to be a sufficient reason not to invest in research on a commodity, Malaysia decided to invest heavily in natural rubber research in the 1950s despite competition from synthetics in order to reduce production costs. This preserved the competitiveness of rubber producers for more than two decades (Barlow, 1978).

Problematically, the characteristics of non-timber forest products and their markets may make them particularly fraught with market failures and susceptible to rent seeking that leads to distortions in policies (Tomich, 1996) that are biased against smallholder producers. On the other hand, selective interventions to improve the markets for agroforestry tree products with benefits for all producers, large and small, could include programs to disseminate information on prices and improve grading standards (without undue bureaucracy); establish named cultivars; and invest in transportation, communications and other market infrastructure.

10.2 Poverty Reduction and Food Security

Products from "Cinderella" trees and other agroforestry species are not in competition with cereal grains or starch from tubers. Instead a number of tree products (especially fruits) provide important vitamins and minerals. Malnutrition, particularly in the most vulnerable groups, women and children, is a major concern in the poor households of the tropics, especially in periods of drought and other disasters. Increasingly, this is a problem in urban populations.

Despite overall success, the Green Revolution also has had some shortcomings, primarily regarding limitations in its impact on marginal lands and under harsh environment conditions. The Green Revolution 'package' of genetic improvement through breeding, fertilizer responsive seeds and water controlled through irrigation simply may not be a feasible means of fully overcoming poverty and household food insecurity in marginal and harsh areas, especially in sub-Saharan Africa where only four percent of the cropland was irrigated in 1990 and economic scope for expansion or irrigation is severely limited (Tomich *et al.*, 1995). On the other hand, agroforestry systems may yield higher payoffs than continuous annual cropping in the rainfed uplands of Asia and Latin America, which also are home to some of the poorer sectors of rural society.

One feature of farming in much of sub-Saharan Africa is that soil fertility is seriously depleted and farmers lack cash to buy fertilizers. What are the solutions to this problem? One would be to subsidize fertilizer; another might be credit schemes, but these solutions have been tried in the past without much success. A more promising solution, increasingly coming to the fore, is agroforestry.

The service role of nitrogen-fixing trees as soil improvers has received considerable attention over the last 20–30 years. It has not always been a great success, although recent economic analysis of agroforestry projects in Central America and the Caribbean was generally positive (Current *et al.*, 1995). Similarly, in Africa, some success stories are also emerging (see review by Cooper *et al.*, 1996). For example, in Eastern Zambia short-rotation fallows with a leguminous tree *Sesbania sesban* are improving soil fertility and increasing maize yields three–four fold from 1-2 t ha^{-1} to 5-6 t ha^{-1}. Standard recommended fertilizer applications improve yields by the same magnitude, although the maize has the genetic capacity to produce yields of over 10t ha^{-1}. These responses to *Sesbania* fallows are currently being achieved on many farms and more and more farmers in Zambia are testing these fallows on their own farms. What these fallows demonstrate is that farmers do have a choice if they want to improve their maize yields, they can use *Sesbania* fallows, or buy fertilizers. A combination of fallows and fertilizer may raise the yields still further. However, to minimize the risks of pest and disease problems, these fallows will ultimately have to become part of a diverse and functioning agroecosystem (Leakey, in press).

But how does this relate to domestication of trees for high-value crops? These farmers in Zambia have *per capita* incomes of only $300–400 per year, much of this in kind rather than cash. If, for example, through the domestication of the indigenous fruit trees for which favorable markets exist, sales of fruit could significantly increase cash income for poor households. When people live on less that $1 per day, even small quantities of cash can have very significant effects on household welfare. In particular, alleviating their cash income constraint may make it possible for them to purchase fertilizer and other inputs and thus be in a position to take better advantage of the maize breeding that has been done in the past. Research on these fruit trees would therefore enhance returns from the research during the Green Revolution.

One potential pitfall of domesticating commercially important fruit trees is that it may induce shifts in benefits from poorer groups of farmers to richer ones, or even multi-national companies. For example, the people who gather the wild product may not be in the best position to undertake production of the domesticated version. Concerns have been expressed, for example, that large-scale monoculture plantations may take over production of domesticated tree products, displacing smallholders although evidence from rubber production in Indonesia is the converse (Barlow *et al.*, 1994). Leakey and Izac (1996) posed the following questions in this regard:

"What are the new conceptual frameworks for policy development to ensure the economic viability of agroforestry for small-scale farmers? Alternatively, under what conditions is small-scale production competitive with large-scale production?"

In the absence of significant economies of scale in production, large-scale production may not be a serious threat unless policy distortions and biased development programs create artificial advantages for large-scale operations (Tomich *et al.* this volume). Probably the greater area of concern is that the households that formerly were small-scale gatherers will not be the same as the households that become small-scale producers and gathering in the wild may cease altogether. This may be associated with a shift from common property sources to private property, and raises complex questions about the impact of domestication on incentives for community-based management and conservation of natural forests. We know of no studies of this potential problem, which may be an important topic for further social science research. The desired impact of domestication on poverty alleviation may therefore benefit farming communities at the expense of forest dwellers, unless ways can be found to address this problem.

Dewees and Scherr (1996) have called for higher priority to be given to non-timber forest products marketing and policy specifically relating to market development. They argue that rather than seeing these issues as a threat to the conservation and management of natural forests and woodlands, this need should be seen as a positive move towards improving human welfare and reducing environmental degradation on already cleared land. This could help to take the pressures off natural forests. Interestingly, the World Food Summit in November 1996 called for governments and forestry institutions to improve the food security of small farmers and the rural poor through the creation of forest policies and forest institutions that will support the needs of households who depend on trees for a significant part of their nutrition. This initiative should be based on the potential of the market for forest products to trigger substantial environmental benefits through the growth of trees on farms. Part of such an initiative would be to increase the funding opportunities for research to domesticate the "Cinderella" trees.

10.3 Environmental Opportunities and Pitfalls

The foregoing discussion of tree domestication for improved products raises several important issues regarding the environmental services from trees. Views of the benefits of research on tree domestication might change dramatically if empirical assessment of economic/social benefits were broadened from its current narrow focus on marketed products; especially if the externalities arising from the many environmental services of trees (Ingram, 1990) could be taken into account. This can be additional to the soil fertility restoring attributes of improved fallows, described above. Difficulty in measuring these environmental externalities and public goods is the main barrier to extending the analysis in this direction. The potential of agroforestry to supply these environmental services is taken up by Tomich *et al.* (this volume). One aspect of this topic fits best in the present chapter, however. The domestication process may alter the stature of trees and affect their suitability for their natural niches. For example, hybrid varieties of durian (*Durio zibenthinus*) flower and fruit much earlier and are shorter than wild trees. Although these characteristics are highly desired by farmers, Michon and de Foresta (1996) have pointed out that these changes also affect the environmental role of the trees. Whereas the wild varieties grow well in complex, multistrata systems (agroforests), which approximate some of the ecological functions of forests, the hybrids will not grow well in the shade of the lower storey. Options could therefore be to grow them in monocultures, in simple agroforestry systems (e.g., orchards, contour hedges, field boundaries, *etc.*), or to similarly modify the other trees forming agroforests. More likely however different cultivars will be produced for different systems. Such options in the domestication process for high value species should be considered among the multiple objectives in public sector research agendas, to ensure that domesticates also provide the important environmental services of trees in farmland. This raises complex questions about balancing farmers' priorities and interests, with broader environmental benefits.

11. CONCLUSIONS

It is clear that there are many important environmental, socio-economic and policy questions associated with the domestication of tropical trees which need to be further considered. The following list is not intended to be comprehensive. Instead, we hope it will provide a useful starting point for expanding the efforts of public research institutions to domesticate tree species in order to promote economic growth with enhanced food security and reduced poverty, while also preserving to the greatest extent possible the environmental services of wild species:

1. What is the scope for expanding the productivity and quality of this tree species to have a positive impact on economic growth, poverty alleviation and enhanced food security?
2. What share of the existing market is supplied by this species? Are there prospects that demand for its products will increase (or decline) in local, regional or international markets? Will expansion of output flood the market and cause prices to collapse? What traits of the species need to be improved to provide incentives to farmers to grow it and the trade to buy it?
3. Is this species suited to production in marginal or severely degraded lands or under harsh environmental conditions, thereby creating profitable options for small-scale farmers in regions where scope for intensification of annual crop production may be severely limited? What agroforestry practices would be most suited to the growth of this species and are there traits for selection that will enhance the performance under the targeted environment?

4. What is the scope for expansion of planting of this species to have an indirect impact on economic growth, poverty alleviation and enhanced food security through environmental services, such as soil amelioration, carbon sequestration, ecological equilibrium through food web expansion, reduced erosion, etc. How can the quantification of these benefits affect decisions about land-use policy, allocation of research funding, *etc*.?
5. Will domestication of this species cause adverse distributional consequences by shifting income away from small-scale gatherers to other people who are in a better position to adopt the new germplasm? Do technical factors, institutions or policies make it likely that production will be dominated by large-scale plantations to the detriment of smallholders? What policy measures should be taken to ensure a level playing field for all producers or to ensure opportunities for small-scale production?
6. Can domestication choices and practices maintain or improve the ecological and environmental services of the species? Do domestication practices lead to an unnecessary sacrifice of genetic diversity? Are there apparent trade-offs between farmers' priorities and conservation of the ecological functions and other environmental services of the species? How can these be reconciled?

REFERENCES

Abbiw, D. K., 1990, *Useful Plants of Ghana: West African Uses of Wild and Cultivated Plants,* Intermediate Technology Publications, London, U.K.

Ahuja, M. R. and Libby, W. J., Eds., 1993, *Clonal Forestry II: Conservation and Application,* Springer Verlag, Berlin, Germany.

Aumeeruddy, Y., 1994, "Local representations and management of agroforests on the periphery of Kerinci Seblat National Park, Sumatra, Indonesia," *People and Plants Working Paper 3,* UNESCO, Paris, 46.

Barlow, C., 1978, *The Natural Rubber Industry: its Development, Technology and Economy in Malaysia,* Oxford University Press, New York.

Barlow, C. and Jayasuriya, S. K. J., 1984, "Bias toward the large farm subsector in agricultural research: The case of Malaysian rubber," *Research and Development in Agriculture,* 1(3), 153–164.

Barlow, C., Jayasuriya, S. K. and Tan, C. K., 1994, *The World Rubber Industry,* Routledge, London, U.K.

Bergh, B. O., 1976, "Avocado: *Persea americana* (Lauraceae)," in *Evolution of Crop Plants,* Simmonds, N. W., Ed., 148–151, Longmans, London and New York.

Bonkoungou, E. G., 1995, "Practiques agroforestieres traditionnelles et gestion des ressources naturelles dans les zones semi-arides de l'Afrique de l'Ouest," in *Proceedings of the International Workshop for a Desert Margins Initiative,* ICRISAT, Hyderabad, India.

Brickell, C. Ed., 1996, *A–Z Encyclopedia of Garden Plants*, Royal Horticultural Society, Dorling Kindersley, London.

Cameron, J. W. and Soost, R. K., 1976, "Citrus *Citrus* (Rutaceae)," in *Evolution of Crop Plants,* N. W. Simmonds, Ed., 261–265, Longmans, London and New York.

Cannell, M. G. R., 1989, "Food crop potential of tropical trees," *Exp. Agri.,* 25, 313–326.

Clement, C. R. and Villachica, H., 1994, "Amazonian fruits and nuts: potential for domestication in various agroecosystems," in *Tropical Trees: The Potential for Domestication and the Rebuilding of Forest Resources,* Leakey, R. R. B. and Newton, A.C., Eds., 230–238, HMSO, London.

Cooper, P. J. M., Leakey, R. R. B., Rao, M. R., and Reynolds, L., 1996, "Agroforestry and the mitigation of land degradation in the humid and sub-humid tropics of Africa," *Exp. Agri.,* 32, 235–290.

Cunningham, A. B., Ayuk, E. K., Franzel, S., Duguma, B., and Asanga, C., in press, "An economic evaluation of medicinal tree cultivation: *Prunus africana* in Cameroon," *Report of WWF/UNESCO/Kew "People and Plants Program,"* No. 6, WWF, Godalming, Surrey, U.K.

Cunningham, A. B. and Davies, G., in press. "Human use of plants," in *Vegetation of Southern Africa*Cowling, R. and Richardson, D., Eds., Oxford University Press, Oxford, U.K.

Cunningham, A. B. and Mbenkum, F. T., 1993. "Sustainability of harvesting *Prunus africana* bark in Cameroon: A medicinal plant in international trade," *Report of WWF/UNESCO/Kew "People and Plants Program,"* No 2, WWF, Godalming, Surrey, U.K.

Current, D., Lutz, E., and Scherr, S., Eds., 1995, *Costs, Benefits and Farmer Adoption of Agroforestry,* Project Experience in Central America and the Caribbean, World Bank Environment Paper No. 14, Washington, DC.

Dewees, P. A. and Scherr, S. J., 1996, "Policies and Markets for Non-Timber Tree Products," *EPTD Discussion Paper No. 16,* IFPRI, Washington, DC.

Evans, J., 1992, *Plantation Forestry in the Tropics: Tree Planting for Industrial, Social, Environmental and Agroforestry Purposes,* Second edition, Clarendon Press, Oxford, U.K.

Falconer, J., 1990, "The major significance of 'minor' forest products. The local use and value of forests in the West African humid forest zone," *Forests, Trees and People, Community Forestry Note No. 6,* FAO, Rome, 232.

Faulkner, R., 1976, "Timber Trees," in *Evolution of Crop Plants,* Simmonds, N.W., Ed., 298–301, Longmans, London and New York.

Franzel, S., Jaenicke, H., and Janssen, W., 1996, "Choosing the right trees: Setting priorities for multipurpose tree improvement," *ISNAR Research Report 8,* 87

Griliches, Z., 1957. "Hybrid corn: An exploration in the economics of technical change," *Econometrica,* 25, 501–522.

Harlan, J. R., 1976, Barley *Hordeum vulgare* (Graminae — Triticinae)," in *Evolution of Crop Plants,* Simmonds, N. W., Ed., 93–98, Longmans, London and New York.

Hartmann, H. T., Kester, D. E., and Davies, F. T., 1990, *Plant Propagation — Principles and Practices.* 5th edition, Prentice Hall, Englewood Cliffs, N.J., 647.

Homma, A. K. O., 1994, "Plant extractivism in the Amazon: Limitations and possibilities," in *Extractivism in the Brazilian Amazon: Perspectives on Regional Development,* Clüsen-Godt, M. and Sachs, I., Eds., *MAB Digest, 18,* UNESCO, 34–57, Paris.

Homma, A. K. O., 1996, "Modernisation and technological dualism in the extractive economy in Amazonia," in *Current Issues in Non-Timber Forest Products Research,* Ruiz Perez, M. and Arnold, J.E.M., Eds., 59–81, CIFOR, Bogor, Indonesia.

ICRAF, 1995, *Annual report 1994,* ICRAF, Nairobi, Kenya, 152–156.

ICRAF, 1997a, *ICRAF Medium Term Plan: 1998–2000,* ICRAF, Nairobi, Kenya, 73

ICRAF, 1997b, *Annual Report 1996,* ICRAF, Nairobi, Kenya.

Ingram, J., 1990, "The role of trees in maintaining and improving soil productivity — a review of the literature," in *Agroforestry for Sustainable Production: Economic Implications,* Prinsley, R.T., Ed., 243–303, Commonwealth Science Council, London.

International Rubber Study Group, 1997, *Rubber Statistical Bulletin,* 51, 6.

Irvine, F. R., 1961, *Woody Plants of Ghana,* Oxford University Press, Oxford, U.K.

Jaenicke, H., Franzel, S., and Boland, D. J., 1995, "Towards a method to set priorities among species for tree improvement research: a case study from West Africa," *J. Trop. For. Sci.,* 7, 490–506.

Kunin, W. E. and Lawton, J. H., 1996, "Does biodiversity matter?" in Gaston, K. G., Ed., 283–308, *Biodiversity: a Biology of Numbers and Difference.*

Ladipo, D. O., Foundoun, J. M., and Ganga, N., 1996, "Domestication of the Bush Mango (*Irvingia* sp.): some exploitable intraspecific variations in West and Central Africa," in Leakey, R. R. B., Temu, A.B., Melnyk, M., and Vantomme P., Eds., 193–205, *Domestication and Commercialization of Non-timber Forest Products in Agroforstry Systems, Non-Wood Forest Products No. 9,* FAO, Rome.

Ladipo, D. O., Leakey, R. R. B., and Grace, J., 1991, "Clonal variation in a four-year-old plantation of *Triplochiton scleroxylon* K. Schum. and its relation to the Predictive Test for Branching Habit," *Silvae Genetica,* 40, 130–135.

Lamien, N., Sidibe, A., and Bayala, J., 1996, "Use and commercialization of non-timber forest products in western Burkina Faso," in Leakey, R. R. B., Temu, A.B., Melnyk, M., and Vantomme P., Eds., *Domestication and Commercialization of Non-timber Forest Products in Agroforstry Systems, Non-Wood Forest Products No. 9,* 51–64, FAO, Rome.

Leakey, R. R. B., 1987, "Clonal forestry in the tropics — A review of developments, strategies and opportunities," *Comm. For. Rev.,* 66, 61–75.

Leakey, R. R. B., 1991, "Towards a strategy for clonal forestry: some guidelines based on experience with tropical trees," in *Tree Breeding and Improvement,* Jackson, J. E., Ed., 27–42, Royal Forestry Society of England, Wales and Northern Ireland, Tring, U.K.

Leakey, R. R. B., 1996, "Definition of agroforestry revisited," *Agrofor. Today,* 8(1), 5–6.

Leakey, R. R. B., in press a, "Potential for novel food products from agroforestry trees," submitted to *Food Chemistry.*

Leakey, R. R. B., in press b, "Agroforestry for biodiversity in farming systems," in *The Importance of Biodiversity in Agroecosystems,* Collins, W. and Qualset, C., Eds., Lewis Publishers, Boca Raton, FL.

Leakey, R. R. B. and Izac, A-M. N., 1996, "Linkages between domestication and commercialization of non-timber forest products: implications for agroforestry," in *Domestication and Commercialization of Non-Timber Forest Products in Agroforesty Systems,* Leakey, R. R. B., Temu, A. B., Melnyk, M., and Vantomme, P., Eds., 1–7, *FAO Non-Wood Forest Products No. 9,* FAO, Rome.

Leakey, R. R. B. and Jaenicke, H., 1995, "The domestication of indigenous fruit trees: opportunities and challenges for agroforestry," in *Proceedings 4th International BIO-REFOR Workshop,* Suzuki, K., Sakurai, S., Ishii, K., and Norisada, M., Eds., 15–26, BIO-REFOR, Tokyo, Japan.

Leakey, R. R. B. and Ladipo, D. O., 1996, "Trading on genetic variation — fruits of *Dacryodes edulis,*" *Agrofor. Today,* 8(2), 16–17.

Leakey, R. R. B., Last, F. T., and Longman, K. A., 1982, "Domestication of forest trees: A process to secure the productivity and future diversity of tropical ecosystems," *Comm. For. Rev.,* 61, 33–42.

Leakey, R. R. B., Mesén, J. R., Tchoundjeu, Z., Longman, K. A., Dick, J. McP., Newton, A. C., Matin, A., Grace, J., Munro, R. C., and Muthoka, P. N., 1990, "Low-technology techniques for the vegetative propagation of tropical trees," *Comm. For. Rev.,* 69, 247–257.

Leakey, R. R. B. and Newton, A. C., Eds., 1994a, *Tropical Trees: the Potential for Domestication and the Rebuilding of Forest Resources,* HMSO, London.

Leakey, R. R. B. and Newton, A. C., 1994b, *Domestication of Timber and Non-timber Forest Products, MAB Digest 17,* UNESCO, Paris.

Leakey, R. R. B. and Newton, A. C., 1994c, "Domestication of 'Cinderella' species as the start of a woody-plant revolution," in *Tropical Trees: the Potential for Domestication and the Rebuilding of Forest Resources,* Leakey, R. R. B. and Newton, A.C., Eds., 3–4, HMSO, London, U.K.

Leakey, R. R. B., Newton, A. C., and Dick, J. McP., 1994, "Capture of genetic variation by vegetative propagation: processes determining success," in *Tropical Trees: The Potential for Domestication and the Rebuilding of Forest Resources.* Leakey, R. R. B. and Newton, A.C., Eds., 72–83, HMSO, London.

Leakey, R. R. B. and Simons, A. J., 1997, "The domestication and commercialization of indigenous trees in agroforestry for the alleviation of poverty," *Agro. Syst.*

Leakey, R. R. B. and Simons, A. J., in press. "When does vegetative propagation provide a viable alternative to propagation by seed in forestry and agroforestry in the tropics and sub-tropics?" In *Problem of Forestry in Tropical and Sub-tropical Countries — The Procurement of Forestry Seed — The Example of Kenya,* 21

Leakey, R. R. B., Temu, A., Melnyk, M., and Vantomme, P., Eds., 1996, *Domestication and Commercialization of Non-Timber Forest Products for Agro. Syst., Non-Wood Products No. 9,* FAO, Rome.

Michon, G. and de Foresta, H., 1996, "Agroforests as an alternative to pure plantations for the domestication and commercialization of NTFPs," in *Domestication and Commercialization of Non-Timber Forest Products for Agro. Syst., Non-Wood Products No. 9,* Leakey, R. R. B., Temu, A.B., Melnyk, M., and Vantomme, P., Eds., 160–175, FAO, Rome,

Michon, G., de Foresta, H., and Aliadi, H., in press. "Reinventing the forest: Damar extraction and cultivation in Sumatra, Indonesia," in *People, Plants and Justice,* Zerner, C., Ed., Rainforest Alliance.

Ndoye, O., 1995, "The markets for non-timber forest products in the humid forest zone of Cameroon and its borders: Structure, conduct, performance and policy implications," *Report to CIFOR,* Bogor, Indonesia, 86

Newton, A. C., Leakey, R. R. B., Powell, W., Chalmers, K., Waugh, R., Tchoundjeu, Z., Mathias, P. J., Alderson, P. G., Mesén, J. F., Baker, P., and Ramnarine, S., 1994, in *Tropical Trees: the Potential for Domestication and the Rebuilding of Forest Resources,* Leakey, R. R. B. and Newton, A.C., Eds., 256–266, HMSO, London.

Ohlendorf, W., 1996, "Domestication and crop development of *Duboisia* sp. (Solanaceae)," in *Domestication and Commercialization of Non-timber Forest Products in Agro. Syst., Non-Wood Forest Products No. 9,* Leakey, R. R. B., Temu, A.B., Melnyk M., and Vantomme, P., 183–187, FAO, Rome.

Okafor, J. C., 1978, "Development of forest tree crops for food supplies in Nigeria," *For. Ecol. Mgt.,* 1, 235–247.

Okafor, J. C., 1980, Edible indigenous woody plants in the rural economy of the Nigerian forest zone," *For. Ecol. Mgt.,* 3, 45–55.

Okafor, J. C. and Lamb, A., 1994, "Fruit trees: diversity and conservation strategies," in *Tropical Trees: the Potential for Domestication and the Rebuilding of Forest Resources,* Leakey, R. R. B. and Newton, A.C., Eds., 34–41, HMSO, London.

Panik, F., in press, "Use of biodiversity and implications for industrial production," in *Proceedings of FORUM BELEM,* November 24–30, 1996, POEMA, Belem, Brazil.

Sanchez, P. A. and Leakey, R. R. B., 1997, "Land use transformation in Africa: Three determinants for balancing food security with natural resource utilization," *Eur. J. Agron.*

Scobie, G. M. and Posada, R., 1984, "The impact of technical change on income distribution: the case of rice in Columbia," in *Agricultural Development in the Third World,* Eicher, C.K. and Staatz, J.M., Eds., John Hopkins University Press, Baltimore, MD.

Shackleton, C. M., 1996, "Potential stimulation of local rural economics by harvesting secondary products: a case study of the central Transvaal lowveld, South Africa," *Ambio,* 25, 33–38.

Shackleton, C. M., Dzerefos, C. M., Shackleton, S. E., and Mathabela, F.R., 1997, "The use and trade in indigenous edible fruits in the Bushbuckridge savanna region, South Africa," *Ecology of Food and Nutrition.*

Simmonds, N. W., Ed., 1976, *Evolution of Crop Plants,* Longmans, London and New York.

Simons, A. J., 1996a, "ICRAF's strategy for domestication of non-wood tree products," in *Domestication and Commercialization of Non-Timber Forest Products in Agro. Syst., Non-Wood Forest Products No. 9,* Leakey, R. R. B., Temu, A.B., Melnyk M., and Vantomme, P., Eds., 8–22, FAO, Rome.

Simons, A.J., 1996b, "Delivery of improvement for agroforestry trees," in *Tree Improvement for Sustainable Tropical Forestry,* Vol. 2, Dieters, M. J., Matheson, A. C., Nikles, D. G., Harwood, C. E., and Walker, S. M. Eds., 391–400, Queensland Forestry Research Institute, Gympie, Australia.

Storey, W. B., 1976, "Fig *Ficus carica* (Moraceae)," in *Evolution of Crop Plants,* Simmonds, N. W., Ed., 205–208, Longmans, London and New York.

Tomich, T. P., 1996, "Markets, policies and institutions in NTTP trade: nothing is perfect," in *Domestication and Commercialization of Non-timber Forest Products in Agroforesty Systems, Non-Wood Forest Products No. 9,* Leakey, R. R. B., Temu, A.B., Melnyk, M., and Vantomme, P., Eds., 235–255, FAO, Rome.

Tomich, T.P., Hastuti, and Bennett, C. P. A., in press. "Agricultural intensification and indigenous technological change: policy lessons from smallholder coffee in highland Sumatra," *Bull. Indonesian Eco. Stud.*

Tomich, T. P., Kilby, P., and Johnson, B. F., 1995, *Transforming Agrarian Economies: Opportunities Seized, Opportunities Missed,* Cornell University Press, Ithaca, NY.

Tomich, T. P., Roemer, M., and Vincent, J., 1994, "Development from a primary export base," in *Asia and Africa: Legacies and opportunities in development,* Lindauer, D. L. and Roemer, M., Eds., 151–193, ICS Press, San Francisco.

Tribe, D. E. 1994, *Feeding and Greening the World: The Role of International Agricultural Research,* CAB International, Wallingford, U.K.

Visser, T., 1976, "Tea *Camellia sinensis* (Camelliaceae)," in *Evolution of Crop Plants,* Simmonds, N.W., Ed., 18–20, Longmans, London and New York.

Watkins, R., 1976, "Apple and pear *Malus* and *Pyrus* spp. (Rosaceae)," in *Evolution of Crop Plants,* Simmonds, N.W., Ed., 247–250, Longmans, London and New York.

Watson, G. A., 1990, "Tree crops and farming systems development in the humid tropics," *Exp. Agri.,* 26, 143–159.

Wilson, E. O., 1994, *The Diversity of Life,* Penguin, London.

Wit, F., 1976, "Clove *Eugenia caryophyllus* (Myrtaceae)," in *Evolution of Crop Plants,* Ed. N.W. Simmonds, 216-218, Longmans, London and New York.

Wycherley, P. R., 1976, Rubber *Hevea brasiliensis* (Euphorbiaceae)," in *Evolution of Crop Plants,* Simmonds, N.W., Ed., 77–81, Longmans, London and New York.

Zobel, B. J., 1993, "Clonal forestry in the Eucalypts," in *Clonal Forestry II: Conservation and Application,* Ahuja, M. R. and Libby, W.J., Eds., 139–148, Springer Verlag, Berlin.

15 Ethnobotanical Perspectives of Agroforestry*

David M. Bates

1. INTRODUCTION

As a discipline, ethnobotany seeks to understand the place of plants in the humanized world by proposing and testing hypotheses concerning the relationships between humans and the plant kingdom. Agroforestry, as one expression of human/plant interrelationships, fits comfortably within the expansive sweep of ethnobotany. The nature and application of ethnobotanical thought and practice are explored here, using this exploration as a means of giving agroforestry context within the broader ethnobotanical environment. The introductory sections give a sense of what is meant by ethnobotany: first by considering its historical development and boundaries, then by emphasizing its unifying principles. The third section builds on this base to integrate agroforestry into an ethnobotanical framework. The foregoing themes are then brought together in the concluding sections where an ethnobotanical perspective is used to evaluate the kinds and extent of plant diversity in existing agroforestry systems, and to offer predictions, in accord with ethnobotanical experiences, concerning the prospects for maintaining or increasing the plant species and genetic diversity of these systems in the future.

2. ETHNOBOTANICAL HISTORY AND BOUNDARIES

The term ethnobotany conveys at least three broad meanings. The first meaning stresses what humans, in the course of their daily lives, know and believe about plants, that is, their plant lore, and by extension, how they grow, manage, harvest, and otherwise use and are influenced by plants.

* This chapter is an outgrowth of the generous support received over the years from Cornell University through the L. H. Bailey Hortorium and more recently, the Cornell International Institute for Food, Agriculture and Development, especially its Philippines program, Conservation Farming in the Tropical Uplands. Special thanks to Erick Fernandes for his review of the chapter and his suggestions concerning it.

1-56670-294-1/99/$0.00+$.50

To varying degrees each of us and the societies of which we are members possess plant lore; hence, humans in all kinds and at all levels of social and economic organization, culture, and technological achievement have ethnobotanical awareness, whether formally recognized or not. The second meaning of ethnobotany places the term in the realm of science. In its most basic sense and its grandest scale, the discipline of ethnobotany encompasses, in its entirety, the study of the interrelationships of plants and humans. The third meaning might best be described as socio-political or socio-economic. Here ethnobotanical information and aura are used to promote social justice, environmental conservation and sustainability, intellectual property rights, and other relevant causes, or to justify particular beliefs or resource management systems. Historically and in practice the boundaries of each meaning are not sharply defined and intergrade variously depending on the circumstances.

As a recognized academic discipline ethnobotany, is young, emerging clearly only in the 19th century (Ford, 1978), but its origins are embedded in human behavior and beliefs and trace to antiquity. Can there be doubt that the earliest humans were shaped by and endowed with curiosity about their surroundings or that they were knowledgeable about the plants and animals that shared their environments? Can there be doubt that our ancestors gathered biotic lore from one another, using it to enhance their chances for survival and to enrich their lives? At what point in the course of human evolution, everyday observation and information exchange became deliberate and systematic is not known, but in the detailed Paleolithic cave portraiture of European animals there is ample evidence of astute human observations of nature. Depictions, for example, those of Grotte Chauvet, France, dating to 20,000 years ago (Fischman, 1995), are more than catalogues of the animal life of the time. Together with other evidence, they provide clues concerning the lifestyles of the peoples who created them, albeit lifestyles in which the roles of plants must largely be inferred from habitat reconstructions and cautious extrapolations from the behaviors of modern hunting-gathering peoples.

In seeking the origin of ethnobotany as a discipline, rather than as a practice, one might arguably turn to our post-Neolithic ancestors of the present day Near East, Egypt, India, China, Mesoamerica, and elsewhere, whose representations of plants in reliefs, murals, artworks, and writings chronicle and provide tantalizing glimpses of the extent of human knowledge, beliefs, and practices related to plants and animals. Egyptian murals are rich in plant lore. For example, some, dating from the 12th dynasty onward, variously depict what is presumed to be oil-seed lettuce (*Lactuca sativa* L.) and the sycamore fig (*Ficus sycomorus* L.) in agricultural and mythological associations, indicative of the intimate roles played by these plants in Egyptian life (Harlan, 1986). The Rig Veda, a collection of verses dating from the 2nd millennium BC of India, reveres a plant known as soma, which Gordon Wasson (1972) argues is the fly agaric (*Amanita muscaria* (L. ex Fr.) Hook.), an hallucinogenic mushroom whose use by ancestral Siberian peoples must presage its written veneration. Wasson further suggests that the Siberian birch (presumably *Betula verrucosa* Ehrh. and/or *B. pubescens* Ehrh.), with which the fly agaric often occurs in mycorrhizal relationship, may be the Biblical tree of life and soma the forbidden fruit. Is this not the ultimate example of agroforestry?

The ancients apparently placed no bounds on their ethnobotanical observations. This holistic view, however, did not become the underlying rationale for the development of modern ethnobotany. Rather, much of ethnobotany came to focus on the utilitarian values of plants to peoples of non-literate, non-western cultures, that is, those considered to be primitive from European and North American perspectives. The use/primitive focus emerged quite naturally in Renaissance Europe, attributable to world-wide exploration from the 14th century onward. Information about plants and plant use by non-European peoples was gathered and chronicled by European explorers and naturalists, primarily as inventories of the potential economic wealth of newly reached lands and their peoples, but also to satisfy a growing desire for knowledge about the world and its biotic inhabitants.

In what was to become the United States, the earliest ethnobotanical records flow from the European connection (Vogel, 1970; Ford, 1978; Davis, 1995). Later the intensive and detailed studies of Native American plant use, during the mid-1800s onward, followed the westward movement of the population and served to solidify the dominance of a use/primitive perspective in ethnobotanical studies, even though these studies also revealed much about culture. The writings of Edward Palmer (1871; 1878) are models of this genre; which, in keeping with the tenor of the time, Stephen Powers (1875) termed "aboriginal botany," defining it as "all the forms of the vegetable world which the aborigines use for medicine, food, textile fabrics, ornaments, etc." The term ethnobotany itself followed some two decades later, when it was proposed by the botanist John W. Harshberger in presenting a paper in Philadelphia before the University Archæological Association.* In his presentation, Harshberger spoke more expansively about the values of ethnobotanical study than his legacy suggests, including among those values the elucidation of culture, the understanding of the past distributions and dispersals of cultivated plants, and the use of dendochronological clues to track climatic changes.

During the 20th century, concepts and definitions of ethnobotany evolved along different lines. Most definitions, however, reflect ethnobotany's 19th century use/primitive perspectives. Schultes and von Reis (1995), for example, defined ethnobotany as "the study of human evaluation and manipulation of plant materials, substances, and phenomena, including relevant concepts, in primitive or unlettered societies." In this definition the concept of use is enlarged, but contextually it remains centered on primitive and marginalized peoples. Other contemporary definitions are broad and purged of both use and primitiveness**. Ford (1978) viewed ethnobotany as "the study of direct interrelations between plants and humans." His use of the word "direct" implies peoples whose lifestyles remain intimately related to the plant environment, rather than peoples of technological societies, whose interactions are more compartmentalized and generally removed from substantive, direct interactions with plants.

Recent papers have extended the reach of ethnobotany by variously stressing concepts of ecology (Prance, 1995), chemical ecology and evolution (Johns, 1990), biobehavior (Etkin, 1990), system processes (Alcorn, 1989), and ecological anthropology (Orlove, 1980), although in fairness, the ecological connection was first explicitly proposed over half a century ago by Volney Jones (1941). Other papers have sought to introduce quantitative methods to ethnobotany (Prance et al., 1987; Moerman, 1991; Phillips and Gentry, 1993a, 1993b; Phillips et al., 1994; Begossi, 1996; Principe, 1996) and/or to improve sampling strategies (Martin, 1995; Alexiades, 1996). Yet, despite the diversity of thought and approach represented by the foregoing studies, they retain a focus on direct interactions, as expressed by Ford.

As human activities have become more and more integrated on a global scale, societies and the factors affecting them are increasingly difficult to view in isolation. The relationships between Brazilian campesinos and the babassu palm (*Attalea speciosa* C. Mart. ex Spreng., but also referred to as *Orbignya phalerata* C. Mart.), for example, reflect many local conditions, but ultimately are dependent on external economic, social, and political events (Anderson et al., 1991). Hence, full understanding of the campesino/babassu relationship requires an integrative, holistic perspective. Distinctions made between direct and indirect factors are often clouded and are not attempted in my view of ethnobotany, which then becomes simply the study of the interrelationships of human and plants.

Acceptance of an all-encompassing or holistic view of ethnobotany does not negate the value of more narrowly defined studies, whether discipline-based and/or use- or primitive-focused. Not only do these kinds of studies remain essential as intimate, in-depth records of human interests

* It was reported (Anonymous, 1895) that Harshberger defined ethnobotany as the study of "plants used by primitive and aboriginal people," but in the published form of the address (1896) no definition of ethnobotany is given. The word is derived from the Greek roots: *ethnos*, meaning nation, race, people, or culture and what is characteristic of or used by them, and *botanikos*, of herbs or botany.

** Broad definitions did have 19th century proponents. Walter Hough (1898), for example, defined ethnobotany as "the study of plants in their relations to human culture."

and accomplishments, but also they will continue to provide the data and insights basic to broadly phrased comparative studies. Rather, a holistic view of ethnobotany serves to bridge the gap that separates studies of primitive or traditional societies from those of the westernized world. Holistic ethnobotany encourages the integration of data from all levels of approach and all societal contexts in broad reference frameworks, which address issues of major human concern, among them the effects of human globalization on the sustainability of the earth's environment.

Because the study of ethnobotany relates human interests, concerns, and activities so intimately to plants, it is not unexpected that its study and data can take on social, economic, and/or political meaning. Such meanings are not new. Queen Hatshepsut, 14th century BC Egyptian ruler, whose expedition to the Land of Punt, perhaps present day Somalia, to seek myrrh (*Commiphora* sp.) and other valuables, is said to have been inspired as a way of currying political support (Wells, 1969). The over exploitation of a species of silphium (*Ferula* sp.), which was endemic to present day Libya and which was used to regulate fertility in the ancient Mediterranean region, led to its extinction in the 3rd or 4th century AD. The use of this silphium and other plant species with abortificient activity engendered the still continuing debate about the morality of abortion and the point at which human life begins (Riddle et al., 1994). And, although designed to reveal details of material culture, 19th century studies of Native American plant use could be seen as social statements, for they contributed to both the romanticized and patronized views of these peoples.

In a more contemporary context, solutions to many of the world's pressing environmental problems are sought in the knowledge and practices of primitive and traditional subsistence peoples. Such knowledge and practice are thought to hold special significance with respect to management of vegetational resources and associated animals (Denevan and Padoch, 1988; Posey and Balee, 1989; Anderson, 1990; Inglis, 1993), maintenance of agroecosystems and preservation of genetic resources (Brush, 1986; Altieri and Merrick, 1987; Altieri and Montecinos, 1993), and discovery of new bioactive compounds (Akerele et al., 1991; Cox, 1993; Balick et al., 1996). In other instances, ethnobotanical focus is on preservation of cultures and the biotic resources on which those cultures depend by seeking alternatives to the conversion of forests to agriculture through maintenance of forests as reserves from which non-timber forest products, including medicinals, can be extracted (Balick and Mendelsohn, 1992; Nepstad and Schwartzman, 1992). While ethnobotanical knowledge and action may be cast as non-political, they underlie issues of broad socioeconomic concern, including those concerned with the ownership of nature, knowledge about it, and rights to the benefits to be derived from it. Such issues may be seen from the perspectives of both the developing (Brush, 1993; Kloppenburg and Balick, 1996) and technological worlds (Baenziger et al., 1993; MacDonald, 1995), and call into question the role of academics in the political arena (Posey, 1990) and the nature and paradigms of science (Cox, 1990; Martin, 1994).

3. UNIFYING PRINCIPLES OF ETHNOBOTANY

The extent of human dependence on and richness of involvement with plants through time and space is so great and so far-reaching that study of it, especially in modern times, has tended to fragment and compartmentalize research along narrow disciplinary lines within the social, biological, and physical sciences. Thus, human and plant interrelationships are seen through an astonishingly wide lens of disciplinary perspectives, which often carry their own ethno appellation, for example, ethnopharmacology, ethnomedicine, ethnomusicology, ethnomycology, and, I suppose, ethnoagroforestry, among others, often without recognition of their ethnobotanical heritage. Even in ethnobiology, which encompasses ethnobotany, the latter holds the central position because of the extent to which plants are determinants of human existence. What then, given its seemingly infinite breadth, gives unity and substance to ethnobotany and distinguishes it from other disciplines? Definitions alone do not seem to provide the answer. Rather, I suggest that ethnobotany's common ground is found in the explicit marriage of two themes: botany, especially as represented by systematics and ecology, and human selection. Taken together and given connectedness and continuity, these themes place ethnobotany at the intersection of the biological and the human realms and give it a perspective in which nature and humanity are intimately intertwined.

Botanical literacy is a cornerstone of ethnobotany. In other words, ethnobotanists must understand the structure, function, and relationships of plants and, if the study warrants it, possess in-depth knowledge of other attributes of plants, for example, biochemistry, if the focus is on plant compounds exhibiting bioactivity. Yet, in all cases the underlying foundation of ethnobotany is found in systematics and ecology. The centrality of systematics to ethnobotany is not a new concept. How else might one refer to the plants that are of human interest? But systematics encompasses far more than placing names on plants. It is the fundamental medium for organizing, evaluating, and communicating information and thoughts about plants, including assessments of biodiversity from both the biological (Humphries et al., 1995) and ethical (Rolston, 1988) perspectives. The central position of ecology in ethnobotany is now equally obvious, perhaps so much so that its role is largely alluded to rather than stated explicitly, although exceptions are noted above (Jones, 1941; Orlove, 1980; Alcorn, 1989; Etkin, 1990; Johns, 1990; Prance, 1995). In what are extensions of ethnobotanical concepts, others have seen in ecology the foundation of ecosystem management and sustainability at every level (Kaufmann et al., 1994; Goodland, 1995; Hartshorn, 1995).

By recognizing humans as integral components of the earth's biosphere and, by extension, embodiments of empirical experiences and philosophical concepts of systematics and ecology, an irreducible framework for ethnobotanical study is established. All human interrelationships with plants can be evaluated in terms of our effects on plants and plant communities and the reverse, even in situations where the interactions might be posited as neutral. While systematics and ecology represent nature's component of ethnobotany, selection is indicative of the human component. Human interactions with plants and plant communities through time and space occur as a continuing series of decisions, which can be cast in the context of selection. Selection, in turn, represents the summation of all of the myriad of factors that influence the decision-making process at a given time and place, that is, needs, environment, culture, economics, technology, government policies, whimsy, and any others. The complexities of selection are poignantly illustrated in attempts to understand the factors that lead to sustainable land use (Kruseman et al., 1996).

The systematic and ecological underpinnings of ethnobotany are most clearly expressed in concepts of the resource base, which is the totality of the plants, animals, and other organisms that constitute the biota in relation to a given physical environment. The resource base can be described in terms of its life forms, its taxonomic or chemical composition, the processes that occur within it, or by any other relevant criteria. It has time and space dimensions, which may be adapted to essentially any scale, for example, the entire course of human existence or only the agricultural period placed at single site or over the world as a whole. The character of the resource base at a specific site represents nature's response to a given suite of environmental conditions. Arguably, the response can be viewed as the optimal or most energetically efficient natural solution to those conditions. The convergence of vegetational physiognomy under similar environmental conditions in different regions of the world, for example, tropical rain forest in South America, Africa, and Asia, illustrates this principle.

The resource base gains ethnobotanical meaning when humans are placed in it. The effects of human presence on the resource base are expressed through selection and result in changes in the character of natural vegetation. The degree to which natural vegetation is modified or replaced by humans can be considered as deviation from the natural state and may be measured in such terms as loss or gains in biodiversity, altered biomass production, and cost of inputs to replace the natural occurring ecological functions. Through time, the interplay of human selection with the resource base has led to the present mosaics of use, indicative of relative states of deviation. As a measure, deviation has neither negative nor positive connotations. It is a way of expressing change, yet conceptually, deviation implies more than a simple description of change. Embedded in it are human dynamics, expressed in terms of scale, integrative processes, and connectedness through time, the study of which is aptly holistic and multidisciplinary.

The framework provided by the foregoing resource base/deviation model is applicable to much of ethnobotanical thought, but is especially relevant in broadly based comparative research, which seeks to understand the patterns of human/plant relationships as seen through their effects on the resource base. Two hypotheses are implicit in these comparative studies: 1) that in ecologically similar habitats, human societies of similar character will react to and exploit the plant resource base in similar ways, and 2) that reaction to and exploitation of the resource base are under selective pressure that optimizes the return on use within that selective regime (Bates, 1985, 1988a). Practically, the problems of testing these hypotheses are great, especially as they are cast in terms of foraging theory (Winterhalder and Smith, 1982; Stephens and Kreb, 1986), for it is difficult to identify truly similar cultures, to determine what optimization should mean and the currencies that will be effective measures of it, and to uncover data sets that are actually comparable. Yet, enough is known of human/plant interactions to give general support to the hypotheses and to suggest the kinds of criteria that might be used in developing comparative models of plant use that have applicability to current human concerns, whether quantitative (Prance et al., 1987; Johns et al., 1990; Moerman, 1991; Phillips and Gentry, 1993a, 1993b; Phillips et al., 1994; Begossi, 1996; Principe, 1996) or qualitative (Bates, 1988b; Alcorn, 1995) in approach.

4. AGROFORESTRY IN AN ETHNOBOTANICAL CONTEXT

Agroforestry, as one complex series of relationships between humans and plants, animals, and other life forms, carries meanings that parallel those of ethnobotany. First, agroforestry is embedded in the knowledge and practices of peoples who integrate trees and shrubs into their subsistence and resource management systems. In their most elemental expressions, for example, among the Kayapó of the Amazon Basin (Posey, 1984; Anderson and Posey, 1989; Parker, 1992; Posey, 1992), the disciplines of ethnobotany and agroforestry display obvious unity. In more complex modern societies, in which disciplines tend to be relegated to discrete tracks, agroforestry may be practiced without reference to ethnobotany, yet unity remains. Second, agroforestry is an academic discipline, which not only can draw on the knowledge and practices held in traditional systems, but also generates new knowledge related to the integration of trees and shrubs into managed landscapes (Sanchez, 1995). Third, and perhaps more generally and explicitly stated than ethnobotany, agroforestry may be cast in positive bio-physical, cultural, socio-economic, and political terms. These aspects of agroforestry are emphasized in definitions such as the those of Bene et al. (1977) "a sustainable management system for land that increases production, combines agriculture crops, tree crops, and forest plants and/or animals simultaneously or sequentially, and applies management practices that are compatible with the cultural patterns of the local population," and Lassoie (1996) "any intensive land management system that optimizes the environmental, social, and economic benefits arising from the biological interactions created when trees and/or shrubs are deliberately grown over space and/or time with crops and/or livestock."

In a less qualified definition, which recognizes the difficulties that may be faced in simultaneously trying to meet social, economic, cultural, and other goals (Raintree, 1987) and which parallels that given to ethnobotany in its simplicity, agroforestry may be considered as any agricultural system that purposely integrates trees and shrubs into managed landscapes. This definition includes practices referred to as agrosilvicultural (agricultural crops among wild forest trees), silvopastoral (trees and shrubs combined with grazing), and agrosilvopastoral (agricultural crops, trees, and grazing) (Pedersen, 1990). The terms "agricultural" and "managed landscapes" refer to lands under the plow as well as lands in fallow, within and beyond those cultivated, on which trees and shrubs are planted or occur spontaneously but which are utilized as sources of goods and/or services. As one leaves the cultivated and obviously managed landscapes and enters forested areas, the boundaries of agroforestry tend to lose definition. Whether or not wild trees and shrubs within the forest, which largely represent sources of extractives (Denevan and Padoch, 1988; Brush, 1993; Godoy and Bawa, 1993), are considered part of an agroforestry system depends on the degree to which they are part of an integrated management and subsistence strategy and one's point-of-view on this matter (see Chapter 17).

In an historical context the development of agroforestry essentially parallels that of ethnobotany. The archeological record over the past ten thousand years or so includes the remains of fruits that apparently had been gathered from local shrub and tree species. In the Near East the record reveals a shift from gathered sources to the probable cultivation, with the cereal and legume complex of that region, of the date (*Phoenix dactylifera* L.) possibly by 6,000 BP, and shortly thereafter, the olive (*Olea europea* L.), grape (*Vitis vinifera* L.), and fig (*Ficus carica* L.) (Zohary and Speigel-Roy, 1975; Zohary and Hopf, 1993). Such prehistoric and protohistoric use and cultivation of trees and shrubs and practices related to them eventually coalesced to form a foundation from which modern discipline-based studies of woody plants, including agroforestry, emerged.

As in ethnobotany, agroforestry can be approached from many perspectives and contexts, ranging from those that relate to basic agricultural concerns and practices, for example, water relationships, soil management, nutrient cycling, pest management, and productivity (Hatfield and Karlen, 1994), to those that enter the socio-economic realm, including issues of land tenure, equitable access to resources, labor allocations, social status, wealth, and gender (Browder, 1992a; Kainer and Duryea, 1992; Röling, 1994). As in ethnobotany, individual studies add to our basic knowledge about the role of trees and shrubs in managed landscapes, but alone they cannot address broader issues/questions concerning the place of trees and shrubs in humanity's future. Finally, as in ethnobotany, agroforestry can find unifying themes in systematics and ecology and in human selection and its derivative aspects. It is in the latter thematic context that ethnobotanical perspectives find their strongest application to agroforestry.

The potential of agroforestry in developing sustainable plant management systems appears great, especially in terms of ecosystem functionality and productivity, but questions of the following kind remain to be addressed:

1. What is meant by sustainability in terms of agroforestry systems?
2. Are agroforestry systems more sustainable than those that are not? If so, or if not, what are the conditions that increase or decrease sustainability?
3. Do agroforestry systems maintain adequate levels of ecosystem and socio-economic functionality? Or, how much plant diversity and of what kinds are necessary to sustain humans in an ecosystem. What is meant by functionality and productivity and what is the relationship between the two concepts?
4. Will the further development of agroforestry systems result in less exploitation of the world's remaining primary forests than will other plant/animal management strategies? Conversely, will the further development and widespread adoption of agroforestry systems actually result in increased exploitation of primary forests?
5. Do agroforestry systems conserve biodiversity within themselves in a significant way? That is, are agroforestry systems effective *in situ* banks of genetic diversity?
6. Do indigenous, traditional agroforestry practices have relevance to modern agroforestry practice?

The formulation of questions concerning the roles of trees and shrubs in the future management of the earth's ecosystems are a way of focusing attention on complex issues. None of the foregoing questions is meant to be answered simply or, given our current state of knowledge, definitively, for each question includes assumptions and variables that will condition responses. Each question includes elements that are not strictly ethnobotanical or confined to agroforestry alone. For example, questions related to ecosystem functionality might be approached solely from any one of several perspectives, that is, social, economic, biological, or physical, but greater value and insights are likely to be realized when questions are considered across disciplines in ways that integrate biological and human imperatives. In the following section, such an approach is taken in considering plant species diversity in agroforestry.

5. EVALUATING PLANT DIVERSITY IN AGROFORESTRY

Issues related to the conservation and preservation of biodiversity on local, regional, and global scales are of compelling human concern (Kinnaird, 1992; Krattiger et al., 1994; McNeely, 1995; Reaka-Kudla et al., 1997). Anthropocentrically, these issues resolve to one as yet unanswerable question: how much biotic diversity and of what kind is necessary to assure the future survival of humanity? Again, context is confounding, but answers can be sought within given sets of assumptions about those factors that affect the future human condition, for example, the future size and distribution of the human population, the availability of and access to resources, and degree to which humans can attain quality and equality in their lives (Myers and Simon, 1994; Myers, 1997; Vincent and Panayotou, 1997). Similarly, concepts of diversity differ (Humphries et al., 1995). Here diversity means the number of plant species in any given system, local to world-wide, employed for a specific use or suite of uses. Implied in species diversity numbers are other measures of diversity, including community, life form, spatial, functional, and genetic diversity.

The general rationale for increasing plant species diversity in agroecosystems, including agroforestry systems, rests on the assumption that diversity per se has positive ecological value in sustaining given systems. If this is true, then human selection ought to work effectively to increase species and other diversity measures in those systems, resulting in increases in useful plant productivity and economic returns to humans, while also improving ecological functions. By this reasoning, increased returns could then lessen the need to expand agriculture into now forested lands or permit the regeneration of forests on previously cleared but marginal lands (Kaoneka and Solberg, 1997). In these ways agroforestry could enhance biodiversity within agricultural systems and promote conservation of biodiversity beyond their boundaries. Whether or not agroforestry will ultimately fulfill these goals is not certain, but by understanding factors that influence the extent and nature of plant diversity in current agroforestry systems, it is possible to gain insights into the prospects for maintaining or increasing biodiversity in the future of agroforestry (Padoch and Peters, 1993; Thiollay, 1995).

Pattern analysis, which in this case derives from the ethnobotanical principles of systematics, ecology, and selection, provides the framework in which evaluation may be undertaken and conclusions reached. The key elements of pattern analysis are found in identifying and understanding the factors that influence human selection of the plant resource base, that is, the determinants of plant utilization, and the patterns that result from selection through time and space. Predictions concerning the future of plant use are based on the nature of the patterns, tempered by assumptions concerning the future. The method has been used to analyze the future prospects for plant use in agriculture (Bates, 1985, 1988a) and for palms in agroecosystems (Bates, 1988b).

5.1 Determinants and Evaluation of Plant Utilization

Determinants of plant use are ultimately expressions of need. For emerging humans, needs were expressed in simple survival terms — foods and perhaps medications that could be harvested and consumed directly, and shelter as it could be found in nature. Beyond the survival state, human needs expand in relation to complex interactions involving technological advancements, cultural changes, social and economic factors, and human aspirations, that is, the totality of factors that affect the human condition. The process is self-aggrandizing, in that meeting existing needs engenders new paradigms that, in turn, create new needs.

In earlier periods of history when the human population was fragmented and societies could develop in relative isolation, change in lifestyles brought on by similar need-cycles occurred in concert in different regions of the world. Under similar human and environmental conditions, needs were essentially equivalent the world over, as were the manner in which they were met, differing principally in the taxonomic character of the biological resources used and cultural peculiarities of the inhabitants. In a selection sense, the world-wide resource base represented a vast repetitive reservoir of opportunities and constraints that determined human responses in relation to needs. In an academic sense, the repetitiveness of needs-based situations world-wide created a richly stocked laboratory in which comparative studies of human responses to needs/resources could be undertaken.

It is apparent from the study of peoples who have limited access to technology and remain in direct contact with their resource base that the choice of plants to meet particular needs is not a random act but rather is influenced by selection pressures of varying intensity. The inevitable outcome of selection is a narrowed number of plant species used for a particular purpose, relative to the number of species in the flora as a whole. Data presented by Moerman (1991), for example, demonstrate the narrowing force of selection by Native Americans as they sought plants for medicinal use. Of a total flora of 16,270 species in the continental United States and Canada, 2,147 have been used as sources of herbal medicines, that is, about 13% of the species present. Selection itself tends to be strongest when focused on those plants that are critical to human existence, as they are in the choice of staple foods, as demonstrated by Lee (1986) and Tanaka (1980), among others, for the foraging San of the Kalahari Desert of southern Africa. In botanically richer environments than those of the Kalahari, choices are greater and the total number of species utilized may be greater (Begossi, 1996), but much of the use is in categories such as medicinals that permit relatively high degree of substitution, where opportunistic harvest from wild plants is part of the subsistence strategy, and selection pressures are low (Alcorn, 1984; Avevedo-Rodriguez, 1990; Boom, 1990, 1996). Cultivated staples, even among peoples such as the Amazonian Yanoama, who also forage and hunt in the forest, are relatively few (Smole, 1976).

Assessment of selection pressures and their outcomes is most direct among societies such as the desert dwelling San, where food choice, for example, may be seen as an expression of energetic efficiency and explained, in large measure, in terms of foraging theory (Winterhalder and Smith, 1982; Stephens and Krebs, 1986). More complex societies, engendered by the development of agricultural, industrial, and informational economies, each with its accompanying globalization, remove increasingly large proportions of their members from subsistence contact with the resource base and make energetic analyses more difficult. Under these circumstances energetic values are usually expressed in economic terms, for the assignment of costs is a convenient way to draw together a large number of relevant variables. The validity of such an approach in evaluating selection relies on the recognition of the true costs associated with each variable, especially those related goods and services provided by the physical and biotic resource base (Costanza and Daly, 1992; Repetto, 1992; Costanza et al., 1997), although there remains disagreement as to the meanings and applications of this approach (Myers and Simon, 1994; Daly, 1995; Sagoff, 1995). For instance, uncertainty exists in attempts to assess the actual costs of agriculture (Pimental et al., 1995, 1997), the tangible value of nature's services (Westman, 1977; Ehrlich and Mooney, 1983; Anderson et al., 1991), and the potential of extractive reserves (Gray, 1990; Balick and Mendelsohn, 1992; Browder, 1992b; Nepstad and Schwartzman, 1992; Godoy and Bawa, 1993; Godoy et al., 1993; Hegde et al., 1996), each of which has implications for agroforestry.

5.2 Patterns and Outcomes of Utilization

The patterns that result from needs and the interrelated processes of selection are considered here in two contexts: first, in terms of utilization pools, and second, in terms of agricultural systems.

Among any group of people, ranging from communities to tribes, nations, and humanity itself, the results of selection array species along a gradient of least to greatest importance or value. A few plant species in any use category, that is, foods, medicines, oils, fibers, among others, are of primary importance and are generally recognized as staples in a given society. Other species tend to be complementary or augmentative to the staples, thus forming groups of secondary importance; while still others are used sporadically and are easily substituted for. These species constitute a third rank or one of tertiary importance. On a world-wide basis and for humanity in total this hierarchical ordering results in a group of one hundred or so species that constitute the staples of humankind, including, as examples, rice (*Oryza sativa* L.), wheats (*Triticum* sp.), maize, (*Zea mays* L.) cottons (*Gossypium* sp.), opium poppy (*Papaver somniferum* L.), and para rubber (*Hevea brasiliensis* (Willd. ex A. Juss.) Müll. Arg.). The primary group caps a broad triangle in which successively lower strata are composed of larger numbers of less valued species.

Although the placement of plant species along use-gradients is more or less continuous, for purposes of analysis it is useful to cluster species by ranks, which I have termed utilization pools (Bates, 1985, 1988a, 1988b). In those studies, three ranks of use by any definable group of people were recognized. These ranks correspond to those given above — primary, secondary, and tertiary. A fourth rank or level could be added to represent those plant species that have no apparent direct use by humans although they are dominant contributors to world-wide ecological functionality and stability. This rank would add a broad base to the triangle.

Placement of a plant species in a particular utilization pool is not based on a single criterion, but rather on a composite of factors that relate to the use category. For example, in ranking food plants not only are energetic, nutritional, and agronomic aspects important, but also social, cultural, economic, culinary, and chance aspects, as well. The method is descriptive and the loosely defined boundaries of categories and ranks make them adaptable to many points-of-view. In general, the method's lack of precision is its strength, for it permits a qualitative distillation of data of many kinds into relational context. The outcomes can be used to make cross-cultural comparisons of the identity, uses, and values of single species or groups of species at any level of human organization at a given time or through a series of time periods. An example of the methodology is found in the application of utilization pool concepts to the use and potential use of species of the palm family, the Arecaceae (Bates, 1988b).

In addition to their value in comparative studies, utilization pool concepts focus attention on the broader aspects of plant use and provide a basis for making generalizations about it, including the following, which are cast in a world-wide frame of reference.

1. Through the course of human history, definition of utilization pools developed first around food plants. Thereafter, as needs and technology expanded, pools in other use categories took on definition. On a world-wide scale the composition of the primary utilization pools for food and fiber plants is essentially established. In other use categories, for example, as sources of forage, timber, or biocrude, primary and secondary pools remain in the process of definition world-wide, although they may be defined locally or regionally.
2. Selection is most strongly expressed at the upper levels of the use-gradient, leading to stability in the membership and boundaries of primary utilization pools. Many species of primary pools are cultigens and their replacement, given their critical value to human societies, is less likely than in secondary and tertiary pools. However, human/plant relationships are not static; hence as selection values change, the character and extent of utilization pools can also change.
3. Geographical, economic, and political isolation of peoples of the world maximizes diversity of plant utilization by creating repetitive pool sequences (primary, secondary, tertiary) of different species composition but of similar function in any given use category.
4. With lessening isolation of human populations, the world-wide tendency has been toward globalization of primary and secondary pools and decreasing dependence on tertiary pool species.
5. Plant species that form the growing inventory of those considered underutilized are largely those of past, locally or regionally defined utilization pools that are now competitively disadvantaged as a result of globalization trends and currently recognized market forces.

The selective forces that lead to patterning of plant use find parallels in the patterning of agricultural systems or, taken more broadly, plant management systems, but give a somewhat different perspective to utilization. Variables affecting agricultural systems, whether internationalized or traditional, involve the same suite of factors expressed in different ways and with different intensity (Conway, 1987; Bates, 1988b; Hatfield and Karlen, 1994; Hoffman and Carroll, 1995; Pitman, 1995; Pretty, 1995; Vandermeer, 1995). Variables are found in the scale of systems; their biodiversity, including their representation of wild plant resources; the distinctiveness of their agronomic, horticultural, pastoral, and silvocultural tracks; their planting strategies, that is, whether to use species, varieties, and landraces adapted to the environment or the converse, or whether to seek to maximize productivity or security; their access to capital, energy, labor, organized research and extension programs, and other external inputs; their integration into market economies, including supportive infrastructure of roads and storage facilities and distribution and marketing systems; their degree of land tenure; and their gender and socio-economic equability.

The foregoing list of factors is not exhaustive. Rather, it conveys a sense of the kinds of variables on which selection operates. The outcomes of selection, however, are not infinite in number, but rather they result in a few basic agricultural tracks, embellished by local variations. For example, the range of variations in slash and burn and slash and mulch systems collectively may be seen as a common, widespread agricultural solution to nutrient and pest problems, which stem from low human population density, abundance of land, and lack of access to external inputs (Inglis, 1993; Thurston et al., 1994). The integration of plant species into such systems is likewise due to selection, not random phenomena. This does not mean that all slash systems evolved around the same species, but rather that local and regional selective forces identified and domesticated species with functional and use characteristics adapted to slash systems, for example, among vegetatively propagated crops, cassava (*Manihot esculenta* Crantz) in the New World, bananas and plantains (*Musa* spp.) in Southeast Asia, and yams (*Dioscorea* spp.) in Africa, together with supporting casts of other adapted crops. Similar functional requirements, coupled with similar human needs for carbohydrates, have now shifted selection from regional to a world-wide basis. The result has been greater homogenization of the staples of slash systems at the primary level.

The interplay of needs-based categories and systems-based categories expands the two dimensional triangular representation of utilization pools into a topographically varied three dimensional landscape. The base of the landscape can be visualized as a mantle of varying depth, that is, deep in areas dominated by native vegetation, shallow in areas of cultivation, and representative of those species not directly utilized by humans. The mantle is overlain, in part, by a thinner crust of varying relief, reflecting differences in use and systems categories that are of marginal or minor human importance and indicative of the tertiary utilization pool species. Peaks which arise from the crust singly and in clusters represent species constituting secondary and primary utilization pools. As their mountain counterparts, some peaks are relatively stable, others more active, changing conformation as selective forces continue to modify and refine their composition. Whatever the dynamics of the model, it is clear that the outcome of human selection on the plant resource base is to define peaks more sharply by focusing attention on select groups of species whose functions and productivity can be channeled to meet human needs.

5.3 Patterns of Plant Utilization and Diversity in Agroforestry

The foregoing, generalized summary of patterns of plant utilization provides a framework for evaluating present and future patterns of diversity and utilization in agroforestry systems. Given the wide range of approaches to and practices in agroforestry, selection intensity within them is predictably variable. This variation is recognized here by considering species diversity in three agroforestry regimes: (1) on lands that are in active agricultural production, (2) on lands that are in long-term fallow and subject to forest regeneration by invasive tree species, and (3) on lands occupied by forest, whether primary or not. As a generalization, selection intensity is greatest on cultivated lands and least in forested situations.

Agroforestry, where it involves the integration of trees and shrubs on lands under intensive cultivation, for instance, alley cropping or fields with hedgerows, on the one hand, and home gardens on the other, is obviously more species and life form diverse than single crop systems. In comparison to the native flora that is replaced, however, agroforestry systems under these conditions are inevitably less rich in species diversity, although they might approach the native flora in life form stratification. However, shrub and tree species diversity is summed for intensive agroforestry systems, whether by cataloguing all species used in them world-wide, by focusing on those restricted to their respective temperate or tropical regions and subzones, or by considering them only in functional or use categories, the number of species used is substantially reduced compared to the number of tree and shrub species that compose the indigenous resource base of a given climatic zone or geographical region. In other words, agroforestry under strong selective pressure behaves as other agricultural systems by narrowing species diversity in ways that are directly related to human needs (Lawrence, 1996).

Despite its recent fall from grace, the ubiquitous occurrence of *leucaena* (*Leucaena leucocephala* (Lam.) de Wit) in tropical hedgerow and other agroforestry systems provides an example of the kinds of factors that lead to dominance. In a single species, *leucaena* meets many of the needs of intense agroforestry management. It fulfills functional roles, that is, erosion control, water sequestering, and nutrient cycling, which when coupled with its intrinsic attributes, such as nitrogen fixation, ease of propagation, rapid growth, and coppicing tendencies; its usefulness as a source of mulch, fodder, wood, and shade; its general adaptability to social conditions; and its strong promotion, created a favorable selection environment for its rapid adoption (National Academy of Sciences, 1977; Bray, 1995). Other woody species that are integrated into intense management situations, for instance, *Gliricidia sepium* (Jacq.) Kunth ex Walp., also have desirable characters that place them at a selective advantage in cultivated regimes. Once functional and use niches are occupied by one or a small number of species, those taxa are displaced only with difficulty, even by species that might have superior qualities, a phenomenon that might be termed the inertia effect.

While individual tree and shrub species, *Gmelina arborea* Roxb., for example, combined with field cultivation, may contribute both ecological services and goods in subsistence and market economies in the tropics, home gardens and fallows provide a greater abundance of niches and opportunities to produce efficiently a wide variety of products for both household use and market (Padoch, 1988a). Home gardens and fallows characteristically are heterogeneous in species composition and vertically layered, often consisting of a ground layer of annual and perennial herbaceous crops overtopped by tree and shrub species of varying heights. The extent of the herbaceous or ground layer depends on the spatial arrangement of trees and shrubs and the density of shade that they produce. These systems are seen as actual or potential storehouses of species diversity, but diversity is owed context. While some systems are characterized by indigenous species dominance (Padoch, 1988a), in most instances the staple species composing intensive systems, ranging from root crops to fruit trees, are widely distributed and recognizable within world-wide primary and secondary utilization pools (Rehm and Espig, 1991; Smith et al., 1992). Diversity among these intensive systems may be expressed largely in variant forms or landraces of common domesticates. A representation of indigenous taxa, including those yielding medicinals, craft materials, and ornamentals, may be added to the cadre of internationalized elements, thereby giving these plantings a local flavor (Owusu et al., 1993).

The foregoing observations of intensive, tropical agroforestry systems are not judgmental concerning desirable levels of species diversity, nor do they mean that the composition of these systems is fully defined. They do, however, illustrate the strength of selection under present conditions and suggest an emerging stability in the primary and secondary species composition of these systems. Paradoxically, as world-wide primary and secondary utilization pools take on stronger form, opportunities emerge to integrate some greater species diversity into local systems that serve local needs or specialty markets. Some of these potential species additions may be indigenous to a given area, but others are likely to be drawn from the tertiary cadre of species already representative of minor, wider cultivation or use.

As agroforestry practices move away from intensively managed landscapes and venture into less purposely managed systems, that is, lands left to long-term fallow, regenerating secondary forest patches, and even primary forest, selective forces remain, but are more variably and for the most part less strongly expressed. In fact, it is in these environments that the dynamics of local selective forces remains most elemental. As a result species diversity is generally greater than that of intensively managed systems and more reflective of the indigenous flora and local practices related to it. Utilization pools often retain much of their local character, although they too are in the process of change. For example, one of the effects of the extractive reserve movement's concern for non-timber forest products is to strengthen focus on the utilization of selected species, thought to have significant economic potential (Nepstad and Schwartzman, 1992). Such directed focus brings into play socio-economic forces, such as equitable return on invested labor and lost opportunity costs, that may outweigh the intrinsic merits of the resource (Gray, 1990; Browder, 1992a; Godoy et al., 1993; Hegde, 1996). As a species gains economic value, humans realize greater return on invested labor if the species is brought into cultivation and domesticated, rather than by continuing to place reliance on gathered wild sources (Farooquee and Saxena, 1996). This sequence of change duplicates the general trends of plant selection and utilization.

Just as the tree legumes are the dominants of intensive agroforestry systems because of their nitrogen-fixing capabilities and agronomic adaptability, palms are the dominants of less intensive agroforestry systems throughout much of the tropics because of their taxon diversity, ubiquitous occurrence and wide ecological tolerances (Uhl and Dransfield, 1987); the incredible richness of goods and services they provide to humankind (Balick and Beck, 1990); and their adaptability to many socio-economic settings (Johnson, 1988; Pedersen and Balslev, 1990). Perhaps more than any plant family, the palms are illustrative of the range of human/plant associations that constitute tropical agroforestry.

As sources of vegetable oils, two palms, the coconut palm (*Cocos nucifera* L.) and the African oil palm (*Elaeis guineensis* Jacq.) rank among the primary staples of humankind. But both species are more than sources of oils. For instance, the coconut palm, which Heiser (1990) calls the world's most useful tree, is a source of beverages, sugar, fiber, and wood, among other products. Both the African oil palm and the coconut palm span the range of agroforestry systems. Both are cultivated in plantation settings, the African oil palm normally so, but it is also a component of indigenous west African agroecosystems. The coconut palm, in the Philippines and often elsewhere, is the dominant low-elevation cultivated and naturalized tree, occurring in essentially every humanized habitat from that of home gardens to agropastoral landscapes. Its presence and importance to rural peoples as a source of income, principally copra, and subsistence products is such that it is integral to life itself. In a similar way, the babassu palm, which occurs in wild, widespread, high density stands in Brazil and Bolivia, where it has regional and some international importance as a source of oil and other useful products, is the principal ecological and socio-economic determinant of the regions in which it dominates the landscape (Anderson et al., 1991). Other palm species which fit this mold, having broad geographical range, ecosystem dominance, and multiple use, include the moriche (*Mauritia flexuosa* L.f.) of the Amazon Basin (Padoch, 1988b; Gragson, 1995), and the sago palm (*Metroxylon sagu* Rottb.) (Ruddle et al., 1978), the palmyra or lontar palm (*Borassus flabellifer* L.) (Fox, 1977), and the Nypa palm (*Nypa frutescens* Wurmb.) (Stanton and Flach, 1980) all of Pacific islands and southeast Asia. The natural and often encouraged range of these palms concentrate resources and limit the presence of other species in ways that are characteristic of monocultures.

Other palm species, which are less dominant and thus components of communities and ecosystems of greater vegetational diversity, may represent management strategies in which mixes of valued species are sometimes planted but often are simply encouraged by removing competing species of less value, both approaches exemplified in the reports of Alcorn (1984) and Anderson and Posey (1989). The palm species constituting this level of agroforestry are diverse, but characteristically are indigenous and of limited ecogeographical range. They yield products of subsistence to international market value, the latter seen in such items as fruits, palm hearts, and vegetable ivory (Bates, 1988b). Examples of palm-based agroforestry systems of this kind are found in Balick (1988) and Pederson and Baslev (1990) and more recently in papers such as those of Jensen and Baslev (1995) and Feil (1996). As a means of conserving indigenous species diversity and ecological functionality, palm-based agroforestry systems have value to the degree that representations of indigenous floras remain part of them (Everett, 1995).

Species diversity in existing agroforestry systems fits classic utilization patterns. In intensively managed, mature agroforestry systems — in-field and home garden systems — in which selection pressures are high and are cast against a regional or internationalized background, the expected and realized outcomes are decreased numbers of and increased homogeneity in species composition and more clearly defined utilization pools. In contrast, under less intensive agroforestry management regimes, in which selection operates more diffusely, is expressed locally, and accommodates large numbers of taxa of no immediate human use or of use that is incidental to the system as a whole, species diversity tends to be high in total, reaching a maximum in systems that combine fallow areas and secondary and primary forest. Depending on the focus and purpose of the system, the diversity of taxa that have recognized value as sources of services and/or goods can be low to moderately high.

6. THE FUTURE OF PLANT DIVERSITY IN AGROFORESTRY

The preceding sections illustrate basic lessons of ethnobotany that apply to plant utilization and species diversity, in general, and to agroforestry, in particular. The lessons drawn indicate that: (1) in any given environment and situation, humans directly utilize less than the total of the plant resources available to them, the outcome of selection; (2) selection of a particular plant species or group of species for a given use is dictated by any or a combination of the totality of factors influencing the human condition; (3) while selection may remain a local event with local outcomes, especially in subsistence economies, it is becoming ever more influenced by external events, often of global proportions; (4) the outcomes of selection fit definable and generally predictable patterns, which can be analyzed hierarchically and cross-culturally through time and space; and (5) as the value and importance of a plant species to humans increases, selection favors bringing it into a managed or cultivated, domesticated state, rather than relying on its presence and harvest from the wild.

In considering species diversity in the future of agroforestry, the foregoing premises are assumed, yet the manner in which they may play out cannot be fully inferred from the past experiences alone. To have relevance, predictions concerning plant utilization need to reflect assumptions about the future directions and intensity of selection. The assumptions and predicted outcomes of those assumptions forecast the future. For example, while there is no doubt that the size of human population will increase in the future, barring some catastrophic events, there is strong disagreement, as expressed in polar fashion by Myers and Simon (1994), about extent of the increase, the point at which it will stabilize, and the effects it may have on humanity as a whole. Here, the assumptions drift toward the pessimism of Myers, for at whatever level the human population eventually stabilizes, one expects that a larger population will have major impact on human lives and the environment. The potential for disaster abounds and can be predicted in many forms, ranging from soil degradation and loss to the accumulation and spread of pollutants to global warming. Whatever one's ethical and practical views of biotic conservation, the future promises continued loss of native flora through conversion of natural ecosystems to agriculture or to other managed systems and an accelerated homogenization of natural ecosystems through the widespread introduction and naturalization of exotic species (Vitousek et al., 1996). While some tracts of natural vegetation will probably be spared, the net expected effect will likely be major extinctions of plants and animals and the germplasm diversity that they represent, thus narrowing the genetic library that serves or has the potential to serve the needs of humanity.

The key to assessing the future, however, is perhaps to be found less in predicting disasters than in coming to terms with what continued humanization of the earth will mean and in formulating strategies to meet human needs in an altered world, a world dominated by increasing globalization of human, capital, and technological resources. Predictably, globalization related to plant use implies increased systems specialization, whether based on polycultures or monocultures, and their wider regionalization in agricultural production, accompanied by stronger definition in the membership and boundaries of primary and secondary plant utilization pools and less reliance on the tertiary pool species, whether cultivated or wild. Furthermore, the future impact of biotechnology, which for all its advances remains in its infancy, could be revolutionary, altering our focus on biodiversity from habitats and species to genes (MacDonald, 1994; Hoffman and Carroll, 1995). Insofar as plant species diversity is concerned, the point at issue is not with the assumptions that are made about the future or what the predicted outcomes of those assumptions may be. Rather, it is in recognition that selection is a dynamic process, which reflects the interplay and relative strengths of each of the factors which influence the outcome. If plant species diversity stands low or cannot be effectively expressed on the scale of selection criteria in agricultural or agroforestry systems, then predictably these values will be reflected in low plant species diversity in the systems. Positive and negative changes in diversity are directly tied to changes in selection values.

The current interest in increasing species diversity in agricultural systems is widespread, much of it engendered by the 1975 National Academy of Sciences publication "Underexploited Tropical Plants with Promising Economic Value" (1975) but is in evidence in more local views, for example, that of Negi (1996) for Uttar Pradesh, India. In effect, there is no shortage of plant species that might be integrated into current agricultural and agroforestry systems, whether to meet existing needs or to meet newly emerging needs, such as biomass conversions for energy and industrial feedstocks. Furthermore, the practical values of increased plant diversity in agricultural systems are known (Coffman and Bates, 1993). Yet, while some species thought to be underutilized are likely to gain greater prominence in agricultural and agroforestry systems in the future, for example, the neem tree (*Azadirachta indica* A. Juss.) (Schmutterer, 1995), and despite the avowed interest in and suggested importance of increasing plant species diversity in agroecosystems, ethnobotanical experience suggests the opposite effect. Historical trends extrapolated to the future indicate that world-wide, plant species diversity in agroecosystems is unlikely to increase significantly and may even decrease, despite seeming local increases in diversity that are outgrowths of introductions from the existing, universal utilization pools. Selection priorities emphasizing functional attributes of plant species and/or their value as sources of goods results in primacy being given to efficiency in terms of labor, capital, and short-term economic return. Under present selection regimes the long-term ecological and economic benefits that might accrue from increased plant species diversity are not expressed or are expressed in ways that lack significant influence.

Increasing species diversity in agroecosystems, including agroforestry systems, is dependent on changing the paradigms of selection by giving greater selective value to diversity. This requires demonstration that increased plant species diversity actually increases the value and sustainability of agroforestry systems, consistent with the value parameters set by society as a whole, a problem related to realistic valuation of natural capital and nature's services (Westman, 1977; Ehrlich and Mooney, 1983; Browder, 1992a; Costanza and Daly, 1992; Repetto, 1992; Myers and Simon, 1994; Daly, 1995; Pimental et al., 1995; Principe, 1996; Costanza et al., 1997; Pimental et al., 1997). Whether or not demonstration of diversity values in agroforestry systems can be effectively made in the current economic arena and thus serve to modify selection parameters is simply unknown at this time, but the problem remains and is a principal challenge facing agroforestry.

While academics forecast the future, much of humanity toils daily for survival. For better or worse, world-wide disparity in rates of population growth and in human, physical, and economic resources assure that variations in agroforestry approaches will persist into the foreseeable future. Among the marginalized subsistence farmers of the tropics and elsewhere, who are sufficiently removed from the dictates of external factors to establish their own selective regimes, the most effective survival strategy is likely to remain the integration of trees and shrubs, both exotic and indigenous at different levels of management intensity, throughout individual farm holdings and the surrounding natural vegetation (Milan and Margraf, 1992). Such variations are among the testing grounds of agroforestry, and should be well studied and documented. As models they may hold promise for the future of agroforestry in economically and agriculturally richer regions of both the tropics and temperate zones. Yet, ultimately, the separation of plant management tracks to create carefully designed mosaics of annual and perennial crops, some undoubtedly in monocultures, integrated with zones of natural vegetation, may represent the most efficient agroecological/agroforestry production strategy, one that is also the most conducive to maintaining desired and defined levels of biodiversity.

7. CONCLUSIONS

Ethnobotanical perspectives, derived from an ever evolving interest in the interrelationships of humans with the plant environment, provide a framework for evaluating issues of critical human concern. One issue is concerned with the degree to which plant species diversity can be preserved or increased in managed agricultural environments, including those of agroforestry. Pattern analysis, an ethnobotanical approach grounded in concepts of systematics, ecology, and human selection within the resource base, provides a framework for understanding the dynamics of plant utilization. The approach demonstrates that plant utilization and the systems in which plants are grown, managed, or collected are expressed in patterns that reflect the wide array of factors basic to human selection, that is, decision making. Changes in patterns of plant utilization depend on changes in the selection factors, including technology, knowledge, and culture, among others. In general, as human needs intensify, selection factors likewise intensify. The results of pattern analysis are descriptive summations of events that have affected plant use and diversity in given contexts in the past and present. Such summations can be used to make predictions concerning the future place of plants in the human context.

When applied to issues of plant species diversity in agroforestry systems, pattern analysis demonstrates that trends in agroforestry are similar to those in other agricultural and managed plant systems. Although agroforestry systems, as expressed in home gardens and cultivated fields, are species rich relative to their monoculture counterparts, they are less diverse than the natural vegetation that they replace. At any given locality, selection leads to a narrowed suite of utilized plant species. These species have ranked value in the system; such rankings may be described in terms of utilization pools. As human interactions with plants have become increasingly global, so too has selection become global, leading to the expectation that world-wide plant species and genetic diversity in agroforestry systems will become increasingly simplified and homogenized in both in total and in functional and/or use categories. However one might view this trend, it is evident that increasing plant species diversity in agroforestry systems is dependent on establishing new selective paradigms that recognize and give values to such increased diversity.

REFERENCES

Akerele, O., Heywood, V., and Synge H., Eds., 1991, *The Conservation of Medicinal Plants*, Cambridge University Press, Cambridge, U.K.

Alcorn, J. B., 1984, "Development policy, forests and peasant farms: reflections on Huastec-managed forests contribution to commercial production and resource conservation," *Eco. Bot.*, 38, 389.

Alcorn, J. B., 1989, "Process as a resource: The traditional agricultural ideology of Bora and Huastec resource management and its implications for research," *Adv. Eco. Bot.*, 7, 63.

Alcorn, J. B., 1995, "The scope and aim of ethnobotany in a developing world," in *Ethnobotany: Evolution of a Discipline*, Schultes, R. E. and von Reis, S., Eds., Dioscorides Press, Portland, OR, 23.

Alexiades, M., 1996, "Selected Guidelines for Ethnobotanical Research: A Field Manual," *Advances in Economic Botany*, Vol. 10, New York Botanical Garden, Bronx, NY.

Altieri, M. A. and Merrick, L. C., 1987, "*In situ* conservation of crop genetic resources through maintenance of traditional farming system," *Eco. Bot.*, 41, 86.

Altieri, M. A. and Montecinos, C., 1993, "Conserving crop genetic resources in Latin America through farmers' participation," in *Perspectives on Biodiversity: Case Studies of Genetic Resource Conservation and Development*, Potter, C. S., Cohen, J. I., and Janczewski, D., Eds., AAAS Press, Washington, DC, 45.

Anderson, A. B., Ed., 1990, *Alternatives to Deforestation: Steps Toward Sustainable Use of the Amazon Rain Forest*, Columbia University Press, New York.

Anderson, A. B., May, P. H., and Balick, M. J., 1991, *The Subsidy from Nature: Palm Forests, Peasantry, and Development on* an *Amazon Frontier*, Columbia University Press, New York.

Anderson, A. B. and Posey, D. A., 1989, "Management of tropical scrub savanna by the Gorotire Kayapó of Brazil," *Adv. Eco. Bot.* 7, 15.

1895, "Some New Ideas," *Philadelphia Evening Telegraph*, Dec. 5.

Avevedo-Rodriguez, P., 1990, "The occurrence of pesticides and stupefactants in the plant kingdom," *Adv. Eco. Bot.*, 8, 1.

Baenziger, P. S., Kleese, R. A., and Barnes, R. F., 1993, *Intellectual Property Rights: Protection of Plant Materials*, Crop Science Society of America, Special Publication 21, Madison, WI.

Balick, M. J., Ed., 1988. "The Palm — Tree of Life: Biology, Utilization and Conservation," *Advances in Economic Botany*, Vol. 6, The New York Botanical Garden, Bronx, NY.

Balick, M. J. and Beck, H. T., Eds., 1990, *Useful Palms of the World: A Synoptic Bibliography*, Columbia University Press, New York.

Balick, M. J., Elisabetsky, E., and Laird, S., Eds., 1996, *Medicinal Resources of the Tropical Forest*, Columbia University Press, New York.

Balick, M. J. and Mendelsohn, R., 1992, "Assessing the economic value of traditional medicines from tropical rain forests," *Cons. Biol.*, 6, 128.

Bates, D. M., 1985, "Plant utilization: patterns and prospects," *Eco. Bot.*, 39, 241.

Bates, D. M., 1988a, "The potential for new crops," *Angewandte Botanik*, 62, 31.

Bates, D. M., 1988b, "Utilization pools: A framework for comparing and evaluating the economic importance of palms," *Advances in Eco. Bot.*, 6, 56.

Begossi, A., 1996, "Use of ecological methods in ethnobotany: Diversity indices," *Eco. Bot.*, 50, 280.

Bene, J. G., Beall, H. W., and Cote, A., 1977, *Food and People: Land Management in the Tropics*, IDRC, Ottawa, ON, Canada.

Boom, B. M., 1990, "Useful plants of the Panare Indians of Venezuelan Guyana," *Adv. Eco. Bot.*, 8, 57.

Boom, B. M., 1996, "Ethnobotany of the Chacobo Indians, Beni, Bolivia," Second Edition, *Adv. Eco. Bot.*, 4, 1.

Bray, R. A., 1995, "*Leucaena, Leucaena leucocephala* (Leguminosae-Mimosoideae)," in *Crop Evolution*, Second Edition, Smartt, J., and Simmonds, N. W., Eds., Longman Scientific & Technical, Burnt Mill, 274.

Browder, J. O., 1992a, "Social and economic constraints on the development of extractive reserves in Amazon Rain Forests," *Adv. Eco. Bot.*, 9, 93.

Browder, J. O., 1992b, "The limits of extractivism," *Biosci.*, 42, 174.

Brush, S. B., 1986, "Genetic diversity and conservation in traditional farming systems," *J. Ethnobiol.*, 6, 151.

Brush, S. B., 1993, "Indigenous knowledge of biological resources and intellectual property rights: The role of anthropology," *American Anthropologist*, 95, 653.

Coffman, W. R. and Bates, D. M., 1993, "History of crop improvement in sustainable agriculture," in *Crop Improvement for Sustainable Agriculture*, Callaway, M. B. and Francis, C. A., Eds., University of Nebraska Press, Lincoln, NE, 19.

Conway, G. R., 1987, "The properties of agroecosystems," *Ag. Syst.*, 24, 95.

Costanza, R. and Daly, H. E., 1992, "Natural capital and sustainable development," *Cons. Biol.*, 6, 37.

Costanza, R., d'Arge, R., de Groot, R., Farber, S., Grasso, M., Hannon, B., Limburg, K., Naeem, S., O'Neill, R. V., Paruelo, J., Raskin, R. G., Sutton, P., and van den Belt, M., 1997, "The value of the world's ecosystem services and natural capital," *Nature*, 387, 253.

Cox, P. A., 1990, "Samoan ethnopharmacology: plants and traditional medicine," *Economic and Medicinal Plant Research*, 4, 123.

Cox, P. A., 1993, "Saving the ethnopharmacological heritage of Samoa," *J. Ethnopharmacology*, 39, 181.
Daly, H. E., 1995, "Reply to Mark Sagoff's 'Carrying capacity and ecological economics'," *Biosci.*, 45, 621.
Davis, E. W., 1995, "Ethnobotany: An old practice, a new discipline," in *Ethnobotany: Evolution of a Discipline*, Schultes, R. E. and von Reis, S., Eds., Dioscorides Press, Portland, OR, 40.
Denevan, W. M. and Padoch, C., Eds., 1988. "Swidden-Fallow Agroforestry in the Peruvian Amazon," *Advances in Economic Botany*, Vol. 5, The New York Botanical Garden, Bronx, NY.
Ehrlich, P. R. and Mooney, H. A., 1983, "Extinction, substitution, and ecosystem services," *Biosci.*, 33, 248.
Etkin, N. L., 1990, "Ethnopharmacology: Biological and behavioral perspectives in the study of indigenous medicines," in *Medical Anthropology: Contemporary Theory and Method*, Johnson, T. M. and Sargent, C., Eds., Praeger, New York, 149.
Everett, Y., 1995, "The *kitul* palm: ethnobotany of *Caryota urens* L. in highland Sri Lanka," *J. Ethnobiol.*, 15, 161.
Farooquee, N. A. and Saxena, K. G., 1996, "Conservation and utilization of medicinal plants in high hills of the central Himalayas," *Environmental Conservation*, 23, 75.
Feil, J. P., 1996, "Fruit production of *Attalea colenda* (Arecaceae) in coastal Ecuador — an alternative oil resource," *Eco. Bot.*, 50, 300.
Fischman, J., 1995, "Painted puzzles line the walls of an ancient cave," *Science*, 267, 614.
Ford, R. I., 1978. "Ethnobotany: Historical diversity and synthesis," in *The Nature and Status of Ethnobotany*, Ford, R. I., Ed., Anthropological Papers, No. 67. Museum of Anthropology, University of Michigan, Ann Arbor, MI, 33.
Fox, J. J., 1977. *Harvest of the Palm*, Harvard University Press, Cambridge, MA.
Godoy, R. A. and Bawa, K. S., 1993, "The economic value and sustainable harvest of plants and animals from the tropical forest: assumptions, hypothesis, and methods," *Eco. Bot.*, 47, 215.
Godoy, R. A., Lubowski, R., and Markandya, A., 1993, "A method for economic valuation of non-timber tropical forest products," *Eco. Bot.*, 47, 220.
Goodland, R., 1995, "The concept of environmental sustainability," *Annual Review of Ecology and Systematics*, 26, 1.
Gragson, T. L., 1995, "Pumé exploitation of *Mauritia flexuosa* (Palmae) in the llanos of Venezuela," *J. Ethnobiol.*, 15, 177.
Gray, A., 1990, "Indigenous peoples and the marketing of the rainforest," *The Ecologist*, 20, 223.
Harlan, J. R., 1986, "Lettuce and the sycamore: sex and romance in ancient Egypt," *Eco. Bot.*, 40, 4.
Harshberger, J. W., 1896, "The purposes of ethno-botany," *Botanical Gazette*, 21, 146.
Hartshorn, G. S., 1995, "Ecological basis for sustainable development in tropical forests," *Annual Review of Ecology and Systematics*, 26, 155.
Hatfield, J. L. and Karlen, D. L., 1994. *Sustainable Agricultural Systems*, Lewis Publishers, CRC Press, Boca Raton, FL.
Hegde, R., Suryaprakash, S., Achoth, L., and Bawa, K. S., 1996, "Extraction of non-timber forest products in the forests of Biligiri Rangan Hills, India. 1. Contributions to rural income," *Eco. Bot.*, 50, 243.
Heiser, C. B., 1990, *Seed to Civilization: the Story of Food*, 3rd edition, Harvard University Press, Cambridge, MA.
Hoffman, C. A. and Carroll, C. R., 1995, "Can we sustain the biological basis of agriculture?" *Annual Review of Ecology and Systematics*, 26, 69.
Hough, W., 1898, "Houses built of fossil trees," *Plant World*, 1, 35.
Humphries, C. J., Williams, P. H., and Vane-Wright, R. I., 1995, "Measuring biodiversity value for conservation," *Annual Review of Ecology and Systematics*, 26, 93.
Inglis, J. T., Ed., 1993, *Traditional Ecological Knowledge: Concepts and Cases*, International Program on Traditional Ecological Knowledge and International Development Centre, Canadian Museum of Nature, Ottawa, ON, Canada.
Jensen, O. H. and Balslev, H., 1995, "Ethnobotany of the fiber palm *Astrocaryum chambira* (Arecaceae) in Amazonian Ecuador," *Eco. Bot.*, 49, 309.
Johns, T., 1990, *With Bitter Herbs They Shall Eat It: Chemical Ecology and the Origins of Human Diet and Medicine*, University of Arizona Press, Tucson, AZ.
Johns, T., Kokwaro, J. O., and Kimanani, E. K., 1990, "Herbal remedies of the Luo of Siaya District, Kenya: Establishing quantitative criteria for consensus," *Eco. Bot.*, 44, 369.

Johnson, D., 1988, "Palms as multipurpose cash and subsistence tree crops," in *Multipurpose tree species for small-farm use*, Withington, D., MacDicken, K. G., Sastry, C. B., and Adamas, N. R., Eds., Winrock International, Arlington, VA.

Jones, V. H., 1941, "The scope of ethnobotany," *Chronica Botanica,* 6, 219.

Kainer, K. A. and Duryea, M. L., 1992, "Tapping women's knowledge: Plant resource use in extractive reserves, Acre, Brazil," *Eco. Bot.*, 46, 408.

Kaoneka, A. R. S. and Solberg, B., 1997, "Analysis of deforestation and economically sustainable farming systems under pressure of population growth and income constraints at the village level in Tanzania," *Agri. Ecosys. Env.*, 62, 59.

Kaufmann, M. R., Graham, R. T., Boyce, D. A., Jr., Moir, W. H., Perry, L., Reynolds, R. T., Bassett, R. L., Mehlhop, P., Edsminster, C. B., Block, W. H., and Corn, P. S., 1994. *An Ecological Basis for Ecosystem Management*, USDA Forest Service General Technical Bulletin RM246, Rocky Mountain Forest and Range Experiment Station, Fort Collins, CO.

Kinnaird, M. F., 1992. "Competition for a forest palm: Use of *Phoenix reclinata* by human and nonhuman primates," *Cons. Biol.*, 6, 101.

Kloppenburg, J. R., Jr., and Balick, M. J., 1996, "Property rights and genetic resources: A framework for analysis," in *Medicinal Resources of the Tropical Forest*, Balick, M. J., Elisabetsky, E., and Laird, S., Eds., Columbia University Press, New York, 174.

Krattiger, A. F., McNeely, J. A., Lesser, W. H., Miller, K. R., St. Hill, Y., and Senanayake, R., Eds., 1994. *Widening Perspectives on Biodiversity,* IUCN, Gland, Switzerland, and International Academy of the Environment, Geneva, Switzerland.

Kruseman, G., Ruben, R., Kuyvenhoven, A., Hengsdijk, H., and van Keulen, H., 1996, "Analytical framework for disentangling the concept of sustainable land use," *Ag. Syst.*, 50, 191.

Lassoie, J. P., 1996, *Planning for the Cornell Program for the Study and Application of Agroforestry (CPSAA)*, unpublished report, Cornell University, Ithaca, NY.

Lawrence, D. C., 1996, "Trade-offs between rubber production and maintenance of diversity: the structure of rubber gardens in West Kalimantan, Indonesia," *Agro. Syst.*, 34, 83.

Lee, R. B., 1986, "What hunters do for a living, or, how to make out on scarce resources," in *Man the Hunter*, Lee, R. B. and DeVore, I., Eds., Aldine, Chicago, IL, 30.

MacDonald, J. F., 1994, *Agricultural Biotechnology and the Public Good*, National Agricultural Biotechnology Council Report 6, Ithaca, NY.

MacDonald, J. F., Ed., 1995, *Genes for the Future: Discovery, Ownership, Access*, National Agricultural Biotechnology Council Report 7, Ithaca, NY.

Martin, G. J., 1994, "Ethnoecology and postmodernism: Friends or foes," *J. Ethnobiology,* 14, 265.

Martin, G. J., 1995, *Ethnobotany: A Methods Manual*, Chapman & Hall, London.

McNeely, J. A., Ed., 1995, *Biodiversity Conservation in the Asia and Pacific Region: Constraints and Opportunities*, Proceedings of a Regional Conference, June 1994, Asian Development Bank and The World Conservation Union, Manila, Philippines.

Michon, G., and de Foresta, H., 1998, "Agroforests: incorporating a forest vision in agroforestry," in *Agroforestry in Sustainable Agroforestry Systems*, Buck, L. E., Lassoie, L. E., and Fernandes, E. C. F., Eds., CRC Press Lewis Publishers, Boca Raton, FL.

Milan, P. P. and Margraf, J., 1992, "Land-use management towards increased biodiversity of ecosystems," in *Proceedings, Workshop on Indigenous Knowledge and Sustainable Development*, IIRR, Silang, Cavite, Philippines, 1.

Moerman, D. E., 1991, "The medicinal flora of native North America: an analysis," *J. Ethnopharmacology,* 31, 1.

Myers, N., 1997, "Consumption: challenge to sustainable development," *Science*, 276, 53.

Myers, N. and Simon, J., 1994, *Scarcity or Abundance: A Debate on the Environment*, Norton, New York.

National Academy of Sciences, 1975. *Underexploited Tropical Plants with Promising Economic Value*, National Academy of Sciences, Washington, DC.

National Academy of Sciences, 1977, *Leucaena: Promising Forage and Tree Crop for the Tropics*, National Academy of Sciences, Washington, DC.

Negi, K. S., 1996, "Wild food plants of Uttar Pradesh hill, India," *Plant Genetic Resources Newsletter*, 108, 33.

Nepstad, D. C. and Schwartzman, S., Eds., 1992, *Non-Timber Products from Tropical Forests*, Advances in *Eco. Bot.*, Vol. 9, New York Botanical Garden, Bronx, NY.

Orlove, B. S., 1980, "Ecological anthropology," *Annual Review of Anthropology,* 9, 235.

Owusu, J. G. K., Quashie-Sam, S. J., Nkyi, K. A., and Oppong, S. K., 1993, *Indigenous African Food Crops and Useful Plants, Their Preparations for Food and Home Gardens in Ghana*, UNU/INRA Natural Resources Survey Series No. B1, United Nations University/Institute for Natural Resources in Africa.

Padoch, C., 1988a, "The economic importance and marketing of forest and fallow products in the Iquitos region," *Adv. Eco. Bot.*, 5, 74.

Padoch, C., 1988b, "Aguaje (*Mauritia flexuosa*) in the Economy of Iquitos, Peru," *Adv. Eco. Bot.*, 6, 214.

Padoch, C. and Peters, C., 1993, "Managed forest gardens in West Kalimantan, Indonesia," in *Perspectives on Biodiversity: Case Studies of Genetic Resource Conservation and Development*, Potter, C. S., Cohen, J. I., and Janczewski, D., Eds., AAAS Press, Washington, DC, 167.

Palmer, E., 1871, "Food products of the North American Indians", *U.S. Commissioner of Agriculture Report*, 1870, 404.

Palmer, E., 1878, "Plants used by the Indians of the United States," *American Naturalist* 12, 593, 646.

Parker, E., 1992, "Forest islands and Kayapó resource management in Amazonia: A reappraisal of the *apêtê*," *American Anthropologist,* 94, 406.

Pedersen, H. B. and Balslev, H., 1990, "Ecuadorean palms for agroforestry," *AAU Reports*, 23, 1.

Phillips, O. and Gentry, A. H., 1993, "The useful plants of Tambopata, Peru: I statistical hypotheses tests with a new quantitative technique," *Eco. Bot.*, 47, 15.

Phillips, O. and Gentry, A. H., 1993, "The useful plants of Tambopata, Peru: II additional hypotheses testing quantitative ethnobotany," *Eco. Bot.,* 47, 33.

Phillips, O., Reynel, C., Wilkin, P., and Galvez-Durand, C. B., 1994, "Quantitative ethnobotany and Amazonian conservation," *Cons. Biol.*, 8, 225.

Pimentel, D., Harvey, C., Resosudarmo, P., Sinclair, K., Kurz, D., McNair, M., Crist, S, Shpritz, L., Fitton, L., Saffouri, R., and Blair, R. D., 1995, "Environmental and economic costs of soil erosion and conservation benefits," *Science*, 267, 1117.

Pimentel, D., Houser, J., Preiss, E., White, O., Fanf, H., Mesnick, L., Barsky, T. Tariche, S., Schreck, J., and Alpert, S., 1997, "Water resources: Agriculture, the environment, and society," *Biosci.*, 47(2), 97.

Pitman, W. D., 1995, "Ameliorating effects of alternative agriculture," in *Soil Amendments, Impacts on Biotic Systems*, Rechcigl, J. E., Ed., Lewis Publishers, CRC Press, Boca Raton, FL, 215.

Posey, D. A., 1984, "A preliminary report on diversified management of tropical forest by the Kayapó Indians of the Brazilian Amazon," *Adv. Eco. Bot.,* 1, 112.

Posey, D. A., 1990, "Intellectual property rights: what is the position of ethnobiology?" *J. Ethnopharmacology,* 10, 93.

Posey, D. A., 1992, "A reply to Parker," *American Anthropologist*, 94, 441.

Posey, D. A. and Balée, W., 1989, *Resource Management in Amazonia: Indigenous and Folk Strategies*, Advances in *Eco. Bot.*, Vol. 7, The New York Botanical Garden, Bronx, NY.

Powers, S., 1875, "Aboriginal botany," *California Academy of Sciences Proceedings*, 5, 373.

Prance, G. T., 1995, "Ethnobotany today and in the future," in *Ethnobotany: Evolution of a Discipline*, Schultes, R. E. and von Reis, S., Eds., Dioscorides Press, Portland, OR, 60.

Prance, G. T., Balée, W., Boom, B. M., and Carneiro, R. L., 1987, "Quantitative ethnobotany and the case for conservation in Amazonia," *Cons. Biol.*, 1, 296.

Pretty, J. N., 1995, *Regenerating Agriculture*, Joseph Henry Press, Washington, DC.

Principe, P. P., 1996, "Monetizing the pharmacological benefits of plants," in *Medicinal Resources of the Tropical Forest*, Balick, M. J., Elisabetsky, E., and Laird, S., Eds., Columbia University Press, New York, 191.

Raintree, J. B., 1987, "The state of the art of agroforestry diagnosis and design," *Agro. Syst.*, 5, 219.

Reaka-Kudla, M. L., Wilson, D. E., and Wilson, E. O., Eds., 1997. *Biodiversity II: Understanding and Protecting Our Biological Resources,* Joseph Henry Press, Washington, DC.

Rehm, S. and G. Espig, G., 1991, *The Cultivated Plants of the Tropics and Subtropics: Cultivation, Economic Value and Utilization,* Verlag Josef Margraf, Weikersheim.

Repetto, R., 1992, "Accounting for environmental assets," *Scientific American*, 266(6), 94.

Riddle, J. M., Estes, J. W., and Russell, J. C., 1994, "Ever since Eve: Birth control in the ancient world," *Archaeology*, March April 1994, 29.

Röling, N., 1994, "Platforms for decision-making about ecosystems," in *The Future of the Land: Mobilising and Integrating Knowledge for Land Use Options*, Fresco, L. O., Stroosnijder, L., Bouma, J., and van Keulen, H., Eds., John Wiley, New York, 385.

Rolston, H., III, 1988, *Environmental Ethics: Duties to and Values in the Natural World*, Temple University Press, Philadelphia, PA.

Ruddle, K., Johnson, D. V., Townsend, P. K., and Rees, J. D., 1978, *Palm Sago: A Tropical Starch from Marginal Lands*, University Hawaii Press, Honolulu, HI.

Sagoff, M., 1995, "Carrying capacity and ecological economics," *Biosci.*, 45, 610.

Sanchez, P. A., 1995, "Science in agroforestry," *Agro. Syst.*, 30, 5.

Schmutterer, H., Ed., 1995, *The Neem Tree: Source of Unique Natural Products for Integrated Pest Management, Medicine, Industry and Other Purposes,* VCH, Weinheim.

Schultes, R. E. and von Reis, S., 1995, "Preface," in *Ethnobotany: Evolution of a Discipline*, Schultes, R. E. and von Reis, S., Eds., Dioscorides Press, Portland, OR, 12.

Smith, N. J. H., Williams, J. T., Plucknett, D. L., and Talbot, J. P., 1992. *Tropical Forests and Their Crops*, Comstock Publishing Associates, Cornell University Press, Ithaca, NY.

Smole, W. J., 1976. *The Yanoama Indians: A Cultural Geography*, University of Texas Press, Austin, TX.

Stanton, W. R. and Flach, M., Eds., 1980. *Sago: The Equatorial Swamp as a Natural Resource,* Proceeding of the Second International Sago Symposium, Martinus Nijhoff, The Hague.

Stephens, D. W. and Krebs, J. R., 1986. *Foraging Theory*, Princeton University Press, Princeton, NJ.

Tanaka, J., 1980. *The San: Hunter-Gatherers of the Kalahari*, University of Tokyo Press, Tokyo.

Thiollay, J.-M., 1995, "The role of traditional agroforests in the conservation of rain forest bird diversity in Sumatra," *Cons. Biol.*,9, 335.

Thurston, H. D., Smith, M., Abawi, G., and Kearl, S., Eds., 1994, *Slash/Mulch: How Farmers Use It and What Researchers Know About It*, CATIE and CIIFAD, Cornell University, Ithaca, NY.

Uhl, N. W. and Dransfield, J. D., 1987, *Genera Palmarum, A Classification of Palms Based on the Work of Harold E. Moore, Jr.*, L. H. Bailey Hortorium and The International Palm Society, Lawrence, KS.

Vandermeer, J., 1995, "The ecological basis of alternative agriculture," *Annual Review of Ecology and Systematics,* 26, 201.

Vincent, J. R. and Panayotou, T., 1997, "[Consumption: challenge to sustainable development]...or distraction," *Science*, 276, 53.

Vitousek, P. M., D'Antonio, C. M., Loope, L. L., and Westbrooks, R., 1996, "Biological invasions as global environmental change," *American Scientist*, 84, 468.

Vogel, V., 1970. *American Indian Medicine*, University of Oklahoma Press, Norman, OK.

Wasson, R. G., 1972, "What was the soma of the Aryans?" in *Flesh* of *the Gods*, Furst, P. T., Ed., Praeger, New York, 201.

Wells, E., 1969, *Hatshepsut*, Doubleday, Garden City, NY.

Westman, W. E., 1977, "How much are nature's services worth?" *Science* 197, 960.

Winterhalder, B. and Smith, E. A., Eds., 1982. *Hunter-gatherer Foraging Strategies: Ethnographic and Archeological Analyses*, University of Chicago Press, Chicago, IL.

Zohary, D. and Hopf, M., 1993, *Domestication of Plants in the Old World*, 2nd edition, Oxford University Press, Oxford, U.K.

Zohary, D. and Spiegel-Roy, P., 1975. "Beginning of fruit growing in the Old World," *Science*, 187, 319.

16 Sustainable Mulch-Based Cropping Systems with Trees

Donald C. L. Kass, H. David Thurston, and Kenneth Schlather

1. INTRODUCTION

Raintree and Warner (1986) have presented a scheme by which increasing pressure on land and increasing availability of labor per unit of land results in the development of different agroforestry systems. Systems in which trees occupy the land permanently are considered to represent the greatest land use intensity but relatively low labor intensity. These systems, which Raintree and Warner (1986) call tree-based economies, home gardens, and other tree based alternatives, are considered to have evolved from a forest fallow, either directly, or by passing through stages in which the fallow was gradually eliminated. Both alley cropping and the enriched fallow are seen as steps in which trees are restored to systems which have become exhausted by the elimination of tree fallows. In 1986, Raintree and Warner considered that alley farming was less labor intensive than annual or multiple cropping and integral taungya, a position that recent experience has not born out. The scheme of Raintree and Warner (1986) indicates that the elimination of trees will eventually lead to their restoration, a sort of Paradise Lost, Paradise Regained scenario which might seem overly Gaian (Joseph, 1990) or optimistic. Still, the resemblance of many tree-based systems to the natural forest has been noted by many authors (Nair, 1993; Sanchez, 1995).

Despite its limitations, we will follow the Raintree and Warner scheme in the organization of this chapter. We will first discuss cropping systems within a forest context and then the systems that have evolved as population pressure and availability of labor led to more intensive farming systems. We will give some attention to coexistence with the original forest or other forms of relatively undisturbed vegetation because hunting and gathering continues to be an important economic activity in many areas and coexists with more intensive land uses in many places in the Americas (Grigg, 1974). There are probably very few areas where the natural vegetation has not

1-56670-294-1/99/$0.00+$.50

to some degree been altered by human activity (Roosevelt *et al.*, 1996). The development of farming in the Americas does not seem to have occurred directly in response to population pressure, although Harlan (1992) feels that the disappearance of game was a significant factor. Recent evidence has shown that people in the Americas might have cultivated squashes at least while their economy was principally based on hunting and gathering (Roush, 1997). The combination of hunting and gathering with more intensive cropping systems in many parts of the world may contribute to the sustainablity of these systems since they reduce the amount of food and income these systems must produce.

Finally, we will discuss the tree-based systems which appear to offer the greatest possibilities for sustainable land use in the tropics.

2. SOME STEPS IN THE DEVELOPMENT OF MULCH-BASED CROPPING SYSTEMS WITH TREES

2.1 Trees and Mulch in Uncropped Areas

Litterfall has long been recognized as a significant process in the maintenance of natural forests and other vegetation formations. Falling leaves and other senescent plant parts not only recycle nutrients but also perform most of the functions of mulch in agricultural systems: they conserve moisture, prevent the growth of potential competitive species, reduce soil temperatures, and provide a substrate for organisms which contribute to their decomposition and release of nutrients. As these organisms often provide food to animals which are hunted, it can be argued that mulch plays a significant role in hunting and gathering systems even though we might not go so far as to call these systems mulch-based.

Harlan (1992) entitles the chapter in which he discusses hunting and gathering "The Golden Age" and provides considerable evidence that not only is the system sustainable but that it provides a much higher standard of living to its practitioners than agricultural systems. Harlan uses a study of the Kung bushman of Botswana to show that not only do these people have a higher standard of living, but they also devote less time to obtaining food than do most agricultural peoples.

Harlan (1992) also points out that Australian aborigines did not farm before the arrival of the Europeans but modified the vegetation in many ways, including burning, to increase the proportion of useful plants available. Apparently, aborigines even diverted water to flood forests in the dry season (Harlan, 1992). As Harlan states, "Perhaps the key difference here between foraging and farming is that no native Australian plant was actually domesticated, otherwise hunter-gatherers do about everything farmers do."

A mixture of hunting and gathering and actual farming was practiced by most North and South American Indian tribes. Some commercially important foodstuffs in the Americas continue to be produced almost exclusively by gathering: wild rice (*Zizania aquatica*), Brazil nuts (*Berthollea excelsa*), and hearts of palm (*Euterpe* spp.). Among the genera gathered by native Americans is *Canavalia*, which was later domesticated as a cover crop (Harlan, 1992).

So, without actually tilling the soil, we have a sustainable, mulch-based system with trees. Various methods are used to keep populations from exceeding food supply. "There is evidence that the diet of gathering peoples was better than that of cultivators, that starvation was rare, that their health status was generally superior, that there was a lower incidence of chronic disease and not nearly so many cavities in their teeth" (Harlan, 1992). The system maintained the human race for more than a million years, a hundred times longer than agriculture has yet been shown able to do.

Some compulsion was probably required to force people to take up farming and provide food to non-productive sectors of the population. In the Judeo-Christian tradition, the need to farm is equated with the expulsion from the Garden of Eden. In all our fairy tales, the woods is presented as a dangerous place, where evil occurs. No wonder most South American hunter-gatherers are characterized as "dropouts from farming" (Grigg, 1974). It may be questioned whether humans were actually the first agriculturists. Lanner (1985) describes how European Jays (*Garrulus glandarius*) select both seeds and sites for burial of acorns. They later pull on the sprouted oaks, consuming those in which the taproot is not fully developed, leaving those which they cannot pull up due to good taproot development. Lanner (1985) also describes how squirrels plant nuts. "Typically, they dig a hole, place the nut in it, cover it with soil, compact the soil, and place some grass or litter on top, probably as camouflage." So mulching was probably practiced long before humans arrived on the scene.

2.2 Slash and Burn Systems: Are They Mulch Based?

With the exception of slash-mulch (Thurston, 1997), homegardens, and shaded plantation systems, most cropping systems are considered to have passed through a slash and burn stage sometime in their development, and slash and burn continues to be important in areas where population densities are sufficiently low for a long enough fallow period to restore fertility and soil properties. Well after the development of more intensive agriculture in Europe, there were areas of low fertility that were only cropped at wide intervals and allowed to revert to forest for periods of twenty or more years after which they were burned and cropped for short periods (Grigg, 1974). Given enough land on which to practice the system, human populations can be maintained at satisfactory levels. However, since the cut vegetation is generally burned, slash and burn agriculture may not qualify as a mulch-based system, unless the cut vegetation and the ash may be considered mulch materials. In a recent review of changes in nutrient levels following burning and subsequent cropping cycles, Juo and Manu (1996) concluded that initial increases in soil pH and levels of nutrients are generally short-lived and decrease rapidly in subsequent cropping cycles; depending however, on soil properties, the frequency of cropping, and the amount of time in which the soil is left bare.

They also feel that such changes in soil properties are to some degree illusory since they represent a movement of nutrients from the above ground biomass to the soil. On the other hand, they do not consider the possibility that a population of species which require and accumulate higher levels of nutrients can be established in the period that the ash remains on the soil surface. Juo and Manu (1996) find a smaller decline in nutrients when pastures and continuous cropping systems using fertilizers were established following the burn although these systems, which are outside the scope of this chapter, eventually led to a decline in soil P availability and soil physical properties respectively. Juo and Manu (1996) conclude that maintenance of the total nutrient stock under shifting cultivation is possible only when the size of the cultivated plot is sufficiently small so that reestablishment of native species is rapid following cropping, the cropping period is less than two years, and length of fallow is sufficiently long.

Several authors have shown recently how farmers have manipulated slash and burn systems to make them more productive. Fruit and other trees are often planted in the cropping period which persist into the fallow when they continue to be harvested (Padoch and De Jong, 1987). The persistence of unburned trees in the fallow will affect the activities of animals (Unruh, 1990). Dufour (1990) has shown that farmers often plant species in the cropping period to attract animals during the fallow. Thus, the biodiversity and productivity of natural forests can be improved by shifting cultivation. Similar benefits to the well-being of the practitioners would contribute to the sustainability of the altered systems.

2.3 Slash and Rot Systems: Are They with Trees?

In very humid areas, the vegetation is generally not burned. Seeds are sown into the slashed vegetation which serves as a mulch. In many areas of the tropics, this system is practiced to produce beans in areas where the woody vegetation has been removed. Recent modifications to include trees in the *frijol tapado* system (Kettler, 1996/1997) will be discussed in the section on alley farming. Several of the systems described by Thurston (1997) include trees. Most of the slash/mulch systems of the Amazon include trees, either left in the original vegetation or planted by the farmers. These systems are generally considered to be sustainable because environmental destruction is minimal. The systems of the Pacific coast of South America would appear to be somewhat intermediate between the *tapado* systems of Central America and the slash/mulch systems of the Amazon with respect to the presence of trees.

Other slash/mulch systems which include trees described by Thurston are those of Cameroon, where the alang (*Crossocephalum mannii*) and Fewim (*Albizia* sp.) trees are left in the vegetation because they are considered to improve the soil. In the Tanzania sunnhemp system, *Crotolaria ochroleuca* is sown between bananas, coconuts, and citrus and cut to provide a mulch. The slash mulch systems of Indonesia include the planting of fruit trees and are considered sustainable but are generally being replaced as part of a governmental development policy (Thurston, 1997). In Papua New Guinea, not only are fruit trees planted in the slash mulch systems but the felling of the trees in the original vegetation is delayed until after taro and bananas are planted. In this way, much of the nutrient loss associated with slash and burn systems is avoided. Thurston (1997) also mentions a coppicing system practiced in Sierra Leone where trees are only gradually removed from the cropped area and some of the larger trees left. The presence of trees through the cropping period is supposed to reduce nutrient loss and aid in the reestablishment of forest vegetation at the end of the cropping period.

2.4 Managed Fallows: Are They Sustainable or Mulch Based?

Managing fallows is generally considered to be a response to the increased pressure placed on the fallow period because of land shortages, caused either by increased population or by some other demand for greater production. Farmers shift fields from crop production back into fallow in part for nutrient replenishment but primarily to reduce weed pressure (de Rouw, 1995). DeRouw (1995) describes the weed break function of fallows as a two stage process which necessarily requires maintaining shade on the fallow field for long time spans. Hence, shortened fallow periods ultimately lead to system instability unless the "cropped" portion of the cycle contains some form of shade or other way of maintaining relatively weed-free conditions. Very often this would mean trees that produce products of value that grow throughout the cropped cycle or are relayed into the crop very early in the crop cycle. In fact, "managed fallows" applies to nearly all of the systems described in the previous section, since in nearly all of them the fallow is manipulated in some fashion, either by planting or favoring the growth of valuable species or by allowing some desirable fallow species to remain at fallow clearing.

Some form of fallow for grazing animals was probably integrated with farming in the Middle East at least 6,000 years ago (Grigg, 1974). By Roman times, fallow areas for grazing animals were already scarce and animals were frequently stall fed, principally with tree fodders, and the manure applied to crops. *Medicago arborea* was intercropped with garlic and onions, lopped periodically, and fed to animals. Woodlands were managed for pig and fodder production, but there is no record of whether these wooded areas were rotated with cropped areas (Robinson, 1985). These manipulations did not include trees or mulches, but the previously mentioned Indonesian fruit forests contained both. Intensification of agriculture in the developed world generally involved a disappearance of the fallow with the nutrient restoring and weed control function of the fallow being taken first by animal manures and tillage and, after World War II, by chemical fertilizers and herbicides (Grigg, 1985).

There are several indigenous cropping systems in the Americas where an almost monospecific stand of trees is allowed to establish itself between cropping periods. Because of seed dormancy, the seeds of these trees germinate in great numbers following a burn. The extent of human intervention in the establishment of such "oligarchic forests" may vary and is not clear in many cases (Peters *et al.*, 1989). During the cropping period which follows, selective weedings ensure that these trees will dominate in the fallow vegetation. Thus, undesirable species are gradually eliminated. The best known and documented of these systems is the bracatinga (*Mimosa scabrella*) fallow of southeastern Brazil but similar systems, using *Mimosa tenuiflora* and *Senna guatemalensis* have been reported in Central America. An unburned variant of the system using *Lippia torresii* has been described in Costa Rica (Kass *et al.*, 1993). There might be some question as to what degree these systems are mulch-based since the mulch component does not really coexist with the crop except for the *Lippia torresii* system. In the bracatinga system, Graça *et al.* (1986) say that maize and bean yields can be maintained at 1600 kg ha^{-1} and 330 kg ha^{-1} respectively, even though cropping is only realized for one year in an eight year rotation. Financial analyses indicated a good return in comparison to more intensive systems and profits could be further increased by raising bees which feed on the Mimosa nectar (Graça *et al.*, 1986). Over a 21-year period, the maize-bean-bee-Mimosa system was shown to be more profitable under a wide range of conditions than planting pine or Eucalyptus trees, although a plantation of erva-mate (*Ilex paraguensis*) might be equally profitable. The study of Graça et al. (1986) is probably the best evidence for sustainability of a managed fallow system.

Farmers in Central America are beginning to plant timber and pulp trees in their cropped fields to serve as a long-term cash crop while fulfilling the same functions of fallow. Large increases in income are reported by Montagnini and Mendelsohn (1997) for native timber species planted in degraded pastures in the humid tropical lowlands of Costa Rica. Annual species like maize and cassava are often planted in the first years of tree establishment, although they are not always mulched. Improving fallows by planting trees for firewood is a traditional practice in some parts of Southeast Asia. *Leucaena leucocephala* stumps that are coppiced regularly throughout the cropping cycles are continuously cut back until just before the initiation of the fallow period in a system called Amarasi in East Nusa Tenggara, Indonesia. Farmers in Cebu province in the Philippines plant *L. leucocephala* as well as other species for firewood production during the fallow period and erosion control during the cropping phase (Bensel, 1995). In another Philippine province (Iloilo), one of the authors (Schlather) observed managed fallows of trees planted at 1 × 1 m or 1 × 2 m spacing and harvested annually for charcoal production. Tree leaves and twigs are left to serve as a mulch for the one maize crop planted each year.

Sanchez (1995) has described a similar relay intercropping system used in southern Malawi which would fulfill the requirements of a mulch-based cropping system with trees. *Sesbania sesban* seedlings are transplanted into a maize crop. Because of their slow growth, there is little interference with the maize crop. The *Sesbania* continues to grow into the dry season, shedding its leaves on the ground, which supply 80 kg N and 6 kg P ha^{-1} The stems are harvested for firewood yielding 1.7 t ha^{-1} of dry matter. The next year, maize is planted again and the *Sesbania* either reseeds itself or is replanted. This cycle has been repeated for five years. (Table 1).

2.5 Palm Based Systems: An Expected Payoff from Burning

The population of palms in natural forests invariably increases following burning. This is generally explained by the fact that palms do not have a vascular cambium but "produce secondary growth from better protected meristems within the stem" (Esau, 1965), and therefore are able to survive a burn better than most dicotyledons. Various seed dispersal mechanisms probably further aid in increasing the population of palms in disturbed areas. Humans were not slow to recognize the benefits of palms even where burning was not practiced, and the date palm (*Phoenix dactylifera*) was one of the first plants to be domesticated (Grigg, 1974. Coconut palms (*Cocos nucifera*) are the basis of many agroforestry systems (Nair, 1993) and are usually the most important trees in

TABLE 1
Performance of mixed intercropping system of maize and *Sesbania sesban* for five consecutive years near Makoka, Zambia. Fertilizer was 28 kg ha^{-1} N, 5 kg ha^{-1} P. All values in t ha^{-1} except as noted

Year	Seasonal rainfall (mm)	Sole maize, no fert.	Sole maize + fert.	Maize + relay, *Sesbania*, no fert.	Maize + relay, *Sesbania*, fert.	Wood relay, *Sesbania*, no fert.
1990	1001	1.00	3.75	1.02	3.27*	2.08
1991	1017	1.33	6.21	2.41*	6.13	0.05
1992	551	0.61	2.93	1.86*	3.54	2.24
1993	962	1.02	4.79	2.33**	5.99*	3.01
1994	522	0.51	2.41	1.16*	3.29**	1.01
Mean	810	0.89	4.01	1.76	4.44	1.68

*,** indicate significant differences at 10% and 5%, presumably from corresponding treatment without *Sesbania sesban* (Author does not state).

home gardens. Slaveowners imported oil palms *Elaeis guineensis*) from Africa to feed imported populations with an accustomed food (Walsh, 1997). The sago palm (*Metroxylon sagus*) in Asia and the buriti palm (*Mauritia flexuosa*) in the Americas were soon recognized as a source of food that could be grown on flooded soils, increasing the value of areas that would otherwise be marginal for food production.

Except for the use of cover crops in oil palm and coconut plantations to be discussed in the next section, the only palm system that could be considered mulch-based is that of the babbassu palm (*Orbignya phalerata*) in the Eastern Amazon where the leaves are pruned every four years, placed on the ground, burned and the area between the palm trees used for the production of annual crops. Whether or not the extensive areas of babassu forest in the Eastern Amazon is a result of human activity remains a mystery (Anderson, 1987). Soils of the area generally have higher base status than the rest of the Amazon but precipitation is also somewhat less. The hard seeds of both the babassu and the related Cohun palm (*Orbignya cohun*) which is common on the drier margins of the Central American rainforest make animal dispersion unlikely, although Janzen and Martin (1982) have speculated that presently extinct animals could have spread the palms in the past. Farmers in Belize have assured the senior author that wild pigs are able to crack the seedcoat of the cohun palm although it requires considerable human effort to do so.

Raintree and Warner (1986) and Anderson (1987) have described an apparently very sustainable system based on the culture of the lontar palm (*Borassus sundaicus* "Beccari") on the islands of Timor and Sumba. The trees are tapped for their sugary sap which is used for pig production as well as human consumption. The palm litter is used as a mulch for vegetable gardens. Raintree and Warner (1986) observed that the returns to labor in this system are high and that the general welfare of the population was generally higher than that of populations practicing swidden cultivation on neighboring islands. Raintree and Warner (1986) believed that the system evolved as a culmination of a long period of swidden cultivation when the farmers realized that the tree which came to dominate the degraded fallow was more useful than the crops it replaced. It is also possible that the palms emerged quite early in the process of agricultural intensification since they would have been one of the first species to dominate burned areas. Coconuts dominate the economies of many Pacific Islands, not as a result of a long period of evolution, but because they are the only species that can survive saline soils and other unfavorable positions.

3. SUSTAINABLE MULCH-BASED CROPPING WITH TREES

3.1 Shaded Plantation Systems

Shaded plantation systems have proved to be one of the most productive and sustainable systems in the humid tropics. In many of those systems, the shade trees are pruned for mulch production and to alter shading intensity. Tropical crops grown under shade include coffee (*Coffea* sp.), cacao (*Theobroma cacao*), tea (*Camellia sinensis*), vanilla (*Vanilla fragrans*), black pepper (*Piper nigrum*), cardamon (*Elettaria cardamomum*) and other spice and medicinal crops, and tobacco (*Tabacum* sp.) for cigar wrappers.

The shade produced by interplanting trees among various tropical crops has diverse effects. Shade trees may reduce crop stress by ameliorating adverse climatic conditions and nutritional imbalances; however, they also may compete for growth resources. Natural litter fall and the residues from pruning shade trees produces significant mulch in many agroforestry systems that has numerous benefits. Benefits include decreasing or increasing air and soil temperatures depending on elevation, buffering soil moisture and humidity, maintaining soil fertility, and reducing soil erosion. Reduced light quantity and quality also may avoid excessive vegetative growth and negative overbearing effects in some crops. Purseglove (1968) and Beer *et al.* (1998) have reviewed the agronomic effects of using shade trees in coffee and cacao plantations. Moreover, the conditions produced by shade trees reduce the incidence of many important diseases, although in some cases disease may be increased (Thurston, 1992). Purseglove (1968) notes that in the absence of both shade and mulching, young coffee and cacao plants bear heavily for one or two crops and then decline and suffer dieback. The labor and cost of weed management is also considerably reduced in shaded and mulched plantations. Purseglove (1968) summarized the benefits derived from shade as follows:

> extends the productive life of the tree; (2) prevents over-bearing and die-back; (3) gives more even annual cropping; (4) reduces temperature of air and soil at high temperatures and raises it at low temperatures, and provides a favorable microclimate; (5) reduces hail damage and may reduce pests and diseases; (6) reduces evaporation and transpiration; (7) leaf fall will provide mulch, humus and nutrients, particularly if deep-rooting by bringing nutrients from the lower levels in the soil; legumes may fix nitrogen; (8) protects the organic matter of the surface layer from breakdown by exposure to the sun; (9) root systems of the shade trees may assist in drainage and aeration; (10) branches may be lopped for mulch; (11) reduces cost of weeding and helps to keep out grass; (12) shade may provide firewood, timber or other useful products.

Shaded coffee has been a sustainable agroforestry system for centuries in the tropics. Coffee is grown primarily in tropical countries, and is the major export crop of many tropical countries. The removal of shade trees to "modernize" coffee production in many of the most important producing countries has produced considerable controversy because of possible effects on nutrient cycling, soil degradation, and biodiversity. Coffee is one of the most commonly shaded tree crops. Shaded coffee has been a sustainable agroforestry system for centuries in the tropics. (See Chapter 4.) Arabica coffee, the most desirable type, is an evergreen shrub or tree which may grow to five meters without pruning. The first crop generally is harvested three to four years after planting, and under ideal conditions trees may continue to bear for several decades. Although today most of the world's coffee is planted without shade (sun coffee), in many countries coffee is still cultivated with shade. Coffee without shade requires a higher level of management.

Recent efforts at modernization of coffee systems have consisted of the replacement of old, traditional coffee cultivars with newer cultivars that respond well to chemical fertilizers. Another feature of modernization involves the elimination or reduction of shade, the goal being to open the coffee up to the sun to deter the spread of fungal diseases, and to increase coffee yields. The transformation also allows increases in the density of coffee plants. There is evidence that coffee plants in the "modern" systems may only last 12 to 15 years, much less than traditional coffee plants. Thus, the sun coffee systems do not appear to be as sustainable as the older traditional shade coffee systems.

In many cases, shade removal and the establishment of high-yielding coffee systems has not achieved the intended objectives. For example, modernization has often entailed planting of *Caturra,* a dwarf mutant coffee cultivar of *Coffea arabica* discovered in Brazil during the last century. Caturra yields about 30% more coffee per plant if supplied with adequate fertilizer. Although initially reported to be resistant to coffee rust, Caturra is, in fact, susceptible to the disease. Not all sun coffee is grown in full sun. Coffee-growing systems vary along a range of full shade to full sun. Some modernized coffee farms retain up to 20% shade, while others are completely devoid of a canopy.

The modernization of coffee production has some negative effects on wildlife that is gaining the attention of ecologists (Perfecto, 1996; Tangley, 1996). Historically coffee, a shade-tolerant shrub or tree, has been grown beneath a canopy of native forest trees often intermingled with fruit trees. More than 40 species of trees have been found on some traditional coffee farms. This combination of coffee and trees mimics the native forests and is attractive to a wide variety of migratory birds such as Baltimore orioles, warblers, and vireos, as well as year-round residents such as parrots and toucans. Few birds actually eat coffee berries. The Baltimore oriole (*Icterus galbula*) actually seems to prefer traditional coffee farms over natural forests. The traditional coffee plantations have a relatively high degree of biodiversity "because of the structural and floristic complexity of the shade trees" (Perfecto, 1996). Various studies have shown that bats, snakes, insects, wild cats, monkeys, and other arthropods and vertebrates were more abundant in the traditional than in modern coffee plantations (Perfecto, 1996). Nevertheless, the habitat of coffee farms is rapidly disappearing. The diverse agroecosystems of traditional coffee farms is giving way to monocultures of high-yielding cultivars of coffee grown in evenly spaced rows under full sun, with little or no forest canopy. These modernized or "sun" coffee plantations offer little habitat for wildlife. Various studies have shown that the diversity of migratory birds drops dramatically when coffee is converted from shade to sun. More than 1.1 million ha of coffee within northern Latin America have been modernized. Future losses to modernization are "likely to include a severe loss of biodiversity in areas where coffee plantations currently provide the last refuge."

Another negative aspect of shade removal from coffee plantations is the effect on nutrient cycling. Unshaded plantations require higher levels of fertilization, especially of nitrogen, and there is some evidence that this nitrogen is accumulating in groundwater. Babbar and Zak (1995) measured nitrate accumulation and denitrification in shaded and unshaded coffee plantations in Costa Rica receiving 300 kg ha^{-1} yr^{-1} of N. Such a rate of N fertilization would seem unnecessarily high but it is quite common in Costa Rica because farmers find it gives an economic return in terms of coffee yield. There was much less nitrate accumulation in shaded plantations but the combination of shade with high fertilization rates led to increased denitrification. Pot studies indicated that considerable N_2O, a greenhouse gas, might evolve from heavy fertilization of shaded plantations but the area of shaded coffee in the world is not enough for the N_2O evolved to have any significant effect on global warming (Beer *et al.*, 1998).

Mulches are often used in coffee production, even with "sun" coffee. Wrigley (1981) cited a number of benefits from mulching coffee with non-living crop residues. Wiley (1975) went so far as to state that all of the benefits from shade could be obtained through mulching. Wrigley (1981) suggested that mulches reduced soil temperatures, protected against rain, conserved rainfall, increased soil nutrients, increased soil organic matter, produced conditions ideal for root growth, reduced weeds, reduced soil acidity, and increased coffee yields. High labor costs were cited by Wrigley as the main disadvantage for the use of mulches. Large quantities of grass and banana trash were commonly carried to the coffee plantations in East Africa (Wellman, 1961). Discussing mulches for coffee, Wellman (1961) noted that during dry seasons mulches provided sufficient soil moisture so that coffee roots could obtain nutrients otherwise unavailable without them. He also noted "I was told in Africa, by researchers and farmers alike, that fertilizers applied to mulch grass, and this mixture then used as mulch, was more profitable than fertilizer applied directly to the soils of their coffee shambas." Many different materials have been used to mulch coffee plantings. Acland (1971) reported the use of various grasses, banana trash, sisal waste, coffee pulp, wheat straw, sawdust, and wood shavings for mulching coffee in East Africa. *Arachis pintoi* (wild peanut) is being used as a live mulch in some coffee plantations of Central America, but its long-term value is yet to be established.

In Costa Rica, poró (*Erythrina poeppigiana*) is a commonly used shade tree for coffee (Kass, 1994). Trees are pruned one to three times a year. The pruned branches provide a mulch and return nitrogen to the soil. Beer (1988) concluded that poró, when pruned two–three times per year, can return, as a litter layer, the same quantity of nutrients as are applied to coffee plantations in Costa Rica via inorganic fertilizers, even at the highest recommended rates of 270 kg N, 60 kg P and 150 kg K/ha/yr. In addition, trees contributed 5,000–6,000 kg organic inputs/ha/yr. The total leaf litter from both coffee and poró was between 5,000 and 20,000 kg/ha/year. This amount is within the range of leaf litter fall reported for tropical forests. Although the nutrient contribution by nitrogen fixation is important, Beer (1988) concluded that, especially in fertilized plantations of cacao and coffee, leaf litter productivity is a more important contribution of the leguminous shade trees than nitrogen fixation. Litter also provides organic material to the soil and shades out weeds. Kass et al. (1997) discuss in detail the role of nitrogen fixation and nutrient supply in coffee and cacao plantations. According to Teketay (1990), in Ethiopia, when nutrient deficiencies are noted in coffee, leaves and branches of *Erythrina burana* are cut and buried around the coffee bushes. Subsequently, farmers claim higher production for several years.

Other trees used as shade in coffee plantation are other species of *Erythrina*, *Gliricidia sepium*, *Leucaena glauca*, and *Inga* spp., especially *Inga edulis*. *Inga* sp. are favored as a shade tree in both coffee and cacao as the mulch produced by *Inga* sp. decomposes very slowly in comparison to other leguminous trees (Fernandes *et al.*, 1993). The fruits of several *Inga* species sometimes reach more than a meter in length and contain large seeds surrounded by a white, succulent, sugar-rich pulp. The fruit is highly prized and is sometimes called "the ice cream bean." The Inca Emperor Atahualpa sent Francisco Pizzaro a basketful of *Inga* beans as a gift (National Research Council, 1989). Some farmers in Costa Rica don't like *Inga* because it attracts people who collect the fruit and damage the coffee. Thus *Erythrina* spp. are preferred because they are not an "attractive nuisance." Shade trees produce many alternative products in addition to fruit. Purseglove (1968) lists the most commonly used shade trees for coffee and cacao.

Purseglove (1968) stated: "in its natural habitat *Theobroma cacao* is a small tree in the lowest story of the evergreen tropical rain-forest of South America." Traditional farmers in tropical Latin America have cultivated cacao, probably almost always under shade, for more than 2,000 years (Purseglove, 1968). Fernandez de Oviedo (1986), writing in the 1600s, wrote that Indians in Nicaragua used shade trees for cacao, and that they pruned the trees to give the proper shade. Most cacao is grown in the tropics in countries between 15°N and S latitude with abundant rainfall. Plants from seed are usually used to establish plantings, but cuttings can also be used although their use is more expensive. Almost all cacao is grown under shade. Plantings are usually established at densities ranging from 500–2,000 trees per hectare.

Cacao can be grown in full sunlight with higher yields, but the long-term results are not promising. Purseglove (1968) noted that permanent overhead shade is necessary for cacao except under the most favorable conditions. A number of cacao diseases are more serious when cacao is grown in full sun than when it is grown under shade (Thurston, 1992). The results of a 17 year shade and fertilizer trial in Ghana was given by Ahenkorah *et al.* (1974). Cacao fertilized and grown without shade yielded three times as much as shaded trees. However, cacao without shade was more susceptible to insects and, according to the authors, "probably due to diseases." Furthermore, after ten years the unshaded trees declined in vigor, due to a high incidence of pests, establishment of mistletoe (*Tapinathus bangwensis*) and mosses, loss of exchangeable bases from the soil, and depletion of reserves. Alvim (1977) suggested that lack of protection from the wind may have been an additional factor. Regarding Ahenkorah's results Alvim wrote:

> The above findings are in agreement with practical observations by cacao farmers in many countries showing that increase in yield following complete removal of shade usually does not last very long and is followed by rapid decline of plantations, with many plants showing severe defoliation and dieback after the third or fourth year, (Alvim, 1977).

In practice some degree of partial shade is probably needed for almost all cacao plantings. The proper degree of shade may not always be easy to obtain, considering the multitude of conditions and objectives under which cacao is managed, varietal differences, soil differences, the large diversity of shade trees utilized, and the different climates where cacao is grown. The long-term effects of growing cacao and other shade-tolerant crop species in full sunlight need to be carefully studied in long-term experiments such as those described by Ahenkorah et al. (1974).

Purseglove (1968) noted that tea was interplanted with shade trees, but that this practice had become controversial. Grown in full sun, tea gives much higher yield when given adequate nitrogen. Also, shade increases the occurrence of blister blight, caused by *Exobasidium vexans*. The fungus causes a destructive disease in Asia and caused more severe losses in shaded than in unshaded tea (Kerr and Rodrigo 1967, Visser et al. 1961).

The long-term sustainability of the various agroforestry systems with some degree of shade and mulch is not known, but their sustainability surely is greater than that of many modern farming systems. Beer et al. (1998) wrote:

> Many coffee farmers and their organizations have demonstrated that they are concerned about medium- to long-term perspectives by implementing erosion control recommendations while ignoring advice to eliminate shade trees. However, they generally lack solid information about the factors which control sustainability and environmental effects.

Developmental and environmental organizations, agronomic scientists, and decision makers concerned with the future of the tropics need to provide farmers with sound information on which to make realistic decisions that will expand and preserve those agroforestry systems that are truly sustainable for the future.

How sustainable is the shaded plantation system? The fact that these systems have existed for as long as 2,000 years might be cited as evidence of sustainablity. Fassbender (1993) has shown considerable nutrient accumulation in such systems, but they were fertilized. Herrera et al. (1987) have maintained that a total nutrient (N+P+K+Ca+Mg) output of 57 and 102 kg/ha/yr for coffee and cacao could be maintained in unfertilized plantations in Venezuela shaded with a mixture of leguminous and non-leguminous trees through nitrogen fixation, mycorrhizal activity, and inputs from rain and dust.

3.2 Cover Crops in Plantations

Growing cover crops under trees is a common practice in tree plantations throughout the world. Many of the cover crops used are leguminous and contribute to the nitrogen economy of the systems. They are also extremely effective in controlling weeds (Samson, 1980). The primary functions of a cover crop in orchards are to improve soil structure, prevent soil erosion, improve nutritional conditions for the trees by rendering nutrients in the soil more available, add nitrogen, and remove excess water from heavy soils in years of excess rainfall (Jones and Embleton, 1973) The use of cover crops is generally recommended in tree plantations in the humid tropics. Where there is a marked dry season, they may compete with the trees for moisture and become a fire hazard (Samson, 1980).

Other possible disadvantages of cover crops is their serving as a source of insect and disease problems, preventing the movement of water during periods of excess rainfall or during irrigation operations. They also restrict movement of air and radiation of heat from the soil in cold periods. Thus, the decision whether or not to use cover crops in plantations depends on the relative importance of the various positive and negative effects (Jones and Embleton, 1973).

Giller and Wilson (1991) describe a succession of species used in oil palm plantations: *Calopogonium mucunoides* germinates first but is later replaced by *Pueraria phaseoloides*. As the trees develop larger crowns and shade the ground more, *Centrosema pubescens* becomes dominant. All three species are sown at the same time (Giller and Wilson, 1991). The same sequence is recommended by Samson (1980) for most perennial crops in the humid tropics. Fassbender (1993) cites data from the Rubber Research Institute of Malaya which show that after five years, understory legumes in a rubber plantation had accumulated 226–353, 18–27, 85–131 and 15–27 kg ha^{-1} of N, P, K, and Mg. After the fifth year, the understory legumes were considered to interfere with harvesting activities and were eliminated. Giller and Wilson (1991) state that the beneficial effects of leguminous cover crops on rubber production may persist for 20 years even if the legumes die out after only four to five years. It was also found that the benefits of the leguminous cover crops was sufficient to justify the application of fertilizer, especially rock phosphate, to aid in their establishment (Giller and Wilson, 1991).

Benefits from the use of cover crops in plantations include increased yield and faster development of the trees as well as erosion control (Giller and Wilson, 1991). In dry areas, cover crops may compete with trees for moisture. As plantation crops often have high value, the system is sustainable as long as measures are taken for a more equitable distribution of earnings as occurs in Malaysian oil palm plantations. Where such provisions are not taken, sustainability may be more limited. Removal of the trees for timber as frequently occurs in rubber plantations also may reduce the sustainability of the system (Gller and Wilson, 1991).

Cover crops are not generally recommended in banana plantations because the bananas generally give adequate soil cover soon after establishment. *Centrosema* and *Indigofera* are reported to be toxic to bananas. Bananas have a more superficial root system than other perennials and may be adversely affected by cover crops (Samson, 1980).

Use of leguminous cover crops in citrus plantations is often recommended as a weed control measure because costs of conventional weed control are high. Where there is a marked dry period, cover crops can be turned under at the end of rainy period so they do not compete for moisture in the dry period (Jones and Embleton, 1973). Some live mulches, as these cover crops are often called, can overgrow the trees if not properly managed (Samson, 1980). This difficulty has prevented the use of cover crops in coffee and cacao plantations but better results have been obtained recently with *Flemingia* (Giller and Wilson, 1991) and *Arachis pinto*i, now widely used as a live mulch in perennial plantations although it is often considered to provide a habitat for snakes.

Leguminous cover crops can also be used in forestry plantations. The senior author has observed mucuna growing vigorously under a two-year-old teak plantation in El Salvador.

3.3 Homegardens and Multistory Tree Gardens: Paradise Regained?

Gardens variously called home gardens, backyard gardens, household gardens, kitchen gardens, village gardens, and dooryard gardens are found around the homes of many traditional farmers in the tropics. Anyone who has traveled outside of the large cities in the humid tropics has seen the almost omnipresent backyard gardens. The gardens are usually multistoried systems dominated by tall trees, and in some tropical locations, they closely mimic the natural forest. Wilken (1987) suggested that the fullest exploitation of vertical (height) as well as horizontal distribution of space was found in home gardens, with their multiple levels of useful plants. He described such a garden in Mexico with a four-tiered vertical arrangement containing more than 24 plant species. Abdoellah and Marten (1986) found 235 useful species of plants grown in household gardens in a survey in

West Java. Beckerman (1983) noted that the Bari or Motilones Bravos, Indians of Northern Colombia and Venezuela, often planted fields or household gardens near their houses. The fields nearest their longhouses received mulches in the form of food scraps, garbage, ashes from fires, and human waste.

The home garden is one of the most widespread cropping systems in the tropics. Home gardens occupy 20% of the arable land on Java (Jensen, 1993a). Its structure mimics the tropical rainforest with almost all niches occupied by species producing products for human consumption. It would thus be an alternative to the intensification of natural systems through clearing, burning, and the planting of annual crops (Raintree and Warner, 1986). Nair (1993) makes a distinction between home gardens and multistory tree gardens in that the former are located close to the house and the latter at a distance from the dwelling, often on communal lands. However, there is no structural difference in the two systems, so the distinction would appear to be socioeconomic rather than biophysical. Sanchez (1995) feels that the complex agroforests of Indonesia are perhaps "the epitome of sustainability" because they provide a wide variety of high value products that have assured markets and that are grown with little nutrient removal from the soil. (See Chapter 17.)

Considerable evidence of the sustainablity and the role of mulch in home gardens has been provided by Jensen (1993b) who measured nutrient inputs and outputs over a 16-week period in a 0.13 ha. Javanese home garden and extrapolated these findings to an annual basis. He found that nutrient balance was slightly negative but was small enough to be provided by soil reserves for a considerable time. Still, the only evidence we have would indicate that the home garden is less than ideally sustainable at least from the point of view of nutrient balances. The mulch based nature of the system was however demonstrated by the amount of nutrients relative to the total storage and outflow that was provided by the mulch. Zech *et al.* (1990) give evidence of carbon accumulation in the soil as home gardens develop but after 25 years, levels of soil carbon were still lower than in an adjacent secondary forest.

Other authors have questioned the sustainability from a socioeconomic viewpoint, indicating that farmers are under much pressure to convert home gardens to more market oriented systems (Jose and Shanmugaratnam, 1993). As the home garden functions by filling gaps in the provision of nutrients and income left by the other household activities, some inputs to the garden from these systems would be justified. A recent study of Nicaraguan and Honduran home gardens, however, found that return to labor in a home garden is much higher than the minimum wage (Marsh and Hernandez, 1996). The multistory tree garden or home garden would appear to have all of the components of sustainability without always demonstrating long-term sustainability. Rather than existing as a separate unit, the garden is invariably tied to other activities of the household which determine the degree of inputs given and amounts of outputs demanded by the garden. When inputs sufficiently balance outputs, the system is sustainable but management of the inputs will also affect the outputs obtained (Wojtlkowski, 1993).

3.4 Alley Farming: Sustainable under Certain Conditions

Alley farming is defined by Kang and Wilson (1987) as the growing of arable crops between hedgerows of planted shrubs and trees, preferably leguminous species, which are periodically pruned to prevent shading the companion crops. Over the past fifteen years, the system has been the subject of a considerable amount of research for which the results are generally described as negative to mixed. Various reasons were given for the poor performance and acceptance of alley farming. Among them, poor selection of the tree and crop component, inadequate nutrient cycling, excessive competition from the tree component, excessive labor demands, and problems with land tenure (Fernandes *et al.*, 1993; Palm *et al.*, 1995; Carter, 1995; Sanchez, 1995). There were some circumstances in which alley farming was shown to have considerable promise, such as on sloping lands where it was an effective erosion control measure (Table 2), where rainfall was adequate throughout the cropping season, where soils were fertile without major nutrient limitations, where

TABLE 2
Soil loss reductions reported due to alley farming

Source	Location and number of years	Soil type	Hedge species	Crops	Soil loss without hedges (t ha^{-1} yr^{-1})	Soil loss with hedge (t ha^{-1} yr^{-1})
Alegre and Fernandes, 1991	Peru — 1	Typic Paleudult	*Inga edulis*	Rice, cowpea	13-bare 2.4-with tillage 0.7-forest	0.6-alley farming
Lal, 1989	Nigeria — 6	Haplustalf	*Leucaena, Gliricidia*	Maize, cowpea	5-without tillage 12-with tillage	0.35 leucaena in 2 0.08 leucaena in 4 0.36 gliricidia in 2 0.39 gliricidia in rows
Ghosh et al., 1989	India — 2	Gravelly clay	*Leucaena, Eucalyptus*	Cassava	8.3-bare 4.54-cassava	1.83-Eucalyptus 3.59-Leucaena
Vega et al., 1987	Colombia — 1	No data, slopes of 45% and 75%	*Gliricidia*	Maize	344-bare, 45% slope 282-bare, 75% slope 37.6-maize, 45% slope 23.5-maize, 75% slope	13.5-Glir.+maize, 45% slope 13.0-gliricidia, 75% slope
Lebauf, 1993	Costa Rica — 2	Acrudoxia Melanudand, 15–35% slope	*Erythrina fusca*	Maize, beans	0.4-bare 0.1- crop	0.2–4m alleys 0.2–6m alleys

labor supply was adequate, coupled with a scarce supply of land, and where land tenure was secure (Sanchez, 1995). Kass *et al.* (1995) found a close relationship between the price of nitrogen fertilizer and labor costs in determining whether applying the loppings of nitrogen-fixing trees to crops was more economical than applying mineral fertilizers.

Whether alley farming is a sustainable system is the subject of some controversy. Nair et al. (1995) provide considerable evidence that it is, based on long-term experiments of up to ten years duration, at least on more fertile soils of the humid tropics. The experience in areas with lower rainfall, lower soil fertility, and other factors contributing to poor tree growth have been less encouraging (Juo and Manu, 1996; Nair et al., 1995; Sanchez, 1995). Under these conditions, where cultivation is on slopes and hedges are planted on contours, alley farming can be a successful soil erosion control measure (Sanchez, 1995). Soil erosion has generally been recognized as the leading cause of soil degradation in the tropics. Low profitability might be the major obstacle to sustainability of this system. With the introduction of high value crops and trees producing valuable products, the system might be made more attractive to farmers (Carter, 1995).

In a study of economic return to agroforestry systems in Central America and the Caribbean, Current and Scherr (1995) found alley farming to be one of the most profitable systems although the number of farms on which this conclusion was based was relatively small. Kass et al. (1995) showed considerable evidence of long-term sustainability of alley farming systems. On an Andic Eutropept, yields of more than 2 t/ha/yr of maize and 1 t/ha/yr of beans were maintained for ten years without the use of nitrogen fertilizer. Some improvement in soil properties were noted, in particular a significant increase in potassium levels, indicating that trees were indeed bringing up nutrients from lower soil depths and making them available to shallow rooting crops. There was evidence though of depletion of readily available organic P reserves when tree prunings were the only phosphorus source (Paniagua et al., 1995).

Carter (1995) suggested the following practices which would make alley farming more acceptable:

Alley farming on sloping land, in the form of contour hedgerows
The use of pigeon pea (*Cajanus cajan*) as the hedgerow species
Widening alley spacing to allow more mechanized cultivation
A form of alley grazing in which widely-spaced hedgerows are grazed directly

Contour hedgerows was a principal component of the Sloping Agriculture Land Technology (SALT) developed with much success in the Philippines. Use of *Leucaena leucocephala* which was later severely attacked by psyllid was a major factor in poor acceptance of alley farming (Carter, 1995). Recently, Kettler (1996/1997) has shown an increase of bean yields in the *tapado* system by introducing rows of trees into the system, a form of alley farming. Other farmers are finding that leaving the area between the hedgerows in fallow for several seasons can increase productivity and reduce weed infestations (Suson and Garrity, 1997). Rather than being an economically and technologically bankrupt system (Sanchez, 1995), it would seem that more and more niches are being recognized where alley farming can contribute to crop productivity and environmental sustainability (Kass et al., 1997).

4. CONCLUSIONS

Sustainable mulch-based systems with trees include many of the most important and economically viable production systems of not only the tropics but also temperate areas where cover crops are also used in tree plantations. Because of their value in erosion control, weed control, and fertility maintenance, such mulch-based systems contribute considerably to the sustainability of the systems. Many of these systems appear to have evolved almost spontaneously as population pressure and labor availability increased although some tree-based systems are not particularly labor-intensive. It would

TABLE 3
Yield and stability indices for maize and beans under different cropping systems on an Andic Eutropept, Turrialba, Costa Rica. (from Kass et al., 1995)

Treatment	Maize Square root of MS	Yield (kg ha^{-1})	Stability index	Beans Square root of MS	Yield (kg ha^{-1})	Stability index
Control –N	1432	2061	0.69	622	757	0.82
Control +N	1776	2885	0.62	793	1018	0.78
Erythrina mulch –N	978	3137	0.31	734	1425	0.51
Erythrina mulch +N	1054	3317	0.32	1146	1613	0.71
Cattle manure –N	1325	2921	0.45	865	1141	0.76
Cattle manure +N	1545	3329	0.46	1194	1493	0.80
Gmelina mulch –N	2028	2525	0.67	948	1222	0.78
Gmelina mulch +N	1461	2938	0.50	760	1506	0.50
Gliricidia mulch –N	1883	2535	0.74	1091	1039	1.05
Gliricidia mulch +N	995	2935	0.34	1213	1248	0.97
Erythrina alley crop –N	1483	2229	0.67	529	1141	0.46
Erythrina alley crop +N	1737	2139	0.79	376	1242	0.30
Gliricidia alley crop –N	888	2412	0.37	602	1318	0.46
Gliricidia alley crop +N	1199	2533	0.47	590	1487	0.40

seem unlikely that all the agricultural areas will eventually turn into labor-efficient and ecologically friendly home gardens, a Gaian (Joseph, 1990) view that Raintree and Warner (1986) seem to support. Although there are examples of inequity in such systems, some of the most equitable systems that have developed in the tropics are mulch-based systems with trees. As more importance is given to the development of sustainable systems, these systems will certainly maintain if not increase their importance in the economies in many countries of the world. There might be some question as to whether or not they are able to provide a significant source of the human food supply. Tree crops account for less than ten percent of total arable land. While an increasing amount of food crops in developing countries will probably be grown on sloping land with barriers on the contour, half of the world's grain supply will continue to come from the temperate zone where agroforestry systems will probably be limited to windbreaks between fields. Whether or not human diets will change to incorporate a greater amount of tree crops is beyond the scope of this chapter. However, chestnuts, walnuts, almonds, and acorns were formerly more important items in European diets, especially in hilly areas of Italy before importation of grains became a common practice (Grigg, 1974).

A combination of shaded plantations of appropriate crops, leguminous cover crops in other perennial fruit and forest species, multistory tree gardens, and modified alley farming on sloping land will certainly contribute to improved human well-being and reduced environmental degradation. Future research will show whether managed fallows can also be considered a sustainable system. However, there is no need to wait until environmental abuse and loss of productivity in other production systems create a greater imperative for wider use of mulch-based cropping systems with trees.

REFERENCES

Acland, J. D. 1971, *East African Crops*, Longman, London, 252

Agus, Fahmuddin. 1997, *Intensification of Indigenous Fallow Rotation Using Leucaena leucocephala*, paper submitted to the workshop "Indigenous Strategies for Intensification of Shifting Cultivation in Southeast Asia, Bogor, June 23–27, 1997,

Abdoellah, O. S. and G. G. Marten, 1986, "The complementary roles of home gardens, upland fields, and rice fields for meeting nutritional needs in West Java," in *Traditional Agriculture in Southeast Asia, A Human Ecology Perspective*, Martin, G.G., Eds., Westview, Boulder, CO, 293–325.

Ahenkorah, Y., Akrofi, G. S., and Adri, A. K., 1974, "The end of the first cocoa shade and manurial experiment at the Cocoa Institute of Ghana, *J. Hort. Sci.* 49: 43–51,

Alegre, J. C. and E. C. M. Fernandes, 1991, "Runoff and erosion losses under forest, low input and alley farming on slopes," in *Tropsoils Technical Report*, Y-443B, pp. 227–228, 1988–1989.

de Alvim, T-P, 1977, *Cacao III. Climate, in Ecophysiology of Tropical Crops* Alvim de T-P. and Kowalski, T.T., Eds. Academic Press, New York.

Anderson, A.B, 1987, "Use and management of native palm forests in Agroforestry: Realities, Possibilities, and Potentials," Gholz, H.L., De Martinus Nijhoff, Dordrecht Netherlands, 169–178

Babbar, L. I. and Zak, D. R., 1995, "Nitrogen loss from coffee agroecosystems in Costa Rica. Leaching and denitrification in the presence and absence of shade trees," *J. Env. Qual.* 24, 227–233.

Beckerman, S, 1983, "Bari swidden gardens: crop segregation patterns," *Hum. Ecol.,* 11, 85–101, 1983

Beer, J, 1988. "Litter production and nutrient cycling in coffee (*Coffea arabica*) or cacao (*Theobroma cacao*) plantations with shade trees," *Agrofor. Sys.* 7: 103–114, 1988

Beer, J., Muschler, R., Kass, D., and Somarriba, E., 1998, "Shade management in coffee and cocoa plantations," *Agrofor. Sys.* 38, 139–164.

Bensel T. G, 1995, "Rural woodfuel production for urban markets: problems and opportunities in the Cebu province, Philippines," *Pac. Asian J. Energy* 5, 9–28.

Carter, J, 1995, "Alley farming: Have resource poor farmers benefitted?" ODI (Overseas Development Institute) *Natural Resource Perspectives*, No. 3, June, 1995.

Current, D. and Scherr, S.J, 1995, "Farmer costs and benefits from agroforestry and farm forestry projects in Central America and the Caribbean. Implications for policy," *Agrofor. Sys.* 30 (1–2), 87–103.

Dufour, D. L, 1990, "Use of tropical rainforest by native Amazonians," *Biosci.* 40(9), 652–659

Esau, K.1965, *Plant Anatomy*, second edition, John Wiley & Sons, New York, 400–401,

Eswaran, H., S.M. Virmani, and L.D. Spivey, 1993, "Sustainable agriculture in developing countries: constraints, challenges, and choices," 7–24, in *Technologies for Sustainable Agriculture in the Tropics,* Ragland, J. and Lal, R., Eds. American Society of Agronomy Special Publication 56; ASA. *Madison, WI.*

Fassbender, H.W, 1993, "Edaphalogical models of agroforestry systems" (in Spanish), *Serie Materiales de Enseñanza*, No. 29, Centro Agronómico Tropical De Investigación y Enseñanza. Turrialba, Costa Rica, 1993, Ch. 4–7.

Fernandes, E. C. M., Davey, C. B., and Nelson, L. A., 1993, "Alley cropping on an acid soil in the upper Amazon: Mulch, fertilizer and hedgerow root pruning effects," in *Technologies for Sustainable Agriculture in the Tropics*, Ragland, J. and Lal, R., Eds. ASA Special Publication 56, American Society of Agronomy, Madison, WI, Ch. 7.

Fernandez de Oviedo, G, 1986, *Summary of the natural history of the Indies* (in Spanish), Ballesteros, M., Ed., Historia 16, Madrid, 181.

Ghosh, S. P., Mohna Kumar, B., Kabeerathumma, S., and Nair, G. M, 1989, "Productivity, soil fertility and soil erosion under cassava based agroforestry systems," *Agrofor. Sys.* 8, 67–82.

Giller, K. E. and Wilson, K. J., 1991, "Nitrogen fixation in tropical cropping systems CAB International," Wallingford, U.K., Ch. 9.

Graça, L. R., Ribas, L. C., and Baggio, A. J, 1986, "The economic profitability of bracatingas in Parana" (in Portuguese), *Boletim de Pesquisa Forestal* (EMBRAPA, Brasil) 12: 47–52.

Grigg, D.B, 1974, *The Agricultural Systems of the World, An Evolutionary Approach*, Cambridge University Press, New York, 1974, Ch. 1–5 and 8.

Harlan, J. R., 1992, *Crops and Man*, second edition, American Society of Agronomy, Madison, WI.

Herrera, R., Aranguren, J., Escalante, G., Accardi, A., Navidad, E., Toro, M., and Cuenca, G, 1987, "Coffee and cacao plantations under shade trees in Venezuela," in *Advances in Agroforestry Research*, Beer, J., Fassbender, H. W., and Heuveldop, J., Eds. CATIE GTZ Serie Tecnico, *Informe Tecnico No, 117*, Centro Agónomico Tropical de Investigación y Enseñanza. Turrialba, Costa Rica.

Janzen, D. H. and Martin, P. S, 1982, "Neotropical anachrononisms: the fruits the gomphomeres ate," *Science* 215, 19–27, 1982.

Jensen, M, 1993a, "Soil conditions, vegetation structure and biomass of a Javanese home garden," *Agrofor. Sys.* 24, 171–186.

Jensen, H, 1993b, "Productivity and nutrient cycling of a Javanese home garden," *Agrofor. Sys.* 24, 187–201.

Jones, W. W. and Embleton, T., 1973, "Soils, soil management, and cover crops," 98–121, in *The citrus industry, Vol. III. Production Technology,* Reuther, W., Ed., University of California Press. Riverside, CA.

Jose, D. and Shanmugaratnam, N., 1993, "Traditional home gardens of Kerala: A sustainable human ecosystem, *Agrofor. Sys.* 24, 203–213.

Joseph, L.E, 1990, *Gaia, the growth of an idea.* St. Martin's Press, New York.

Juo, A. and Manu, A., 1996, "Chemical dynamics in slash and burn agriculture," *Agriculture, Ecosystems, and Environment* 58: 49–60.

Kang, B.T. and Wilson, G.F, 1987, "The development of alley cropping as a promising agroforestry technology," in *Agroforestry: A decade of development*, Steppler, H.A. and Nair, P.K.R, Eds., ICRAF, Nairobi, Kenya.

Kass, D. C. L., Foletti, C., Szott, L. T., Landaverde, R., and Nolasco, R., 1993, "Traditional fallow systems of the Americas," *Agrofor. Sys.* 23, 207–218.

Kass, D. L, 1994, "Erythrina species — Pantropical multipurpose tree legumes," in *Forage Tree Legumes in Tropical Agricultural Development*, Gutteridge, R. C. and Shelton, H. M., Eds., CAB International, Wallingford, U.K.

Kass, D. L., Araya, J. S., Sanchez, J., Pinto, L. S., and Ferreira, P., 1995, "Ten years experience with alley farming in Central America," in *Alley Farming Research and Development*, Kang, B. T., Osiname, A. O., and Larbi, A., Eds. International Institute of Tropical Agriculture, Ibadan, Nigeria, 379–392.

Kass, D. L., Sylvester-Bradley, R., and Nygren, P., 1997, "The role of nitrogen fixation and nutrient supply in some agroforestry systems of the Americas," *Soil Biol. Biochem.* 29: 775–785.

Kass, D., Jiménez, J., and Schloenvoigt, A., 1997, "How to make alley farming more productive, sustainable, and acceptable to small producers" (in Spanish), *Agroforestería en Las Américas*, 4(14): 21–23.

Kerr, A. and Rodrigo, W. R. F., 1967, "Epidemiology of tea blister blight (*Exobasidium vexans*). III. Spore deposition and disease prediction," *Trans. Br. Mycol. Soc.* 50, 49–55.

Kettler, J. S, 1996/1997, "Fallow enrichment of a traditional slash/mulch system in Southern Costa Rica: Comparisons of biomass production and crop yield," *Agrofor. Sys.* 35: 165–176

Lal, R., 1989, "Agroforestry systems and soil surface management of a tropical Alfisol. II Water runoff, soil erosion, and nutrient loss," *Agrofor. Sys.* 8: 192–215.

Lanner, R. M, 1985, "Some attributes of nut-bearing trees of temperate forest origin," in *Attributes of Trees as Crop Plants*, Cannell, M. C. R. and Jackson, J. E., eds. Institute of Terrestrial Ecology, Titus Wilson & Sons, Kendal, U.K.

Larson, W. E. and Pierce, F. J., 1974, "The dynamics of soil quality as a measure of sustainable management," 37–51, in *Defining Soil Quality for a Sustainable Environment,* Doran, J. W., Coleman, D. C., Bezdicek, D. F., and Stewart, B. A., Eds., Soil Science Society of America Special Publication 35, ASA, Madison, WI

Leakey, R., 1996, "Definition of agroforestry revisited," *Agrofor. Today* 8(1): 5–7.

Lebeuf, T., 1993, "Agroforestry systems with *Erythrina fusca* and their effect on soil loss and surface runoff on hillslopes, San Juan Sur, Turrialba, Costa Rica," in *Erythrina in the New and Old Worlds*, Westley, S. B. and Powell, J. M., Eds. NFTA, Paia, U.S., 333–343.

Marsh, R. and Hernandez, I., 1996, "The role of the traditional home garden in the economy of the household: case study from Honduras and Nicaragua," *Agroforestería en las Américas* 3(9–10): 8–16.

Montagnini, F. and Mendelsohn, R. O., 1997, "Managing forest fallows: improving the economics of swidden agriculture, *Ambio*, 26: 118–123.

Nair, P. K, 1993, *An introduction to agroforestry*, Kluwer Academic Publishers, Dordrecht. Netherlands, Ch. 2, 7.

Nair, P. K. R., Kang, B. T., and Kass, D. C. L., "Nutrient cycling and soil-erosion control in agroforestry systems. Agriculture and the Environment: bridging food production and environmental protection in developing countries," Juos, A. S. R. and Freed, R. D., Eds. *ASA Special Publication 60*, America Society of Agronomy, Madison, WI, 117–138.

National Research Council, 1989, *Lost Crops of the Incas: Little Known Plants of the Andes with Promise for Worldwide Cultivation*, Nat. Acad. Press, Washington, DC.

Padoch, C. and de Jong, W., 1987, "Traditional agroforestry practices of native and riberino farmers in the lowland Peruvian Amazon," in *Agroforestry realities possibilities and potentials.*, Gholz, H., Ed., Martinius Nijhoff, Dordrecht. Netherlands, 179–194.

Palm, C. A, Salazar, A. A., Szott, L. T., and Fernandes, E. C. M., 1995, "Long term performance of alley cropping on acid soils of the Amazon Basin," in *Alley Farming Research and Development*, Kang, B. T., Osiname, A. O., and Larbi, A., Eds. International Institute of Tropical Agriculture, Ibadan, Nigeria, 379–392.

Paniagua, A., Mazzarino, M. J., Kass, D., Szott, L., and Fernandes, C., 1995, "Soil phosphorus fractions under five tropical agro-ecosystems on a volcanic soil," *Aus. Journal Soil Res.*, 33, 311–320.

Paningbatan, E. P., 1990, "Alley cropping for managing soil erosion on sloping lands," *Transactions 14th International Congress of Soil Science*, Kyoto, Japan, 376--377

Perfecto, I., Rice, R. A., Greenberg, R., and Van de Voort, M. W., 1996, "Shade coffee: A disappearing refuge for biodiversity," *Biosci.* 46: 598–608.

Peters, C. M., Balick, M. J., Kahn, F., Anderson, A. B, 1989, "Oligarchic forests of economic Plants in Amazonia: utilization and conservation of an important tropical resource," *Cons. Biol.* 3: 341–361

Purseglove, J. W., 1968, *Tropical Crops. Dicotyledons 1*, Longmans, Green & Co., London. 719

Raintree, J. B. and Warner, K., 1986, "Agroforestry pathways for the intensification of shifting agriculture," *Agrofor. Sys.* 4: 39–54.

Robinson, P. J., 1985, "Trees as fodder crops," in *Attributes of Trees as Crop Plants*, Cannell, M. C. R and Jackson, J. E., Eds., Institute of Terrestrial Ecology. Titus Wilson & Sons. Kendal, U.K., 281–300

Roosevelt, A. C. et al., 1996, "Paleoindian cave dwellers in the Amazon: The peopling of the Americas," *Science* 272: 371–383.

Roush, W., 1997, "Squash seeds yield new view of early American farming," *Science*, 276: 894–895.

de Rouw, A., 1995, "The fallow period as a weed-break in shifting cultivation (tropical wet forests)," *Agriculture, Ecosystems & Environment*, 54, 31–43.

Samson, J. A, 1980, *Tropical Fruits*, Longman, New York.

Sanchez, P, 1995, "Science in agroforestry," *Agrofor. Sys.* 31, 1–55.

Soil Science Society of America, 1997, *Glossary of Soil Science Terms*, Soil Science Society of America. Madison, WI.

Somarriba, E., 1992, "Revisiting the past: an essay on agroforestry definition," *Agrofor. Sys.*, 19, 233–240.

Steen, E.B., 1971, *Dictionary of biology*, Barnes & Noble, New York.

Suson, P. and Garrity, D, 1997, "Pruned-tree hedgerow fallow systems. 28, Indigenous Strategies for intensification of shifting cultivation in southeast Asia," *Proceedings of a workshop sponsored by ICRAF, IDRC, and CIIFAD* in Bogor, Indonesia, June 23–27.

Tangley, L., 1996, "The case of the missing migrants." *Science*, 274, 1299–1300.

Teketay, D., 1990, "*Erythrina burana.* Promising multipurpose tree from Ethiopia," *Agrofor. Today* 2: 13.

Thurston, H. D, 1992, *Sustainable Practices for Plant Disease Management in Traditional Farming Systems.* Westview, Boulder, CO.

Thurston, H. D, 1997, *Slash/mulch systems. sustainable methods for tropical agriculture*, Westview Press, Boulder, CO.

Tilander, Y., 1996, "Competition for and conservation of water and nutrients in agroforestry systems in semi-arid West Africa," in *Rapport — Institutionen for Ekologioch Miljovard*, Sveriges Lantbruksuniversitet. No. 86, 140.

Unruh, J. D., 1990, "Iterative increases in economic tree species in managed swidden-fallows of the Amazon," *Agrofor. Sys.* 11, 175–197.

Vega, L. E., Van-Eijk Bos, C., Moreno, L.A., 1987, "Alley cropping with *Gliricidia sepium* (jacq.) *Walp.* (Mata raton) and its effect on soil losses on the hillslopes in Uraba, Colombia, in *Gliricidia sepium. (Jaca) Walp. Management and Improvement*, Withington, D., Glover, N., and Brewbaker, J. L. Eds. Nitrogen Fixing Tree Association, Waimanlo, U.S., pp. 68–69.

Visser, T., Shanmuganathan, N., and Sabanayagam, J. V., 1961, "The influence of sunshine and rain on tea blister blight, *Exobasidium vexans Massee*," *Ann. Appl. Biol.* 49: 306–315.

Walsh, R., 1997, "Pods of the gods," *Natural History* 106(6), 70–73.

Wellman, F. L, 1961, *Coffee: Botany, Cultivation and Utilization*, Interscience, New York.

Wiley, R. W., 1975, "The use of shade in coffee, cocoa, and tea," *Horticultural Abstracts* 45,791–797.

Wilken, G. C, 1987, *Good Farmers. Traditional Agricultural Resource Management in Mexico and Central America*, University of California Press, Berkeley, CA.

Wojtlkowski, P. A., 1993, "Toward an understanding of tropical home gardens," *Agrofor. Sys.* 24: 215–222.

Wrigley, G, 1981, *Tropical Agriculture. The Development of Production*, Longman, London.

Zech, W., Haumaier, L., and Hempfling, R., 1990, "Ecological aspects of soil organic matter in tropical land use," in *Humic Substances in Soil and Crop Sciences, Selected Readings*, MacCarthy, P., Clapp, E. E., Malcolm, R. L., and Bloom, P., Eds. American Society of Agronomy, Madison, WI, 187–202.

17 Agro-Forests: Incorporating a Forest Vision in Agroforestry

Genevieve Michon and Hubert de Foresta

1. INTRODUCTION

Though extensively practiced throughout the tropics by indigenous farmers, agroforestry as a science-based technology was first introduced through forestry, not agriculture. It developed in the mid-19th century, when professional foresters strove to improve the economic efficiency of forest

1-56670-294-1/99/$0.00+$.50

plantation establishment through the technology that later became known as the "taungya system" (King 1987). This first development of modern agroforestry was not concerned with farmers, nor was it considered a system that could improve global land utilization patterns in forest areas. In the early 1970s, when global concerns for the degradation of forested lands increased, agroforestry was reassessed as a system of land management applicable to, and with great potential for, both farmlands and forests. This new brand of agroforestry was primarily targeted at improving the conditions of the rural poor. It did not fundamentally change perceptions about farmers and farming in forestry sciences, but it did contribute to a broader vision of agricultural sciences in general. Suddenly, trees in agricultural landscapes, that had remained quite invisible to agronomists, became valued as important elements of the agricultural system itself. But how far did this reassessment of trees in agriculture translate into a better integration of forestry and agriculture?

Probably to balance a history that for more than one hundred years had ignored the importance of farms and farmers to promote the forest itself, modern agroforestry science as developed through the International Centers emphasized its agricultural side at the expenses of its forest dimension. But even after more than a decade of development, in many parts of the world agroforestry remained more "agro-" than "-forestry" oriented. With the exception of the taungya system, agroforestry research perceived studies related to forest resource management as outside of its mandate. See for example, Steppler and Nair 1987; Nair 1989; Kartasubrata et al. 1989; Vergara and Briones 1987. While focusing on integration of trees into farmers fields, agroforestry research neglected the forest itself.

Ignoring forests seems inappropriate, for in the real rural world of the tropics many forms of forest management directly interact with the management of farmlands. Hunting, gathering and extractivism are essential complements to field culture in forest margins. Farmers often manage more or less artificial forests, either evolved from natural vegetation or purposefully established within their farmlands, as central elements of their farming systems. Is that not real agroforestry too? This hidden face of agroforestry has recently been unveiled in the American tropics where present research programs include major components on forest-related systems, but it remains ignored in agroforestry programs developed in Africa and in Asia.

The importance of forest resource utilization by indigenous communities is a well recognized fact, but how forest management relates to field culture should be more systematically considered. What does the forest actually contribute to agricultural development at local and national levels? What is the extent and the role of indigenous forest gardening and how does it affect land development in general? What mechanisms and strategies have farmers developed to integrate forest resources into their farming systems? How do farmers compensate for the depletion of natural forests in their direct environment? All these questions, and many others, should logically fall into the scope of agroforestry research. But they commonly do not, as this is considered more the field of forestry research. By neglecting its forest-related side, modern agroforestry science is probably losing considerable understanding of the present dynamics of farming systems in "forest areas." It is also drastically reducing its potential impact on the future development of the so-called forest lands.

By undermining the forest dimension of their vision, the main institutions for agroforestry research are also failing to address important issues that stand at the crossing between agriculture and forestry and represent new challenges in rural development. How to orient smallholder farmers to forest resource management, and especially how to integrate forest production on farmlands through domestication and cultivation of timber as well as non-timber forest resources are the questions? Also, how to involve farmers in conservation, especially for biodiversity conservation outside protected areas? Can agroforestry help reduce the present trend toward segregation between forest and agriculture? Can it help balance the planned expansion of obviously unsustainable forms of agriculture at the expenses of natural forests, and the related shrinkage of the forest resource base of smallholder farmers? Can it positively address the roots and dynamics of forest clearing by smallholders and help restore forest functions through adequately designed systems? Finally, can it achieve a real integration between agriculture and forestry?

This chapter focuses first on those indigenous examples of forest management that achieve a true integration of forest resources and structures into farmlands — cyclic and permanent cultivation of artificial forest in swidden agricultural systems in an Indonesian context. Agroforesters often deny the "agroforestry label" to these indigenous systems, as their "forest" component is so immediately visible that it obscures what pertains to the "agro" part of it. However, a close look at these systems reveals how, more than the plot itself, it is the whole conception of resource management and land development that concerns the interface between agriculture and forestry. The paper then highlights the qualities of these reconstructed forests from an agroforestry perspective and proposes initial approaches for their analysis. Based on these defined qualities, it examines how these indigenous systems can inspire new models for the "forest end" of the agroforestry continuum, and how their extrapolation to other areas can be accomplished. Finally, we discuss the potential benefits of incorporating a larger "forest vision" in the agroforestry approach.

2. INDIGENOUS FOREST FARMING AS TRUE AGROFORESTRY OPTIONS

Studies on indigenous forest management have multiplied during the last 15 years. Few of them, however, have emphasized the importance of basic differences between the various systems encountered throughout the tropics. Among these differences, the most significant might be the distinction between extraction — passive resource management in natural ecosystems — and production — active management of resources, usually occurring through the establishment of artificial forests into farmlands. It is essential to overcome the classic confusion between these two strategies for forest resource management for, despite obvious similarities in the biological and ecological nature of harvested forests and artificial forest-gardens, their historical justification, socio-cultural and institutional foundations, and political implications diverge. And whereas managed forests may not have their place in agroforestry research, cultivated forests that clearly associate forest resources with agricultural management touch directly upon basic agroforestry issues.

Southeast Asia, and Indonesia in particular, is rich in examples of indigenous forest resource management. However, if Indonesian farmers are often cited as skilled managers of natural forests, it is less often acknowledged that an essential part of this indigenous forest resource management is carried out outside natural forests in complex tree gardens established within agricultural lands for active production of forest commodities. These gardens are often viewed as scattered home gardens. They are not! They usually form large blocks that extend beyond villages and their permanent open fields, like the damar gardens in the south of Sumatra that form an obvious 25,000 hectares forested unit in agricultural lands. They are sometimes referred to as "managed forests." This is confusing as a "managed forest" most commonly refers to natural vegetation, which forest gardens are not. They are always born from forest clearing and tree plantation. They are regarded as devoted to personal consumption through the provision of complementary foods and materials. They are much more than that. Most of them assume a determining role in the farm economy, as the main income-generating unit of the production system. Indonesian forest gardens are in fact closely related to plantation agriculture, as the main incentive for their development is the production of commodities for trade. The fact that these commodities are forest resources like fruits and spices, rattan, resins or latex might be misleading, but one should not forget that those forest resources have long been and still are essential commodities in forest trade. Forest gardens are also commonly regarded as anecdotal components of backwards traditional agriculture. But, again, they are not. In Indonesia, they are so diverse, so dynamic and so important that they are a major element of 20th century smallholder agriculture. They can be found, in one form or another, in almost every farming system evolved from former forest lands.* They cover altogether several million hectares

* Rubber-based forest gardens cover between 2.5 and 3 million hectares in the lowlands of Sumatra and Borneo (Dove 1993; Gouyon et al 1993), and the total of fruit-dominated forest gardens in Sumatra alone approximates 3 million hectares (Laumonier 1983; Laumonier et al 1986; 1987).

FIGURE 1. Some forest commodities such as damar resin in Indonesia, though considered harvested from natural forests are actually produced in complex agroforestry systems. Illustrator: G. Michon.

(Michon 1993). They provide 80% of the rubber latex consumed and exported by Indonesia, roughly 95% of the various fruits marketed in the country, and 80% of the Dipterocarp resins traded in and outside the country, a significant part of bamboos and small cane rattans, an immense part of the firewood used in the country, and the majority of such items as medicinal plants and handicraft material (Figure 1).

Active production of forest commodities in Indonesia is a modern fact, but it is by no means a recent strategy in indigenous resource management. According to some authors, it probably launched the bases of plant domestication that began, in this part of the world, not from annual food plants but through the transfer of forest trees that produced materials essential to subsistence, such as tannins for fishing nets or bark fibers for clothes, to artificial environments nearby dwellings (Sauer 1952; Barrau 1967). Forest cultivation really expanded under two complementary dynamics (Michon et de Foresta, 1996). One was linked to subsistence resources in the forest and led to the domestication and cultivation of more than a hundred fruit and nut species. In shifting cultivation areas, this subsistence-oriented tree gardening established cultivated forests around settlements, which further remained as semi-managed "fruit islands" disseminated along the successive migration routes. These fruit forests remain along roads and river banks all over the archipelago. The other forest production dynamics evolved with the expansion of long-distance trade in forest products and created large areas of forest-gardens along the forest margins of the great islands: Sumatra, Kalimantan, Sulawesi, the Mollucas. This commercial forest gardening probably emerged between the 1st and the 5th century through the domestication of spices and stimulants — tea, clove, nutmeg and local species of pepper — traded from the lowland forests to China and the Middle East. It really expanded, however, during colonial times, with resources as varied as cinnamon, illipe nut, rattan, damar or rubbers. These indigenous dynamics of forest cultivation remained distinct from the plantation movement imported by the Dutch colonizers. The variety of forest garden systems developed by local people had nothing in common with the uniform colonial estate model. Except for rubber, they used different forest species. Colonial planters exclusively brought species from other parts of the world; oil palm and coffee from Africa, and rubber, cinchona and cocoa from Southern America, whereas indigenous people dealt mainly with local species. These examples of commercial forest cultivation emerged continuously in the course of recent centuries.

One of the oldest traded forest products in the region is benzoin, a fragrant resin that is produced by *Styrax* sp., Styracaceae, entering incense mixtures and perfumery or pharmaceutical preparations. Benzoin was produced in permanent agroforestry systems in North Sumatra, probably before the 18th century (Marsden 1783). Cinnamon production, the aromatic bark of several species of *Cinnamomum*, Lauraceae (Indonesian cinnamon is traded as *Cassia vera* and is mainly provided by the cultivated *Cinnamomum burmanii*), became established two centuries ago in the central highlands of Sumatra (Heyne 1927; Burkill 1935). In western Borneo, the illipe nut forests developed at the beginning of the 19th century. *Tengkawang*, or illipe nut, is a fatty nut produced by roughly 15 *Shorea* species, Dipterocarpaceae, that provides a substitute for cocoa butter in the manufacture of chocolates, margarines and cosmetics. In South Sumatra and Central and East Kalimantan, rattan cultivation emerged 100 years ago (Heyne 1927). Rattan includes 600 different species of climbing palms, 12 to 15 of which are used regularly in Indonesia. Only two are cultivated: *rotan sega*: *Calamus caesius*, and *rotan irit*: *Calamus trachycoleus*. At the beginning of this century, in the south of Sumatra, swidden farmers started cultivating damar trees (Michon and Bompard 1987). Damar trees are resin-producing Dipterocarps, from the genus *Shorea*. The cultivated species is *Shorea javanica*, which produces *damar mata kucing*. Their colleagues in the eastern lowlands began appropriating the Amazon rubber tree (Pelzer 1945). Nowadays, new types of commercial forest gardens are still being developed with fruit and timber species (de Jong 1994; Salafsky 1994).

Integration of forest resources into farmlands occurred not only during various periods in history, but also through different strategies related to complex sets of circumstances. Differences in the intensity of agricultural practices, in the level of forest-related knowledge, in the strength of socio-cultural habits or in the power of local institutions resulted in the creation of different types of structures. Present indigenous forest gardens can be grouped into two main categories. The first set consists of gardening practices of various intensities that remain integrated into forest structures and dynamics. It includes forest-gardens evolving from localized enrichment planting or gradual substitution of species in natural forests, as well as gardens established as productive fallows, integrated in the cycle of shifting cultivation. The second set of forest-gardens implies a more or less permanent replacement of the natural ecosystem by cultivated forests, established after total removal of the original vegetation and intensive replanting of useful trees.

2.1 Forest Production Integrated into Forest Structures and Dynamics

Some indigenous forest gardens rely on an *in situ* replacement of the few and scattered individuals of a given commercial species in the wild by many planted individuals of that same species, coupled with practices that locally modify the forest to the benefit of these planted species. These include selective slashing of competing vegetation and a slight opening of the canopy. This enrichment planting is integrated into existing forest structures, but never destroys or replaces them.* The forest is artificial in spots, to more or less important degrees, but without any meaningful impact on its biology, structure, ecology or function. This practice is presently limited to some forms of rattan production in Central and East Kalimantan, where rattan is planted in late successional forests after slashing of the undergrowth bushes and vines (Godoy et Feaw 1989). Rattan canes use the standing canopy trees as support. The resulting "rattan garden" does not differ from an old successional vegetation structurally nor floristically but the astonishingly high density of rattan clumps clearly indicates the artificial nature of this seemingly "secondary" forest. The rattan

* This practice has been mainly documented in the Amazon, with the management of cultivated stands of *Euterpe* palms — for palm heart and juice production — in swamp forests (Anderson 1987), or of Brazil nuts (Lescure 1995).

FIGURE 2. Some semi-permanent forest gardens, like rattan gardens in Central and East Kalimantan, are included in the cycle of shifting cultivation. Illustrator: Wiyono, ORSTOM.

stand is managed until cane quality and growth performances of the rattan clumps decrease, which may take several decades. A new cycle can then be initiated through selective slashing of remnant vegetation and replanting of rattan seedlings (Figure 2).

Most cases of rattan production in Kalimantan, however, are associated with younger stages of forest succession and are fully integrated within shifting cultivation cycles (Sevin 1983; Weinstock 1983; Fried 1995). Rattan seeds are co-planted with rice in the swidden, and the canes develop with the fallow vegetation that they use as a support. These early "rattan fallows" are not visited during the whole period of rattan growth, and the rattan canes can be harvested 8 years after the plantation. The commonly cultivated rattan species grow in clumps so that harvest of individual canes does not entail the death of the productive plant. The rattan garden can yield canes for the following 20 to 35 years, which usually goes with a notable enrichment of garden plots with fruit trees. The garden can also be completely harvested after two or three "crops" — 3 to 6 years after the first harvest — for a new rice-and-rattan cycle.

In this rotational system, the forest production phase is well integrated within existing agricultural cycles, under the form of a planted fallow. This refers to what has been defined as "rotational agrisilvocultural systems" in classical agroforestry classifications (Nair 1990). The forest phase of this agro/forestry cycle is ecologically equivalent to early successional vegetation, with dense stands of intertwined pioneer trees and climbers. Its main function is the production of commodities for the market. Depending on the degree of planting and control in the garden, however, it can also fulfill several secondary economic functions from occasional subsistence production, gathering of wild vegetables, firewood and fruits, to intended commercial ones, mainly fruits.

Another outstanding example of rotational forest production is represented by rubber gardens in the lowlands of Sumatra and Kalimantan. Though the cultivated rubber tree is not a native species, swidden cultivators in Sumatra and Borneo have adopted it in their swidden production system not long after its introduction in the colonial estates (Pelzer 1945; Dove 1993; Gouyon et al. 1993). Rubber trees, sown in a rice swidden and growing among a classical fallow vegetation, can be tapped after 8 to 10 years. The "normal" cycle for this smallholder rubber production is 35 to 40 years, but some rubber gardens are more permanent, with gradual replacement of decaying trees by self-established rubber seedlings. After 70 to 80 years, however, yields are definitively declining and the vegetation has to be slashed for a new cycle. These native rubber gardens have overcome rubber estates in terms of surface area and volumes of production. In the lowlands of Sumatra and Kalimantan they represent the major land-use systems in smallholder farming.

Rubber gardens tend to be more permanent than the rattan gardens discussed above, and the tree density is higher. However, as for rattan gardens, their structure is similar to that of classic successional vegetation and, although rubber trees are exotics, rubber gardens are often confused with and classified as "natural secondary forests." Due to their relative perennial nature, combined with tending practices that leave a major role to natural processes, they harbor a considerable number of plant species. This half-managed richness allows the provision of secondary products — plant foods and material, timber, game meat — which compensates for the relative low productivity in rubber. Beside their economic importance to farmers, rubber gardens play a determining role in the conservation of plant and animal biodiversity in the lowlands. This role is dramatically increasing with the depletion of the last unlogged dipterocarps forest of this ecotone (de Foresta 1992).

2.2 Cultivated Forests in Farm Lands

Other forest gardening practices in Indonesia have resulted in the establishment of permanent structures closer to those of late successional or old growth forests, which constitute examples of true "forest culture" (Michon et Bompard 1987). These permanent tree gardens are also established through classic subsistence slash and burn as tree seedlings are directly planted in the swiddens. The management of the establishment phase constitutes a complex process of forest reconstruction that can be illustrated by the example of damar gardens in Sumatra (Michon et Bompard 1987; de Foresta et Michon 1993; Michon et al. 1995).

The plantation begins as a classic "taungya system." Damar seedlings, usually raised in nurseries, are introduced in an already planted rice swidden or, preferably, a young coffee or pepper plantation established after rice production. This coffee-damar association is maintained for up to eight years. It allows seedlings to grow in the best possible conditions in terms of micro-environment and concurrence. However, the parallel with more conventional tree plantations does not go further, and the consecutive phases are conceived more in a logic of connivance with the forest ecosystem than of environmental confrontation. Once the crop phase is abandoned, damar trees freely develop with the natural pioneer vegetation that establishes spontaneously. During this period of relative abandonment, through natural processes of dispersion and niche colonization, the young plantation gradually acquires the structure and composition typical of any secondary forest. This successional forest-garden becomes more complex over the years due to a combination of free functioning — development of natural silvigenetic processes — and integrated management — selective cutting and enrichment planting. This management pattern does not fundamentally change when the damar garden begins producing. In the mature plantation, a balance between natural dynamics and appropriate management of individual trees helps maintain a system which produces and reproduces without disruption in structural or functional patterns. It also allows further diversification through the establishment of more climactic forest species among the cultivated stand. Once established, damar gardens usually reproduce without any major disruption, as decaying trees are replaced whenever needed. Unlike plantations that evolve through

FIGURE 3. For damar resin production farmers have developed a system of highly complex forests cultivated in the middle of farmlands. Illustrator: G. Michon.

cycles of total harvest, damar gardens remain permanent without reverting to a phase of massive regeneration (Figure 3).

After 40–50 years, the damar plantation reaches its full production period. From a socio-economic point of view, it is not fundamentally different from any specialized commercial plantation. It provides the majority of household income and constitutes an essential complement to ricefields in the farming system (Mary 1987; Levang 1992). From a biological point of view, however, the mature phase finally resembles more the forest it replaced than a conventional tree plantation. As a natural forest, it is characterized by a high canopy, dense undergrowth, high levels of biodiversity and perennial structures. Apart from damar trees that form the frame of the garden, the damar plantation shelters several tens of commonly managed tree species, and also several hundreds of additional species of trees, treelets, liana and epiphytes that are spontaneously established and often used. As natural lowland and hill dipterocarp forests in the area are severely depleted, the damar gardens constitute the major habitat for many true forest plants and animals, among which are represented some highly endangered mammal species including Sumatran rhinoceros, Sumatran goat, tigers, tapir, gibbons and siamangs (Sibuea et Herdimansyah 1993; Michon et de Foresta 1995; Thiollay 1995).

The "agroforestry nature" of damar gardens is more visible in the establishment phase, which constitutes taungya system associating tree seedlings and annual crops, than in the mature phase that can be characterized as a forest. However, it is the mature phase where forest and agriculture really intersect. Damar gardens have the ecological integrity of a forest, but rely on agricultural practices and are managed mainly as an agricultural enterprise in the middle of farmlands. They epitomize what integrating forest into farming systems can look like.

Benzoin gardens of the Tapanuli regency in North Sumatra represent another original practice of indigenous forest plantation (Watanabe 1990; Simanullang 1988; Katz et al. 1997). In this case, the silvicultural pattern does not result in the establishment of a forest garden that remains permanent, but rather integrates a semi-cyclic phase of actual forest production in a global continuum of old-growth forest. Benzoin trees are first planted in the undergrowth of montane forest. Canopy trees and undergrowth species are then selectively cut as the benzoin develop, in order to maintain a balance

FIGURE 4. Spice and nut trees in forest gardens of the island of Saparua, Moluccas. Illustrator: Wiyono, ORSTOM.

between high light and low temperature in the micro-environment of the benzoin trees. The resulting diversity of large tree species associated with benzoin is not as high as for damar gardens — only good hardwood species and pines in the canopy are effectively favored. But the undergrowth retains many bushes and epiphytes typical of the surrounding montane forest. As long as benzoin trees are tapped for resin — from year 10/12 to 25/45, the garden is more or less carefully maintained and its structure remains somewhat open. The gardens are then gradually abandoned, after a maximum of 65 years, with less and less control over the self-established tree species. This increasing lack of maintenance allows the abandoned garden to quickly revert back to a typical high growth forest. After several decades, the resulting montane forest can be reutilized for benzoin production if needed.

Many other examples of permanent or semi-cyclic forest gardens have been reported. In West and central Sumatra some cinnamon gardens integrate the production of cinnamon plus coffee or nutmeg below a high canopy of large cultivated trees targeted at timber and fruit production. The cinnamon stand is usually completely harvested after 8 to 10 years and then replanted. However, some gradual harvesting can occur if needed. Self established vegetation is usually conserved, but due to the high density of the cinnamon stand, biodiversity concentrates mainly on epiphytes on the canopy trees, small lianas, and undergrowth herbs (Michon 1983; Aumeeruddy 1993). Many types of fruit-based forest gardens can be found throughout Sumatra and Kalimantan. In North Sulawesi or in Lombok, forest garden systems are centered around a palm producing sugar, whereas in the Moluccas, they associate coconut trees and tall Canarium in the canopy with tahitian chestnut, nutmeg or clove, or a mixture of both, plus banana groves in the lower levels. The most remarkable systems in terms of plant richness and similarity to old growth forest are to be found on the island of Borneo, with illipe-nut gardens in West Kalimantan (Momberg 1993; Padoch et Peters 1993; Sundawati 1993) and "lembo" fruit forests in East Kalimantan (Bompard 1988; Seibert 1989; Sardjono 1992) (Figure 4).

3. AGROFORESTS: THE QUALITIES OF EXISTING EXAMPLES

In the large range of agroforestry systems and practices, the Indonesian forest gardens exhibit very specific qualities linked to their structure and management. To understand these particularities it is useful to introduce an artificial, but clear distinction that allows us to differentiate the existing agroforestry models based on physiognomic criteria (de Foresta et Michon 1991). This distinction leads to the definition of "simple" and "complex" agroforestry systems. "Simple" and "complex" refer here to the structure and biodiversity of the system without any pejorative connotation. The terms do not refer to the establishment and management processes which can be indeed quite simple for "complex agroforestry systems" and sometimes very complex for "simple agroforestry systems."

3.1 Complex and Simple Agroforestry Systems

"Simple agroforestry systems" refer to associations involving a small number of components arranged with obvious, usually well-ordered patterns: one or a couple of tree species, either as a continuous canopy, in equally distant lines or in edges, and some annual species for ground cover. The tree component can be of major economic importance such as coconut, rubber, clove, teak, or have a more service-oriented role such as *Erythrina*, *Leucaena*, or *Calliandra,* planted for fodder as well as for soil fertility. The annual species is almost always important economically, as with paddy, maize, vegetables, and forage herbs. The ground species can also be a semi-perennial like banana, cocoa, or coffee. These simple agroforestry associations represent the classical agroforestry model, which is favored in research and development programs of most institutions dealing with agroforestry (Steppler and Nair 1987; Nair 1989). Most agroforestry experimentation and extension projects until now have also concentrated on simple tree/crop associations, which are important as improved technologies, such as alley-cropping, hedgerows and improved fallows. These are also practiced in plantation agriculture as with coffee under *Erythrina*, coconut and cocoa or rubber and rattan. In traditional agriculture in Indonesia, most dry fields include trees either as true components, such as coconut with maize, or as borders of teak, mahogany or rosewood. Trees are also commonly associated with irrigated ricefields, which sometimes reflect natural constraints, such as the association between coconuts and ricefields in the swamp areas of Sumatra East Coast. Or they characterize areas of high population density, such as kapok trees cultivated for centuries on the dikes of Javanese ricefields or the recent trials of introduced clove and lime trees on mounds in the middle of Sundanese ricefields.

Complex agroforestry systems are tree-based with a forest-like configuration. They associate a high number of components, among them trees as well as treelets, lianas, and herbs. Their physiognomy and functioning are close to those observed for natural ecosystems. Complex systems are encountered most commonly in peasant agricultures of the humid tropical world. Except for home garden systems, a particular form of complex agroforestry association that has been well documented all over the world, complex systems, although common components of indigenous farming systems, have been ignored and poorly investigated.

This striking physiognomic, physiological and functional reference to a natural forest ecosystem — the "forest preference" in agroforestry — is one of the main features separating "complex" from "simple" agroforestry systems. It is also the main determinant of their diverging ecological as well as economic functions. To emphasize this forest affinity, we have introduced the term "agroforest" (Michon 1985; Michon et Bompard 1987). This term tends now to be widely appropriated. It is often conveniently applied to any type of agroforestry association, be it a complex multistoried garden of forest trees, a simple association between trees and herbs or even a taro garden overshadowed by papaya "trees." This recent but increasingly widespread confusion will soon obscure the uniqueness and the potential impact of the original concept. As the term "forest" refers to a complex association of plants dominated by large trees, "agroforest" applies only to those multistrata systems that exhibit typical features of a true forest ecosystem, i.e., a closed canopy structure, a multilayered configuration, a great diversity, a dominant forest origin of plant components and the predominance of natural processes in vegetation development and regeneration. The agroforest concept also implies a relative continuity in time and space. Even if it consists of a mosaic of different management units, often exhibiting different silvigenetic units, an agroforest constitutes a large and easily recognizable forest block in agricultural landscapes. A mosaic of open fields and small tree gardens can be compared to a fragmented forest ecosystem. Even though complex in structure and composition, it usually does not allow the full development of forest functional processes, especially those that relate to species regeneration and, more globally, biodiversity (Simberloff 1992). Though related to "complex agroforestry systems," this kind of agroforestry mosaic should not be considered an "agroforest," unless a highly fragmented agroforest.

The implications of the concepts, simple and complex agroforestry systems, go far beyond this physiognomic description. Simple and complex agroforestry systems relate to two different, though potentially complementary, conceptions of land development. One refers to field management: simple agroforestry systems address the integration of trees in agricultural lands. The other refers to resource management: complex agroforestry systems address the integration between forests and agriculture. This difference not only embraces important ecological aspects but, has essential socio-political implications especially concerning the global role and interest of smallholder farmers in the management of forest lands and resources.

3.2 Rationale and Outputs of Agroforest Development in Indonesia

Existing agroforests hold a high degree of local particularity. They also exhibit qualities that are of central interest in the framework of sustainable development, and that are quite different from those commonly acknowledged for simple agroforestry combinations. These include, among others, the simplicity of establishment and maintenance techniques, the conservation of plant and animal biodiversity, the protection of soils, high economic profitability, diversity and flexibility.

The rationale and the qualities of the Indonesian examples consistently feature the ambiguity of their nature and their development: neither really a forest, nor totally a plantation. As exhibited in the examples, agroforests begin as specialized plantations, as an attempt to switch from the management of wild resources in traditional extractive systems to their adoption as new crops in farming systems. These plantations diversify into a mixed stand of planted tree crops and useful resources which have similarities to a climax formation: the evergreen rainforest. This diversification process is not totally planned by the farmers, but it is commonly accepted and purposefully retained. Agroforest management maximizes the benefits of this forest similarity. This "forest preference" in agroforest development is essential, as it allows a combining of the ecological qualities of a forest and the economic benefits of a diversified, multipurpose plantation.

3.2.1 Agroforests as an Original Process of Land Development: Ecological Qualities

As a process of land development, indigenous agroforests constitute an outstanding example of forest conversion through plantation establishment that changes the original forest cover, but in the final analysis does not drastically threaten the forest quality or its biodiversity. Even if it does not revert to the pre-disturbance state, this "forest preference" allows an original process of guided reforestation to develop that restores a good range of economic forest resources as well as essential forest functions. As in a forest, agroforests provide soil protection against erosion, leaching and landslides, and ensure the control of water flows on slopes at a watershed level. The observance of integrated management practices that respect natural processes in vegetation development allows a fairly good level of conservation of plant and animal biodiversity. Even in cyclic agroforests, where mass regeneration is used to rejuvenate the system, the significant natural regeneration of tree-to-tree replacement enables the farmers to wait longer before felling and replanting. Agroforests, either cyclic or permanent, are the only plantation system that allows the restoration of a fair proportion of original forest plants and the capture of wild animals. Besides conserving wild species, the agroforest plays an important role in genetic conservation of useful plants by ensuring the maintenance of a wide range of genotypes of economic trees, especially fruit, spices and nut species.

3.2.2 Agroforests as An Integrated Management Unit: Technical Qualities

The forest similarity also maximizes the benefits of "minimized intervention" that gives the major role to natural processes in the evolution and shaping of the cultivated ecosystem (Michon et Bompard 1987). It avoids resorting to intensive labor, sophisticated techniques or costly technology, while maintaining labor and chemical inputs at low levels. As the production of basic commodities relies mainly on primary production of the vegetation, the necessary inputs are generated by the agroforest itself. The strict economy of minerals between the agroforest components allows the maintenance of natural fertility potential. The active reproduction of productive structures through anticipated interventions is enhanced by natural regeneration processes. Through the "forest preference," the agroforest strategy not only achieves a simple transfer of forest resources and structures, it also guarantees the renewability of these resources and structures, thus assimilating the long-term aspect linked to the management of forest species.

3.2.3 Agroforests as an Original Model of Commercial Tree Gardening: Economic and Socio-Cultural Qualities

From an economic point of view, agroforests are a determining element in the generation of cash income (Mary 1987; Levang 1992). At household level, they ensure the economic independence and welfare of farmers, being the main source of income for the family, a standing reserve similar to monetary savings for urban people, and an important family patrimony that is transferred from one generation to the next. The village community also benefits from agroforests either through their direct production or from their derived activities: harvesting, transport, sorting, processing and marketing of the agroforest produce raise additional value and create job opportunities that are especially important for the landless members of the community. At the national level, agroforests provide important quantities of essential commodities for both national markets and export. The originality of this example of commercial tree gardening is that it is not conceived as an exclusive enterprise: while focusing on income generation, as estates do, the agroforest allows the maintenance of numerous other economic functions that help, as the forest does, to diversify the farmer's income and reinforce his economic stability. Through the diversification of income sources and rhythms, the agroforest serves as a "bank" that allows farmers to cover both every-day expenses with regularly harvestable products such as rubber latex, resin, coffee, cinnamon bark and annual expenses, at least partly, with seasonal products as fresh fruits, clove, and nutmeg. Other commodities such as timber provide occasional, but important sums of cash for exceptional expenses. This diversity of income sources is essential in areas where habits of storing cash are not developed and where credit is very expensive or unavailable — which represents the bulk of rural areas in the tropics. The "multifunctional" strategy also follows the age-old model of multipurpose forest use: the forest preference in the agroforest contributes to maintaining a large variety of wild foods and materials for immediate consumption or sale, or "just in case." By allowing a certain economic as well as ecological flexibility in the management of the main crop, this forest preference also constitutes an insurance against risk.

It is important to stress that agroforests never occur in isolation in native production systems: their main quality is that they are highly compatible with other activities. They are usually associated with permanent food production systems, shifting cultivation practices, specialized plantation practices, and sometimes with extra-agricultural activities.

Socio-cultural benefits associated with the agroforest system are also important. Besides maintaining an obvious coherence between practices and the cognitive base of the society, most of the agroforest management rules ensure a good distribution of benefits through several mechanisms, among them seasonal employment, processing and marketing of agroforest produce and free gathering of wild products such as small fruits, leaves, firewood, medicinal plants.

3.2.4 Agroforests as a Strategy in Forest Resource Management: Socio-Political Qualities

The socio-political consequences of the agroforest development process might be the most important, from the farmers' point of view. Agroforests secure a better grip on forest resources for farmers. Transferring wild resources from the forest to the field has occurred under different types of dynamics, including the need to increase natural levels of production or to escape from natural competitors and pests, to avoid the depletion of commercial resources in the wild, or any kind of stimulus from outsiders.* However, by switching from the management of wild resources in traditional extractive systems to their adoption as new crops in farming systems, Indonesian farmers also often aimed at maintaining or reestablishing their traditional authority over important forest resources and at capturing more efficiently the benefits of their management. There is historical evidence that the transfer of wild resources into cultivation has immediate as well as indirect effects on the bases of authority over these resources. Colonial plantations have clearly demonstrated how domestication can deprive indigenous collectors not only of the benefits of forest collection, but also of their basic rights to forest resources.**

Many examples of agroforest development were carried out as an answer to a politically induced dispossession of native collectors, as in the case of rubber or damar at the turn of this century (Dove 1994; Michon et de Foresta 1996). In many areas in Indonesia, most relations between local populations and the forest resources do not occur anymore through the forest itself, but through one or another type of agroforest. Through agroforests, farmers could secure reappropriation of the coveted resource, but this process had an unexpected consequence: the restoration of numerous other resources that were not initially encompassed in the domestication effort. This allowed farmers not only to reshape bits and pieces of the former forest economy, but also to reroot this economy in an agricultural context that allowed the establishment of new rules and regulations for the use and the control of forest resources. Establishing agroforests is a tentative strategy for legal land and resource appropriation, based on agricultural claims over forest lands through tree plantation. It could constitute a powerful weapon in political contexts where legal rights of farmers over forest resources are abused.

Indigenous agroforests, in spite of their relative success in the areas where they were conceived, are still to be improved. However, they should be considered now as models of utmost interest for the development of sustainable forms of land-use which could combine both agricultural and forest qualities in an economic as well as ecological perspective. These models clearly do not concern Indonesia only, but globally any area where forest and agriculture need true integration. The Indonesian agroforest model not only can, but should be extensively extrapolated. It is essential to

---

* Chinese traders played a determining role in stimulating the cultivation of rubber by smallholders, by providing rubber seedlings to local farmers. Local rulers — the Sultan of Kutai in East Kalimantan- —or missionaries — in Bengkulu, Sumatra- reportedly recommended or actively supported rattan cultivation in their respective regencies.

** Homma (1992) argues that the high productivity in forest plantations unavoidably leads to the fall in prices of natural products and to the economic collapse of any collection business. In the case of rubber, intensive rubber cultivation in colonial estates in Southeast Asia has clearly and quickly entailed the spoilage of indigenous forest collectors mainly in the Amazon but also in Southeast Asia and Africa.

understand that extrapolation does not mean transposition, and it would be useless to try replicating damar or benzoin gardens of, let us say, Congo or Brazil.

It needs to be stressed that even if the Indonesian examples do retain some "Indonesian" specificities, these are not determining factors in the establishment or reproduction of the agroforest systems. In particular, the common assumption that such systems owe their success to the high fertility of volcanic soils in Indonesia is wrong. Such highly fertile soils occur mainly in Java where complex agroforestry systems occur exclusively in the form of small home gardens. True agroforests like the damar gardens in Sumatra or fruit forest-gardens in Kalimantan have developed on low quality ferralitic soils. Benzoin gardens are reproduced on highly infertile podzols while rubber agroforests occupy the less fertile soils developed through agriculture in the lowlands in Sumatra and Kalimantan.

In the same way, the early development of trade links between Indonesia and the outer world might have been an important historical criterion for development of agroforests, but as markets are presently developing all over the world, it is no more a local specifity. Indonesian agroforest models globally hold more "generalizable" qualities — like technical simplicity, good adoptability in shifting cultivation systems, or integration of local tree species and commodities — than local specifities. If the model is now more or less defined and understood, more precise criteria for extrapolation have to be worked out. These concern mainly extra-sectorial factors like market accessibility, global land tenure security, product taxation, local institutions, etc. The most promising arena for extrapolation is that where integration of forest resources into agricultural systems is preferable to segregation between forestry and agriculture, such as the integrated development of forest margins, integration rather than exclusion of farmers in the management of permanent forest areas, or forest production through forest domestication for and by farmers. In this arena, Indonesian examples represent an invaluable source of inspiration for the future.

4. AGROFORESTS AS A NEW PARADIGM IN AGROFORESTRY

In attempting to systematize the conditions for extrapolating the agroforest examples, there is a common danger of focusing exclusively on the concept of a technological package. This could too easily obscure the real importance of agroforest as a new paradigm in agroforestry.

The paradigm of agroforestry itself is founded on tree management in general (Bene et al. 1977), which includes farms as well as forests and single trees as well as complex tree formations. However, it focuses on trees as components of a wider agricultural, forest or agroforestry system. Agroforests do not represent agroforestry in general. In the wide agroforestry landscape, it is clear that they do not address the management of scattered trees on farmlands, nor the introduction of trees in open fields as a support for agricultural production.

Agroforests are only that part of agroforestry that focus on the management of "forests" in farmlands. The implications of the agroforest paradigm, however, might concern agroforestry as a whole. Though narrowing the agroforestry focus towards its forest end, the paradigm enables a reevaluation of agroforestry systems and practices as integral biological and social processes in the context of global resource management by farmers. More than just proposing models of complex plant associations, the agroforest paradigm address fundamental logics in resource development that touch the very interface between forests and agriculture.

In forestry, plantation intensification follows the agricultural model through extreme specialization and homogenization typical of the development of grain crops. In agriculture, they simplify the structure and the function of the cultivated "forest" to the limit. From the perspective of sustainable development, these production-oriented tendencies have demonstrated their limits. Further development of forests and agriculture urgently need a fair dose of fresh imagination.

Existing agroforests synthesize centuries of positive interdependence between farmers and forests in a model that has not been sufficiently explored, but constitutes an invaluable source of inspiration for devising systems for better agro/forestry harmonization. The derived agroforest paradigm enables a view of the original alliances between agriculture and forestry through the incorporation of forest resources, structures and logics in farmlands. Agroforests represent an ecological, cultural and socio-economic replicate of forest structures and features in agriculture that allow an optimum balance between production and conservation functions and maintain continuity with local representation and knowledge systems that have evolved from former forest traditions. What are the merits of this forest preference in agriculture?

4.1 Ecological Benefits of the "Forest Preference" in Agroforestry Systems

4.1.1 A Farmer-Oriented Technology

The main quality of the agroforest model for smallholder farmers' development, as epitomized in the Indonesian examples, lies in the simplicity and the low costs of the techniques involved. This is in sharp contrast to intensive plantation agriculture based on trees, or with plantation forestry, which both depend heavily on specialized techniques, genetically altered plant material and capital and energy intensive processes of crop establishment and maintenance. Modern tree-based plantations either lie far beyond smallholders' financial and technical capacities, and therefore involve the participation of large firms, or require a heavy dependence on credit. Agroforest development, on the other hand, is affordable by local populations as it relies on simple techniques that all shifting cultivators in humid tropical countries have at their disposal. It is based on local knowledge shared by all farmers and does not imply high energy or capital inputs (de Foresta et Michon 1993).

The establishment phase requires technical skills in maintaining personal nurseries, in planting and transplanting tree seedlings on swidden fields, in managing intercropping phases and in carrying-out selective clearing to sustain the growth of the commercial plants. The management of mature agroforests mainly involves relevant selection skills when managing the undergrowth, and above all enough anticipation to allow efficient and timely regeneration of the productive structures (Michon et de Foresta 1995). The ecological knowledge required for the maintenance of agroforests is more important than simple technical skills, and mainly addresses the ecological qualities of the plants and animals components, rather than the fundamentals of ecological processes. Farmers do not need to know how natural recovery in forest gaps operates, but they do need to know which plants will — and which won't — grow in the sunlight of this gap. This kind of basic knowledge of species is usually quite well shared among shifting cultivators.

4.1.2 The Ecological Benefits of a Forest?

The ecological qualities of agroforests are basically those of a natural rainforest. They range from soil protection against erosion and leaching and water conservation through improved drainage of rains, to relatively high levels of carbon sequestration in trees, and conservation of a good portion of the original forest biodiversity. The extent to which these qualities are maintained is not fully comparable to what is happening in an undisturbed rainforest. But these are not targeted outputs, simply benefits derived from an original conception of crop structure and management. The main function of an agroforest is not to restore a forest, but to produce selected commodities. The originality of the model is that these forest qualities apply to agricultural lands that were cleared for economic development. There is no other system in agriculture that allows forest qualities and functions to be preserved to such an extent. The present models used in plantation forestry, although producing forest material and rehabilitating some of the forest ecological functions, do not replace a true forest ecosystem. Though forest cover is shrinking at an unprecedented rate throughout the tropics, the purpose of agroforest development is certainly not to compete with active forest conservation measures. Besides providing an original model for multipurpose forest development, however, it can help in reducing the ecological costs of natural forest conversion for intensive production.

4.1.3 Restoring a Forest? Ecological Integrity in Agroforests

The most original and fundamental feature of the agroforest strategy is how the reestablishment of forest features becomes a tool in agroforest establishment and management. The construction of a forest not only reinstates structure and diversity, but also restores integral biological and ecological processes that are vital for both commodity production and reproduction of structures. In existing examples, farmers do not try to simulate natural processes, they just let them happen and intervene only slightly at key points. The benefits of these processes to the farmer are not expressed in the form of resource production only, but also in the form of ecosystem development. Dynamic processes of vegetation recovery after disturbance are first channelled to speed up and secure the integration of slow-growing trees in the cultivated system. This has proved quite efficient for rapid restoration of not only the resource, but also ecosystem integrity (Figure 5).

Farmers in Sumatra and Kalimantan have succeeded in what remains a dream for most foresters: establishing, maintaining and reproducing, at low costs and for large areas, a healthy dipterocarp plantation. These plantations rely on selected and planted forest trees and exhibit high density stands and a good productivity. But they also exhibit ecological sustainability, low cost establishment and easy regeneration over the years, which is uncommon in conventional plantation forestry. The mimicry of natural structures then enables the use of internal dynamics to sustain the production process and maintain a continuous balance between "obsolescence" and regeneration of the cultivated stand. Plants and animals that colonize the agroforest are not weeds, but support essential ecological processes that are crucial in the maintenance of this "natural" integrity of the agroforest as ecosystem. These natural processes schematically replace the complex technical nature and high energy costs that sustain forest plantations. In this sense agroforests constitute an attempt at "ecosystem domestication" through the full integration of economic and non-economic components, and the utilization of natural ecosystem dynamics to the benefit of a selected, artificially established population of trees.

4.2 Promoting Forest Diversity to Increase Economic Benefits

Among the economic qualities related to this forest preference in agricultural development, the most important ones are related to economic diversity, flexibility and reversibility.

4.2.1 Agroforest as a Sustainable Economic Strategy

Risk management through diversification is a common strategy for smallholder farmers. The development of commercial strategies in agriculture, especially through intensive perennial crop cultivation, usually assumes the complete conversion of existing production systems to monocultures with high inputs of external energy, capital and labor. This increases risks of ecological failure as well as vulnerability to market fluctuations. The agroforest model rests on diversification of structures and components, and this harmonizes intensification and risk management at the plot level through a fusion of commercial production with a wide range of other economic and ecological functions.

Short and long term economic flexibility is also directly linked to the forest qualities of the agroforest. An important principle in agroforest management is that the harvest of the main commercial commodity can be delayed in case of necessity: the concerned species can simply be neglected in the garden for a while, until its exploitation becomes profitable again. This process does not involve any disruption of the system itself. The agroforestry plot is maintained intact and still productive while the species concerned can survive in the structure and will be ready for further

FIGURE 5. The restoration of a dipterocarp forest though planting, maintenance, and reproduction by farmers. Illustrator: G. Michon.

exploitation. The complementary principle is that the agroforest harbors other resources that can be harvested if needed to balance the temporary weakness of the main crop. This aspect of temporary relay in harvesting commercial products can also be turned into a more general process, with the possibility of switching from existing commodities to more profitable ones as the market changes, as often happened in the history of forest product collection for trade.

4.2.2 Agroforests: A Reversible Process

The main economic significance of the "forest end" of the agroforest lies in the maintenance of high levels of reversibility that are apparent in the establishment process, but are also present in the mature agroforest. If analyzed in the context of forest management, the agroforest represents an example of forest development for commercial purposes that entails a total transformation of the original ecosystem, but succeeds in restoring most of its resources and retaining its biodiversity. The process of forest conversion through agroforest enables reversion to forest in a more significant way than through conventional forest plantations. This should offer a lesson for foresters who rarely attempt to manage forests as a global ecosystem. If encompassed in the framework of agricultural plantation strategies, agroforest development represents a process of forest conversion for intensive production which does not conform to economic reductionism and does not irreversibly close economic potentials formerly linked to the presence of natural forest. Agroforests shelter potential economic resources which could be developed if the main economic crop fails, as new tree crops can easily be integrated without disrupting the overall structure of the production system. Agroforests can also generate valuable inputs — material, fertilizer, genetic resource, and capital — for further evolution if needed.

4.2.3 Agroforests and Land Development: In Context

These qualities acquire a new dimension if examined in the context in which agroforests are currently developed: that of forest conversion processes for commercial agriculture through shifting cultivation. This conversion process represents a real, and dual transformation of production systems as it leads from swidden agriculture in temporary fields to more or less permanent field cropping, and from forest extractivism to forest plantation.

Agroforests embody an intensification process of shifting cultivation through the introduction of long-lived crops in the traditional swidden. The commercial importance of these crops prevents any further use of the fields in the short term. The ecological advantages of such a rupture of the traditional cycle in favor of permanent or semi-permanent forest structures are obvious: agroforest establishment practically removes these fields from the realm of slash and burn cycles, thus preventing probable reduction of site fertility through shortening of fallow periods. It also helps to reduce the global extent of slash and burn for food-cropping as it provides income that will be used to purchase food. The long-term economic advantages are also important as agroforest-based farming systems can accommodate much higher population densities than subsistence shifting cultivation.

4.2.4 Agroforests and Capitalization

Agroforest establishment also represents a process of land development that relies on capitalization. This is a major revolution in livelihood strategies. Wealth in strict shifting cultivation systems cannot be generated through agriculture as the cultivation process, in spite of fairly good levels of labor productivity, does not enable the production of significant surpluses. Investments in field preparation are mostly converted into food production for subsistence, being consumed more than capitalized. The organization of the slash-burn-and-harvest cycle allows farmers to carry out income-generating activities during certain periods, mainly through forest collection. Wealth is then created through tapping the forest rent. Planting forest trees on the swidden first constitutes the best way to increase returns to initial labor investment with minimal additional inputs. At the same time, agroforests consistently minimize labor inputs, they allow extra-agricultural activities for income generation if needed. Wealth is then generated through tree production. Besides generating goods for consumption and sale, agroforests can provide surpluses for saving or converting into "luxury" consumption. Above all, they create assets that will increase the production capacity of the farm, and that can be transferred to children. Agroforests constitute the basis of a family or lineage patrimony that comprises the capital that feeds further production as well as the rent itself. More than an income-generating strategy, agroforest development constitutes an integral appropriation process involving switching from the exploitation of natural stocks to the establishment of production structures and property rights that will be transmitted to future generations. This pertains to more global strategies that touch sensitive socio-political issues.

4.3 Agroforest and the Reappropriation of Forest Resources by Smallholder Farmers

A main feature of the forest preference in agroforest establishment is the restitution of an integral forest in a context that is totally different from that currently in use in forestry. The transfer of wild resources to cultivated lands, from the sphere of Nature to that of Agriculture, is an essential process not only for domestication purposes, but for empowering smallholders in the "forest business." This empowering process relies on practical, conceptual and legal aspects of forest lands and resources development.

The strong preference of governments in many tropical countries for a production-oriented open-field model in plantation forestry might have obvious advantages and well established economic foundations, but it has socio-political attributes that may lead to the marginalization of smallholder forest farmers, or even to their total exclusion from the future management of forest resources. In the classic plantation model, increasingly specific technical knowledge and domesticated plant material are developed through research institutes and become available to farmers only through markets or credit. Besides the increased risks due to ecological as well as economic fragility, processes of crop establishment and maintenance are so capital intensive that they lie far beyond smallholders' financial and technical capacities. For these financial and structural reasons, plantation forestry is developing under the control of private or state corporations that physically and legally replace local farmers on forest lands.

Agroforests originate from the transfer of selected forest species to agricultural lands, but they rely on processes that restitute a whole range of resources. As they maintain the pre-existing resource and knowledge bases of indigenous management systems, rely on techniques based on local knowledge shared by every farmer, and do not imply high energy inputs, they allow local communities to maintain control and authority over these resources. Moreover, the agroforestry strategy enables restitution of the former forest resource base in a "cultivated" form which might be more beneficial to the farmers' rights and interests. In most ideologies and political regimes of tropical countries, the "agricultural" context secures better legal structures for land or resource appropriation by smallholders than does a forestry context. Indigenous forest management, however sustainable or profitable, has never been seriously considered by the governing elites, as reflected in the widespread lack of legal recognition of native rights and traditional property regimes concerning forest lands and resources in tropical countries. Moreover, most forest lands in the tropics, and the resources they bear, are under State control. The status of "public goods" facilitates tacit "appropriation" of resources traditionally controlled by indigenous people by those who are close to the seat of national power. On agricultural lands, private property, for either collective or individual owners, is more easily recognized and implemented. In such a context, cutting the forest and planting trees might help native resource management systems to gain official support and be enough to secure property rights themselves, or at least the right to claim for such legal rights. The agroforest strategy therefore constitutes both a symbolic and a political act of appropriation.

5. THE LESSONS FOR AGROFORESTRY RESEARCH

The recognition of agroforests as a new paradigm in agroforestry introduces several alternative dimensions into classic agroforestry approaches. First, it involves the integration of forest perspectives into logics that have remained basically agricultural. Second, it concerns the application of these agricultural logics to the management of forest resources. Insight from this forest perspective for agroforestry research applies to various complementary levels, beginning with the choice of the research object or the definition of the research approach, to the refinement of the model itself and its implications for development.

5.1 The Objects and Approaches of Agroforestry Research

Field experimentation and testing of new agroforestry technologies has long been considered the "noble way" of agroforestry research. This technocratic approach has softened over the years, but still influences the scientific perception of other approaches. In particular, the study of existing systems remains viewed as journalistic works of "description" rather than valid research, especially if it concentrates on non-quantifiable aspects and does not provide files of hard data for statistical analysis. Understanding the value of local agroforestry examples as potential models for innovative agroforestry development, as illustrated by the agroforest story, is developing. What is currently underestimated is the value of including local examples of forest management in the scope of agroforestry-related research. This is incongruous with the current aims of agroforestry to mitigate deforestation and land depletion. Understanding local forest management systems not only will reveal practices that are of utmost importance for land development, but will help in clarifying the interrelations between what is commonly perceived as "agriculture" and as "forestry." It is also likely to generate new insights into the underlying causes of deforestation as well as possible remedies to land depletion. To reach this objective, such examples need to be analyzed in an evolutionary and comparative perspective that puts them "in context." More than focusing on the present patterns and qualities of existing systems, it is important to understand when and why they emerged, how they have succeeded in solving local problems and how and why they are surviving or disappearing. Characterization of determining factors and driving forces that have led to various present situations is essential in assessing not only their impact, but also their future. In this analytical process, bio-ecological or economic factors should be placed in parallel with political or socio-institutional ones, as the current importance of indigenous systems reflect not only the technical or economic validity of the models to which they refer, but, ultimately, the enforcement of socio-political choices from the governing spheres.

Including local forest management examples in agroforestry research will also directly benefit agroforestry experimentation. Until now, the economic and political necessity of providing tangible results in a reasonable time span has forced the experimental approach to concentrate on fast-growing tree species rather than dealing with long-lived forest trees. Does this mean that practical agroforestry research is condemned to work with short-gestation tree crops, keeping true forest resources out of its scope? Indigenous examples based on long-lived trees could be understood as experiments in their own right especially in the arenas of domestication, tree crop association, and system design, conclusions of which could immediately benefit global agroforestry research. Existing structures could also be used to foster further scientific experimentation dealing either with plant material — like selecting genetically superior strains or testing existing, high yielding varieties of forest plants — or with management techniques. This type of experimental research is presently being carried-out by CIRAD and ICRAF-SEA in Sumatra and Kalimantan with the testing of highly productive rubber clones under the technical and environmental conditions of the "jungle rubber" agroforests (Penot and Wibawa 1996).

Dealing with forest resources and forest structures in agroforestry will also require new types of experimentation that can integrate the principle of "forest preference" into development. The most promising field for such experimentation concerns domestication. It implies, of course, widening the range of domestication candidates to the bounty of forest resources, especially commercially important ones. But, once species have been prioritized it implies that the development of improved plant material should be designed for a complex, forest-like environment, rather than for conventional plantation conditions. Present selection and breeding techniques should be adaptable to both forest-like ecological conditions and farmers, technical as well as energetic preferences. Instead of trying to adapt wild species to homogenous open field conditions, plant selection effects could try to draw from "forest" characteristics of the species for both ecological and economic benefits. Low branching varieties of canopy fruit trees have no particular appeal in an agroforest environment, and transforming undergrowth forest species into light-demanding cultivars, as happened for coffee or cocoa, is not the best way to allow the integration of the species into a domesticated forest ecosystem. (See Chapter 16.) Nor is relying on sophisticated hybridization techniques the best way to empower local farmers in the domestication process.

The main feature of the agroforest paradigm for agroforestry approaches is the new focus it puts on restoration or maintenance of ecological processes, rather than on land-use systems.* Introducing forest models in agroforestry not only widens the range of agroforestry technologies and mixture designs, but also provides an ecosystem perspective.

5.2 New Fields for Agroforestry Implementation and Development

The agroforest paradigm addresses the classical fields of agroforestry implementation, rehabilitation of lands degraded through unsustainable agricultural practices, or prevention of foreseen degradation, but with a different perspective. In particular, it can reorient agroforestry action, from promoting the introduction of several universally recognized "agroforestry" species of either fast-growing or multipurpose trees in agricultural lands where original trees have long disappeared, to maintaining local forest resources traditionally known, used and managed by peasants in more productive domesticated forests. It also offers new fields for implementation, namely the management of lands traditionally classified as "forests" rather than "agricultural lands." For multipurpose forest production it offers challenges to the poorly defined model of "extractive reserves." In forest areas where the development perspective rests on intensive forest plantations, the agroforest paradigm provides an ecologically as well as socially interesting alternative to the common introduction of firm-managed fields of *Acacia* or *Eucalyptus*. For conservation areas, it can help devise useful models for buffer zone management. On forest margins, it can provide a smooth transition from shifting cultivation to permanent agriculture. Lastly, for rehabilitation of forest lands through reforestation, it might help in shaping acceptable models for social forestry. In this field, the participation of agroforestry research is essential as smallholder farmers all over the tropics are increasingly involved in the rehabilitation of deforested lands. Due to their "smallholder" dimension, agroforests can enable the promotion of new forms of reforestation by local people. Another promising field for further agroforest development is the production of good quality timber on agricultural lands. As it is vanishing from natural forests, timber might well be the principal next crop for smallholders, and agroforests offer interesting technical as well as social models for its management.

The main task of the agroforest paradigm for applied agroforestry activity is to undermine the still pervasive dichotomy between forest resources and agricultural commodities, between agricultural systems and forests and between forest and agricultural lands. Agroforestry until now has been defined for agricultural lands and agricultural systems. The challenge now is to find new crops for new agroecosystems. This means investigating how forest resources and forest structures can be transferred to agriculture, through new concepts, logics and strategies.

* This input has been recently integrated by R. Leakey in a new definition of agroforestry. Leakey, who has visited Indonesian agroforests, claims that agroforestry should be more than "prescriptions for land use," but instead a land-use development process which, similar to natural ecological processes, increases in complexity over time (Leakey 1996).

5.3 A revision of Development Models for Agriculture

The most meaningful questions raised by the agroforest paradigm relate to the very conception of agriculture and agroforestry. Does modern agriculture, which is usually perceived as the ultimate, least reversible step of an artificialization process, necessarily imply a total partition from the original environment? Should patterns of tree cover in agricultural lands be quite different from those found in natural forests? Should agroecosystems escape natural, ecological laws through intensive human control (Michon and Bouamrane 1997)?

These basic questions can be stated in terms of continuity or partition with the initial "natural" state, of confrontation or connivance with Nature (Henry 1987). This confrontation/connivance perspective relates to two main models defined by ethnobotanists for the development of agriculture, based on differences in patterns of plant manipulation and ecosystem design for cultivation (Haudricourt and Hedin 1943; Geertz 1966; Barrau 1970). One concerns the development of grain crops. It epitomizes "agriculture" in its narrow sense, the cultivation of *ager*, the open field (Barrau 1967; Michon and de Foresta 1996). It rests on a clear distinction between the cultivated field and the natural ecosystem through oversimplification of structural patterns and adaptation of the wild plant to these artificially simplified open fields through domestication. This clearly reflects a "confrontation" approach. It relies on heavy and uniform human interventions and control. In the modern version of this "grain model," homogenization and control extend to the genes, with a generalized use of hybrids and clones which depend on man for production and reproduction. Artificialization culminates in intensive use of chemical and mechanical inputs, associated with high energy consumption, aiming at maximum yields while overcoming natural constraints. The grain model reflects on the productivist mentality which sustained the development of modern agriculture. As it has achieved incredible results in raising food production throughout the world, it tends to be considered the only valuable model for efficient (agri)cultural development. Though it was initially devised for annual crops, the grain model has deeply influenced modern horticulture and silviculture in the tropics: forest plantations and plantation agriculture based on tropical forest trees, such as *Acacia*, *Eucalyptus*, rubber, oil palm, or cocoa, replicate the biological model and the technical options of a corn field.

The second model concerns the development of trees and tuber crops. It is commonly referred to as "horticulture," the cultivation of *hortus*, the garden. This retains the complexity of the natural ecosystem, trying to replicate its structure to accommodate the exigencies of the cultivated. The garden model reflects "connivance" approaches. Diversity is the key word ranging from plant types, herbs, tuberous perennials, trees and lianas, to species and genotypes and including architectural diversity that mimics natural ecosystems, as well as functional diversity, from production aspects — foods or various plant materials — to social functions. Though their evolution "gardens" have integrated many exotic species, but even their modern version maintains these basic patterns of diversity and complexity. The garden, devised for multipurpose production as well as for optimal management of ecological and economic risks, does not comply with the strict exigence of short term productivity in agriculture. Whereas "gardens" are still a major component of indigenous agricultures in the tropics, swiddens, home gardens, anthropogenic forests and forest-gardens are all variations of the "hortus" model — they are relatively absent of modern, scientific agriculture.

This opposition between two diverging models is essential to understand. On the one hand is a technically complex but biologically homogenous and structurally oversimplified model whose deviation from a "natural" model depends on complex sets of knowledge and inputs and on heavy interventions and control. On the other hand is a biologically and structurally complex, but technically simple model requiring management that relies not on segregated knowledge or inputs, but on the free development of functional processes existing in natural forest vegetation. If "mastering" is the key word in the grain model, "taming" is more appropriate for the agroforest strategy. The question should be not which strategy expresses the greatest achievements in transforming and channelling ecological processes to the benefit of humankind, but which appears as the most adapted to present constraints of the tropics, ecological as well as socio-economic? The open field model has proven its success for immediate development, but its long term negative consequences are obvious. Reverting to less productive, but more sustainable models is currently acknowledged. But what does it mean for the tropics? In humid tropical environments, the open field model implies a total partition with the climax formation: the evergreen rainforest. The agroforest, on the contrary, rehabilitates it, not in its undisturbed stage, but in its basic dynamics and functions.

The agroforest paradigm also addresses the integration of agriculture and forestry. It achieves this amalgam between a long term perspective of forest ecology and development, and short-term imperatives of production in agricultural systems. Elaborating on this original association between forests and agriculture may allow the design of alternative strategies for forest domestication and culture.

6. CONCLUSIONS

The most recent important development in agroforestry was the formulation of new research priorities that address more widely the integration of forestry and agricultural problems: mitigating deforestation, alleviating poverty, devising sound alternatives to slash and burn agriculture, helping biodiversity conservation through improved agroforestry practices or carbon sequestration through tree growing. In this new strategic context, agroforestry would benefit substantially by including a broader vision of its "forestry" mandate. This means an integration not only of trees, but of forests in farmlands. The aim is to capture not only forest components, but also forest functions, dynamics and benefits for agriculture, and to translate the socio-cultural dimensions of forest resource management into the agricultural domain. There are many obvious and immediate justifications for such an enlargement of the "forest side" of agroforestry. Ecologically, tree crops are more sustainable than intensive food cropping on marginal soils, and forest-like systems are more likely to succeed than specialized tree plantations. Economically, pure forest crops, such as timber, have bright economic prospects in agriculture. From a socio-political point of view, forestry and agricultural issues are more and more interwoven in the race for sustainable development, and a large part of the lands available for further agricultural development in the tropics are under the jurisdiction of forestry services. But there is another essential reason for agroforestry research to be more involved in forestry issues, a reason that draws upon the socio-political dimension of devising new alliances between forests and agriculture.

The agroforest paradigm can help in revising the relations between central State power and farmers' communities in forest areas. Smallholders or landless farmers all over the tropics have often been largely deprived of their traditional forest resources by forest development projects and expelled from forest lands by coercive forestry regulations. Present forestry laws are often too rigid to accommodate significant changes related to the devolution of forest management to local communities. Or if they allow it, it is often because forest resources are so depleted that foresters urgently need the help of farmers to replenish them. This call for participative approaches in forestry might well be a poisoned gift; farmers are urged to succeed where foresters have so obviously failed. They certainly will be blamed if they fail. They also may be dispossessed if they succeed. However, this new political orientation of many tropical countries also opens unexpected opportunities for smallholders. In addition to fostering participative approaches to conservation and conventional social forestry, the agroforest concept may help to enable the development of a legal framework for forest resource management by local farmers through the transfer of management responsibility into an agricultural context.

REFERENCES

Anderson, A. B., 1987, "Management of native palm forests: a comparison of case studies in Indonesia and Brazil," in *Agroforestry: Realities, Possibilities and Potentials*, Gohlz, H. L., Ed., Martinus Nijhoff Publishers, Amsterdam 155–167.

Aumeeruddy, Y., 1993, *Agroforests and protected forest areas*, University of Montpellier II, Montpellier, (Translated from French).

Barrau, J., 1967, "From Man the Gatherer to Man the Cultivator," *Notebooks of History X* (2): 275–292, (Translated from French).

Barrau, J., 1970, *Man and His Plant Environment in the Humid Tropics, The Malayo Oceanian example,* Faculty of Arts, Museum of Natural History, Paris, (Translated from French).

Bene, J. G., Beall, H. W., and Cote, A., 1977, *Trees, Food and People: Land Management in the Tropics*, International Development and Research Center (IDRC), Ottawa, ON, Canada.

Bompard, J. M., 1988, *Wild Mangifera species in Kalimantan (Indonesia) and in Malaysia*, IBPGR–IUCN–WWF, Bogor.

Burkill, I. H., 1935, *A Dictionary of the Economic Products of the Malaya Peninsula,* Crown Agents for the Milbank Colonies, London.

de Foresta, H., 1992, "Botany's contribution to the understanding of smallholder rubber plantations in Indonesia: an example from Sumatra," in *Sumatra: Environment and Development: its Past, Present and Future*, Seameo-Biotrop, Bogor.

de Foresta, H. and Michon, G., 1991, *Indonesian agroforestry: systems and approaches. Paper presented at the workshop, Which Agroforestries?*, ORSTOM, Paris, September 1991, (translated from French).

de Foresta, H. and Michon, G., 1993, "Creation and management of rural agroforests in Indonesia: potential applications in Africa," in *Tropical Forests, People and Food: Biocultural Interactions and Applications to Development*, Hladik, C. M., Pagezy, H., Linares, O. F., Hladik, A., Semple, A., and Hadley, M., Eds., UNESCO and the Parthenon Publishing Group, Paris, 709–724.

de Jong, W., 1994, "Recreating the forest: successful examples of ethno-conservation among land-Dayaks in central West Kalimantan," in *Management of Tropical Forests: Towards an Integrated Perspective,* Sandbukt, O., Ed., Center for Development and Environment, Oslo, Norway, 295–304.

Dove, M. R., 1993, "Smallholder Rubber and Swidden Agriculture in Borneo: A Sustainable Adaptation to the Ecology and Economy of the Tropical Forest," *Eco. Bot.* 47(2): 136–147.

Dove, M. R., 1994, "Transition from native forest rubbers to Hevea Brasiliensis (Euphorbiaceae) among tribal smallholders in Borneo," *Eco. Bot.* 48 (4): 382–396.

Fernandes, E, C. F., 1997, Personal communication.

Fried, S. T, 1995, *Writing for Their Lives: Bentian Dayak Authors and Indonesian Development Discourse*, Cornell University, Ithaca, N.Y.

Geertz, C., 1966, *Agricultural Involution: The Process of Ecological Change in Indonesia*, University of California Press, Berkeley and Los Angles.

Godoy, R. and Feaw, T. C., 1989, "The profitability of smallholder rattan cultivation in Central Borneo," *Hum. Ecol.* 16(4): 397–420.

Gouyon, A., and de Foresta, H., 1993, "Does 'jungle rubber' deserve its name? An analysis of rubber agroforestry systems in southeast Sumatra," *Agro. Syst.* 22: 181–206.

Haudricourt, A. G. and Hedin, L.,1943, *Man and the Cultivated Plants*, Gallimard, Paris, (translated from French).

Henry, C., 1987, *Confrontation or Connivance? Nature, the Engineer and the Tax-Payer*, Multigraph, Paris, (Translated from French).

Heyne, K., 1927, *Useful Plants of the Dutch Indies*. Ruygrok, Batavia, (translated from Dutch)

Homma, A. K. O., 1992, "The dynamics of extactivism in Amazonia: a historical perspective," *Adv. Eco. Bot.* 9: 23–31.

Kartasubrata, J., S.S. Tjitrosomo, and R. Umaly, 1990, *Agroforestry Systems and Technologies*, Center for Tropical Biology (BIOTROP), Bogor.

Katz, E., Gouloubinof, M., Ruiz, M., and Michon, G., 1997, *Benzoin in North Sumatra*, paper presented at the International Conference on Gums and Resins in Africa. Nairobi, September 1997.

King, K. F. S., 1987, "The history of agroforestry," in *Agroforestry: A Decade of Development,* Steppler, H. and Nair, P. K., Eds., Nairobi, Kenya, ICRAF, 3–11.

Laumonier, Y., 1983, *International map of the vegetation: Southern Sumatra*, 1:1,000,000, International Institute for Vegetation Mapping (ICIV) and Center for Tropical Biology (BIOTROP), Toulouse.

Laumonier, Y., et al., 1986, *International map of the vegetation: Central Sumatra*, 1:1,000,000, International Institute for Vegetation Mapping (ICIV) and Center for Tropical Biology (BIOTROP), Toulouse.

Laumonier, Y., et al., 1987, *International map of the vegetation: Northern Sumatra*, 1:1,000,000, International Institute for Vegetation Mapping (ICIV) and Center for Tropical Biology (BIOTROP), Toulouse.

Leakey, R., 1996, "Definition of agroforestry revisited," *Agrofor. Today* 8(1): 5–7.

Lescure, J. P., 1995, *Extractivism in the Amazon: Viability and Development*, ORSTOM-INPA-AARHUS University, Manaus, Paris, (translated from French).

Levang, P., 1992, *Pahmungan, Penengahan, Balai Kencana: Socio-economic interviews in Krui, Lampung*, ORSTOM/Institute for Tropical Biology (BIOTROP), Bogor.

Marsden, W., 1986, *The History of Sumatra*, Oxford University Press, London.

Mary, F., 1987, *Agroforests and Societies. Socio-economic Analysis of Indonesian Agroforestry Systems*, School of Agriculture, National Research Center for Agriculture, National University, Montpellier, (translated from French).

Michon, G., 1983, *Village-Forest-Gardens in West Java*, Plant Research and Agroforestry. International Center for Research in Agroforestry (ICRAF), Nairobi, 13–24.

Michon, G., 1985, *From forest people to tree farmers: Indonesian agroforestries*, U.S.T.L., Montpellier, France.

Michon, G. and Bompard, J. M., 1987, "Indonesian agroforestries: farmers contributions to the conservation of natural forests and of their resources," *Earth and Life J. of Ecology* 42: 3–37, (translated from French).

Michon, G. and Bouamrane, M., 1997, *Anthropisation and artificialization: continuity or break-up with a natural state?*, manuscript prepared for workshop on "Dynamics and Use of Natural Resources," ORSTOM, Paris, (translated from French).

Michon, G. and de Foresta, H., 1993, *Indigenous Agroforests in Indonesia: Complex Agroforestry Systems for Future Development*, International Center for Research in Agroforestry (ICRAF) — Asian-Pacific Agricultural Network (APAN) — Food and Agriculture Organization (FAO), Bogor.

Michon, G. and de Foresta, H., 1995, "The Indonesian agro-forest model," in *Conserving Biodiversity Outside Protected Areas: The Role of Traditional Ecosystems*, Gilmour, D. and Halladay, P., Eds. International Union for the Conservation of Nature and Natural Resources (IUCN), Gland, Switzerland.

Michon, G., de Foresta, H., Levang, P., 1995, "Farmers agroforestry strategies: damar agroforests in Sumatra," *Nature, Science and Society* 3(3): 207–221, (translated from French).

Michon, G. and de Foresta, H., 1996, "Agroforests as an alternative to pure plantations for the domestication and commercialization of NTFP's," in *Domestication and Commercialization of Non-Timber Forest Products in Agroforestry Systems*, Leakey, R. R. B., Temu, A. B., Melnyk, M., and Vantomme, P., Eds., FAO, Rome, 160–175.

Michon, G., de Foresta, H., and Aliadi, A., 1996, "Damar resins, from extraction to cultivation: An 'agroforest strategy' for forest resource appropriation in Indonesia," in *Ethnobiology in Human Welfare*, Jain, S.K., Ed., Deep Publication, New Delhi, 454–459.

Momberg, F., 1993, *Indigenous Knowledge Systems Potentials for Social Forestry Development: Resource Management of Land-Dayaks in West Kalimantan*, Polytechnical University, Berlin.

Nair, P. K. R., Ed., 1989, *Agroforestry Systems in the Tropics. Forestry Science*, Kluwer Academic Publishers/ICRAF.

Nair, P. K. R., 1990, "Classification of agroforestry systems," in *Agroforestry Classification and Management*, McDicken, K. and N.K. Vergara, Eds., John Wiley & Sons, Inc., New York., 31–57.

Padoch, C. and Peters, C., 1993, "Managed forest gardens in West Kalimantan, Indonesia," in *Case Studies of Genetic Resource Conservation and Development, Perspectives on Biodiversity*, Potter, C. S., Cohen, J. I., and Janczewski, D., Eds., AAAS, Washington, DC, 167–176.

Pelzer, K. J., 1945, *Pioneer Settlement in the Asiatic tropics*, International Secretariat Institute of Pacific Relations, New York.

Penot, E. and Wibawa, G., 1996, *Improved Rubber Agroforestry Systems in Indonesia: an Alternative to Low Productivity of Jungle Rubber Conserving Agroforestry Practices and Benefits*, report presented at the Annual Meeting of the International Rubber Research and Development Bureau (IRRDB), Beruwala, Sri Langka.

Salafsky, N., 1994, "Forest gardens in the Gunung Palung region of West Kalimantan, Indonesia," *Agro. Syst.* 28: 237–268.

Sardjono, M. A., 1992, "Lembo culture in East Kalimantan: A model for the development of agroforestry land-use in the humid tropics," *GFG-Report* 21: 45–62.
Sauer, C. O., 1952, *Agriculture Origins and Dispersals,* American Geographical Society, New York.
Seibert, B., 1989, *Indigenous fruit trees of Kalimantan in traditional culture,* First PROSEA International Symposium on Plant Resources of South-East Asia, Jakarta, Indonesia, Pudoc Wageningen.
Sevin, O., 1983, *The Dayak of Central Kalimantan,* ORSTOM, Paris, (translated from French).
Sibuea, T., and Herdimansyah, D., 1993, *The Variety of Mammal Species in the Agroforest Areas of Krui (Lampung), Muara Bungo (Jambi) and Maninjau (West Sumatra),* report for the ORSTOM/SOFT Project, ORSTOM, Bogor.
Simanullang, B., 1988, *Benzoin Farmers: An Anthropological Study About Benzoin Production in Desa Aek Nauli, North Sumatra, Indonesia,* North Sumatra University, Faculty of Political Sciences, (translated from Bahasa Indonesia).
Simberloff, D., 1992, "Species-area relationships, fragmentation, and extinction in tropical forests," in *Harmony with Nature,* Yap, S. K. and Lee, S. W., Eds., Malayan Nature Society, Kuala Lumpur, 398–413.
Steppler, H. A. and Nair, P. K. R., Eds., 1987, *Agroforestry: A Decade of Development,* ICRAF, Nairobi, Kenya.
Sundawati, L., 1993, *The Dayak Garden Systems in Sanggau, West Kalimantan: An Agroforestry Model,* Georg-August University, Göttingen, Germany.
Thiollay, J. M., 1995, "The role of traditional agroforests in the conservation of rain forest bird diversity in Sumatra, *Cons. Biol.* 9(2): 335–353.
Vergara, N. T., and Briones, N. D., 1987, *Agroforestry in the Humid Tropics,* East-West Center, Honolulu, HI.
Watanabe, H., 1990, "Lessons from traditional forest management in which non-wood forest products are mainly harvested in Southeast Asia," *International Agriculture* 1: 140–143.
Weinstock, J. A., 1983, "Rattan: ecological balance in a Borneo rainforest swidden," *Eco. Bot.* 37 (1): 58–68.

Index

A

F

G